PRAISE FOR
ZINN & THE ART OF ROAD BIKE MAINTENANCE

"*Zinn & the Art of Road Bike Maintenance* can help you remedy any problem that might arise while working on a road bike. It's packed with in-depth explanations and useful diagrams."

—*VeloNews* magazine

"*Zinn & the Art of Road Bike Maintenance* is the gold standard textbook for aspiring home mechanics. From simple tasks such as fixing a flat tire to advanced overhauls of drivetrains or brakes, this book's step-by-step guides explain the tasks and tools your newbie will need to get the job done right."

—RoadBikeReview.com

"This smartly organized guide shows how to repair new and old bicycles from top to bottom. *Zinn & the Art of Road Bike Maintenance* is essential cycling gear for all road and cyclocross riders."

—CrossBikeReview.com

"Lennard Zinn is an institution in the bicycle world—a legend. Legions of cyclists have learned to repair bikes from him, ridden bicycles he's built, or used his advice as guidance on how to better enjoy the world on two wheels."

—*Bicycle Times* magazine

"Today's bicycles are complicated machines that can be expensive to maintain and repair. Zinn has written this book to help both the leisure bike rider and expert mechanic handle almost any problem associated with road bikes."

—*Library Journal*

"Lennard Zinn really is the world's most helpful and comprehensive human when it comes to bicycle repair and maintenance."

—*Bike* magazine

"*Zinn & the Art of Road Bike Maintenance* has instructions on anything an aspiring wrench would want to know. What impresses most is Lennard's overall approach of simplifying a task and reminding us how rewarding it is to perform our own service."

—Podium Café

"Lennard Zinn is a veritable cycling Einstein and, as a naturally gifted teacher, he has the unique ability to explain even the most difficult mechanical task. So unless you currently ride on a high-profile pro team with your own mechanic (and maybe even then), *Zinn & the Art of Road Bike Maintenance* is an absolute 'must-have' book."

—Davis Phinney, Olympic medalist, national champion, and Tour de France stage winner

ZINN & THE ART OF
ROAD BIKE
MAINTENANCE

ZINN & THE ART OF
ROAD BIKE
MAINTENANCE

The World's Best-Selling Bicycle Repair and Maintenance Guide

5TH
EDITION

LENNARD ZINN

Illustrated by Todd Telander and Mike Reisel

VELO press

BOULDER, COLORADO

Zinn & the Art of Road Bike Maintenance, 5th Edition

Text copyright © 2016 by Lennard Zinn

Illustrations copyright © 2016 by VeloPress

3002 Sterling Circle, Suite 100

Boulder, CO 80301–2338 USA

VeloPress is the leading publisher of books on endurance sports. Focused on cycling, triathlon, running, swimming, and nutrition/diet, VeloPress books help athletes achieve their goals of going faster and farther. Preview books and contact us at velopress.com.

Distributed in the United States and Canada by Ingram Publisher Services

Library of Congress Cataloging-in-Publication Data

Zinn, Lennard, author.

Zinn & the art of road bike maintenance: the world's best-selling bicycle repair and maintenance guide / Lennard Zinn; illustrated by Todd Telander and Mike Reisel.

Includes bibliographical references and index.

ISBN 978-1-937715-37-3 (pbk.: alk. paper)

1. Bicycles—Maintenance and repair. 2. Road bicycles—Maintenance and repair.

TL430.Z557 2015

629.28'772—dc23

2015039577

Illustrations by Todd Telander and Mike Reisel

Cover and interior design by Erin Farrell/Factor E Creative

Cover, bike, and author photos by Brad Kaminski

Custom paint on cover bike by Alchemy Bikes

1984 Olympic Murray by Serotta courtesy of The Pro's Closet, Boulder, CO; www.theproscloset.com

Title font Good Headline Pro; body text Proxima Nova

This paper meets the requirements of ANSI/NISO Z39.48-1992 (Permanence of Paper).

17 18 / 10 9 8 7 6 5 4 3

CONTENTS

A TIP OF THE HELMET TO . . .

Mike Reisel. A picture is worth a thousand words, adding up to many thousands that Mike has added in this edition to the hundreds of thousands of illustrative "words" from Todd Telander's capable hand over the years. Mike's and Todd's drawings make my written words more intelligible and this book more useful and beautiful.

My everlasting appreciation goes to the late Bill Woodul for teaching me much of what I know about working on road bikes. I have also learned tricks from thousands of other people too numerous to count, every one of whom I greatly appreciate. I have incorporated suggestions from Scott Adlfinger, Paul Ahart, Sheldon Brown, Peter Chisholm, Saul Danoff, Skip Howat, Paul Kantor, Calvin Jones, Dan Large, Nick Legan, Paul Morningstar, Nate Newton, Chris Rebula, Tom Ritchey, Dag Selander, Daniel Slusser, Wayne Stetina, Stu Thorne, and many others, including innumerable readers of my Q&A column on velonews.com who have written to me with great tips. Thanks!

Many thanks to Ted Costantino for his gentle nudging of me toward completion and his fine editing, to Kara Mannix for keeping this book moving smoothly through the editorial process, and to Charles Pelkey for his editing contributions and content suggestions. My appreciation to Dave Trendler and Renee Jardine for keeping the fires burning under this book and for their contributions to making VeloPress such a fine organization to work for.

And thanks to Felix Magowan and John Wilcockson for creating VeloPress in the first place, and to Competitor Group for keeping it going.

And, last, thanks to my family for providing endless support and a nice environment at home for me to work in, as well as the freedom to work away from it.

INTRODUCTION

First things first, but not necessarily in that order.

—DOCTOR WHO

ABOUT THIS BOOK

So, you want to learn how to care for your road bike? Congratulations. You will be glad you took this step. Although it is nice to learn about your bike from friends or shop employees who know more about bicycles than you do, you don't want to depend on them for routine maintenance or fixing basic mechanical problems. And the exhilaration of riding with the wind in your hair will be enhanced by understanding the structure of the mechanical system on which you are sitting and to which you are entrusting your life.

Even the purest romantic can follow the simple step-by-step procedures and exploded diagrams in this book and discover a passion for spreading new grease on old parts. And, I hope, everyone will develop an appreciation for how infusing love into the work will guarantee success at bike maintenance. If not, frustration will take over, you will use less care, and your riding enjoyment will be compromised.

Zinn & the Art of Road Bike Maintenance allows you to pick maintenance tasks appropriate for your level of skill and confidence. However, I firmly believe—and my experience with the repair classes I have taught confirms this—that anyone can perform the repairs illustrated on these pages. It takes only a willingness to learn and the appropriate tools.

This book is intended for everyone from experienced shop mechanics to those who only want to know about the most minimal maintenance their bike requires. Chapter 2 is for those whose interest is limited to the latter; the rest of the book is for those who choose to go to greater lengths to make everything work optimally and look clean and beautiful. Even for those who wish to focus on Chapter 2, the information in Appendix C on fitting your bike to you instead of the other way around will increase your riding pleasure and safety.

WHY DO IT YOURSELF?

There are a number of reasons for learning to maintain your bike. Obviously it is a lot cheaper to fix a bike yourself than to pay someone else to do it. Once you have some skill and experience, it is also faster. And home-based maintenance is a necessity for most racers and others who live to ride and have no visible means of support.

As your income increases, economic necessity ceases to be a significant issue. However, you may find that you enjoy working on your bike for reasons other than just saving money. Unless you have a trusted mechanic who services your bike regularly, you are not likely to find anyone who cares as much about your bicycle's smooth operation and cleanliness as you, or who will make your bike a priority when you need to have it the next day or in the next few hours. Furthermore, if you love to ride, you need to be able to fix mechanical breakdowns that occur on the road, especially if you ride alone.

If time is your biggest issue, having someone else work on your bike might seem like a no-brainer. But in reality, even finding the time to drop off your bike and pick it up from the shop, while coordinating with the shop's schedule, can be hard. You may be able to perform a simple repair faster or more conveniently than you can make a trip to the bike shop during working hours. And you won't like missing a ride during beautiful weather while your bike sits in a shop that is backed up with repairs. Finding out that you can't just drop off your ailing bike during high season and expect anything faster than a three-week turnaround on a minor repair can ruin your day. Even arranging and adhering to a repair appointment with a shop can be a hassle. Finally, a shop slammed with summer work may return your bike in less than optimal condition because too little time was devoted to the repair or the mechanic was inexperienced. Ultimately, you may decide that having someone else work on your bike creates more aggravation than it alleviates.

Working on your bike can be fun. Bicycles are the manifestation of elegant simplicity. Bicycle parts, particularly high-end components, are a fantastic value. They are made to work well and last a long time. With the proper attention, they can shine in appearance and performance for many years. Satisfaction can be found in dismantling and cleaning a filthy, barely functional part, lubricating it with fresh grease, and reassembling it so that it works like new. Knowing that you made those parts work so smoothly—and that you can do it again when they next need it—is rewarding. You will be eager to ride hard and long to see how your work holds up, rather than being reluctant to get far from home for fear of breaking down.

It is liberating to go on a long ride confident that you can fix just about anything that may go wrong. Armed with this confidence and the tools to put it into action, you will have the freedom to explore new roads and go farther than you may otherwise have gone. You may also find yourself more willing to share your love of the sport with riders who are less experienced. You will enjoy riding with them more if you know that you can fix their questionably maintained bikes, and you can bask in their appreciation after you have eliminated an annoying squeak or a skipping chain.

HOW TO USE THIS BOOK

Skim through the entire book. Look at the table of contents and the exploded diagrams, and get the general flavor of the book and what's inside. When it's time to perform a particular task, you will know where to find it, and you will have a general idea of how to approach it. Illustrators Mike Reisel and Todd Telander and I have done our best to make these pages as understandable as possible. The exploded diagrams show precisely how each part goes together. Nevertheless, the first time you go through a procedure, you may find it easier to have a friend read the instructions out loud as you perform the steps.

Obviously, some maintenance tasks are more complicated than others. I am convinced that anyone with an opposable thumb can perform any repair on a bike. Still, it pays to spend some time getting familiar with the simple tasks, such as fixing a flat, before throwing yourself into a complex job, such as building a wheel.

Tasks and the tools required to accomplish them are divided into three levels indicating their complexity or your proficiency. Performing level 1 tasks demands level 1 tools and requires of you only an eagerness to learn. Level 2 and level 3 tasks also have corresponding tool sets and are progressively more difficult.

All suggested tools are shown in Chapter 1. At the end of Chapter 2 is the must-read section "A General Guide to Performing Mechanical Work" (2-18); it states general policies and approaches that apply to all mechanical work. (Note that the cross-references list chapter first and then the section within the chapter; for example, 2-18 indicates material found in section 18 of Chapter 2.)

Each chapter starts with a list of suggested tools in the page margin. If a section demands more than basic experience and tools, there will be an icon designating the difficulty. Tasks and illustrations are numbered for easy reference.

If you're wondering what to do first, a routine maintenance schedule is included at the end of Chapter 2 (2-19). A troubleshooting section is included at the end of some chapters. This is the place to go to identify the source of a certain noise or particular malfunction in the bike. There is also a comprehensive troubleshooting index in Appendix A.

For those into cyclocross, almost every chapter includes a specific cyclocross maintenance section.

Many tasks will be simplified or improved by using the information presented in the appendixes. Appendix B is a complete gear chart and includes instructions on how to calculate a gear if you're using nonstandard-size wheels or tires. Appendix C is an extensive section on selecting the proper-size bike and positioning it to fit you. It includes information about setting up your bike for time trials or triathlons, as well as road and cyclocross. Appendix D, the glossary, is an inclusive dictionary of bicycle technical terminology. Appendix E lists the tightening (torque) specifications of almost every bolt on the bike. I can't emphasize enough how useful it is to use a torque wrench to tighten bolts as tightly as the component manufacturer intended, but no tighter. Flag Appendix E so you can flip to it easily whenever you work on your bike.

The Internet can be a useful supplement to this book. For instance, bikeschool.com, dtswiss.com, and other sites have spoke-length calculators to use when you are building wheels.

THE ROAD BIKE

This is the creature (Fig. i.1) to which this book is devoted. All of its parts are illustrated and labeled. Take a minute to familiarize yourself with these parts now, and then refer back to this diagram whenever necessary.

The road bike comes in a variety of forms, from road racing (Fig. i.1) to time trial or triathlon (Fig. i.2), to longer-wheelbase touring models, which are rigged for carrying luggage (Fig. i.3), to models with front—and even rear—suspension. Some cousins are the track bike (Fig. i.4) and the cyclocross bike (Fig. i.5).

THIS MEANS YOU!

Because this book clearly spells out the steps necessary to properly maintain and repair a road bike, even those who see themselves as having no mechanical skills will be able to tackle problems as they arise. With a willingness to learn and a little practice, you will find that your bicycle will become transformed from a mysterious contraption too complicated to tamper with to a simple machine that is a delight to work on. Just allow yourself the opportunity and the dignity to follow the instructions, take your time, and trust yourself.

So, if you think you are not mechanically inclined, set that opinion aside, along with any other factors that may stand in the way of rolling up your sleeves to improve your bike's performance. The bicycle is one of our greatest inventions. Another is the book. Here is a chance to use them both. See you on the road!

I.1 The object of our attention (and affection), racing version

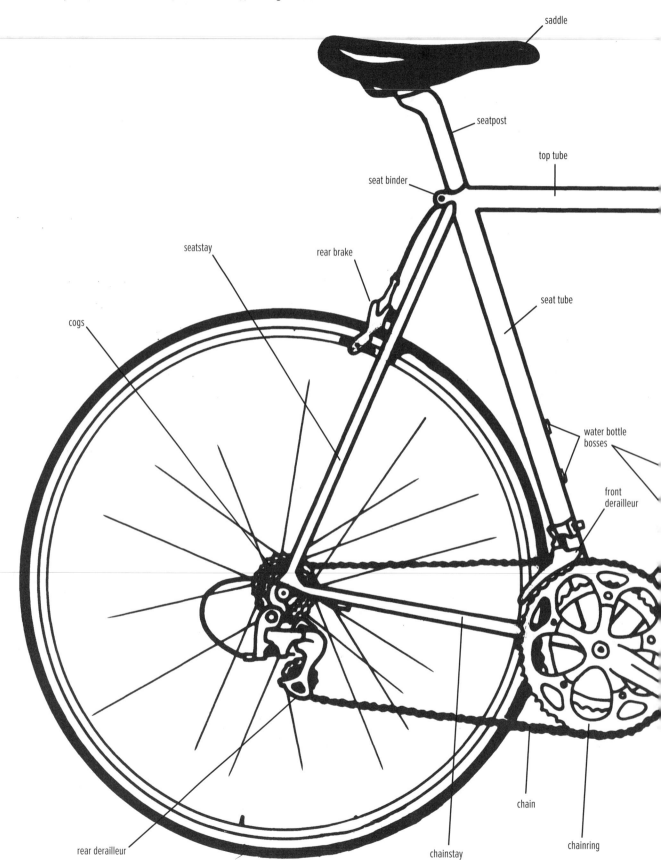

saddle

seatpost

top tube

seat binder

seatstay

rear brake

seat tube

cogs

water bottle bosses

front derailleur

chain

chainring

rear derailleur

chainstay

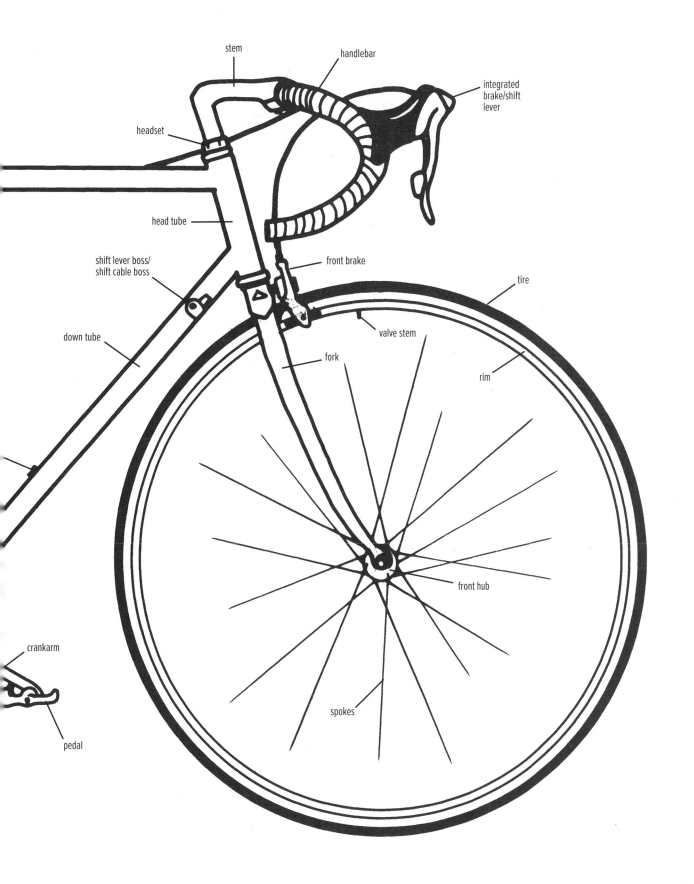

stem

handlebar

integrated brake/shift lever

headset

head tube

shift lever boss/ shift cable boss

front brake

tire

down tube

valve stem

rim

fork

crankarm

front hub

spokes

pedal

I.2 Triathlon or time trial bike

I.3 Touring bike

I.4 Track bike

I.5 Cyclocross bike

TOOLS

> *If the only tool you have is a hammer,*
> *you tend to see every problem as a nail.*
>
> —ABRAHAM MASLOW

You can't do much work on a bike without a basic tool assortment. Bicycles—like other evolved machines such as automobiles and watches—have specific fasteners and threads that require specific tools to fit them. This chapter will clarify which tools you should consider owning, based on your level of mechanical experience and interest.

As I mentioned in the introduction, the maintenance and repair procedures described in this book are classified by degree of difficulty. Nearly all repairs are classified as level 1, because most bicycle repair jobs are pretty easy to complete once you understand the principles involved. The tools for levels 1, 2, and 3 are pictured in Figures 1.1A, 1.2, and 1.3, respectively, and described on the following pages. In addition, the tools you may need for a specific repair are listed in the margin at the beginning of each chapter.

For the novice, there is no need to rush out and buy a large number of bike-specific tools. The Level 1 Tool Kit (Fig. 1.1A) consists of standard tools, many of which you may already own. This is almost the same collection of tools, in a more compact and lightweight form, that I recommend for carrying on long rides (Fig. 1.6).

The Level 2 Tool Kit (Fig. 1.2) contains several bike-specific tools, allowing you to do more complex work on the bike. Level 3 tools (Figs. 1.3, 1.4) are extensive (and sometimes expensive) and ensure that your riding buddies will show up not only to ask your sage advice but also to borrow your tools. If you are willing to lend tools, you may want to mark your collection and keep a list of who borrowed what, to help recover items that may otherwise take a long time finding their way back to your workshop. I have yet to consistently take this advice and am missing some favorite tools. . . .

1-1

LEVEL 1 TOOL KIT

LEVEL Level 1 repairs are the simplest and do not require a workshop, although a well-lit, comfortable workspace is nice to have. For everyday repairs, you will need the following tools (Fig. 1.1A):

- **Tire pump with a gauge** and a valve head to match your tubes (either Presta or Schrader valves; see Fig. 1.1B).
- **Standard screwdrivers**: small, medium, and large.

- **Phillips-head screwdrivers**: one small and one medium.
- Set of three plastic **tire levers**.
- At least two **spare tubes**—or **tubulars** (see Chapter 7)—of the same size and valve type as those on your bike.
- Container of regular **talcum powder** for coating tubes and the inner casings of tires. Do not inhale this stuff; it's bad for your lungs.
- **Patch kit**. Choose one that comes with sandpaper instead of a metal scratcher. Every year, check that the glue has not dried out.
- One 6-inch **adjustable wrench** (also called a "Crescent wrench").
- **Pliers**: regular and needle-nose.
- Set of **metric hex keys** (or Allen wrenches) that includes 2.5mm, 3mm, 4mm, 5mm, 6mm, and 8mm sizes. Folding sets are available and keep wrenches organized. I also recommend buying extras of the 4mm, 5mm, and 6mm sizes, and a long-handled 8mm hex key for removing and installing some pedals and crankarms.
- **Torx keys**, which look like hex keys with star-shaped tips; they fit some brake bolts and chainring bolts. Torx T25 and T30 are common sizes on road bikes.
- A 15mm **pedal wrench**. This wrench is thinner and longer than a standard 15mm wrench and thicker and longer than a cone wrench, to fit into the space between the pedal and crank (Fig. 13.4). A pedal wrench is not necessary for pedals with only a hex-key hole and no wrench flats on the spindle (Fig. 13.5).
- **Chain tool** for breaking (opening) and reassembling chains. If you have a 9-, 10-, or 11-speed system, you may need a narrower chain tool to avoid bending the center prongs of the tool. Shimano's TL-CN34 (Fig. 4.19) works for 7-, 8-, 9-, 10-, and 11-speed chains. Many other chain tools work as well (Figs. 4.20–4.23); you can ask your bike shop for the brand of tool that matches the brand and size of the chain on your bike.
- **Chain-elongation gauge**. This handy little item helps you determine whether a chain needs to be replaced (Figs. 4.5, 4.6). An accurate 12-inch ruler will substitute adequately.

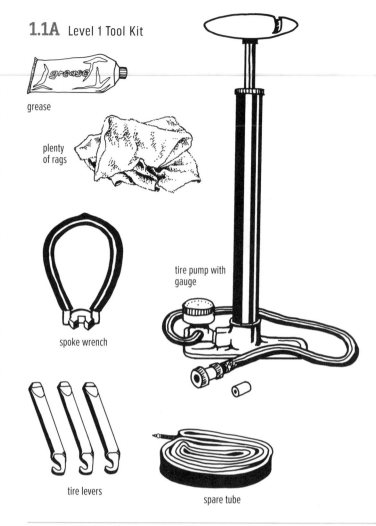

1.1A Level 1 Tool Kit

grease

plenty of rags

spoke wrench

tire pump with gauge

tire levers

spare tube

- **Spoke wrench** to match the size of the spoke nipples on your bike's wheels.
- Tube or jar of **grease**. I recommend using grease designed specifically for bicycles, but standard automotive grease is okay.
- Drip bottle or can of **chain lubricant**. Choose a nonaerosol; it is easier to control, uses less packaging, and wastes less in overspray.
- **Rubbing alcohol** for cleaning brake tracks on rims and discs, doing other light cleaning, and removing and installing handlebar grips, if you have them instead of handlebar tape.
- **Electrical tape** for taping off the end of the handlebar tape, marking your seat height, covering frame holes, etc.
- A lot of **rags**! Old T-shirts work fine.
- **Safety glasses**.
- **Rubber gloves** or a box of cheap latex gloves.

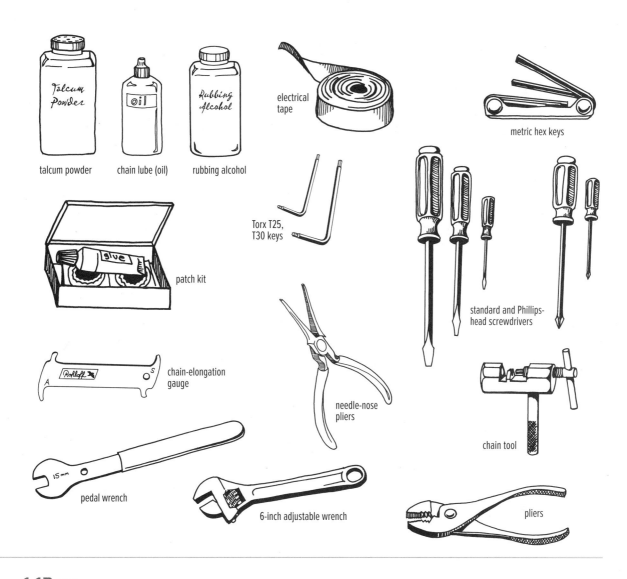

talcum powder

chain lube (oil)

rubbing alcohol

electrical tape

metric hex keys

patch kit

Torx T25, T30 keys

standard and Phillips-head screwdrivers

chain-elongation gauge

needle-nose pliers

chain tool

pedal wrench

6-inch adjustable wrench

pliers

1.1B Valve types

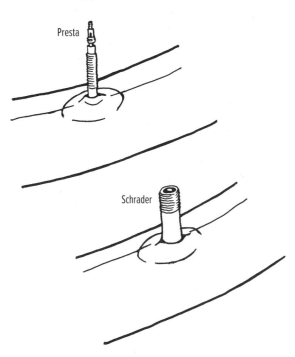

Presta

Schrader

- A **bucket**, **dish soap**, **large brushes**, and **sponges** (Fig. 1.7). These will serve you well for cleaning a dirty machine rapidly.

1-2

LEVEL 2 TOOL KIT

LEVEL Level 2 repairs are a bit more complex, and I recommend that you attack them with specific tools and a well-organized workspace with a shop bench. Keeping your workspace well organized is probably the best way to make maintenance and repair easy and quick. You will need the entire Level 1 Tool Kit (Fig. 1.1A) plus the following Level 2 Tool Kit tools (Fig. 1.2), which of course you can buy as you need them:

- **Portable bike stand**. The stand must be sturdy enough to remain stable when you're really cranking on the wrenches. If you have a bike with an aero seatpost or integrated seat mast, you won't be able to clamp it in a standard bike stand, which is designed to clamp round tubes. In that case, you will need a bike stand that holds the bike by the bottom bracket and the front or rear end with one wheel out. It will have a cradle for the bottom bracket to sit in and a sliding clamp on a long horizontal arm with a quick-release mount to clamp either the fork ends or rear dropouts (see the race mechanic's bike stand in Fig. 1.4). Alternatively, you can use an adapter that holds the frame and can be clamped itself in a standard bike stand.

- **Shop apron** to keep your nice duds nice.
- **Hacksaw** with a fine-toothed blade or a sintered blade for hard materials and composites.
- Set of **razor blades** or a sharp **shop knife** (box cutter).
- **Files**: one round and one flat, with medium-fine teeth.
- **Cable cutter** for cutting coaxial shift cable housing without crushing as well as for cutting brake and shift cables without fraying the ends. If you purchase a SRAM, Shimano, Park, Pedro's, or Jagwire housing cutter, you won't need to buy a separate cable cutter, because all of these cut cables as well as housings.
- Set of **metric socket wrenches** that includes 7mm, 8mm, 9mm, 10mm, 13mm, 14mm, and 15mm sizes. This assumes you have a ratchet handle; if not, get one.
- **Chainring-nut tool** for holding the nut while you tighten or loosen a chainring bolt.
- Medium **ball-peen hammer**.
- Medium-size **bench vise**, bolted securely to a sturdy bench.
- **Cog lockring tool** for removing cogs from the rear hub (Figs. 8.21, 8.22). Note that Campagnolo lockrings require a different tool than Shimano, SRAM, and Mavic.
- **Chain whip** or **Pedro's Vise Whip** for holding cogs while loosening the cassette lockring. The Vise Whip will hold the cog more firmly (Fig. 8.21) than will the chain whip (Fig. 8.22).
- **Bottom-bracket tools**. For external-bearing cranks (Figs. 11.2, 11.8, 11.19), you'll need an oversized splined wrench (external-bearing tool) to remove the cups (Fig. 11.19) and, for Shimano Hollowtech II cranks, a

1.2 Level 2 Tool Kit

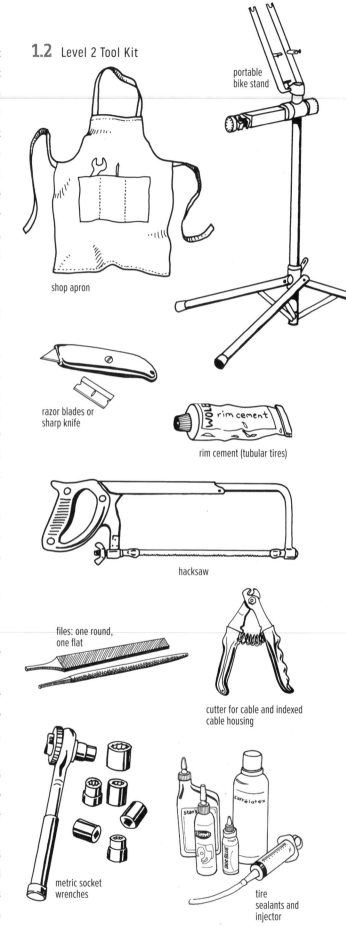

portable bike stand

shop apron

razor blades or sharp knife

rim cement (tubular tires)

hacksaw

files: one round, one flat

cutter for cable and indexed cable housing

metric socket wrenches

tire sealants and injector

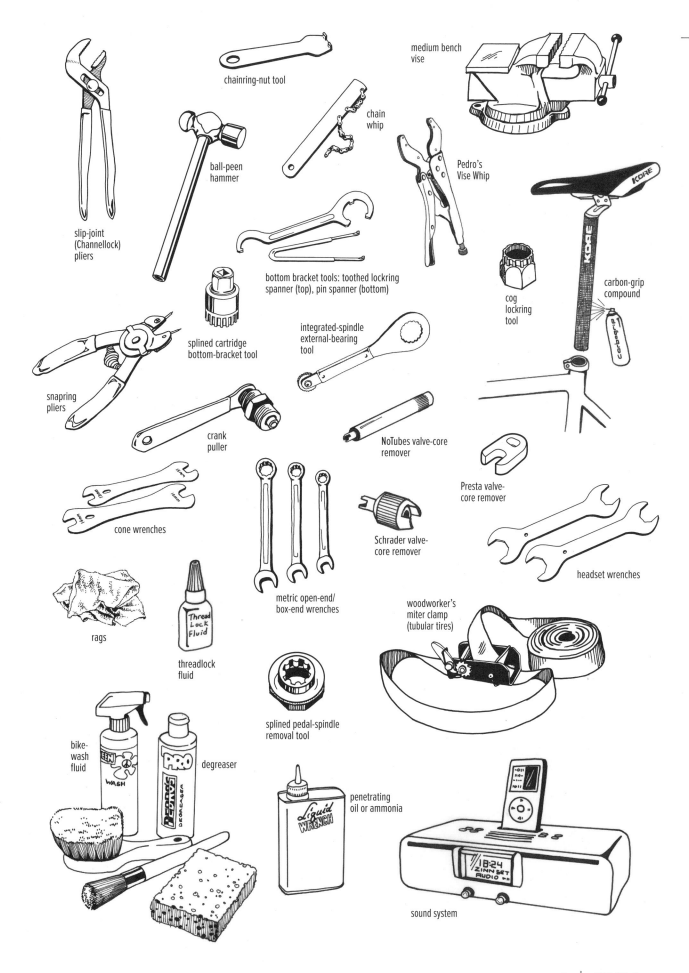

chainring-nut tool

medium bench vise

chain whip

ball-peen hammer

Pedro's Vise Whip

slip-joint (Channellock) pliers

KORE

bottom bracket tools: toothed lockring spanner (top), pin spanner (bottom)

cog lockring tool

carbon-grip compound

splined cartridge bottom-bracket tool

integrated-spindle external-bearing tool

snapring pliers

crank puller

NoTubes valve-core remover

Presta valve-core remover

cone wrenches

Schrader valve-core remover

headset wrenches

metric open-end/ box-end wrenches

rags

threadlock fluid

woodworker's miter clamp (tubular tires)

splined pedal-spindle removal tool

bike-wash fluid

degreaser

penetrating oil or ammonia

sound system

little splined tool to tighten the left crank's adjustment cap (Fig. 11.3). For Campagnolo Ultra-Torque integrated-spindle cranks (Fig. 11.8), you'll also need a long, 10mm hex key to tighten the bolt in the middle of the axle. For sealed cartridge bottom brackets (Figs. 11.21, 11.25), you'll need the splined cartridge bottom-bracket tool (Fig. 11.34). Note that if you have an ISIS or Octalink splined-spindle bottom bracket, you need a splined tool with a bore large enough to swallow the fatter spindle (Fig. 11.25). And for cup-and-cone bottom brackets (Fig. 11.22), you'll need a lockring spanner and a pin spanner to fit the bottom bracket (Fig. 11.37).

- **Snapring pliers** for BB30 cranks (Fig. 11.17) and other unthreaded bottom brackets with snapring grooves (Fig. 11.33). Also useful for removing snaprings from pedals, PRO seatposts for Shimano Di2 batteries, and other parts.

- **Crank puller** for removing crankarms (Fig. 11.7). This tool is only necessary for older cranks; it is not needed for integrated-spindle cranks (Figs. 11.2, 11.8, 11.15–11.19) or for cranks with self-extracting crank bolts. The pushrod of this tool is sized for either square-taper spindles (Figs. 11.21–11.23) or ISIS/Octalink spindles (Figs. 11.24, 11.25) but not both, so get the right one for your crankset.

- **Cone wrenches** for loose-bearing hubs (Figs. 8.8, 8.12). The standard sizes are 13mm, 14mm, 15mm, and 16mm, but check which size you need before buying.

- **Slip-joint pliers** (also called "Channellock" pliers).

- **Splined pedal-spindle removal tool**. Note that Shimano's plastic tool (Fig. 13.14) is different from Look's, although high-end Shimano and Look pedals no longer require either (they take standard 20mm and 19mm wrenches).

- For older bikes, a set of **metric open-end wrenches** that includes 7mm, 8mm, 9mm, 10mm, 13mm, 14mm, 15mm, and 17mm sizes.

- Two **headset wrenches** (only needed for older bikes with threaded headsets; Figs. 12.1, 12.8, 12.11, 12.23). Check the size of the headset on your bike before buying.

- **Rim cement** for tubular tires, if you have them. Use Continental clear glue or Vittoria Mastik'One for aluminum rims, but stick to Mastik'One for carbon-fiber rims. See Chapter 7 for more options.

- **Valve-core remover** for Presta or Schrader valves. Some tools have separate ends for both, while other tools like Stan's NoTubes can remove both Presta and Schrader valves.

- **Carbon-grip compound** for the clamping areas of carbon seatposts and handlebars to keep them from slipping.

- **Tire sealants** and **sealant injector syringe** to prevent or fill small punctures and to set up tubeless tires.

- **Woodworker's miter clamp** for gluing tubulars (optional).

- **Sound system**, if you plan on spending a lot of time working on your bike.

- **Threadlock fluid**, **specialty bike-wash fluid**, **degreaser**, **penetrating oil** (or **ammonia** for breaking free stuck parts), **antiseize grease** for titanium bolts will all come in handy, along with a small stack of shop rags.

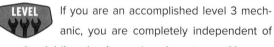

LEVEL 3 TOOL KIT

LEVEL If you are an accomplished level 3 mechanic, you are completely independent of your local bike shop's service department. You can even build a complete bike from a bare frame. By now, you have a well-organized, separate space intended solely for working on your bike. Some elements of the Level 3 Tool Kit (Fig. 1.3) are more accurate or heavy-duty replacements for tools in the Level 2 Tool Kit.

- **Parts washing tank**. Use an environmentally safe degreaser. Dispose of used solvent responsibly; check with your local environmental safety office.

- **Fixed bike stand** (optional). Be sure it comes with a clamp designed to fit any size of frame tube.

- Large **bench-mounted vise** so that you can free stuck parts.

- **Master-link pliers** (Fig. 4.25).

- A **tire pressure gauge** separate from the one on a floor pump is more accurate at low pressures and can save time.

- A **telescoping** or **articulating magnet** for picking up dropped parts or small tools.

- Large **ball-peen hammer**.

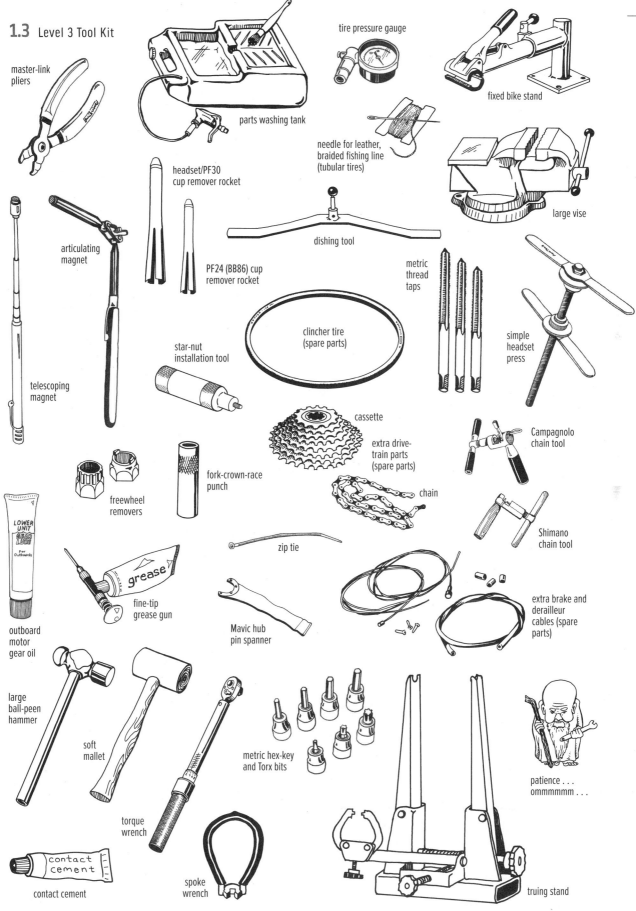

1.3 Level 3 Tool Kit

master-link pliers

parts washing tank

tire pressure gauge

fixed bike stand

needle for leather, braided fishing line (tubular tires)

large vise

headset/PF30 cup remover rocket

dishing tool

articulating magnet

PF24 (BB86) cup remover rocket

clincher tire (spare parts)

metric thread taps

simple headset press

telescoping magnet

star-nut installation tool

cassette

Campagnolo chain tool

extra drive-train parts (spare parts)

freewheel removers

fork-crown-race punch

chain

Shimano chain tool

zip tie

outboard motor gear oil

fine-tip grease gun

grease

Mavic hub pin spanner

extra brake and derailleur cables (spare parts)

large ball-peen hammer

soft mallet

metric hex-key and Torx bits

patience . . . ommmmmm . . .

torque wrench

contact cement

spoke wrench

truing stand

- **Soft mallet**. Choose leather, rubber, plastic, or wood to prevent damage to parts.
- **Headset press** used to install headset bearing cups (Fig. 12.44) and bearings and bearing cups into threadless bottom-bracket shells (Fig. 11.29). The press should not push on the bearing's inner race; if it does, get the appropriate insert to adapt your headset press to the particular headset and bottom-bracket bearings and cups you will be installing.
- **Headset-cup and PF30 bottom-bracket remover**. Often called a "rocket" tool due to its shape, this tool expands inside the head tube (Figs. 12.37, 12.38) or bottom-bracket shell behind the bottom bracket or headset cup.
- **Small press-fit PF24 (BB86) bottom-bracket remover**. Identical to a headset remover rocket but smaller for the press-fit bottom brackets for cranksets with 24mm integrated spindles.
- **Fork-crown race punch** (a.k.a. "slide hammer") for installing the fork-crown headset race (Fig. 12.43). Thin crown races require a second support tool to protect the crown race during installation.
- **Star-nut installation tool** for threadless headsets in metal fork steerer tubes.
- **Freewheel removers**. If you will be working on retro stuff, you'll need these for unscrewing freewheels from threaded hubs. Different freewheels take different removers, so only buy as you need them.
- **Torque wrenches** for checking proper bolt tightness. Most component manufacturers provide torque specs to prevent parts from stripping, breaking, creaking, or falling off while riding. There is a torque specification list in Appendix E of this book. You need a long torque wrench that goes to high torque for big items like crank bolts, bottom-bracket cups, and pedals, and a short torque wrench accurate at low torque settings for small items like stem bolts, shoe cleat bolts, and cable-clamp bolts. Also get a set of **metric hex-key and Torx bits** to fit the wrenches.
- Set of **metric thread taps** that includes 5mm × 0.8mm, 6mm × 1mm, and 10mm × 1mm. These will thread bottle bosses, seat binder clamps, derailleur hangers, and cantilever bosses (on touring or cyclocross frames).

- **Fine-tip grease gun** for parts with grease fittings and for Campagnolo headsets with grease holes.
- **Outboard motor gear oil** for lubricating some freehubs.
- **Truing stand** for truing and building wheels.
- **Dishing tool** for checking whether wheels are properly centered.
- **Spoke wrenches** for the wheels you own; you may require a splined one for splined nipples or a socket-type for wheels with internal nipples.
- **Pin spanner** for adjusting Mavic hubs.
- **High-quality chain tool** for chain installation. A rudimentary chain-breaker tool will be fine for occasional use, but you'll be much better off with a good one, particularly for 10- and 11-speed chains; see Pro Tip on chain tools in 4-12.
- **Needle** made for sewing leather, braided **fishing line**, and **contact cement** for patching tubular tires.
- One healthy dose of **patience** and an equal willingness to work and rework jobs until they have been properly finished.

Other Stuff

- **Spare parts** to save you from having to make a lot of last-minute runs to the bike shop for commonly used parts. Any well-equipped shop really requires several sizes of ball bearings, bolts, spare cables, cable housing, housing ferrules (cylindrical housing end caps), cable-end caps, valve extenders, and zip ties. You should also have a good supply of spare tires, tubes, chains, and cogsets.

1-4

NOW, IF YOU REALLY WANT A WELL-STOCKED SHOP . . .

The following tools (Fig. 1.4) go well beyond the Level 3 Tool Kit and sure come in handy when you need them.

- **Bottom-bracket tap set**. This tool cuts threads in both ends of the bottom bracket while keeping the threads in proper alignment. English threaded taps are required for most modern road bike frames. Most Italian-made metal frames, however, have Italian threads and will require appropriate taps. French threading and Swiss threading are separate standards and are extremely rare.

- **Pre-set torque keys** are faster for tightening stem bolts, cleat bolts, and other small bolts. These torque key handles accept standard hex-drive tool bits and are pre-set to a given torque setting. Most of them click over when the pre-set torque is reached; CDI pre-set torque keys ratchet when the torque setting is reached (like an automotive gas cap) and won't allow over-tightening.

- **Bottom-bracket facer.** This tool cuts the faces of the bottom-bracket shell so that they are parallel.

- **BB30 reaming cutter and backing plate (base plate)** that fit on bottom-bracket tap handles.

- **Bearing press/puller** for popping cartridge bearings in and out of external bottom-bracket cups.

- **Park BBT-39 bearing remover** for BB30 and BB90 bottom brackets.

- **Bushings** for pressing in bottom-bracket bearings and cups.

- **Park CBP-5 and CBP-3 bearing puller** for Campagnolo/Fulcrum Ultra-Torque bottom brackets, **arm puller**, **plug and pads** for Campagnolo Power Torque cranks, and **bearing puller extension** for removing Campagnolo Power Torque drive-side bearing.

- **Electric drill** with drill bit set for customizing. A cordless drill with an adjustable torque collar and a Torx T25 bit will install 6-bolt disc rotors in a jiffy.

- **Dropout-alignment tools** (a.k.a. "tip adjusters").

- **Derailleur-hanger-alignment tool** to straighten the derailleur hanger after you shift the derailleur into the spokes or crash on it. If your bike has a replaceable derailleur hanger, keep an extra hanger around.

- **Chain keeper** attaches to dropout to hold chain while cleaning drivetrain with wheel off.

- **Cog-wear-indicator gauge** determines whether cogs are worn out.

- **Three-way internal-nipple spoke wrench** with square-drive, 5mm, and 5.5mm sockets for the purpose of tightening spoke nipples internal to a deep rim. A specialty wrench may be needed for a non-standard internal nipple.

- **Spoke nipple screwdriver** with bent, free spinning shaft for quicker wheelbuilding.

- **Hub bearing press.** This tool has bushings for all bearing sizes and ensures that cartridge bearings press in aligned with each other.

- **Antitwist tool** for preventing bladed (aero) spokes from twisting during truing; Mavic and DT make good ones.

- **Spoke-tension gauge.** Brings spoke tension up to precise specs for long-lasting, stable wheels.

- **Shimano TL-EW02** plug-in/out tool for Shimano Di2 electrical connectors (Fig. 6.13A).

- **Magnetic wire-fish tool.** The magnet on the end of the fishing cable pulls electronic wires, cables, housings, and hydraulic hoses through the frame by means of a mating magnetic wire that grabs the connector. Campagnolo offers the tool shown here, specific to its EPS electronic wires. Park Tool's IR-1 Internal Cable Routing Kit (Fig. 5.30) has three magnetic wires and works for EPS and Di2 wires, as well as for standard shift and brake cables and for cable housings and hydraulic hoses.

- **Head-tube reaming and facing tool.** This tool keeps the two ends of the head tube parallel and bored out to the right size.

- **Park universal fork-crown-race remover.** This hefty tool can remove a fork-crown race from any shape of fork without using a hammer and screwdriver and suffering consequent collateral damage to the fork.

- **Heavy-duty headset press** installs headset bearing cups (Fig. 12.44) and bottom-bracket bearings and bearing cups into threadless bottom-bracket shells (Fig. 11.29) more quickly and accurately than a simple threaded headset press can. The press should not push on the bearing's inner race; if it does, get the appropriate insert to adapt your headset press to the particular headset and bottom-bracket bearings and cups you will be installing so that the insert presses only on the outer bearing race.

- **Digital chain-elongation gauge.** Precise monitoring of the gradual increase in chain length allows timely chain replacement without overdoing it.

- **Hydraulic hose cutter.** Hydraulic disc-brake hoses that are cut off cleanly and straight are less likely to leak.

- **Rotor-alignment dial indicator.** Rapidly finds out exactly where a disc brake's rotor is out of true.

1.4 Tools for the well-stocked shop

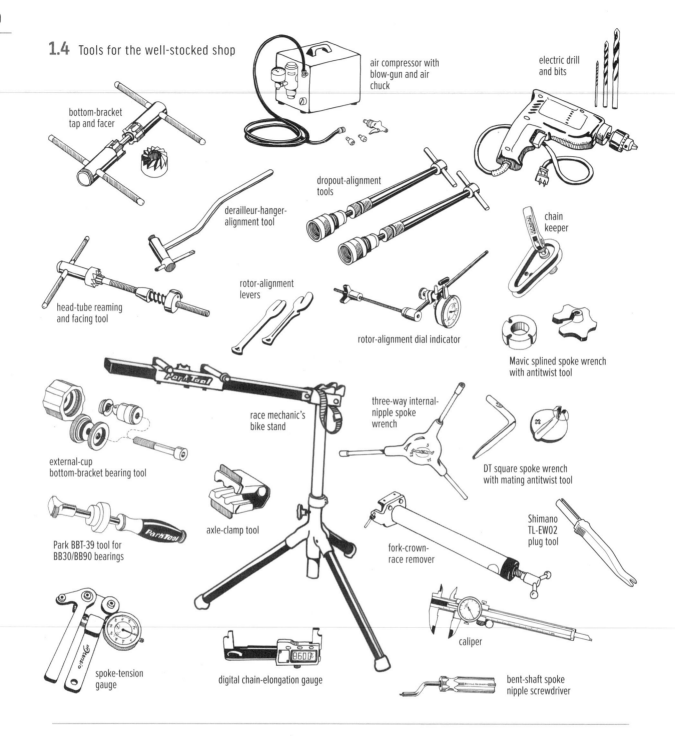

air compressor with blow-gun and air chuck

electric drill and bits

bottom-bracket tap and facer

derailleur-hanger-alignment tool

dropout-alignment tools

chain keeper

head-tube reaming and facing tool

rotor-alignment levers

rotor-alignment dial indicator

Mavic splined spoke wrench with antitwist tool

race mechanic's bike stand

three-way internal-nipple spoke wrench

external-cup bottom-bracket bearing tool

axle-clamp tool

DT square spoke wrench with mating antitwist tool

Park BBT-39 tool for BB30/BB90 bearings

fork-crown-race remover

Shimano TL-EW02 plug tool

spoke-tension gauge

digital chain-elongation gauge

caliper

bent-shaft spoke nipple screwdriver

- **Rotor-alignment levers**. Use in conjunction with dial indicator to precisely bend the rotor into alignment.
- **Cutting guide for threadless steering tubes**. The guide slot keeps the hacksaw blade lined up perpendicular to the steerer. Park makes a good one.
- **Crowfoot sockets**. Turn big nuts and bottom-bracket cups, including splined cups, to precise torque settings. Have the crowfoot at 90 degrees to the torque wrench handle to achieve torque setting shown on

the wrench handle (i.e., if the crowfoot is extended straight out or back, it multiplies or reduces the torque setting shown on the wrench handle).

- **Feeler gauges**. Measures precision of disc-brake pad spacing from the rotor.
- Measuring **caliper** with a vernier dial, or an electronic gauge, to precisely measure parts.
- **Axle-clamp tool** for clamping the end of a hub axle in a vise.

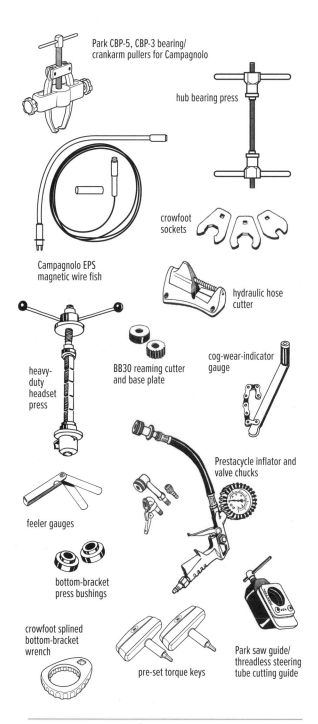

Park CBP-5, CBP-3 bearing/
crankarm pullers for Campagnolo

hub bearing press

crowfoot
sockets

Campagnolo EPS
magnetic wire fish

hydraulic hose
cutter

heavy-
duty
headset
press

BB30 reaming cutter
and base plate

cog-wear-indicator
gauge

feeler gauges

Prestacycle inflator and
valve chucks

bottom-bracket
press bushings

crowfoot splined
bottom-bracket
wrench

pre-set torque keys

Park saw guide/
threadless steering
tube cutting guide

- **Air compressor** to make quick work of mounting tires.
- **Prestacycle tire inflator**. Hook this baby up to your air compressor to inflate to exact pressure quickly.
- **Prestacycle valve chucks**. Inflate accurately regardless of the type or geometry of access of tire valves.
- A European-style **race mechanic's bike stand**, which supports the bottom bracket and has a long arm with a quick-release clamp to hold the fork ends or the rear dropouts, can be the only way to work

efficiently on a bike with an integrated seat mast or an aero seatpost, as there is no way to clamp such a bike in a conventional work stand. Alternatively, **Hirobel's Carbon Frame Clamp adapter** for carbon frames that do not have round tubes or a round seatpost allows you to safely hold them with a standard repair stand.

1-5

SETTING UP YOUR HOME SHOP

Make your shop clean, well organized, and comfortable, and you'll find that the speed and quality of your work will improve. Hanging tools on pegboard or slatboard or placing them in drawers, bins, or trays helps maintain an organized work area. Being able to lay your hand on the tool you need will increase the enjoyment of working on a bike. It is hard to do a job with loving care when you can't find the cable cutter. Placing small parts in a bench-top organizer, one with several rows of little drawers, is another good way to keep chaos at bay.

1-6

TOOLS TO CARRY ON A RIDE

a. For everyday rides

You can keep everything you need for light repairs (Fig. 1.5) in a **small bag** under your seat, in a backpack, or even in a jersey pocket. Look for tools and parts that are light and serviceable. I suggest getting the ones you want in a combination "multitool"; you can even get a lot of them combined into a rear quick-release hub skewer with the Pedro's Tülio! Test all tools at home before taking them on the road.

- **Spare inner tube** or **tubular**. Always carry one. Make sure the valve matches the ones on your bike and is sufficiently long for your rim depth. If rarely needed, keep it in a plastic bag to prevent deterioration.
- **Tire pump** or **air cartridge**. Longer is better for pumping but heavier for carrying. Road bike pumps need to be thin to attain high pressures. Minipumps are compact, but they're slow. Make sure the pump head matches the tire valves (Presta or Schrader). If you

1.5 Tools to take on all rides

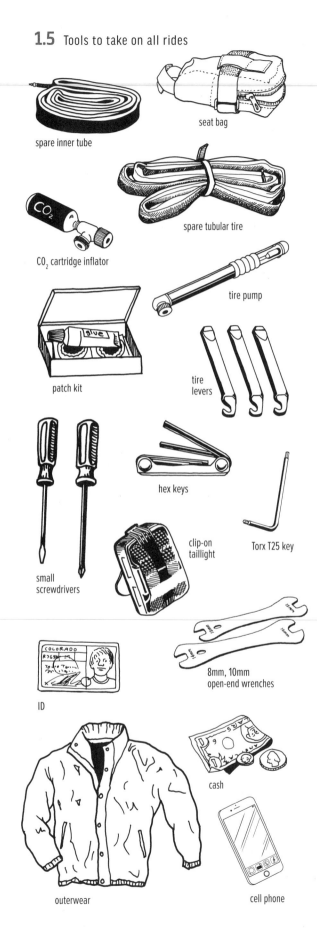

spare inner tube

seat bag

CO₂ cartridge inflator

spare tubular tire

tire pump

patch kit

tire levers

hex keys

small screwdrivers

clip-on taillight

Torx T25 key

8mm, 10mm open-end wrenches

ID

outerwear

cash

cell phone

prefer CO_2 cartridges, get the correct volume for the spare tube or tubular (probably 12g [grams], unless you are filling a huge touring tire, in which case you may need a 16g cartridge). Keep in mind that if you are running sealant inside of a tubeless tire, using CO_2 cartridges to inflate them will tend to solidify the sealant.

- At least two plastic **tire levers**, preferably three (for clinchers and tubeless tires).
- **Patch kit**. You'll need this if you puncture your spare tube. Check it at least once a year to make sure the glue has not dried out. You can also carry glueless patches.
- **Small screwdrivers** for adjusting derailleurs and other parts; ideally on a multitool.
- Compact set of **hex keys** that includes 2.5mm, 3mm, 4mm, 5mm, 6mm, and 8mm sizes; a folding set or multitool is a good investment.
- **Torx T25 key**, ideally on a multitool, if the disc brake bolts have Torx heads.
- For pre-1980s bikes, bring **8mm and 10mm open-end wrenches**. These are often included on some older multitools, eliminating the need to bring separate wrenches.
- Small clip-on **taillight**.
- Warm **outerwear**. Arm warmers, knee warmers, nylon vest or jacket, and a cap for a ride in the mountains or on any cool day. In the mountains and in questionable weather, thin gloves and shoe covers are also a good idea.
- **Identification**.
- **Cash** for food, phone calls, and to boot sidewall cuts in tires.
- **Cell phone**. As if I had to remind you to bring your phone! It can come in handy, although it can also interrupt your rides.
- **GPS computer** to find your way when lost, if your phone doesn't have this capability.

b. For long or multiday trips

Carry the items in Figure 1.6, as well as all of the items in Figure 1.5.

- **Spare folding clincher tire** and a second **spare inner tube**; if you ride tubulars, bring two spares of those.

- **Rain gear**.
- **Spoke wrench** that fits your wheels' nipples (can be on a multitool).
- **Chain tool** in case you break the chain. Chain tools (or "chain breakers") are often included in compact multitools, eliminating the need to bring a separate chain tool (as well as screwdrivers, hex keys, and even box-end or open-end wrenches and spoke wrenches). Try the chain tool at home on spare links you removed when installing new chains to make sure that you can repair the chain 100 percent of the time. This testing is important insurance on long solo rides that include extended stretches away from civilization or cell phone coverage.
- **Spare chain links** from your chain. If you are using a Shimano 8- or 9-speed chain, bring at least two subpin rivets or master links of that width. For 10- or 11-speed chains, bring master links for that width chain.
- **Spare spokes** of the right lengths for your wheels. Or, if you prefer, FiberFix sells a cool folding spoke made from Kevlar. It's worth getting one for emergency repairs on a long ride. See Chapter 3 for directions on how to use it.
- **Small plastic** bottle of **chain lube**.
- **Sealant-filled compressed-air tire inflators**. Especially if you're using tubulars (which you can't patch on the road), an inflator with sealant can get you home if you've ridden through thorns or over a tack.
- Small tube of **grease**.
- Small amount of **duct tape**.
- Small amount of **wire** and/or **zip ties**.
- Compact **15mm pedal wrench** if your bike requires it. Be sure to get one with a headset wrench on the other end, if your bike requires that as well.
- **Headlight**. This can be a lightweight unit to clip onto the handlebar or a headlamp with a strap that will fit over your helmet.
- **Matches**.
- A lightweight, aluminized, folding **emergency blanket**.

NOTE: *Read Chapter 3 on emergency repairs before embarking on a lengthy trip. If you are planning a bike-centered vacation, be sure to take along at least a Level 1 Tool Kit in your car, some headset wrenches (if* your bike has a threaded headset), and incidentals like duct tape and sandpaper. Also pack a few extra tires, a sturdy floor pump, and a spare derailleur hanger.

1.6 Tools for extended trips on the road

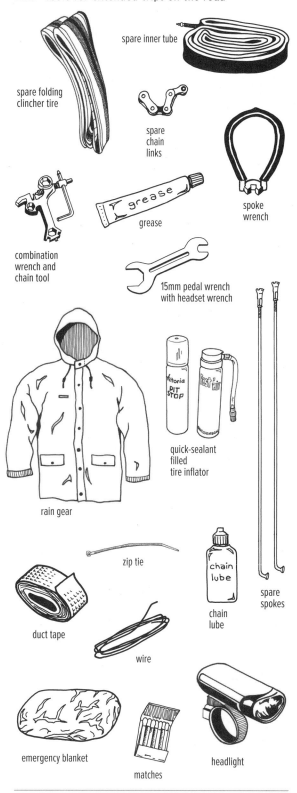

spare inner tube

spare folding clincher tire

spare chain links

grease

spoke wrench

combination wrench and chain tool

15mm pedal wrench with headset wrench

quick-sealant filled tire inflator

rain gear

zip tie

chain lube

spare spokes

chain lube

duct tape

wire

emergency blanket

matches

headlight

1-7

TOOLS FOR CYCLOCROSS RACING

Cyclocross races always have a "pit" where mechanics clean bikes and perform repairs and riders exchange dirty bikes for clean ones. Often the pit is set up so that riders pass it in two directions, meaning that they can pick up a clean bike every half lap. If you are performing mechanic service for a fast friend (maybe you are switching off, each doing service for the other if you race in different categories), you may have less than five minutes to clean the bike (and fix anything that your buddy yelled wasn't working when dropping it off) before he or she is back expecting a clean bike again. In a muddy race, you have to be efficient, which means having the correct tools (Fig. 1.7) as well as the right clothing and a calm demeanor.

- **Digital tire pressure gauge**. Accuracy is critical at low pressures with low-volume cyclocross tires.
- **Waterproof pants**.
- **Rubber gloves**.
- **Rubber boots**.
- Warm and/or waterproof **jacket and hat**, as conditions mandate.
- **Spare bike** set up the same as the bike the rider starts on (i.e., same pedals, same saddle and handlebar position, same type and number of cogs).
- **Spare wheels** with the same type cogs (i.e., same or compatible brand, same speeds) and same brake discs as the bikes have.
- **Spare shift and brake cables, chain links**, and **master links**.
- **Spare saddle, seatpost**, and **seat binder clamp**. You'd be amazed how often these parts break in cyclocross.
- Two or three large, stable, reusable **buckets**, ideally that nest together.
- Large **sponge**(s).
- Large **brush**.
- Small, stiff **cylindrical brush** and/or **narrow brush** with long, thin bristles.
- Curved plastic **cog pick**.
- Environmentally friendly **bike cleaner** or dish soap.
- Environmentally friendly **degreaser**.
- **Chain lube**.

- 3mm, 4mm, 5mm, 6mm, 8mm, and 10mm **hex keys** and **Torx T25 and T30 keys**, two of each in case you lose some in the mud. Don't waste precious minutes searching for lost tools; find them after the race.
- **Long, thin screwdriver** for derailleur adjustment and cleaning mud out of tight spaces.
- **Large screwdriver**.
- **Scissors**.
- **Pliers**.
- **Cable cutter**.
- **Floor pump**.
- **Chain keeper** attaches to dropout to hold chain while cleaning drivetrain with wheel off.
- **Crank puller**, if the bike requires more than a hex key to remove the crank.
- **Spoke wrench**.
- **Duct tape**.
- **Chain tool**.
- **Fresh water**. Lots, if it's a muddy race.
- **Rags**. Lots.

One quick way to equip yourself for the task is with Pedro's Pit Kit: a bucket containing a sponge, brushes, tools, bike cleaners, degreasers, and lubricant. Its lid allows you to toss other tools in it and keep everything together in the car and when walking to the pit. Use earth-friendly lubes and cleaners; after all, you are generally doing this in a public park, an open field, or some generous institution's lawn, and you don't want to despoil it with diesel fuel or other toxic solvents.

Other stuff

- A **portable bike stand** can come in handy.
- If you expect a muddy race and your buddy is a superstar, you can bring a portable **pressure washer**, either rechargeable battery-powered or gas-powered. Races often supply a communal pressure washer (or a hose), which is great unless it's frozen or the wait is too long to get the bike clean before your rider comes back around. Make sure you have enough water; if there's no pond or lake to pull from at the race site, you'll need a large rain barrel with a hose fitting at the bottom, available at garden stores. Also be sure you have enough battery power or gas to run the compressor through the race.

1.7 Tools for cyclocross racing

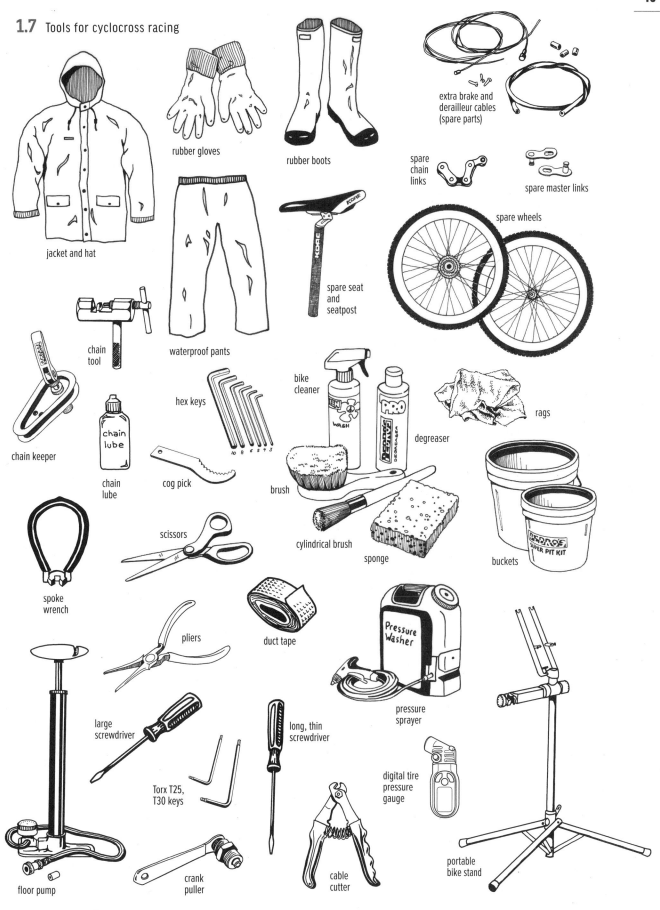

jacket and hat

rubber gloves

rubber boots

extra brake and derailleur cables (spare parts)

spare chain links

spare master links

spare wheels

waterproof pants

spare seat and seatpost

chain tool

bike cleaner

degreaser

rags

chain keeper

hex keys

chain lube

cog pick

brush

cylindrical brush

sponge

buckets

spoke wrench

scissors

pliers

duct tape

pressure sprayer

large screwdriver

long, thin screwdriver

pressure washer

digital tire pressure gauge

Torx T25, T30 keys

floor pump

crank puller

cable cutter

portable bike stand

> *Basic research is what I am doing when I don't know what I am doing.*
> —WERNER VON BRAUN

BASIC STUFF

PRERIDE INSPECTION, WHEEL REMOVAL, GENERAL CLEANING, AND MECHANICAL METHODS GUIDE

TOOLS

Chain lubricant
Rags

OPTIONAL

Solvent (citrus-based)
Chain-cleaning tool
Old water bottle
Bucket(s)
Large sponge
Large and small brushes
Dish soap
Cyclocross pit kit (see 1-7)

Always check your bike before heading out on a ride. This inspection can help you avoid injury as well as getting stranded far from home due to parts failure. You should know how to remove and reinstall a wheel so that you can deal with minor annoyances like flat tires or jammed chains. And even if you do nothing else to your bike, keeping its chain clean and properly lubricated, as outlined in this chapter, will make every ride smoother and quieter.

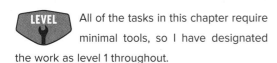 All of the tasks in this chapter require minimal tools, so I have designated the work as level 1 throughout.

2-1

PRERIDE INSPECTION

1. **Check that the wheel quick-release levers or axle nuts (which secure the hub axle to the dropouts) are tight.**
2. **Check the brake pads for excessive or uneven wear.**
3. **Grab and twist the brake pads and brake arms to make sure that the bolts are tight. With disc brakes, check that the caliper is tight.**

4. **Squeeze the brake levers.** A good squeeze should bring the pads flat against the rims (or brake rotor) without hitting the tires. Make certain that you cannot squeeze the levers all the way to the handlebar. See Chapter 9 for brake cable adjustment (Chapter 10 for disc brakes).
5. **Spin the wheels while eyeing the rims, not the tires.** Check for wobbles. Make sure that the rims or rotors do not rub the brake pads.
6. **Spin the wheels again, this time eyeing the tires.** Check for wobbles. If a tire wobbles excessively on a straight rim, it may not be fully seated in the rim. There is usually a mold line or an edge of a tape strip on the tire that should be parallel to the rim edge all the way around. Look for areas where the tire bulges and/or the mold line or tape edge is higher above the rim or deeper into the rim than the rest of the way around the tire. To fix an improperly seated tire, you need to completely deflate the tire, carefully seat it uniformly all the way around, and then inflate it to the proper pressure.
7. **Check the tire pressure.** On most road bike tires, the proper pressure is between 80 and 120 pounds per square inch (psi). Look to see

that there are no foreign objects sticking in the tire. If there are, you may have to pull the tube out and repair or replace it. If you have an aversion to fixing flats, turn to the section on tire sealants (i.e., goop inside the tube that fills small holes) in Chapter 7 (7-15).

8. **Check the tires for excessive wear, cracking, bulges, or gashes.**

9. **Make certain that the handlebar and stem are tight and that the stem is lined up with the front tire.**

10. **Check that the gears shift smoothly and the chain does not skip or shift by itself.** Ensure that each indexed (click) shift moves the chain one cog, starting with the first click. Make sure that the chain does not overshift the smallest or biggest rear cog or the inner or outer front chainring, which would throw the chain off to one side or the other.

11. **Check the chain for rust, dirt, stiff links, or noticeable signs of wear.** It should be clean and lubricated (but not overlubricated; gooey chains pick up lots of dirt). The chain should be replaced on a road bike about every 1,500 to 3,500 miles of paved riding. See 4-6 to accurately evaluate chain wear.

12. **Apply the front brake and push the bike forward and back.** The headset (fork bearings; see Fig. 12.1) should be tight and not make clunking noises or allow the fork any fore-aft play.

13. **Ensure that the crank bearings and crankarms are not loose.** Grasp one crankarm and push and pull it laterally, toward and away from the frame, checking for play.

14. **Ensure that the hub bearings are not loose.** Check for bearing play by grasping each wheel around the rim and tire and pushing it side to side.

15. **Check that the saddle is tight and straight.** Make sure it does not wobble or twist easily.

16. **If all this checks out, go ride your bike!** If not, check the table of contents, go to the appropriate chapter, and fix the problems before you go out and ride.

2-2

REMOVING AND INSTALLING THE FRONT WHEEL

You can't fix a flat if you can't remove the wheel. Front wheel removal is also generally required for placing a bike on a roof rack or in a car. As outlined in the following sections, wheel removal usually involves releasing a rim brake (in most cases) before opening the hub quick-release, bolt-on skewer, or the axle nuts. A through-axle requires also pulling out the axle.

To install the front wheel, leave the caliper rim brake open (or the tire deflated, if there is no brake quick-release and the tire won't fit through the brake while inflated) and slide the tire past the brake pads while lowering the fork onto the wheel so that the bike's weight pushes the top of the dropout slots down onto the hub axle. This action will seat the axle fully into the fork and center the rim between the brake pads. The procedure is the same with a disc brake except you don't have to worry about opening the brake; you merely slip the disc rotor between the brake pads in the caliper and follow the rest of the earlier instructions.

If the fork or wheel is misaligned, you will need to readjust the brake or hold the rim centered between the brake pads when securing the hub (and as soon as you can, true the untrue wheel—see 8-2—or get the bent fork fixed or replaced).

2-3

RELEASING THE BRAKE

Most rim brakes have a quick-release (QR) mechanism to open the brake arms so that they spring away from the rim, allowing the tire to pass between the pads. Most road bike sidepull brake calipers have a lever that you flip up to open the brake (Fig. 2.1). Alternatively, Campagnolo Ergopower systems have a pin near the top of the brake lever that you push outward to allow the lever (and consequently the caliper) to open wider (Fig. 2.2). Cheap sidepull brakes on cheap bikes—as well as early sidepull brakes on classic racing bikes from the late 1960s and early 1970s—don't have a quick-release for the brake. The same is often true with time trial bikes and triathlon bikes, as well as bikes with Campagnolo brake calipers coupled with non-Campagnolo aerodynamic brake levers on the ends of bullhorn bars (see Fig. i.2).

If the brake has no quick-release and the tire is not skinny enough to slip past the brake pads as the wheel is removed, you need to deflate the tire to avoid damag-

2.1 Releasing the brake

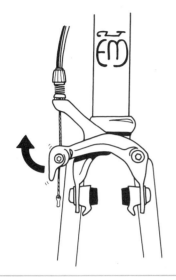

2.2 Releasing a Campagnolo brake

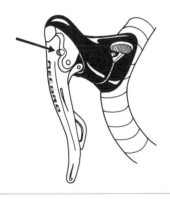

ing it and perhaps even dislodging the brake pads when pulling the wheel out.

Center-pull brakes (rare now, but common on pre-1975 bikes; see Fig. 9.3) have a cable-hanger yoke that must be pulled down to release it from the straddle cable that connects to the brake arms; squeeze the brake pads against the rim with one hand and release the yoke from the cable with your other hand.

Some cyclocross bikes and touring bikes have cantilever-type brakes (Figs. 9.23–9.43) that mount on pivots attached to fork legs or seatstays. Most standard cantilever brakes are released by holding the pads against the rim with one hand and pulling the enlarged head of the straddle cable out of a notch in the top of the brake arm with the other hand. Really old cantilever brakes are released like the center-pull brakes mentioned in the previous paragraph.

Another type of cantilever brake, less common on road and cyclocross bikes because of limited compatibility with the brake levers on drop bars, is commonly called a "V-brake" (Fig. 9.47), after a popular Shimano design. Mountain-bike V-brakes are incompatible with drop-bar brake levers because more cable pull is required to move the long brake arms than they can provide. Short-arm road/cyclocross V-brakes work better with road levers, but care is still required when pulling the front brake due to their high leverage. A V-brake is released by pulling the end of the curved cable-guide tube (the "noodle") out of the horizontal link atop one of the brake arms while squeezing the pads against the rim with the other hand.

Disc brakes (Figs. 10.1–10.2) require no release of the pads; the rotor should simply drop straight down out of the slot in the caliper.

2-4

REMOVING AND INSTALLING A FRONT WHEEL WITH A QUICK-RELEASE SKEWER

Removal

You don't need a tool for this task.

1. **Pull the lever outward to open it (Fig. 2.3).**
2. **Unscrew the nut on the opposite end of the quick-release skewer's shaft until it clears the fork's wheel-retention tabs.** Older forks often don't have these tabs, thus making unscrewing the nut unnecessary.

2.3 Opening a quick-release skewer

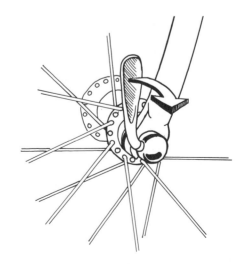

Installation

The quick-release skewer is not a glorified wing nut and should not be treated as such.

1. **Hold the quick-release lever in the "open" position.**

2. **Tighten the opposite end nut until it snugs up against the face of the dropout.** If there are no wheel-retention tabs on the fork and you did not unscrew the skewer nut, this step is unnecessary because the skewer will still be in adjustment.

3. **Push the lever over (Fig. 2.4) to the "closed" position (it should now be at a 90-degree angle to the axle).** It should take a good amount of hand pressure to close the quick-release lever properly; the lever should leave its imprint on your palm for a few seconds.

4. **If the quick-release lever does not close tightly, open the lever again, tighten the end nut a quarter turn, and close the lever again.** Repeat until tight.

5. **If, on the other hand, the lever cannot be pushed down flat, then the nut is too tight.** Open the quick-release lever, unscrew the end nut a quarter turn or so, and try closing the lever again. Repeat this procedure until the quick-release lever is fully closed and snug. The lever should leave an imprint on the palm of your hand for a few seconds. When you are done, it is important to have the lever pointing straight up or toward the back of the bike so that it cannot hook on obstacles and be accidentally opened.

6. **Hit the top of the tire with your open palm to check that the wheel is not loose and you cannot bang it out.**

2-5

REMOVING AND INSTALLING A WHEEL WITH AXLE NUTS

Removal

1. **Unscrew the nuts on the axle ends (usually with a 15mm wrench) until they allow the wheel to fall out (Fig. 2.5).** Really old road bikes may have wing nuts for finger tightening instead.

2. **Loosen the nuts enough to clear the retention tabs on the fork ends.** Your bike may have some type of wheel retention system consisting of nubs or bent tabs on the fork ends, or an axle washer with a bent tooth hooked into a hole in the fork end. These systems prevent the wheel from falling out if the axle nuts loosen. Removing the nuts completely from the axle is not usually necessary.

3. **Pull the wheel out.**

Installation

Snug up the nuts clockwise (opposite direction in Fig. 2.5) with a wrench (usually 15mm) a little from each side

---

2.4 Tightening the quick-release skewer

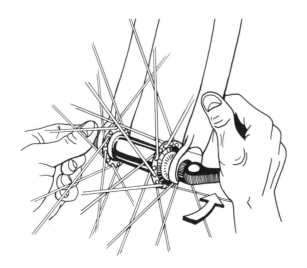

2.5 Loosening the axle nut

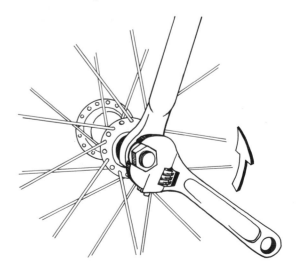

2.6 Bolt-on skewer

until they are quite tight. In the case of wing nuts, the procedure is the same: The tools are your fingers.

NOTE: *Some bikes—usually those of lightweight fanatics or track bikes with hollow axles—have bolt-on skewers (Fig. 2.6), often made of titanium to save weight. The wheel is removed by unscrewing the skewer with a 5mm hex key.*

When installing the wheel, hold the end nut with one hand and tighten the bolt-on skewer (Fig. 2.6) with a 5mm hex key. One maker of these skewers recommends 65 inch-pounds (in-lbs) of tightening torque for steel bolt-on skewers and 85 in-lbs for titanium ones. You can approximate the accurate tightening torque by using a short hex key and tightening as tight as you can with just your fingers, not your palm. These skewers can be overtightened; avoid that problem by being conscious of how much pressure a quick-release skewer applies and do not go higher than that. But make sure it is tight enough to securely hold the wheel and will not loosen on its own.

2-6

REMOVING AND INSTALLING A WHEEL WITH A THROUGH-AXLE

A feature appearing on many bikes with disc brakes is the "through-axle," a motorcycle-style means of attaching the front wheel. Front through-axles are extra-long, removable front-wheel axles, either 12mm or 15mm in diameter, which fit through the hub cartridge bearings and are clamped directly into the fork ends. They stiffen the fork against lateral and torsional flex, and they offer a higher degree of safety against the wheel falling out of the fork than a quick-release skewer. In particular, the closed ends of the fork oppose the braking force on a disc brake, which is forcing the hub axle downward and could push it out of a standard dropout.

2.7 Loosening a through-axle

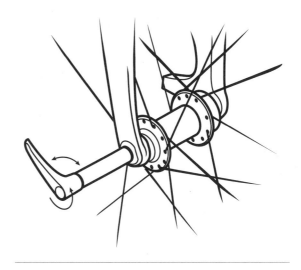

Road through-axles (Fig. 2.7) have a lever that you use as a handle to unscrew the axle from the opposite fork end. Many of them flip open and closed like a standard quick-release lever. A through-axle can be quicker to remove and install than a standard hub with a quick-release, when you include the time it takes to loosen and tighten a conventional skewer enough to clear the fork-end tabs, although a through-axle does complicate mounting the fork on a roof rack's fork mounts.

Removal

1. **Flip open the lever.** It may need to flip into a notch. Skip this step with a DT Swiss through-axle.
2. **Rotate the lever counterclockwise and unscrew the entire axle from the opposite leg of the fork (Fig. 2.7).**
3. **Pull the axle out.**
4. **Let the wheel drop out of the fork.**

Installation

1. **Stick the wheel into the fork (sans axle) with the brake rotor sliding up between the pads**

of the disc-brake caliper. The fork ends usually have small inboard flanges that sit on the hub end caps with everything lined up to insert the axle; if not, you'll have to carefully hold the bike up as you line up the holes through the hub and fork ends to install the axle.

2. **Push the axle through from the side with the bigger hole in the fork end.**

3. **Screw the axle (clockwise) into the opposite fork leg (Fig. 2.7) using the lever as a handle.**

4. **Flip the lever to the closed position.** The feel for proper adjustment is similar to that of a quick-release skewer lever (2-4); rotate the axle in or out until you get a firm, but not too firm, feel as you push the lever over. Note that a DT Swiss through-axle's lever does not flip over; it stays fixed at 90 degrees to the axle, but it does have DT's RWS ratchet wheel-mounting system, which allows you to reposition the lever. After you tighten a DT axle (by turning the lever), pull outward on the RWS lever (you're pulling against a spring) and turn it to whatever position you want the lever to point (so it's less vulnerable to catching on shrubbery, for instance). Release it, and it will snap down and stay fixed in position, engaged in its splines.

Automobile roof racks and truck-bed carrying racks present a challenge for a through-axle fork. If the rack clamps the fork ends, you must first install a through-axle adapter like the Hurricane Components Fork Up, which is a tube welded to two slotted plates. The slotted plates clamp onto the fork mount like standard slotted fork ends, and, once you set the fork over it, you push the axle through the tube and secure it as in steps 3 and 4.

2-7

CLOSING THE BRAKES

The steps required to close the brakes are always the reverse of what you did to release them.

1. **For sidepull road brakes:** With most road bikes, closing the brake caliper is simply a matter of flipping closed the quick-release lever on the sidepull brake caliper (Fig. 2.1 in reverse). With Campagnolo Ergopower, pull the brake lever and push the pin

inward to engage the shallower notch in the lever body (Fig. 2.2).

2. **For cantilever, center-pull, V-, and disc brakes:** With many cantilever brakes (cyclocross and some touring bikes; Figs. 9.23–9.43), hold the brake pads against the rim with one hand and hook the enlarged end of the straddle cable back into the end of the brake arm with your other hand. On older bikes with center-pull brakes (Fig. 9.3), as well as with some cantilevers, hook the straddle cable yoke under the straddle cable.

3. **For V-brakes:** Squeeze the pads against the rim with one hand and pop the end of the curved cable-guide tube ("noodle" in Fig. 9.5) back into the horizontal link atop one of the brake arms with your other hand.

4. **For disc brakes:** There is no quick-release to close on a disc brake (Figs. 10.1–10.2); simply slide the rotor up between the pads. Make sure with hydraulic disc brakes that you don't squeeze the lever when there is no rotor or spacer between the pads, because it may push the pads out too far to allow the rotor to fit in between them, and then you will need to push the pistons back in their bores before you can install the wheel.

5. **Check that the brake cables are connected securely by squeezing the levers.** Lift the front end of the bike and spin the front wheel, gently applying the brakes several times. Check that the pads are not dragging. If they are, center the wheel (or adjust the brakes as described in Chapter 9 or 10). If everything is reconnected and centered properly, you're done. Go ride your bike.

2-8

REMOVING THE REAR WHEEL

Removing the rear wheel is just like removing the front, with the added complication of the chain and cogs.

1. **Open the brake (or deflate the tire), as outlined in 2-3.**

2. **Shift the chain onto the smallest rear cog:** Lift the rear wheel off the ground, turn the cranks, and shift.

2.8 Moving the rear derailleur and chain to remove or install the rear wheel

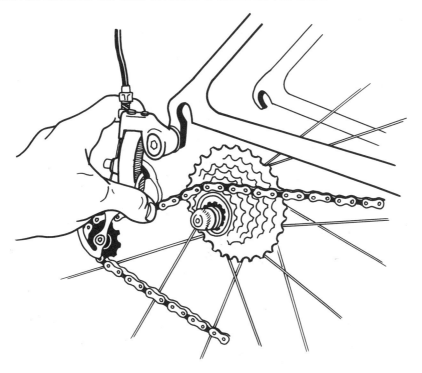

3. **To release the wheel from the rear dropouts and the brakes, follow the same procedure as for the front wheel.** When you push the wheel out, you will need to move the chain out of the way. Unless your bike has rear-entry dropouts (see 2-9 next), this maneuver is usually a matter of grabbing the rear derailleur and pulling it back so that its jockey wheels (pulley wheels that guide the chain onto the cogs) move out of the way, while you push forward on the quick-release or axle nuts with your thumbs and let the wheel fall as you hold the bike up (Fig. 2.8). If the bottom half of the chain catches the wheel as it falls, lift the wheel to free it.

A SRAM CX-1 rear derailleur is made for a single front chainring and a wide-range cassette; getting the rear wheel out requires overcoming its chain retention clutch system. Pull the cage down as in Figure 2.8, and then push in the cage-lock button extending from the derailleur's lower pivot (the button would be under the thumb in Fig. 2.8) until it locks in; the cage will now be locked out straight in line with the derailleur parallelogram plates.

2-9

REMOVING THE REAR WHEEL FROM REAR-ENTRY DROPOUTS

Some time trial and triathlon bikes have tight clearance at the seat tube; there may even be a cutout in the seat tube for the rear wheel to "hide" from the wind (Fig. i.2). These bikes may employ rear-entry dropouts to allow the rear wheel to fit very tightly behind the seat tube. The dropouts work well but can be daunting if you've never dealt with them. Prepare to get your hands dirty.

1. **Open the brake (or deflate the tire), as outlined in 2-3.** If the bike has a disc brake, you can ignore this step.
2. **Shift the chain onto the smallest cog:** Lift the rear wheel off the ground, turn the cranks, and shift. It's a good idea to shift onto the small chainring as well, to create more chain slack.
3. **Open the rear quick-release skewer.**
4. **You must move the chain out of the way before you can release the wheel from the rear-facing dropouts.** Grab the chain with your fingers and pull it back (off the rear cog) and to the right (Fig. 2.9).

2.9 Removal and installation of rear wheel from rear-entry dropouts

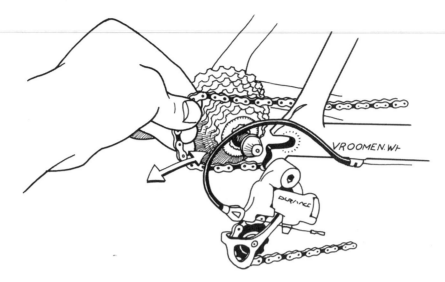

5. **With the chain now out of the way, grab the rear wheel and pull it straight back and out of the bike (Fig. 2.9).**

2-10

INSTALLING THE REAR WHEEL INTO STANDARD DROPOUTS

1. **Check to make sure that the rear derailleur is shifted to its outermost position (over the smallest cog).**
2. **Slip the wheel between the seatstays and between the brake pads; with a disc brake make sure the brake rotor slips between the pads in the caliper.** Maneuver the upper section of chain onto the smallest cog (Fig. 2.8).
3. **Set the bike down on the rear wheel.**
4. **As you let the bike drop down, pull the rear derailleur back with your right hand and pull the axle ends back into the dropouts with your index fingers.** Use your thumbs to push forward on the rear dropouts, which should now slide over the axle ends. If the axle does not slip into the dropouts, you may need to unscrew the nut on the quick-release skewer further. If the spacing of the dropouts and wheel axle length don't match, you may need to spread the dropouts apart or squeeze them toward each other as you pull the wheel in.

With a SRAM CX-1 rear derailleur, push forward on the cage to release the cage-lock button once the wheel is in.

5. **Check that the axle is fully seated in the dropouts;** this should result in the rim being centered between the brake pads. If it is not, readjust the brake or hold the rim in a centered position as you secure the axle. This procedure should not be necessary if the wheel and frame are both aligned and the brakes are centered. Some dropouts have adjuster screws to adjust the depth to which the wheel slides back into the dropouts (see Fig. 2.10); adjust these as needed to center the wheel between the chainstays.
6. **Tighten the quick-release skewer, axle nuts, or bolt-on skewer as explained for the front wheel.**
7. **Reconnect the rear brake the same way as you did on the front wheel.** Check the brake action. If all's well, you're done. Go ride your bike.

2-11

INSTALLING THE REAR WHEEL INTO REAR-ENTRY DROPOUTS

Installing the wheel is the reverse of the removal procedure.

1. **While holding the chain off to the side as you did in 2-9, put the wheel straight into the rear-entry**

2.10 Horizontal dropouts with adjuster screws to center the rear wheel

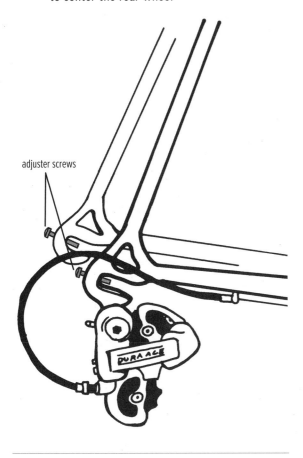

adjuster screws

INSTALLING THE REAR WHEEL WITH A THROUGH-AXLE

1. **With the axle out, follow steps 1–5 in 2-10, while sliding the disc rotor up between the brake pads.** The shelves on the rear dropouts should rest on the hub end caps.
2. **Slide the axle in and tighten it.** Screw it in and flip the lever over to proper adjustment as in 2-6.
3. **Check the brake action.** All's well? Ride on.

FIXING A FLAT TIRE

Now that you can get the wheels in and out, you can fix a flat tire when it happens to you. To do so, follow the directions at the beginning of the next chapter (3-3).

CLEANING YOUR BICYCLE

Most cleaning can be done with soap, water, sponges, and brushes. Soap and water are easier on you and the earth than stronger solvents, which are generally needed only for the drivetrain, if at all.

While using high-pressure car washes to clean your bike is sometimes unavoidable when your bike gets covered with mud while traveling, be careful when using them. The high pressure forces water into bearings and frame tubes, causing extensive damage over time. If you do use a pressure washer, never point it toward hubs and bottom brackets from the side, as it can blow the bearing seals inward; instead, always point it in the plane of the bike. Be careful about pointing the high-pressure water at electronic derailleur connections as well.

The best way to set up your bike for cleaning is to put it in a bike stand. Alternatively, you can lean it against a fence or the like. After initial washing, you can remove the front wheel, stand the bike on the fork and handlebar, and lean it against something.

1. **If the bike is really dirty, start by washing it with a hose while the wheels are on.** A car-washing

dropouts (Fig. 2.9). Slide it forward until it hits the front of the dropout.

2. **Tighten the quick-release to lock the wheel in place.**
3. **Check to see if the wheel is centered in the chainstays and rear brake.** If the wheel is dished correctly and the frame is straight, the wheel should be centered in the frame. If it's not, loosen the quick-release and center the wheel.

Many bikes with rear-entry dropouts have adjusters to set the depth for the axle. Use these to center the wheel before tightening the skewer.

4. **Grab the chain, pull it backward slightly, and put it back onto the small cog. Retighten the skewer.**
5. **Close the rear brake quick-release (as in 2-7).**
6. **Operate the rear brake several times to be sure the wheel is centered between the pads.** Also pedal the bike in the stand and shift the rear derailleur to ensure that the shifting is normal.

brush that screws onto the end of the hose can come in handy.

2. **Remove the wheels.** The wheels can be cleaned easily while they are on the bike, but even more easily when off. Remove the wheels to clean the frame, fork, and components.

3. **Support the chain.** If the bike has a chain hanger, hook the chain over it (the chain hanger is a little nub attached to the inner side of the right seatstay, a few centimeters above the dropout). If not, pull the chain back over a dowel rod (Fig. 2.11), an old rear hub secured in the dropouts, or a chain keeper (Fig. 1.4 or 1.7).

4. **Fill a bucket with hot water and dish soap.** With a stiff nylon-bristle brush and a big sponge, scrub the entire bike and wheels. Leave the chain, cogs, chainrings, and derailleurs for last; scrub those with a different brush. You can also try bicycle-specific cleaners like Pedro's Green Fizz, Finish Line Super Bike Wash, ProGold Bike Wash, etc.

5. **Rinse the bike with water by hosing it off (low pressure!) or wiping it with a wet rag.** Avoid get-

ting water in the bearings of the bottom bracket, headset, pedals, or hubs. Note that most metal frames and forks have tiny vent holes in the tubes; these were drilled at the factory to allow hot air to escape during welding. The holes on the seatstays, fork legs, chainstays, and seatstay and chainstay bridges can let water in. Avoid it by taping over the vent holes. Leaving them permanently taped to keep water out when riding is a good idea.

2-15

CLEANING THE DRIVETRAIN

The drivetrain consists of an oil-covered chain running over gears and through derailleurs. Sounds messy, doesn't it? Because the whole affair is generally exposed to the elements, it picks up lots of dirt.

However, the drivetrain transfers your energy into the bike's forward motion, and so it should be kept clean so that it can move freely. Frequent cleaning and lubrication will keep it rolling well and extend the life of the drivetrain.

2.11 Looping the chain over a dowel rod for cleaning

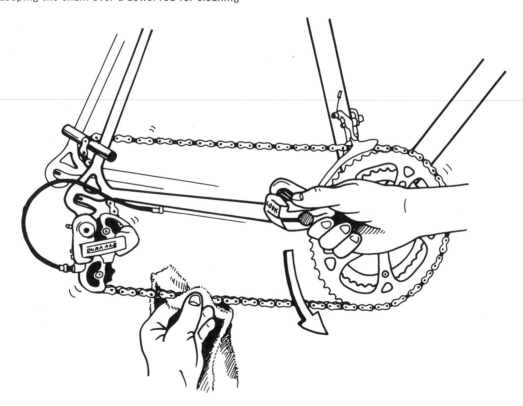

Fortunately, the drivetrain rarely needs to be completely disassembled for intensive cleaning. If you keep after it, regular maintenance can be confined to wiping down the chain, derailleur pulleys, and chainrings with a dry rag and then lubricating. I recommend wearing rubber gloves to keep your hands clean during this job.

1. **To wipe the chain, turn the cranks while holding a rag in your hand and grabbing the chain (Fig. 2.11).**

2. **Clean the rear derailleur's jockey wheels.** Holding a rag, squeeze the teeth of the rear derailleur's jockey wheels between your index finger and thumb as you turn the cranks (Fig. 2.12). This procedure will remove grease and dirt that have built up on the jockey wheels.

3. **Slip a rag between each pair of rear cogs and work it back and forth until each cog is clean (Fig. 2.13).** An even better method is to use Finish Line Gear Floss, a soft, absorbent rope.

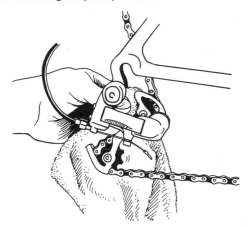

2.12 Cleaning the jockey wheels

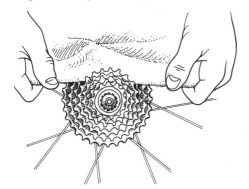

2.13 Cogset flossing with a rag

4. **Wipe down the derailleurs and the front chainrings with the rag.**

The chain will last much longer if you perform this sort of quick cleaning regularly, followed by dripping chain lube on the chain and another light wipe-down. You will also be able to avoid the kind of heavy-duty solvent cleanings that become necessary when the chain and cogs get really grungy from lack of regular cleaning.

You can remove packed-up road grit from derailleurs and cogs with the soapy water and scrub brush. Note, however, that the soap may not dissolve the dirty lubricant that is all over the drivetrain; rather, the brush may smear it all over the bike if you're not careful. Use a different brush from the one you use for cleaning the frame. Follow the drivetrain cleanup with a cloth wipe-down of the frame. Follow the lubrication steps 6–7 in the next section.

2-16

CLEANING THE CHAIN WITH SOLVENT

When a chain gets really dirty, the only way to rescue it is with an immersion in solvent—a nasty task worth avoiding by performing the regular maintenance just described. In fact, if you are sparing with the chain lube—that is, if you only drip it on the chain rollers where it is needed, rather than spraying it all over the chain—you can minimize, if not avoid, the need for solvent cleaning with its associated disposal and toxicity problems.

If you cannot avoid using a solvent, work in a well-ventilated area, use as little solvent as necessary, and pick an environmentally friendly mixture. There are many citrus-based solvents on the market that will reduce the danger to your lungs and skin and create less of a disposal problem. If you are using a lot of solvents, organic ones such as diesel fuel can be recycled, which may be preferable to citrus solvents, as long as you protect yourself from the fumes with a respirator. All solvents suck the natural oils from your skin, so be sure to wear rubber gloves, even with "green" solvents.

A self-contained chain cleaner with internal brushes and a solvent bath is a quick and convenient way to clean a chain (Fig. 2.14). See instructions given in 4-3.

2.14 Solvent cleaning of the chain

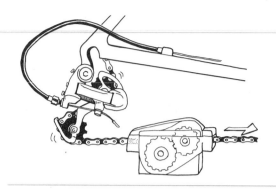

2.15 Dripping oil only where it is needed

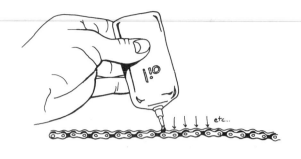

A nylon brush or an old toothbrush dipped in solvent is good for cleaning cogs, pulleys, and chainrings, and it can be used for a quick cleanup of the chain as well. Unless the chain has a master link, do not remove it to clean it in a solvent bath. Modern 11-speed, 10-speed, 9-speed, and even 8-speed chains should not be taken apart for cleaning; each chain rivet is so short that, while riding, it can pop out of a hole enlarged by removal and reinstallation of the rivet. However, chains with master links, which are available all the way up to supernarrow 11-speed models, can be removed for cleaning without being damaged.

1. **Follow the directions in Chapter 4 (4-7) for removing the chain.**
2. **Put the chain in an old water bottle about one-quarter full of solvent.**
3. **Shake the bottle vigorously to clean the chain.** Hold the bottle close to the ground, in case it leaks. Do not leave the chain to soak in the solvent.
4. **Hang the chain to dry completely.**
5. **Install the chain on the bike, following the directions in Chapter 4.**
6. **Drip chain lubricant into each of the chain's links and rollers (Fig. 2.15) as you turn the cranks to move the chain past the drip bottle.** Though it gets more lube on the chain to attract dirt, it's most time-efficient to drip lube on the moving chain by gently squeezing the bottle with the tip on each top edge of the chain for a couple of turns of the crank on each side.
7. **Lightly wipe down the chain with a clean rag.** You want to remove excess lubricant on the outside, where it is not needed.

2-17

CYCLOCROSS BIKE CLEANING IN THE "RACE PIT"

If you are supporting a cyclocross racer, you will be working in the service pit, the section set aside by the race organizer where racers exchange bikes and get other service. You will be cleaning and lubricating the bike similarly to the method described in 2-14 through 2-16, except very quickly, as you will only have a few minutes to do it before your rider is back, trading a dirty bike for the one you've just cleaned and lubed.

You will need the equipment and clothing listed in 1-7 and the tools shown in Figure 1.7. You will have either a hose or a pressure washer supplied by the race (shared with other mechanics) or only water that you brought, perhaps supplemented by buckets of water from a nearby pond. If you are really a pro, you may have brought a portable pressure washer, along with a barrel of water with a hose fitting at the bottom to feed it.

Give yourself enough space to work by keeping your stuff organized in the small area you've claimed along the barricades set up throughout the pit. Give other mechanics room to work so they will be more apt to reciprocate in the heat of the battle. This is especially important when the riders come through early in the race, as they will be bunched up, and conditions will be quite chaotic. You will be busy handing the clean bike to your rider, so you will need somebody else to grab the dirty bike as he or she jumps off of it. Later on, when the riders are spread out, this will no longer be imperative, though still preferable. The rider is not allowed to run without a bike, so he or she must drop the dirty bike close to you.

As soon as you have the dirty bike, get out of the way of incoming riders and other mechanics.

a. If you have a hose or pressure sprayer

1. **Hang the bike on a bike stand or lean it against whatever structure the race has provided.** Orient yourself so you won't be dousing riders, spectators, or other mechanics as you blast water at the bike. When cleaning near the hubs and crank, always spray in the plane of the bike, not from the side, which can force water and grime into the bearings.

2. **First spray right at the tire treads and pedals.** Blast the mud out of them (and all over you and the bike as they spin faster and faster). You did wear waterproof boots, jacket, and pants, right?

3. **Clean from top to bottom.** Once the wheels and pedals are clean, spray down from above, first cleaning the handlebar, saddle, and top tube, working down to the brakes, drivetrain, and bottom-bracket area (which you will spray up at from the bottom to complete, as well as the undersides of the down tube, saddle, and handlebar). Spraying into the bearings is a no-no, hence working in the plane of the bike and from the top and bottom. If there is a waiting line for the pressure washer, this may be all you have time for, and all that is really necessary, other than a check-over for mechanical function and a quick squirt of lube on the chain (skip to step 10).

4. **If you have time, degrease the drivetrain.** Though not usually necessary, if there is time available, you can spray some environmentally friendly degreaser (don't dump anything in this field that you wouldn't want on your own lawn) on the chain, cogs, and derailleurs. Let it sit while you sponge off the rest of the bike.

5. **Clean the handlebars.** If you have time, carefully sponge off the handlebars and brake levers using your bucket of soapy water and your big sponge, remembering that the only thing the rider will see is the front of the bike. To give your rider confidence, you want him or her to have the impression that the bike is completely clean, even if you did not have the time to clean all of it.

6. **Sponge off the frame, fork, saddle, and other non-sharp, nongreasy areas.**

7. **Scrub the bike where muddy.** With the big scrub brush and soapy water, scrub the brakes (especially the front if you are rushed; see step 5), pedals, cranks, derailleurs, and any areas where mud and foliage may have accumulated. Keep your brushes and sponges separated into those you use on greasy areas and those you don't. Brush off the chain with a stiff brush.

8. **Rinse.** Wipe the chain and the jockey wheels with a rag while turning the crank, as in Figures 2.11 and 2.12.

9. **Pick out gunk from the rear cassette.** Using a stiff cylindrical brush; a flat, narrow stiff brush; a curved plastic cog pick; or a thin screwdriver, flick out mud and grass jammed between the cogs and chainrings while turning the crank. Dig out anything stuck in the pedals too.

10. **Lubricate.** Run lube into the chain by squeezing the bottle with the tip against it as you turn the cranks (Fig. 2.15). Besides seeing the front of the bike, the rider can feel the smoothness of the drivetrain, the brakes, and the entry into the pedals, so focus on those if time is short.

11. **Dry.** Move to the front of the pit and, while waiting for your rider, quickly dry off the top of the bike, especially the saddle and handlebars, with a rag.

12. **Check the wheels, brakes, and shifting.** Spin the wheels, checking for wobbles and rubbing brakes. Run through the gears and check that the stem is still lined up straight with the front wheel. Correct as need dictates and time allows. If it's below freezing, shift the gears to keep the cables moving.

13. **Hand off the bike when your rider comes through for another bike change.** Roll it along the ground and give the saddle a push as the rider mounts if it's a riding section next, or hand it onto the rider's shoulder if a running section is up next.

14. **Get out of the way and go back to step 1.** Repeat.

b. If you have no hose or pressure sprayer or bike stand but do have water nearby for your buckets

1. **Start with the wheels.** For the sake of speed, scrape the mud off the tire sidewalls first, using

only your (gloved) thumb and forefinger as you pull the tire through with your other hand. Don't worry about mud left in the tread unless you have time later to work at it with a brush.

2. **Clean the front of the bike.** Grasping the front wheel between your knees, sponge off the stem, handlebar, levers, saddle, fork, and front frame tubes. See 2-17a, step 5, for the psychological explanation.

3. **Clean the rear of the bike.** Grasping the rear wheel between your knees, sponge off the saddle, seat tube, seatpost, and rear stays.

4. **Scrub the pedals with a brush.**

5. **Pick dirt out of the cassette and chainring(s).** To get the big hunks of crud out of the rear cogs and front chainrings, stick a narrow screwdriver or curved plastic cog pick between them as you spin the crank backward.

6. **Unclog the jockey wheels.** Grab them with your fingers or a rag (Fig. 2.12) as you turn the crank, and wipe off the chain as well (Fig. 2.11).

7. **Continue with steps 10–14 in 2-17a.**

c. If you have no water or time is very short

1. **Focus only on important items like free-spinning wheels, functioning brakes and derailleurs, and pedals that will clip in quickly.** Scrape the mud off the tire sidewalls with the side of a big screwdriver or hex key, and poke dirt, grass, and leaves out from around the brakes.

2. **Pick dirt out of the cassette and chainring(s).** With a thin screwdriver or curved plastic cog pick, poke and drag dirt out from the pedals, chainrings, and cogs.

3. **Continue with steps 6–7 in 2-17b.**

2-18

A GENERAL GUIDE TO PERFORMING MECHANICAL WORK

a. Threaded parts

All threads must be prepped before tightening. Depending on the bolt in question (see descriptions in the following list), prep with lubricant, threadlock com-

pound, or antiseize compound. Clean off excess prep compound to minimize dirt attraction.

1. **Lubricated threads.** Most threads should be lubricated with grease or oil. If a bolt is already installed, you can back it out, smear grease on it, and tighten it back down. Bolts that appreciate lubrication include quick-release skewers, bottom-bracket cups, seatpost bolts, stem bolts, crank bolts, pedal axles, cleat bolts on shoes, derailleur- and brake-cable anchor bolts, and control-lever mounting bolts.

2. **Locked threads.** Some threads need to be locked in order to prevent them from vibrating loose. These are bolts that need to stay in place but are not tightened down fully, usually to avoid seizing a moving part, throwing a part out of adjustment, or stripping fine threads. Examples include derailleur limit screws, jockey-wheel center bolts, brake-mounting bolts, spoke nipples, and sometimes crank bolts. Use Loctite, Finish Line Threadlock, or the equivalent on bolts; use Wheelsmith SpokePrep or the equivalent on spokes.

3. **Antiseize threads.** Some threads have a tendency to bind up and gall, making full tightening as well as extraction problematic. Use antiseize compound on them to prevent galling. Any bolt threaded into a titanium part (including any parts mounted to titanium frames, like bottom-bracket cups), as well as any titanium bolt, must be coated with antiseize compound. Use Finish Line Ti-Prep or the equivalent.

CAUTIONARY NOTE: *Unless specified by the manufacturer, never thread a titanium bolt into a titanium part. These can gall and rip apart when you try to remove them, even with antiseize compound. If you must break this rule, use a liberal coating of antiseize compound on the threads. Every six months, unscrew the bolt, clean it, and reapply the compound.*

Wrenches (see Fig. 2.16 for various types) must be fully engaged before tightening or loosening.

1. **Hex keys and Torx wrenches must be fully inserted into the bolt head.** Otherwise, the wrench and/or bolt hole will round off. Shallow bolt heads, such as those used on shoe cleat bolts, are especially susceptible, so be careful; tapping the hex

key in may be necessary. Be sure to clean dirt and debris from bolt heads before inserting the hex key, and make sure the hex key is inserted all the way before turning the bolt. Do final tightening with the straight end of a hex key, not the ball end, and ideally with a torque wrench set to the recommended torque for the part.

2. **Open-end, box-end, and socket wrenches must be properly seated around a hex bolt.** Otherwise, you will round off the bolt head.

3. **Splined wrenches must be fully engaged.** If they are not, the splines will be damaged or the tool will snap. Be especially careful when removing a cog lockring; if you strip the spline teeth, you've got a real problem on your hands.

4. **Toothed lockring spanners need to stay lined up on the lockring.** If the teeth slide off (of a bottom-bracket adjustable cup, for example; see Fig. 11.37), they not only will tear up the lockring but also will damage the frame paint.

5. **Pin spanners need to be fully seated in the holes.** This prevents slipping out and damaging the part. You'll find holes for a pin spanner in some bottom-bracket adjustable cups (Fig. 11.37), hub-adjustment collars, and crank bolt collars.

Tightening torque

Appendix E has a list of specific tightening torques. To best understand torque values, it helps to know a little about metric bolt sizes, particularly as they are used on bikes. If your torque wrench has a head that clicks over, DO NOT continue to tighten after the head clicks.

The designation M in front of the bolt size number means millimeters and refers to the bolt shaft, not to the hex key that turns it; an M5 bolt is 5mm in diameter, an M6 is 6mm, and so on, but the M designation may not have any relationship to the wrench size. For instance, an M5 bolt usually takes a 4mm hex key (or, in the case of a hex-head style, an 8mm box-end wrench). However, M5 bolts on bicycles often accept wrench sizes that are different from the ones normally used on M5 bolts. Bolts that attach bottle cages to the frame are M5, and although some accept the normal 4mm hex key, many have a rounded cap head and take a 3mm hex key. The bolts that clamp a front derailleur around the seat tube or that anchor the cable on a front or rear derailleur are also M5, but they often take a hex key size that is bigger than standard, namely a 5mm. And when you get to the big single-pinch bolts found on old stems, you find lots of different bolt sizes (M6, M7, and even M8), but usually only one wrench size (6mm hex key).

PRO TIP ──ASSEMBLY LUBES

ANTISEIZE PASTE IS FINE IN ITS PLACE, but there's no need to spread it and its zillions of copper flakes all over your bike.

Normal bicycle grease is fine for aluminum and steel threads. You can grease parts that move and creak, or parts that can get frozen in place: front-derailleur band clamps, seat binder clamps, seatposts, stem clamps, and of course on all rolling and sliding parts like bearings and bushings.

However, antiseize paste must be used on all titanium threads. You can also use it on steel and aluminum threads, but grease is sufficient for them.

Use threadlock compound on bolts that won't be fully tightened to high torque but which need to stay put:

Cantilever brake mounting bolts (where overtightening would swell the cantilever post and cause the arm to bind); mounting bolts for rim brakes, disc-brake calipers and rotors; and rear-derailleur jockey-wheel bolts. For spoke threads, use a spoke-specific compound. You can also use spoke nipples that already have threadlock compound inside.

Carbon assembly paste is key for parts that have slipped or may slip in the future. Definitely use it on a carbon seatpost that has slipped in the past. It can't hurt to put it inside the stem clamp around the handlebar, and it is a must if the bar or seatpost slips even after the clamp bolts have been tightened to the specified torque.

2.16 Types of wrenches

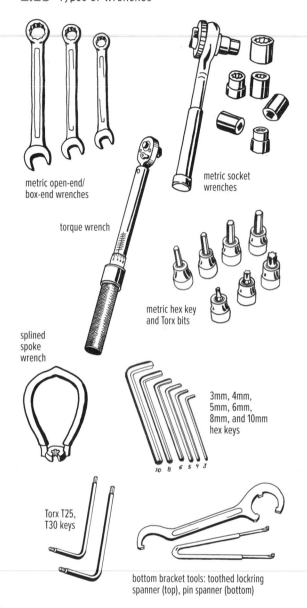

metric open-end/
box-end wrenches

torque wrench

metric socket
wrenches

metric hex key
and Torx bits

splined
spoke
wrench

3mm, 4mm,
5mm, 6mm,
8mm, and 10mm
hex keys

10 8 6 5 4 3

Torx T25,
T30 keys

bottom bracket tools: toothed lockring
spanner (top), pin spanner (bottom)

Generally, tightness can be classified in four levels:

1. **Snug (10–30 in-lbs, or 1–3 N-m [Newton meters in SI units]):** Small setscrews (such as computer-magnet mounting screws and brake bleed screws), bearing preload bolts (such as on the top cap for a threadless headset), and screws going into plastic parts need to be snug.

2. **Firmly tightened (30–80 in-lbs, or 3–9 N-m):** Small bolts, often M5 size, such as shoe cleat bolts, cable anchor bolts on brakes and derailleurs, small (M5) stem bolts, some seatpost binder bolts,

chainring bolts, brake rotor bolts, and brake-lever-clamp bolts need to be firmly tightened.

3. **Tight (80–240 in-lbs, or 9–27 N-m):** Wheel axles, old-style single-bolt stem bolts (M6, M7, or M8), and stem-quill wedge bolts, brake-caliper mounting bolts, some seatpost binder bolts, and seatpost saddle-rail clamp bolts need to be tight.

4. **Really tight (300–600 in-lbs, or 31–68 N-m):** Crankarm bolts, cassette lockrings, and bottom-bracket cups need to be really tight.

b. Cleanliness

1. **Do not think you can get parts to work by just squirting or slathering lubricant on them.** While you're patting yourself on the back for maintaining your bike, the lube will pick up lots of dirt and get very gunky.

2. **Unless otherwise instructed, don't lubricate ball bearings with oil.** They generally require grease.

3. **Do not expect parts to work if you wash them but do not lubricate them.** They will get dry and squeaky.

c. Test riding

Always ride the bike—slowly at first, and then harder—after adjusting in the bike stand. Parts behave differently under load.

2-19

PERIODIC MAINTENANCE SCHEDULE

If you follow this guide, your bike will last longer, and you will have less need for the emergency repairs in Chapter 3.

The interval periods are not written in stone; they depend on the bike, the conditions, and how you ride. In case it is not obvious, a bike used in wet conditions will require more frequent maintenance than one used only in dry, clean conditions. And a bike that's in bad shape will need more frequent attention to provide hassle-free riding than one that is in good shape. Don't add to the stress that your bike riding is intended to relieve by worrying that you're already 30 miles past your 250-mile maintenance interval, and early tomorrow morning you're heading out on a 100-mile ride with some buddies. If you've kept up with it in the past, it probably can wait another 100 miles.

Each maintenance task on this schedule is followed by the section in this book where you can find the instructions for doing it.

You may have other tasks that you want to add to this list; that's why I've added extra lines.

BEFORE EVERY RIDE

1. **Pull the brake levers.** Make sure each brake is working, hits the rim properly, and its quick-release is closed.
2. **Check that quick-release hub skewers are tight.**
3. **Inspect the tires.** Look for cuts, bulges, and worn-through tread.
4. **Check tire pressure.** Ideally use a tire gauge, but at least squeeze the tire to ensure that it has adequate pressure.
5. **Look over the entire bike.** Scan for anything out of the ordinary, like a loose saddle, paint cracks or bulges, frayed cables, rust.
6. _____
   _____
7. _____
   _____

AFTER EVERY RIDE (OR THREE)

1. **Wipe the chain, chainrings, derailleurs, and cogs with a rag, and lubricate the chain (2-15).**
2. **Wipe off the bike.** Look for damage to the frame or fork.
3. **Adjust derailleurs (Chapter 5, 6) and/or brakes (Chapter 9, 10),** if they were not working ideally.
4. **Look for the source of noises.** Investigate any rattles, rubbing noises, or creaks you may have noticed during the ride.
5. **If you've ridden in wet conditions, remove the seatpost and turn the bike upside down to drain water.** Let the bike dry out and then grease and reinstall the seatpost the following day.
6. _____
   _____
7. _____
   _____

EVERY 250 MILES (400 km)

1. **Check chain wear with a chain-elongation gauge (4-6).** Replace chain if wear exceeds acceptable elongation.
2. **Inspect brake pads for wear.** Replace if needed (Chapter 9, 10).
3. **Clean bike and drivetrain (2-14 through 2-16).**
4. **Replace tire if tread wear is excessive or you see other tire damage (7-1 through 7-12).** Inspect rim strip whenever the tire has been removed.
5. **Push rims back and forth to check for play in hub bearings.** Correct if loose (8-6 through 8-7).
6. **Check crank bolt(s) with torque wrench.** Tighten as needed (Chapter 11).
7. **Push crankarms laterally to check for bearing play.** Adjust or replace bottom bracket as needed (Chapter 11).
8. _____
   _____
9. _____
   _____

EVERY 1,000 MILES (1,600 km)

1. **Check that frame pump works.** Or check that CO_2 cartridges and inflator are in good condition.
2. **Check condition of spare inner tube.** Also check for presence of appropriate tools in seat bag (1-6).
3. **Drip chain lube on front- and rear-derailleur pivots.**
4. **Overhaul derailleur jockey wheel bushings and seals (5-27).** If the derailleur has cartridge-bearing jockey wheels, check for smooth action and regrease if needed after removing bearing covers (5-28 through 5-29).
5. **Check wheel trueness.** Correct as needed (8-2).
6. **Check rim brake-track wear.** Replace rim if wear indicator dictates it (Chapter 15).
7. **Check rims for cracks, particularly at the spoke holes.** Replace rim if cracks exist (Chapter 15).
8. **Check shoe cleats.** Replace if worn (13-2).
9. **Lubricate shift and brake cables (5-15, 9-3).**
10. _____
    _____
11. _____
    _____

EVERY 4,000 MILES (6,400 km)

1. **Remove and regrease seatpost.**
2. **Overhaul loose-ball bearings.** These can be in hubs (Chapter 8), pedals (Chapter 13), bottom bracket (Chapter 11), and headset (Chapter 12). Replace or grease cartridge bearings if they are worn, tight, or grinding or they exhibit play.
3. **Replace shift and brake cables and housings (5-6 through 5-14; 9-4).**
4. _____
   _____
5. _____
   _____

EVERY 20,000–30,000 MILES (32,000–50,000 km)

1. **Replace handlebar.**
2. **Replace stem.**
3. **Replace fork.**
4. **Replace seatpost.**
5. **Replace saddle.**

3

TOOLS

Take-along tool kit shown
in Figure 1.5 (and
Figure 1.6 for epic or
multiday rides)

*Eat a live toad the first thing in the
morning and nothing worse will
happen to you the rest of the day.*

—ANONYMOUS

EMERGENCY REPAIRS

If you ride your bike a fair distance from home, sooner or later you are likely to encounter a situation that has the potential to turn into an emergency. The best way to avoid an unpleasant surprise is to plan ahead and be prepared before it happens, which is what this chapter on emergency repairs is all about. Proper planning involves steps as simple as bringing along a few tools, spare tubes, food, water, and extra clothes. And, of course, a little knowledge.

 This chapter will acquaint you with ways to deal with most breakdowns, whether or not you have all the tools you need. Generally, any problem you're likely to encounter will involve only one component on the bike—a flat tire, a broken derailleur cable, or something similar—and in most cases it is pretty easy to find a workaround that will get you home. True, you always have the option of walking, but this chapter is designed to help you avoid that miserable fate.

Always carry a cell phone on long solo rides, just in case something does break in a big way. On the other hand, you may find yourself with a perfectly functioning bicycle and a fully charged phone (with service) and still be in dire straits because you're either lost, cold, dehydrated, bonking (i.e., your body has run out of fuel), or injured. Carefully read the final part of this chapter for pointers on how to avoid these possibilities and what to do if the worst does happen.

3-1

RECOMMENDED TOOLS

The take-along tools to keep in your seat bag are described in 1-6. If you're going to be a long way from civilization, take along the extra tools recommended for longer trips.

3-2

FLAT TIRE PREVENTION

The best way to avoid flats is to keep good tires and rim strips on your bike. Check them regularly for wear, cracking, and cuts. Coat tires you don't ride often with 303 Aerospace Protectant or ArmorAll to prevent ozone cracking. Steer clear of potholes, broken glass, and nails, and you'll rarely have a problem.

Flat tires can be minimized with the use of tire sealant. Tire sealants usually come either as a viscous liquid containing chopped fibers (Slime is the most common brand), or as a thin solution containing liquid latex or the like that hardens into a rubbery glob to plug holes in the tube as air blows past (Stan's is a common brand). Sealant can be poured into a tubeless tire on installation or injected into an inner tube that has a Schrader valve; liquid latex sealants can be injected into Presta valves if they have removable cores (sealant use is covered in 7-15). All Schrader valves have removable cores, and a core-remover tool often comes with the bottle of sealant. Most Presta valves, on the other hand, do not have removable cores, and so you cannot inject sealant through the valve except for thin, slow-solidifying sealants like Hutchinson Protect'air or Effetto Mariposa Caffélatex. Even on Presta valves with removable cores (you take the core out with an adjustable wrench or a specific core-removal tool), you can only inject a thin, liquid-latex-type sealant, because the valve stem is so thin that a chopped-fiber sealant would clog it up. You can get aerosols with liquid latex designed for Presta valves (Fig. 1.6), or you can use a syringe to inject liquid sealant. You can also purchase tubes (both Schrader and Presta) with sealant already inside.

In a pinch, you can use evaporated milk in any Presta valve tube or tubular tire as a sealant. Just pour some canned evaporated milk into a pump you no longer care about, and pump it right into the tube or tubular tire. It actually works quite well to seal small leaks (it can be a lifesaver with a tubular tire with a slow leak), but if you ever get a blowout, boy, does it stink!

If you do have tire sealant in the tube and the tire gets low owing to a puncture, pump in more air and turn the wheel so the hole is at the bottom, or ride the bike for a couple of miles to get the sealant to flow to the hole. Sealant will not fill a puncture if the hole in the tube is on the rim side, because the liquid will be thrown to the outside when the wheel turns.

Sealants cannot fill large punctures and blowouts, although big holes can be plugged sufficiently to get you home if you locate where the sealant is squirting out through the tire. Rotate the wheel so the spot is at the bottom and wait. The sealant may pool up and plug the hole. Add more air and continue. If you have Caffélatex in your tire and it is shooting out through a big hole, you can squirt Effetto Mariposa's ZOT! Nano instant polymerization catalyst at the sealant flowing out of the hole to harden the Caffélatex instantly and plug the leak so you can get home.

Plastic tire liners that fit between the tire and tube are often promoted to ward off flats, but I don't recommend them. Most are so stiff that they roll roughly, decrease traction and cornering ability, and can slip around inside the tire.

3-3

FIXING FLAT TIRES

a. If you have a spare or a patch kit

Simple flat tires are easy to deal with. The first flat you get on a ride is easily fixed by installing a spare tube (7-1 through 7-5). If you have tubeless tires, simply remove the valve stem from the rim and install a new inner tube, as in 7-5.

1. **Remove the tube (7-1).**

2. **Remove whatever caused the flat.** You may see it sticking up from the tread. Feel around the inside of the tire for any other sharp objects. It's important to know that you have eliminated the cause of the puncture before continuing riding, or it could happen right away again with the new tube.

   Sometimes inner tubes just fail, particularly near the valve on the rim side. If this happens and you can't see a cause like a sharp edge on the valve hole, the solution is to shrug your shoulders, replace the tube, and hope it doesn't happen again (perhaps a different make of tube or a less ancient, dried-out one will do the trick).

3. **If you can't find the cause of the puncture, check the rim strip.** The rim strip is the plastic or rubber tape that covers the spoke holes in the well of the rim, and you depend on it to prevent punctures to the underside of the tube. See if the flat was caused by the end of a spoke protruding up into the tube, a metal shard from the rim, or the edge of a spoke hole protruding through or around a worn, dislodged, or insufficiently wide rim strip.

4. **If the rim strip is the culprit, cover whatever bored into the tube.** Use an energy bar wrapper or whatever works to supplement the rim strip; once you get home, replace the rim strip before riding again. Replace it even if the rim strip did not cause the puncture but is damaged or is a limp, narrow strip of soft rubber or cloth. Use either adhesive-backed rim tape or a high-quality plastic rim strip loop sized for your rim, or in lieu of either, apply layers of reinforced packing tape (the kind that has lengthwise fibers inside) to cover the spoke holes.

5. **If you can't find the leak, put air in the tube to see where it's coming out.** You may need to spit on suspicious spots or hold the tube under water and look for bubbles if you can't hear the air hissing out (see Fig. 7.9).

You have a pinch flat if you see two adjacent holes on the top and bottom of the tube (sometimes called a "snake-bite"). Hitting a sharp edge, like a train track, curb, rock, or pothole, with insufficient air pressure in the tire can result in a pinch flat; the inner tube gets pinched between the tire and the rim. Prevent this in the future with sufficient pressure and/or tire width and by avoiding sharp impacts by steering clear of them or standing on the pedals as you cross them to let the bike move freely under you.

Metal shards left from the drilling of rims during manufacture can create rim-side holes in the tube. Shake out any metal fragments inside the rim.

6. **Replace the tube as in 7-5.**

7. **After you run out of spare tubes, rely on your patch kit.** Patching punctures is described in 7-2 through 7-4.

b. No more spare tubes or patches

If you're out of spare tubes and patches, you can actually tie a knot in the inner tube, pump the tube back up, and ride it home. You'd be amazed how well this works.

1. **Fold the tube at the puncture.**

2. **Tie an overhand knot with the folded end (Fig. 3.1).** To maximize the length of inflatable inner tube, minimize the length of the folded end sticking out of the knot.

3.1 Tying a knot in a punctured inner tube to seal off the hole

Riding with a knotted tube works better the larger the tire is. It can work only if there is a single puncture in the tube, a snake bite (pinch flat), or multiple punctures all within a few inches of each other that you can seal off with a single knot. Obviously, multiple knots would seal off sections of the tube from air, leaving them flat!

If tying off the tube won't work because the hole is at the valve, the tube has more than one hole in it, or the tire is too narrow, continue to the next section.

c. No way to inflate the tube or a section of a knotted tube

If you have neither a pump nor air cartridges, if you have multiple knots in the tube or a hole at the valve and no patches, or if the valve is broken, you will have to ride home without air in the tube or in a section of it. However, riding a flat for a long way will destroy the tire and probably damage the rim too. You can minimize that damage by filling the space in the tire with grass, leaves, or similar materials. Pack the stuff in tightly, and then remount the tire on the rim. This "fix" should make the ride a little less dangerous by minimizing the flat tire's tendency to roll out from under the bike during a turn.

d. Torn sidewall

Rocks and glass can cut tire sidewalls. The likelihood of sidewall problems is reduced if you avoid old tires with rotten and weak cords. If the tire's sidewall is torn or cut, the tube will stick out. Just patching or replacing the tube isn't going to solve the problem. Without reinforcement, the tube will work past the cut and blow out again very soon.

First, you have to look for something to reinforce the tire sidewall (Fig. 3.2). Energy bar wrappers and dollar bills work to "boot" a tire. (I told you that cash can get

3.2 Fixing torn tire casing (temporarily) with an energy bar wrapper

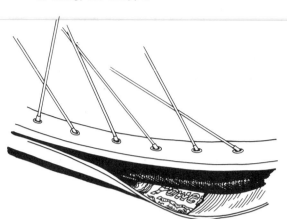

3.3 Freeing a jammed chain

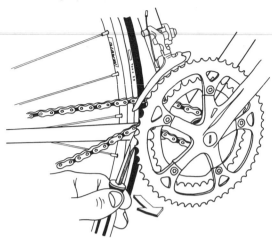

you out of bad situations; credit cards are not acceptable for this purpose.) Business cards are a bit small but work better than nothing. You might even try a piece of a plastic soda bottle. A small piece of lawn-chair webbing or a piece of old tire sidewall cut in an oval might be a good addition to your patch kit for this purpose. You get the idea.

1. **Lay the energy bar wrapper, cash, etc. inside the tire over the gash, or wrap it around the tube at that spot.** Place several layers between the tire and tube to support the tube and prevent it from bulging out through the hole in the sidewall.

2. **Put a little air in the tube to hold the makeshift reinforcement in place.**

3. **Mount the tire bead on the rim.** You may need to let a little air out of the tube to do so.

4. **Make sure that the tire is seated and the boot is still in place.**

5. **Inflate the tire.** Don't pump it up too hard or leave it too soft. Low pressures will allow the boot to move around or may lead to a pinch flat, and higher pressures could blow through your boot.

Check the boot periodically on the ride home to make certain that the tube is not bulging out again.

3-4

JAMMED CHAIN AND TWISTED LINK

When the chain gets jammed between the chainrings and the chainstay, it can be difficult to extract. You may

tug and tug on the chain, and it won't come out. Well, chainrings are flexible, and if you apply some mechanical advantage, the chain will come free quite easily.

1. **Insert a screwdriver or similar thin lever between the chainring and the chainstay.**

2. **Pry the space open while pulling the chain out (Fig. 3.3).** You will probably be amazed at how easy it is to free the chain this way, especially since lots of tugging would not free it.

If you still cannot free the chain

1. **Disassemble the chain.** You will probably need to use a chain tool (Fig. 3.4; instructions in 4-7), but you're really in luck if you have a removable master link (Fig. 3.5; instructions in 4-13).

2. **Free it from the jam.**

3. **Reconnect it (4-9 through 4-13).**

3.4 Disassembling or assembling a chain with a chain tool

3.5 SRAM or KMC master link

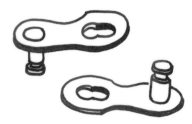

3.6 Twisted chain link

3.7 Untwisting a twisted chain link without tools

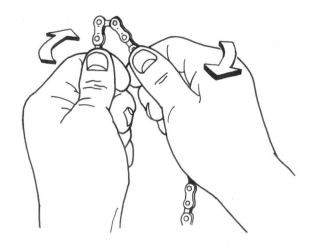

NOTE: *You can push out a pin on any derailleur chain, push it back in (Fig. 3.4), and get home. But I would not recommend riding a reassembled 9-, 10-, or 11-speed chain longer than that, because that link will be very weak due to minimal rivet protrusion; the plate will be prone to pop off the end of the rivet upon shifting.*

You need a master link in order to disassemble and reassemble (4-13) the thin chain required for nine or more rear cogs and to continue to ride it worry-free. If you already have a master link in the chain, fine; otherwise, bring one along that is for the same number of speeds. SRAM, KMC, and Wippermann (Figs. 3.5, 4.26) make them for each chain width.

Once you get rolling again, the chain may skip. You may be experiencing a common side effect of jamming the chain and continuing to pedal a split second too long: twisting a chain link (Fig. 3.6). Once you find the twisted link, it will be obvious why it was popping out of gear.

Untwisting a twisted chain link with tools

With two pairs of pliers, two adjustable wrenches, or one of each, grab the links on either side of the twisted one at the rivet pins, and twist them to straighten the link.

Untwisting a twisted chain link without tools

1. **Shift to the smallest rear cog.**
2. **Flip the chain off of the inner chainring.** Let it drop around the bottom-bracket shell with no tension on it.

3. **Fold the chain at the twisted link.** Have the twisted link alone at the top, horizontal (Fig. 3.7).
4. **Grasp the vertical sections of chain with each hand.**
5. **Pull one hand toward you and push the other hand away from you (Fig. 3.7).** This should untwist the link.
6. **Repeat until the twist is gone.**
7. **Replace the chain when you return home (4-7 through 4-13).**

3-5

BROKEN CHAIN

Chains seldom break on road bikes, although the problem is becoming more common with supernarrow chains. Chains can be overstressed by worn cogs, which cause the chain to skip and then suddenly retighten. The chain breaks when a chain plate pops off the end of a rivet, often during a front shift. As the chain rips apart, it can cause collateral damage as well. The open chain plate can snag the front-derailleur cage, bending it or tearing it off.

When a chain breaks, the end link is certainly shot, and some adjacent ones may be as well. Unless you install a master link, a broken chain is unsuitable for further use, although it can often be repaired well enough to ride home carefully, pedaling gingerly.

1. **Remove the damaged links with the chain tool.** (You or your riding partner did remember to bring a

chain tool, right?) Again, the procedures for removing the damaged links and reinstalling the chain are covered in Chapter 4, 4-7 through 4-12. If the outer chain plates of a single link are damaged, remove them by pushing out the remaining rivet attaching them to the chain, and install a master link (4-13); note proper orientation with a Wippermann master link (Fig. 4.26).

2. **If you have brought along extra chain links, replace the same number you remove.** If not, you'll need to use the chain in its shortened state; it will still work, but don't use the largest cogs when the chain is on the big chainring.

3. **Join the ends and connect the chain (Fig. 3.4);** the procedure is described in Chapter 4, 4-9 through 4-12 (4-13 with a master link). Some lightweight chain tools and multitools are more difficult to use than a shop chain tool. Some flex so badly that it is hard to keep the pushrod lined up with the rivet. Others pinch the plates so tightly that the chain link binds up. It's a good idea to try yours before you need the tool on the road.

3-6

BENT WHEEL

If the rim is banging against the brake pads—or, worse yet, the frame or fork—pedaling becomes very difficult. If you haven't hit a pothole or something similar that has bent the rim, the cause is probably a loose or broken spoke. Another culprit could be a broken rim—fairly obvious with aluminum rims but more difficult to detect with carbon-fiber rims.

If the wheel wobble is so bad that you can't loosen the brake (3-9) and ride home with it that way, you'll have to perform a temporary fix; see the following sections.

3-7

LOOSE SPOKES

If a wheel has a loose spoke or two, the rim will wobble all over the place, often too much to ride home with.

1. **Find the loose spoke (or spokes) by feeling all of them.** The really loose ones, which would cause a

3.8 Tightening and loosening spokes

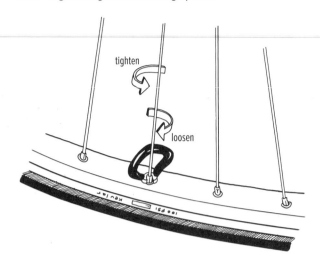

wobble of large magnitude, will be obvious. If you find a broken spoke, skip to the next section (3-8). If there are no loose or broken spokes, skip ahead to 3-10.

2. **Get out the spoke wrench that you carry for such an eventuality.** If you don't have one, skip to 3-9.

3. **Mark the loose spokes, if necessary.** You can tie blades of grass, sandwich bag twist ties, tape, or the like around them so that you can remember which ones were the culprits as the wheel becomes more true.

4. **Tighten the loose spokes (Fig. 3.8) and true the wheel.** Follow the procedures in 8-2.

3-8

BROKEN SPOKES

If you break a spoke, the wheel will wobble so wildly that the tire will hit the chainstay, making it hard to ride, while also wearing away the chainstay.

1. **Locate the broken spoke.**

2. **Remove the remainders of the spoke, both the piece going through the hub and the piece threaded into the nipple.** If the broken spoke is on the freewheel side of the rear wheel, you may not be able to remove it from the hub because it will be behind the cogs. If so, skip to step 6 after wrapping it around a neighboring spoke to prevent it from slapping around (Fig. 3.9).

3.9 Wrapping a broken spoke

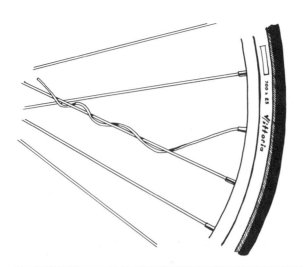

3.10 Loosening the brake cable

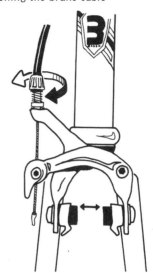

3. **Get out your spoke wrench.** If you have no spoke wrench, skip to 3-9.

4. **If you brought along a spare spoke of the right length or the FiberFix Kevlar replacement spoke mentioned in 1-6b, you're in business. If not, skip to step 6.** Put the new spoke through the hub hole, weave it through the other spokes the same way the old one was, and thread it into the spoke nipple that is still sticking out of the rim. Mark it with a pen or a blade of grass tied around it. With the Kevlar emergency spoke, first thread the head (the "cam piece") attached to the fixed end of the Kevlar cord into the nipple a few turns. Now thread the free end of the Kevlar cord through the hub hole, attach the free end by weaving it through the bars on the cam piece attached to its included spoke stub, pull it tight, and tie off the end. Tighten the spoke by holding the nipple while rotating the cam piece; this will twist the Kevlar spoke and tighten it further.

5. **Tighten the nipple on the new spoke with a spoke wrench (Fig. 3.8).** Check the rim clearance with the brake pad as you go. Stop when the rim is reasonably straight, and finish your ride.

6. **If you can't replace the spoke and you have a spoke wrench, make the wheel more laterally true by loosening the spoke on either side of the broken one.** These two spokes come from the opposite side of the hub and will let the rim move toward the side with the broken spoke as they are loos-

ened. A spoke nipple loosens counterclockwise when viewed from its top (i.e., from the tire side; Fig. 3.8). Don't loosen more than two turns. Ride home conservatively, as this wheel will rapidly get worse.

7. **Once at home, replace the spoke.** Follow the procedure in 8-3, or take the wheel to a bike shop for repair. If you break a spoke more than once on a wheel, relace the wheel with new spokes (Chapter 15). The rim may need replacement as well.

3-9

OPENING YOUR BRAKE TO GET HOME

If the rim is banging the brake pads but the tire is not hitting the chainstays or fork blades, simply open the brake so that you can get home, as detailed here. If the tire is hitting the frame or fork, you may need more extreme measures to temporarily straighten it; see the next section.

1. **Open the brake-caliper quick-release lever as far as necessary for the pads to clear the rim.** If the pads still rub, loosen the tension on the brake cable by screwing in the barrel adjuster on the caliper (Fig. 3.10); it may screw in clockwise or counterclockwise, depending on brand. Remember that braking effectiveness on that wheel will be greatly reduced or nonexistent, so ride slowly and carefully.

2. **If the rim is still banging the brakes and you have a wrench to loosen the brake cable, do so, and**

then clamp it back down. The bolt will probably require a 5mm hex key. You now have no brake on this wheel; ride carefully.

3. **If this still does not cut it, you can remove the brake.** Disconnect the cable and remove the brake caliper from the fork or brake bridge, put it in your pocket, and pedal home slowly. You will usually need a 5mm hex key for this task.

3-10

BENT RIM

If the wheel is too bent to turn even with spoke truing and/or with removing the brake, you can beat it straight as long as the rim is not broken.

1. **Find the area that is bent outward the most and mark it.**
2. **Leaving the tire on and inflated, hold the wheel at the bottom with the bent-outward part at the top facing away from you.**
3. **Smack the bent-outward section of the rim against flat ground (Fig. 3.11).**
4. **Put the wheel back in the frame or fork, and see if anything has changed.**

3.11 Fixing a bent rim

5. **Repeat the process until the wheel can be ridden.** You may be surprised how straight you can get a wheel this way.

3-11

DAMAGED FRONT DERAILLEUR

If the front derailleur is mildly bent, straighten it with your hands or leave it until you get home.

If the front derailleur has simply rotated around the seat tube or twisted in the frame tab that it is attached to (the chain, your foot, or a pants leg can catch it and turn it), reposition it so that the cage is just above (Fig. 3.12) and parallel to the chainrings (Figs. 5.13, 5.14). Tighten the derailleur in place (usually with a 5mm hex key).

If the derailleur is broken or so bent that you can't ride, or if the frame tab for the derailleur is bent, you will need to remove the derailleur or route the chain around it as described next. (If the tab is bent, trying to straighten it will either break it, pull it off, dent or crack the seat tube, or cause a crack to form in the near future. You will need to have a frame builder or the factory put on a new one.)

a. If you have only a screwdriver

1. **Get the chain out of the derailleur cage.** To do this, open the derailleur cage by removing the screw at its tail (Fig. 3.12). If the derailleur cage is riveted closed, you'll have to open the chain with a chain tool (4-7) or by hand at a master link (4-13).
2. **Bypass the derailleur by putting the chain on a chainring that does not interfere with it.** Either shift the derailleur to the inside and put the chain on the big chainring, or vice versa.

b. If you have hex keys and a screwdriver (or a chain tool or master link)

1. **Remove the derailleur from the seat tube, usually with a 5mm hex key.**
2. **Remove the screw at the tail of the derailleur cage with a screwdriver, if it has one.**
3. **Pry open the cage and separate it from the chain.** You can also disassemble the chain, pull it out of the derailleur, and reconnect it (Chapter 4).

3.12 Opening the front-derailleur cage

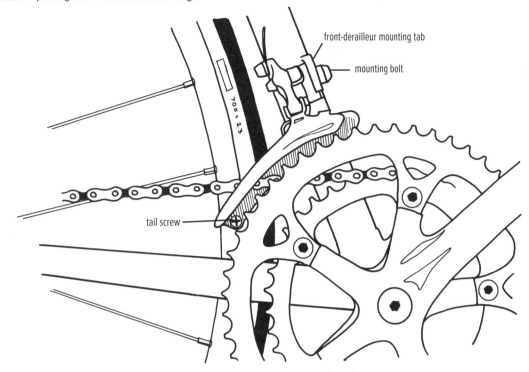

front-derailleur mounting tab

mounting bolt

tail screw

4. **Manually put the chain on whichever chainring is most appropriate for the ride home.** If in doubt, put it on the inner one (or middle one, if you have a triple).

5. **Tie up the derailleur cable so that it won't catch in your wheel.**

6. **Stuff the derailleur in your pocket and ride home.**

3-12

DAMAGED REAR DERAILLEUR

If the upper jockey wheel gets lost, put the lower one on top and thread a wire or zip tie through three threaded Presta valve collar nuts (off your inner tube valves) as a lower wheel. If one of the jockey-wheel bolts gets lost, and you found the jockey wheel, try replacing the bolt with one of the water bottle cage bolts or a zip tie. If the return spring on the rear-derailleur cage breaks, the chain will hang loosely. If you have a bungee cord, hook it to the lower cage, around the skewer (put the lever on the drive side), and up to the seat-tube bottle cage.

If the rear derailleur gets bent just a bit, you can probably straighten it enough to get home. If it gets

3.13 Bypassing a damaged rear derailleur

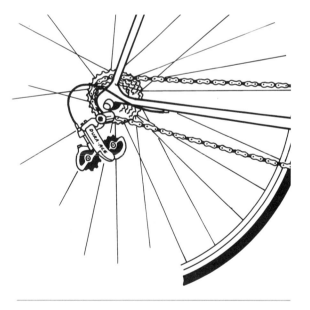

really bent or broken or one of the jockey wheels falls off, you will need to bypass the derailleur, effectively turning your bike into a single-speed for the remainder of your ride (Fig. 3.13).

1. **Open the chain with a chain tool or by hand at a master link (4-7 or 4-13) and pull it out of the derailleur.**

2. **Pick a cog in the middle of the cassette or free-wheel, and set the front derailleur over the small or middle chainring.** Be aware that the chain will tend to fall off the chainring or move down to smaller cogs, unless it is really tight and is lined up as straight as possible with the direction of the bike (i.e., is not crossed at an angle from big to big or small to small chainring and cog).

3. **Wrap the chain over the chainring and the rear cog you have chosen.** Bypass the rear derailleur entirely.

4. **Remove any overlapping chain with the chain tool (4-7).** Make the chain as short as you can while still being able to connect the ends.

5. **Connect the chain with the chain tool or by hand at a master link as described in 4-9 or 4-13.**

6. **Ride home carefully.**

3-13

BROKEN FRONT-DERAILLEUR CABLE

The chain will be on the inner chainring, and you will still be able to use all of your rear cogs. Leave it on the inner ring and ride home.

3-14

BROKEN REAR-DERAILLEUR CABLE

The chain will be on the smallest rear cog, and you will still be able to use both (or all three) front chainrings. You have three options:

1. **Leave the chain on the small cog and ride home.**

2. **Lock the derailleur in one gear.** Move the chain to a larger cog, push inward on the derailleur with your hand, and tighten the high-gear limit screw on the rear derailleur (usually the upper one of the two screws) until it lines up with a larger cog (Fig. 3.14). Move the chain to that cog and ride home. You may have to fine-tune the adjustment of the derailleur stop screw to get it to run quietly without skipping.

3. **Jam the derailleur in one gear.** If you do not have a screwdriver, you can push inward on the rear derailleur while turning the crank with the rear wheel off the ground to shift to a larger cog. Jam a stick between the derailleur cage plates to prevent

3.14 Tightening a high-end limit screw

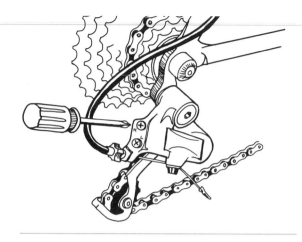

3.15 Wedging the rear derailleur into an easier gear

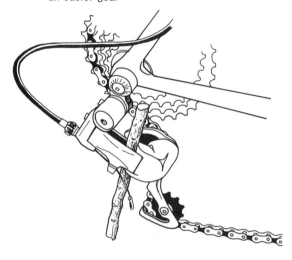

the chain from moving back down to the small cog (Fig. 3.15).

3-15

NONFUNCTIONING ELECTRONIC DERAILLEUR

If the battery is dead or a connection is faulty, your bike's electronic derailleurs will not work. Check the battery level (6-1), check all of the visible wire connections to ensure none have come unplugged, check that the battery is clipped in correctly, and do the standard adjustment procedures (6-3) with the mode buttons on the stem-mounted interface and the shifters in an attempt to get it working again. If shifting still is not happening, you can manually move the rear derailleur into a chosen gear, and it will stay there. Push the derailleur into a gear you can get home with, lifting the rear wheel and turning the crank to get the chain to engage. Don't do this too often; you can wear out the saver clutch designed to protect the derailleur in the case of a crash, and it will then no longer hold a gear.

3-16
BROKEN BRAKE CABLE

Ride home slowly and carefully. Very slowly. Very carefully.

3-17
BROKEN SEAT RAILS OR SEATPOST CLAMP

If you can't tape or tie the saddle back on, try wrapping your gloves or some clothing over the top of the seatpost to pad it. Otherwise, remove the seatpost and ride home standing up.

3-18
BROKEN SEATPOST SHAFT

Ride home standing up.

3-19
BROKEN HANDLEBAR

It's probably best to walk home (or call for a ride). You could splint it by jamming a stick inside and ride home very carefully, but the stick could easily break, leaving you with no way to control the bike. A sudden impact of your face with the road would follow.

If you decide to splint the handlebar, hold the pieces together with duct tape. If the break is adjacent to the stem, slide the bar left or right into the stem so that both pieces are clamped.

3-20
FROZEN PARTS

Riding in snow or freezing rain can freeze shift cables where they pass under the bottom bracket or can freeze the derailleurs and fill the cogs you are not using with ice. You will just have to stay in the gear you are frozen in. But if the freehub mechanism freezes, you won't be able to coast for even a second. You may be able to free it by applying any hot liquid available (even urine!) and hitting the freehub with a stick until it rotates counterclockwise again.

3-21
PREPARING FOR EVERY RIDE

1. **Always take plenty of water and food.**
2. **Tell someone where you are going and when you expect to return.** Consider wearing a wrist ID. If you know of someone who is missing, call the police or sheriff.
3. **Take extra food for any ride over an hour.**
4. **Take a road map or a GPS unit if you don't know the area.** Be willing to ask for directions.
5. **Take a cell phone.**
6. **For long rides in uninhabited areas, take extra supplies.** Carry matches, extra clothing, and perhaps a flashlight and an aluminized emergency blanket, in case you have to huddle under a tree for the night.
7. **Ride carefully and attentively.** Pay special attention to wet roads, gravel-covered turns, turns covered with moss or fallen leaves and other plant debris, and areas with lots of traffic, especially traffic turning into and out of side roads.
8. **Wear a helmet.** It's hard to ride home with a cracked skull.
9. **Don't ride beyond your limits if you are a long way from home.** Take a break. Get out of the hot sun. Avoid dehydration and bonking by drinking and eating enough.
10. **Have your bike in good working order before you leave.**

In short, make appropriate decisions when taking long rides. Prepare well. Just because you have a $4,000+ bike and are riding on paved roads, you are not immune to mechanical problems or becoming exhausted, cold, bonked, injured, lost, or caught out in the dark.

> *Take care of the luxuries and the*
> *necessities will take care of themselves.*
> —DOROTHY PARKER

TOOLS

Chain lubricant
12-inch ruler
Chain tool ("chain breaker")
Lots of rags
Rubber gloves

OPTIONAL

Chain-elongation indicator
Master-link pliers
Solvent (citrus-based)
Self-contained chain
    cleaner
Old water bottle
Caliper
Pliers
Solvent tank
Rohloff cog-wear indicator

THE CHAIN

The bicycle chain is one of those technological wonders that we take for granted. Without it a bike would be a clumsy and inefficient contraption. The chain is nothing more than a simple series of links connected by rivets (also called pins). Rollers surround each rivet between the link plates and engage the teeth of the cogs and chainrings. Nothing to it, and yet it efficiently transmits mechanical energy from the pedals to the rear wheel. In terms of weight, cost, shift-ability, and efficiency, the bicycle chain has no equal, although people have tried endlessly to improve on it.

Perhaps because it is so simple and familiar, the chain is often ignored. To keep your bike running smoothly, though, you have to pay some attention to it. It needs to be kept clean and well lubricated in order to utilize your energy most efficiently, shift smoothly, and operate noiselessly. And because its length increases as it wears, thus contacting gear teeth differently than intended, it needs to be replaced regularly to prolong the working life of more expensive drivetrain components.

To address specific chain problems, check out "Troubleshooting Chain Problems" at the end of this chapter (4-14 through 4-16).

4-1

LUBRICATION

LEVEL For best results, use a lubricant intended for bicycle chains. Most lubes sold for this purpose work reasonably well at the basic task of keeping the chain protected and happy. However, be careful with wax-based lubricants as some don't protect well. Chain life with them can be short (1,000–1,500 miles).

If you want to get fancy about it, you can assess the type of conditions in which you ride and choose a lubricant intended for those conditions. Some lubricants are "dry," formulated to pick up less dirt in dry conditions. Other lubes are "sticky" and therefore less prone to wash off in wet conditions. Still others claim to be "metal conditioners" that actually penetrate and alter the surface of the metal.

Chain lubes are generally sold in spray cans and squeeze bottles. Avoid sprays for regular maintenance chores because they tend to spew oil over everything, and droplets from the spray can end up in your lungs. The chain only needs a reservoir of oil inside each link; on the outside, a thin film is sufficient to keep corrosion at bay. Extra oil on the

47

outside will only attract dirt and gunk; it does nothing to improve the function of the chain.

1. **Ideally, drip a small amount of lubricant across each roller one by one (Fig. 4.1).** Periodically move the chain to give easy access to the links you are working on. Regular application is the most important thing you can do with the chain, and to speed the process, you can turn the crank slowly while dripping lubricant onto the chain as it goes by. This method will cause you to apply excess lubricant, which will pick up more dirt. But overlubricating is preferable to not lubricating, and if you wipe and lube the chain after each ride or two, it won't build up excessive grime.

4.1 Dripping oil only where it is needed

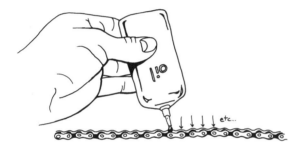

4.2 Wiping the chain

2. **Wipe the chain lightly with a clean rag to remove excess oil.**

3. **If you want to do a champion job, perform this task before you put the bike to bed, and then wipe the chain clean again the next morning or before the next ride.** That way, you'll remove additional oil that has seeped onto the outside of the links, where it isn't needed.

If you're riding in wet conditions, you'll need to apply lubricant frequently (after every ride, or even during a long, rainy ride).

4-2

CLEANING BY FREQUENT WIPING AND LUBRICATION

The simplest way to maintain a chain is to wipe it down frequently and then lubricate it. If you follow this scheme prior to every ride, you will never need to clean your chain with a solvent. The lubricant softens the old sludge buildup, which is driven out of the chain when you ride.

The problem is that the fresh lubricant also picks up new dirt and grime, but if this gunk is wiped off before it is driven deep into the chain, and the chain is relubri-

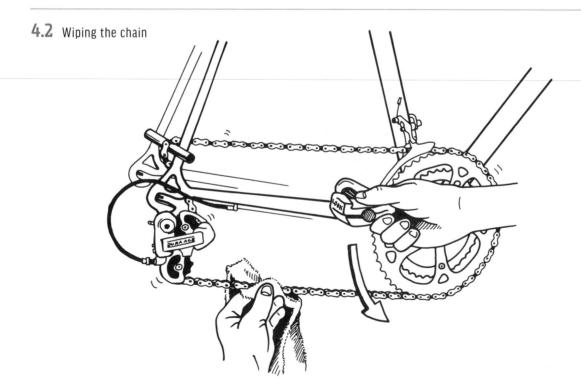

4.3 Wiping the jockey wheels with a rag

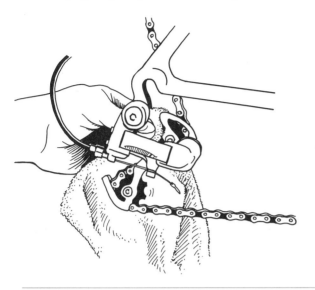

4.4 Chain cleaning on the bike

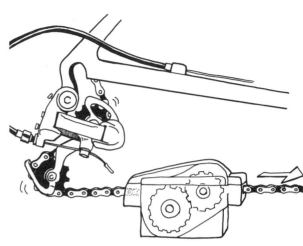

cated frequently, it will stay relatively clean as well as supple. Chain cleaning can be performed with the bike standing on the ground or in a bike stand.

1. **With a rag in your hand, grasp the lower length of the chain (between the bottom of the chainring and the lower jockey wheel of the rear derailleur).**

2. **Turn the crank backward a number of revolutions, pulling the chain through the rag (Fig. 4.2).** Periodically rotate the rag to present a clean section to the chain.

3. **Lubricate each chain roller carefully as in Figure 4.1.** Or take the faster method of running the chain past the dripping bottle tip.

To simplify this procedure, I recommend leaving a pair of rubber gloves, a rag, and some chain lube next to your bike. Whenever you return from a ride, put on the gloves, wipe and lube the chain, and put your bike away. It takes maybe a minute, your hands stay clean, and your bike is ready for the next ride. Wipe the chainrings, cogs, front derailleur, and jockey wheels (Fig. 4.3) while you're at it, and the entire drivetrain will always run smoothly.

4-3

USING CHAIN-CLEANING UNITS

Several companies make chain-cleaning gizmos that scrub the chain with solvent while the chain is on the bike. These types of chain cleaners are gener-

ally made of clear plastic and have two or three rotating brushes that scrub the chain as it moves through the solvent bath (Fig. 4.4). Regularly removing the chain is inadvisable with 9-, 10-, or 11-speed chains unless you use a master link. Not heeding this can result in breaking the chain under high load, driving your foot, and perhaps your entire body, into the asphalt. Cleaning the chain on the bike is a reasonable alternative.

Most chain cleaners are supplied with a nontoxic, citrus-based solvent. For your safety and other environmental reasons, I strongly recommend that you continue to use nontoxic citrus solvents with your chain. Some mechanics prefer to use diesel fuel, which is okay only if you recycle the used fuel (do NOT throw it in the trash). In either case, wear gloves and glasses when using any sort of solvent.

Citrus-based chain solvents often contain some lubricants as well, so that they won't dry out the chain. The lubricant carried with the solvent is one reason diesel fuel has had a following as a chain cleaner. A really strong solvent without lubricant (acetone, for example) will displace the oil from inside the rollers. The solvent will then evaporate, leaving a dry, squeaking chain that is hard to rehabilitate. The same thing can happen with a citrus-based solvent without an included lubricant, especially if the chain is not allowed to dry sufficiently before it is relubricated.

Here's the procedure for cleaning the chain with a chain-cleaning unit:

1. **Remove the top of the chain-cleaner case and pour in solvent up to the fill line.**

2. **Place the unit against the bottom of the chain, and reinstall the top of the unit so that the chain runs through it (Fig. 4.4).**

3. **Turn the bike's crank backward.**

4. **Remove the unit, wipe off the chain with a clean cloth, and let it dry.**

5. **Lubricate the chain as described in 4-1.**

4-4

REMOVAL AND CLEANING

You can also clean the chain by removing it from the bicycle and cleaning it in a solvent. I do not recommend this approach unless the chain has a master link. Even with a wide 5-, 6-, or 7-speed chain, repeated disassembly by pushing rivets in and out weakens the chain by expanding the size of the rivet hole in the outer chain plate, allowing the rivet to pop out more easily. Opening and re-riveting a 9-, 10-, or 11-speed chain almost guarantees that it will break during a ride.

A hand-opened master link can avoid the chain weakening caused by pushing pins out. Master links are standard on SRAM, KMC, Wippermann, and Taya chains. An aftermarket master link can also be installed into any chain, as long as the master link is for the same width chain (same number of speeds).

If you do disassemble the chain (see 4-7 or 4-13 for instructions), you can clean it well, even without a solvent tank. Just drop the chain into an old jar or water bottle half filled with solvent, and agitate. Using an old water bottle or jar allows you to clean the chain without touching or breathing the solvent—even citrus solvents.

Here's the procedure for cleaning the chain if you don't have or don't want to use a chain-cleaning unit:

1. **Remove the chain from the bike (4-7 or 4-13).**

2. **Drop it in a water bottle or jar.**

3. **Pour in enough solvent to cover the chain.**

4. **Shake the bottle vigorously (low to the ground, in case the top pops off).**

5. **Hang the chain to air-dry.**

6. **Reassemble it on the bike (see 4-8 through 4-13).**

7. **Lubricate it as described in 4-1.**

Don't soak the chain for extended periods in citrus-based solvents, since these have a water base and will cause the chain to oxidize (rust), making it move with more friction and making it more prone to breakage. (Some people have two chains they rotate on and off the bike, leaving one soaking in solvent while the other one is on the bike. While this would work with diesel fuel as the solvent, you're asking for a broken chain with water-based solvent.) You gain nothing by soaking it anyway, so just don't do it.

After removing the chain, allow the solvent in the bottle to stand for a few days, decant the clear stuff, and use it again. I'll say it again: Use a citrus-based solvent. It is safer for the environment, gentler on your skin, and less harmful to breathe. Wear rubber gloves when working with any solvent, and use a respirator meant for volatile organic compounds if you are not using a citrus-based solvent. There is no sense in fixing your bike so that it goes faster if you end up becoming a slow, sickly bike rider from breathing solvent fumes.

4-5

CHAIN REPLACEMENT

As the rollers, pins, and plates wear out, the chain will lengthen and hasten wear and tear on the other parts of the drivetrain. An elongated chain will concentrate the load on each individual gear tooth, rather than distributing it over all of the teeth that the chain is wrapped around. The concentrated load will cause the gear teeth to become hook-shaped and the tooth valleys to lengthen.

If such wear has already occurred, a new chain will not solve the problem, because it will not mesh with the deformed teeth and will skip off them whenever you pedal hard. The only cure is to replace the chain, the chainrings, and the rear cogset, at considerable expense. So before all that extra wear and tear hits your pocketbook, get in the habit of checking the chain on a regular basis (4-6) and replacing it as needed.

(Some believe that it's superfluous to replace any part of the drivetrain, choosing instead to let it all wear out together. Since the chain gets longer and the teeth on the cogs and chainrings and jockey wheels all become hook-shaped, everything will tend to still work together, after a fashion at least. This method, which I don't ascribe to, only works if you never interchange cogs, so racers, or anyone with multiple wheelsets or cogsets, would be foolhardy to adopt it. Switch wheels or cogsets for a particular ride destination, or get a wheel change during a race, and your chain will be jumping and skipping like crazy while simultaneously ruining any cogs it lands on.)

Chain life varies depending on chain type, maintenance, riding conditions, and strength and weight of the rider. A heavy rider can count on replacing the chain every 1,000–1,500 miles if the bike is ridden in dirty conditions or with infrequent lubrication (or with wax-based lubricants). Lighter cyclists riding mostly on clean, dry roads can extend the replacement time to 2,000–3,000 miles with poor maintenance and up to 5,000 miles with frequent high-quality lubrication.

4-6

CHECKING FOR CHAIN ELONGATION

a. Chain-elongation gauges

The most reliable way to see whether the chain is worn out is to employ a chain-elongation gauge. Make sure you check a number of spots on the chain; you'll find variation. The Rohloff and ProGold ProLink gauges (Figs. 4.5, 4.6) are quick and easy to use.

The Rohloff is a go/no-go gauge; brace the hook end against a chain roller, and if the opposing curved tooth falls completely into the chain so the length of the tool's body contacts it, the chain is shot. That's it. There's nothing to squint at to determine whether or not to replace the chain. You are supposed to use the tooth marked S for steel cogs (Fig. 4.5) and the tooth marked A for aluminum and titanium cogs, but I only use the A side. I find that if the A edge comes down to the chain and I replace it right then, I get very long life out of my chainrings and cogs, even titanium ones. That's worth it to me.

With ProGold's ProLink chain gauge (Fig. 4.6), brace the hooked end against a chain roller and let the long tooth drop into the chain. If it drops in close to the 90-percent-worn mark, that is equivalent to the A side of the Rohloff dropping down flush with the chain.

The Park CC-32 chain-wear indicator works like the Rohloff; one side indicates 0.5 percent of chain elongation, and the other side indicates 0.75 percent elongation.

Shimano's chain-elongation gauge, Pedro's Chain Checker, and Topeak's Chain Hook & Wear Indicator also check for 0.75 percent chain elongation, which is between 1/16 and 1/8 of an inch of chain growth over 12 inches. They are different from the others in that they do not push opposing rollers apart, and so do not actually measure the distances between trailing sides of rollers, which are what push on the cogs and chainring teeth. Instead, you don't drop their go/no-go tooth into the chain until you have already braced two other teeth between rollers. This ensures that you're actually measuring the spacing between trailing sides of rollers. On the other hand, they are less quick and require touching the chain with your fingers. The Pedro's Chain Checker and Topeak Chain Hook & Wear Indicator have an additional feature that serves as a chain-assembly tool; it consists of a chain hook on the opposite side—two hooked prongs that pull the chain ends and release the tension at the connection point so you can more easily push in a connector pin or assemble a master link.

Feedback Sports makes a digital chain gauge and also offers it under the KMC brand. With it you can monitor chain wear precisely over time.

4.5 Checking chain wear with a Rohloff gauge

4.6 Using a ProGold chain gauge

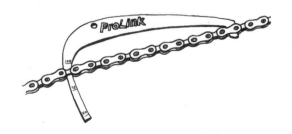

4.7 One complete chain link

b. Ruler method

An accurate ruler offers another way to measure for elongation. Bicycle chains are on an inch standard, and they measure a half-inch between adjacent rivets (and nominally have $3/32$-inch-wide rollers on derailleur chains and $1/8$-inch-wide rollers on single-speed or internal-gear bicycles). There should be exactly 12 links in one foot, where each complete link consists of an inner and outer pair of plates (Fig. 4.7).

1. **Set one end of the ruler on a rivet edge and measure to the rivet edge at the other end of the ruler.**

2. **The distance between these rivets should be 12 inches exactly.** If it is 12 $1/8$ inches or greater, replace the chain; this indicates the chain is one percent longer than when new. If it is 12 $1/16$ inches or more, it is a good idea to replace it (and a necessity if you have any titanium or aluminum cogs or an 11-tooth small cog). The 12 $1/16$-inch measurement indicates 0.5 percent of chain growth and is equivalent to the Rohloff A or Park CC-32 0.5 edge indicating a no-go, and to the ProGold gauge indicating 90 percent worn out.

    If the chain is off the bike, you can hang it next to a new chain; if it is a third of a link or more longer for the same number of links, replace it.

If you always replace the chain as soon as it becomes elongated beyond the spec I've indicated on these chain-elongation gauges, you will replace at least three chains before needing to change the cogs.

4-7

CHAIN OPENING

LEVEL The following procedure applies to all derailleur chains when new and when you are shortening them to length. It also applies to removing a chain from a bike, except for those chains with a master link hand-openable without a chain tool, a.k.a. "chain breaker" (Figs. 4.24–4.27). Wippermann, Taya, SRAM, and KMC chains snap open by hand at the master link (see 4-13), although they can also be opened at any other link with a chain tool, as described next. First-generation Campagnolo 10-speed chains have a master link that cannot be opened.

1. **Place any link over the back teeth on a chain tool (Fig. 4.8).**

2. **Tighten the chain tool handle clockwise to push the pin most of the way out.** For 5-, 6-, 7-, and 8-speed chains other than Shimano or chains with a master link, be careful to leave a millimeter or so of pin protruding inward from the chain plate to hook the chain back together when reassembling. **However, if you have a chain with a master link or a Shimano or Campagnolo chain and a new connector pin for it, go ahead and drive the pin**

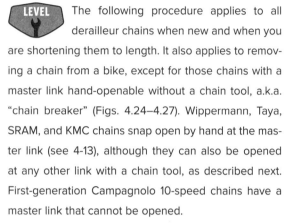

PRO TIP — CYCLOCROSS DEMANDS FREQUENT CHAIN REPLACEMENT

IF YOU ARE RACING CYCLOCROSS, particularly in the Midwest, Southeast, Northwest, or East Coast (or in northern Europe), the chain is going to wear very quickly due to mud. Replacing the chain frequently under these conditions is critically important, or it will skip or ruin cogs. Cyclocross tends to mandate frequent wheel changes, and a worn chain will screw up the cogs on all the wheels you use on that bike. Chains are pricey, yes, but not nearly as pricey as multiple cogsets.

If you're riding in the mud a lot, check the chain for elongation weekly. Replace the chain as soon as (or before) the ProGold indicator goes beyond 75 percent (Fig. 4.6) or the Rohloff A side drops fully into the chain (Fig. 4.5).

4.8 Pushing out the pin (link rivet)

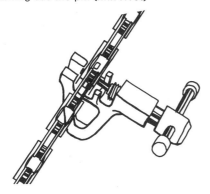

4.9 Proper double-chainring setup

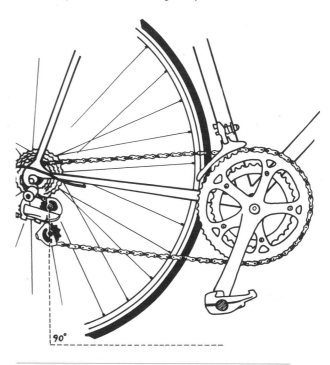

all the way out. Campagnolo requires inserting its special chain-assembly pin only through "virgin" holes in new outer chain plates of its 10-speed chains, requiring that you buy a section of new chain with virgin outer plates at either end and two assembly pins. Then you remove that many links from the chain and insert this new section of chain. Campagnolo 11-speed chains don't require this but do require a special pin and a tool to mushroom out the rivet head after installation.

3. **Separate the chain.** If you pushed the pin all the way out, the two ends will just pull apart. Otherwise, flex it away from the pushed-out pin if you left the stub in.

4-8

DETERMINING CHAIN LENGTH

If you are putting on a new chain for double cranks (including compact doubles), determine how many links you will need in one of the following four ways. Methods 2 and 4 are approximately equivalent, and both work for standard double-chainring setups as well as for compact-drive (smaller) double chainrings. If you have a triple crankset (three chainrings up front) and a long-cage rear derailleur on your bike, however, you should use method 3. For a single-chainring system, you should also use method 3.

Method 1

Under the assumption that your old chain was the correct length, compare the new with the old chain and use the same number of links.

Method 2

With a standard double-chainring setup, route the chain through the derailleurs and over the large chainring and smallest cog. The jockey wheels in the rear derailleur should then align vertically (Fig. 4.9). This method will not work if you are using a large cog that is larger than 27 teeth or beyond your rear derailleur's rated capacity; the chain will end up too short and won't reach over the big chainring and big cog simultaneously. Use method 3 instead.

Method 3

Wrap the chain around the big chainring and the biggest cog without going through either derailleur. Bring the two ends together until the ends overlap; one full link (Fig. 4.7) should be the amount of overlap (Fig. 4.10). This method works for triples and standard doubles and single-ring setups with standard rear derailleurs, and it is a must if you are using a double with a big cogset (like an 11–32 or 11–34). Use one and a half additional links of overlap for a 1 × 11 system using a SRAM X-Horizon roller clutch "straight parallelogram" rear derailleur (such as the Force CX1, Force 1, or Rival 1). In other words, you'll overlap at least two full

4.10 Determining chain length using the big chainring and the biggest cog

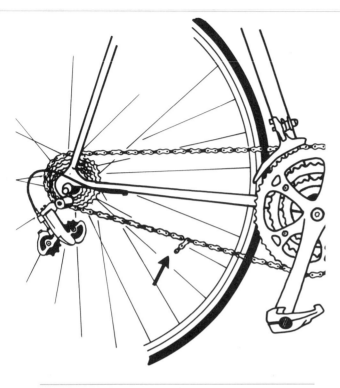

4.11 Determining chain length using the Campagnolo method

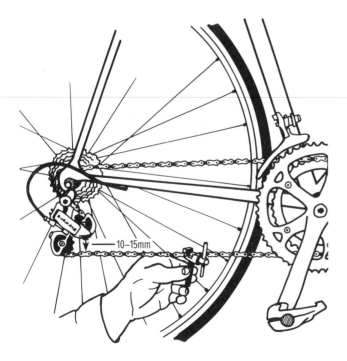

10–15mm

links, but both links will end in an inner link, because you'll be connecting them with a SRAM master link, thus making the actual overlap 2.5 links.

Method 4

Campagnolo suggests, with a double crank, routing the chain over the inner chainring and the smallest cog, as shown in Figure 4.11. Check for about 10–15mm of clearance between the chain wrapped around the upper jockey wheel and the lower run of chain (Fig. 4.11). This is not easy to measure; push the bottom of the ruler up against the chain wrapped around the upper jockey wheel.

Remove the excess links (4-7, Fig. 4.8) and save them in your spare-tire bag so that you have spares in case of chain breakage on the road.

Method 5

Another method is to make sure the chain is still under tension in the small-small combination. When the chain is on the smallest cog and smallest chainring, the chain should not drag on itself where it doubles back through the rear derailleur. There should be 8–15mm between chain sections wrapped around the guide pulleys.

4-9

ROUTING THE CHAIN PROPERLY

1. **Shift into small-small.** Shift the derailleurs so that the chain will rest on the smallest cog in the rear and on the smallest chainring up front.
2. **Guide the chain up through the rear derailleur.** Start with the rear-derailleur pulley that is farthest from the derailleur body (this will be the bottom pulley once the chain is taut). Go around the two jockey pulleys. Make sure the chain passes inside of the prongs on the rear-derailleur cage.
3. **Guide the chain over the smallest rear cog and through the front-derailleur cage.**
4. **Wrap the chain around the smallest front chainring and bring the chain ends together (Fig. 4.11).** When connecting the chain, it is easiest if you completely remove the tension from it by pushing it off of the inner chainring to the inside.

CONNECTING A 5-, 6-, 7-, OR 8-SPEED CHAIN (WITHOUT A MASTER LINK OR A SPECIAL CONNECTOR PIN)

NOTE: *If you have a Shimano chain, a 9-, 10-, or 11-speed chain, or a chain with a master link, go to the appropriate section. Except as a short-term emergency fix to get you home, don't connect it by using the original rivet (as described in this section). Not heeding this warning could result in injury if the chain breaks.*

FURTHER NOTE: *This section applies only to wider chains, such as 5-, 6-, 7-, and non-Shimano 8-speed chains. Never use the same pin (except in an emergency on a ride) on a 9-, 10-, or 11-speed chain or on any Shimano or Campagnolo chain. Connecting a chain that has no special connector pin or link is much easier if the link rivet (pin) that was partially removed when the chain was taken apart is facing outward (toward you). Positioning the link rivet this way allows you to use the chain tool (Fig. 4.11) in a much more comfortable manner (driving the rivet toward the bike, instead of back at you). Be sure that the chain length allows about 10–15mm of clearance between the upper jockey wheel and the lower length of chain.*

1. **Hook the ends together.** Snap the end link over the little stub of pin you left sticking out to the inside between the opposite end plates. You will need to flex the outer chain plates open as you push the link in to get the pin to snap into the hole.

2. **Push the pin through with the chain tool (Fig. 4.12) until the same amount protrudes on either end.** If you have a 9-, 10-, or 11-speed system, or any Shimano chain, you shouldn't be using the original rivet in the first place. But if you are, and the chain tool prongs seem to be getting bent as you push the rivet, see the note in 4-11, step 7.

3. **If there is a stiff link (Fig. 4.13), free it.** You can flex it laterally with your fingers (Fig. 4.14) or, better, you can use the chain tool's back teeth, as illustrated in Figure 4.15. Push the pin a fraction of a turn to spread the plates apart.

4.12 Pushing in the pin (link rivet) with a Park CT-5 chain tool

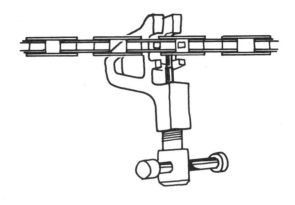

4.13 A stiff link

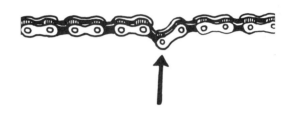

4.14 Freeing a stiff link with your fingers

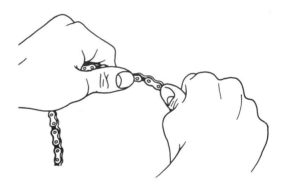

4.15 Freeing a stiff link using a Park CT-5 chain tool

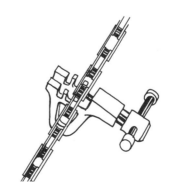

CONNECTING A 7-, 8-, OR 9-SPEED SHIMANO CHAIN

1. **Make sure you have the appropriate Shimano connector pin.** It is more than twice as long as a standard chain rivet and has a pointed tip. It has a breakage groove at the middle of its length. Two connector pins come with a new Shimano chain. If you are reinstalling an old Shimano chain, use a new connector pin, and make sure it is the right length for the chain (10- and 11-speed connector pins are shorter than 9-speed ones, which are shorter than 7- or 8-speed connector pins; see 4-12). If you don't have a connector pin and are going to connect a 7- or 8-speed chain anyway, follow the procedure in 4-10, but be aware that the chain is now more likely to break than if it had been assembled with the proper connector pin. With a 9-speed chain, this is an extremely dangerous approach—don't do it. With a 10- or 11-speed chain, it is a complete no-no to connect the chain without a new connector pin; the chain will likely break immediately. A broken chain is no fun; it can wreck other parts, and you can get injured. For assembling a 10- or 11-speed chain, read 4-12.

2. **Remove any extra links, pushing the appropriate rivet completely out (4-7, Fig. 4.8).**

3. **Line up the chain ends.**

4. **Drip some oil on the connector pin and push it in with your fingers, pointed end first.** It will go in about halfway.

5. **With the chain tool (Fig. 4.16), push the connector pin through until there is only as much left protruding at the tail end as the other rivets in the chain.** It will feel hard, then easy, and then very hard to tighten the tool as you move the pin past its various ridges and valleys. Stop when it gets very hard, and check for binding. Go by feel more than by sight to determine when the pin is seated correctly.

6. **Break off the leading half of the connector pin with the hole in the end of a Shimano chain tool or with a pair of pliers (Fig. 4.17).**

7. **The individual links should move freely when the pin is correctly seated.** If not, depending on which end protrudes more, either push the link rivet in a little deeper (Fig. 4.12) or push it back a hair from the other side with the chain on the tool teeth closest to the screw (Fig. 4.15). As a (poor) last resort not recommended with 9-speed or narrower chains, flex the chain back and forth with your thumbs at the stiff rivet (Fig. 4.14).

NOTE: *If you have a 9-speed chain and an older chain tool, you may find that the prongs in the tool to hold the chain are too far from the backing plate of the tool and will get bent. Shimano tools TL-CN23 (Fig. 4.18) and TL-CN33 (Fig. 4.19) work on all Shimano chains, from 6-speed to 11-speed. Many other brands also have tools for Shimano chains.*

4.16 Pushing a connector pin into a Campagnolo or Shimano chain with a chain tool

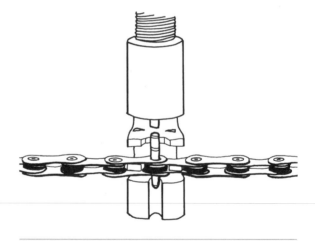

4.17 Breaking off a Shimano or Campagnolo 11-speed connector pin

CONNECTING 10- AND 11-SPEED CHAINS

 LEVEL Breakage can be an issue with 10- and 11-speed chains due to their narrowness. They are narrow because of the tight spacing between cogs required to fit so many of them onto a hub that is the same width as an 8-speed hub. In going from 5-speed to 6-, 7-, 8-, 9-, 10-, and 11-speed systems, the chain width (outside-to-outside) has come down substantially, from 7.3mm to 7.1mm, 6.6mm, 6.1mm, 6.0mm, 5.9mm, and now 5.4mm. But because the cog and chainring teeth did not get significantly narrower, neither did the inner spacing (the roller width) of the chains.

Since the spacing between the inner plates has not decreased, the width decrease has come from shortening the rivets and thinning the steel plates. Imagine how thin the chain plates must be on an 11-speed chain; the spacers on either side of each cog are only 2.2mm thick, and the chain has to be able to run at the most extreme big-to-big or small-to-small cross angles without touching the adjacent cog.

Minimal protrusion of the chain rivets out of the plates means the chain is more prone to breakage under high side loads. Consequently, following recommended assembly procedures is absolutely critical. To install the short connector pin on a 10- or 11-speed chain properly, you need a really good chain tool that is intended for use on 10- or 11-speed chains (see Figs. 4.18–4.22, and read the Pro Tip on chain tools). The only exceptions are chains with hand-openable master links (Figs. 4.24–4.27) that allow opening and reconnecting the chain without using a chain tool. If your chain has a master link, skip to 4-13.

When a chain breaks, it is usually under high pedaling load during shifting. Shifting creates lateral stress. When the rear derailleur shifts with a modern cogset, the chain is simultaneously engaged on two cogs. And when the front derailleur pushes the chain from one chainring to another, it obviously flexes the links sideways. Easing off on the pedals when you shift will greatly decrease the possibility of a broken chain. However, a chain can break under any high load if a link plate is just barely hanging on to the end of a rivet. Breaking a chain is dangerous because when the tension on the chain is suddenly removed, your weight drops straight down, as there is no longer anything supporting the pushing foot. You can easily fall hard. Very hard.

So, pay attention to the extra steps required with a 10- or 11-speed chain to ensure its security. To reuse a Campagnolo 10-speed chain, you must have a new section of chain with a "virgin" outer link at either end and two new connector pins. Shimano connector pins can be installed in used Shimano chains, including 10- and 11-speed chains, and the same goes for Campagnolo 11-speed chains and connector pins.

a. Connecting a Campagnolo 10-speed chain

Original Campagnolo 10-speed chains, when introduced in 1999, came with a separate closing link with two pins, called a PermaLink, and it required an expensive installation tool unique to that link. Fortunately, Campagnolo soon abandoned that closing method and replaced it with a pin system similar to Shimano's.

Campagnolo supplies a 10-speed connector pin that is a two-piece unit that pulls apart, rather than a pin with a break-off end like Shimano's. Remember, you get only one connector pin with each Campagnolo chain (as opposed to the two you get with a Shimano chain), so don't lose it! Replacing one ain't cheap!

Campagnolo, like Shimano, has a special chain tool and highly recommends that you use it. Its main feature is a wire retainer that you slide into the tool to hold the chain down (like the 11-speed tool in Fig. 4.23, except it doesn't need the flip-down peening gate). The Campy tool also has a locator stop for the drive pin specific for its chains.

1. **If you need to shorten the chain, cut excess length (4-7) only from the end that terminates in an inner link.** That way, the holes in the plates of the outer link you will be connecting will never have had a rivet through them and consequently will not have been enlarged and distorted by the insertion and removal of a rivet, ensuring the strongest possible connection. Campagnolo calls these the "virgin holes," and it ships its chains with a zip tie through this pair of holes at the end of the final outer link so that you will make sure to close the chain at these links.

2. Remove the zip tie and install the connector pin as in 4-11. Be careful to insert the connector pin from the wheel side of the chain outward (the opposite direction from that shown in Fig. 4.11!). If you don't have a Campy chain tool (Fig 4.23), hold the chain down in the chain tool so that the tool's drive pin lines up exactly with the connector pin. Note that the Campy connector pin is in two pieces, and you can just insert the leading "guide pin" segment by hand, with or without the other half of the pin on it. The trailing segment that you will press into the chain has two hole sizes in it, so that only one end will fit on the guide pin, thus ensuring proper orientation. Since it is two pieces, it is flimsy laterally; if you push on it with a non-Campy tool, you need to hold the link down in the tool really well to keep the link from sliding up and the connector pin from buckling.

Make sure the same amount of the connector pin is sticking out of both ends (it will only be about 0.1mm) and that you have no stiff link (Fig. 4.13). If you're sure you've pushed it in far enough and it's still binding, free it by carefully using the back teeth on a chain tool. Set the stiff link over the back teeth closest to the screw handle (Fig. 4.15), and push the pin a fraction of a turn to spread the link.

b. Connecting a Shimano 10-speed chain

IMPORTANT NOTE: *In 2009, Shimano introduced an asymmetrical 10-speed chain; it has a different shape on each side. To obtain the correct orientation, make sure that the outer link plates with the Shimano brand and model stamped on them face away from the wheel.*

1. If you need to shorten the chain, cut excess length (4-8) only from the end that terminates in an inner link. That way, the holes in the plates of the outer link you will be connecting will have never had a rivet through them and consequently will not have been enlarged and distorted by the insertion and removal of a peened rivet, ensuring the strongest possible connection.

2. Ensure that the connector pin you insert is the pin leading the outer link over the top of the chainring or cog. Accomplish this link orientation by making

4.18 Shimano TL-CN23 6-speed to 10-speed consumer tool; the TL-CN22 (6s–9s) and TL-CN21 (6s–8s) tools look identical.

4.19 Shimano TL-CN34 6-speed to 11-speed shop tool; the TL-CN33 and TL-CN32 (6s–10s), TL-CN31 (6s–9s), and TL-CN30 (6s–8s) tools look identical.

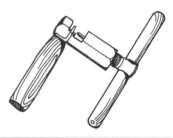

4.20 Pedro's Pro Chain 1.0 single-speed to 10-speed tool

4.21 Rohloff Revolver chain tool

4.22 Park's CT-3 chain tool

sure that, if you are connecting the chain at the bottom as in Figure 4.11, the inner link on one end is to the left (toward the rear derailleur) of the outer link on the other end you are connecting it to.

3. **Lubricate the pin with chain oil prior to installation.** Make sure that you are using a Shimano 10-speed connector pin. Double check the chain orientation before proceeding (see previous Note).

PRO TIP — PROPER CHAIN TOOLS

IF YOU RIDE A LOT, you will change the chain frequently, and it will be worthwhile to have a good chain tool. If you currently just have a cheap little chain tool, you will be glad you made the investment to upgrade.

Most shop chain tools are backward-compatible with the wider chains that came before. Shimano's consumer chain tools work not only on the chain they are contemporary with, like the TL-CN23 tool (Fig. 4.18) and Shimano 10-speed chains, but also on 6-speed chains and every chain in between. Shimano's pricier shop chain tools (Fig. 4.19) share this feature and additionally have four locating teeth in a row to hold the chain, rather than just two, as most chain tools have. These extra two teeth, one extending out on either side of the tool, hold the chain well and do much of what Campagnolo's chain tool (Fig. 4.23) accomplishes with its wire loop to hold the chain down. Pedro's Pro Chain tool (Fig. 4.20) also has these extra two locating teeth extending on either side. Pedro's Tutto (meaning "everything") chain tool works on all single-speed through all 11-speed chains and incorporates an anvil to flare the connector pin on Campagnolo 11-speed chains. Park's CT-4.3 and CT-6.3 and Topeak's All Speeds chain tools also share these features.

Holding the chain in place is nothing new. Before special connector pins came into use, chains still needed to be lined up properly and held in place tightly, because there was no leading tip on a connector pin to slip in by hand and hold the chain together. The Rohloff Revolver tool (Fig. 4.21), which has been around since the 1990s, has a thumbscrew that tightens against the chain and really holds it in place. It also has a revolving plate with different patterns on it to peen the end of the rivet in whatever style you choose. Park's CT-3 (Fig. 4.22) is a standard shop chain tool, with both a front set

of teeth and a back set of teeth, for prying a link apart a bit to free a stiff link, as in Figure 4.15.

If you strictly use Campagnolo chains and don't ride mountain bikes, you might as well get the (very pricey) Campy C11 (UT-CN300) 11-speed tool (Fig. 4.23).

I used a Shimano TL-CN31 (9-speed tool) for years on every kind of 7-, 8-, 9-, and 10-speed chain from Shimano, Campagnolo, Wippermann, and SRAM. As the chains became narrower, the supporting center section on newer Shimano chain tools was moved closer to the chain-locating teeth, to fully support the rear outer link plate while driving the pin in. If you use an older-generation chain tool (for wider chains) with a one-generation newer (narrower) chain, eventually you will damage the tool as it puts too much lateral load on the chain-locating teeth. A two-generation older chain tool will not seat the chain connector pins, so don't use it. Ideally, the latest tool will do the best job with the latest chains, and it will also be compatible with all the older (wider) chains.

If you are careful and are not using a Campagnolo 11-speed chain, you only need one good tool that is at most one generation old (i.e., it is meant for at least 10-speed chains), and you can use it on any chain up through 11-speed. By careful, I mean that you must make sure that the connector pin and the holes are all lined up perfectly (which the Campy, Rohloff, Shimano, Pedro's, and Topeak professional chain tools definitely help guarantee). Shop-quality tools are heavier-duty than consumer tools and often have spare driver pins in the handle.

Furthermore, you must stop at the right point: Do not go too far or not far enough. For this, you must have a feel for the loose-tight-loose-tight pressure changes as you push a Shimano or Campagnolo connector pin into place, as well as an eye for when the pin is protruding (or recessed) the same amount on both faces of the chain.

4. **Follow the instructions in 4-11, with one exception:** Rather than having both ends of the connector pins protruding a bit from their plates (as they are for 7-, 8-, and 9-speed chains), you want the 10-speed connector pin to be slightly recessed, below flush. Go by feel rather than by how much pin you see exposed. Keep tightening beyond where you would normally stop with a 9-speed chain because it feels hard. It will feel easy and then very hard while you continue to push. At this point, stop to check for binding. The chain is connected perfectly when there is no binding at the connector pin, without the need for you to flex the link sideways to free it. Back off on the chain tool screw, check the link for freedom of movement, and retighten the tool if you feel any binding. According to Shimano, if the link is still binding, there is a 99 percent chance that the pin isn't pushed in far enough to be fully seated. As usual, finish by breaking off the tip of the connector pin, as shown in Figure 4.17, or by slipping the hole at the end of the Shimano chain tool over the end of the pin and using that to snap it off.

c. Connecting a Campagnolo 11-speed chain

Campagnolo's hollow 11-speed connector pin is a break-off type, very different from the pull-apart version it uses for 10-speed chains. Install it through the virgin holes in the outer link on the end of the chain, which sports a tag zip-tied to it.

Installing the pin from the wheel side out (the opposite direction from that shown in Fig. 4.11!) makes the extra length protrude toward smaller cogs, so it can't hang up on the face of a larger cog.

You only get one connector pin with a new Campy chain, so don't waste it! It can easily buckle and break at the center if the chain slides up while pushing the pin in. If you use Campagnolo's special (and expensive) UT-CN300 11-speed chain tool (Fig. 4.23), its wire retainer will hold the chain down while pushing in the pin and preclude this from happening.

1. **If you need to shorten the chain, remove excess length (4-8) only from the end that terminates in an inner link.** Campagnolo ships its chains with

4.23 Campagnolo UT-CN300 11-speed chain tool

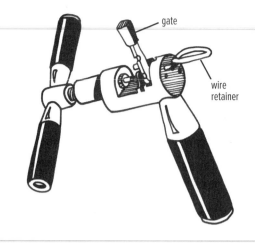

a zip tie through the holes at the end of the final (laser-etched) outer link so that you will close the chain at these holes. These "virgin holes" will not have been enlarged and distorted by the insertion and removal of a rivet. When removing excess links with the Campagnolo 11-speed chain tool, make sure the tool's swinging gate is open (Fig. 4.23) so the old chain rivet can push completely through.

2. **Install the connector pin by hand.** Remove the zip tie and slip the connector pin, pointed end first, from the wheel side of the chain outward through the lined-up holes of the inner and outer end links to hold the chain together.

3. **Flip open the gate and pull out the wire retainer on Campagnolo's 11-speed chain tool (Fig. 4.23), back out the tool's driving pin (spin the T-handle counterclockwise), and set the end links into it.** Since you will be driving the pin outward, away from the wheel, the tool's T-handle will be toward the wheel, making it inconvenient to turn.

4. **Insert the wire retainer through the holes in the end of the tool (Fig. 4.23), over the chain, and through the holes in the tool's back set of teeth.** The chain is now held down.

5. **Push the pin in, going by feel.** You will feel resistance at first, then almost no resistance, and finally a second resistance followed by solid resistance, at which point you should stop pushing. The pin should protrude by 0.1mm; don't push it in flush or beyond.

6. **Open the tool and release the chain.** Check that the link moves easily and that the tail end of the pin is flush with the chain plate.

7. **Turn the tool around, slip its break-off hole (at 3 o'clock on the tool's face) over the pilot pin, and snap it off.** Or use pliers (Fig. 4.17).

8. **With the tool facing in toward the wheel as in Figure 4.11, put it back on the same link, lined up with the assembly pin.** Flip the gate of the tool down behind the chain so that the gate's anvil pin backs up the tail end of the connector pin. Insert the wire loop into the tool to hold the chain in position.

9. **Tighten the chain tool until it flares out (peens) the head of the assembly pin to lock it in place.**

Unlike Campy 10-speed chains, Campagnolo 11-speed chains can be opened (at any spot other than the connector pin) and reconnected with a new connector pin.

d. Connecting a Shimano 11-speed chain

Connecting a Shimano 11-speed chain requires an 11-speed chain tool, like the TL-CN34 (Fig. 4.19) shop tool. Shimano's 11-speed chain tools are backward compatible with Shimano 6- to 11-speed chains. The connecting procedure is the same as with a Shimano 10-speed chain; follow the instructions in 4-12b, except that either side of the chain can face out. When finished, the link pin should be flush with the chain on the side you pushed in, and protruding slightly on the other side where you broke off the lead pin.

4-13

CONNECTING AND DISCONNECTING A MASTER LINK

a. SRAM (Sachs) PowerLink, Lickton's SuperLink, and KMC MissingLink

These links are the same; SRAM (which purchased Sachs) licensed the Lickton's SuperLink design (Fig. 4.24), and the MissingLink works the same way. The master link is made up of two symmetrical link halves, each of which has a single pin sticking out of it and a keyhole on the opposite end.

4.24 SRAM PowerLink, Lickton's SuperLink, and KMC MissingLink

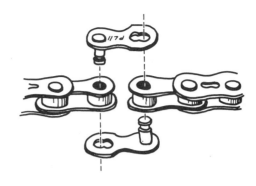

NOTE: *The SRAM 11-speed and 10-speed PowerLock Link and the discontinued KMC MissingLink II are not supposed to be openable.*

ANOTHER NOTE: *SRAM 10-speed PowerLock Links that are stamped with an "M" or an "N" were recalled in late 2009.*

Connecting

1. **Put the pin of each half of the link through the hole in each end of the chain.** One pin will go down, and one up (Fig. 4.24).

2. **Pull the links close together so that each pin goes through the keyhole in the opposite plate.**

3. **Pull the chain ends apart so that the groove at the top of each pin slides to the end of the slot in each plate.** If it won't snap in place by hand, rotate the crank to bring the master link to the top section of chain just behind the front derailleur, and bang down on one pedal with your hand (in the pedaling direction) to snap the master-link pins into the ends of their slots. Another way to do this is to pry two rollers surrounding the master link apart until the link locks by using the notches on the back side of the jaws of Park MLP-1.2 master-link pliers.

Disconnecting

1. **While squeezing the master-link plates together to free the plate from the groove in the pins, push the chain ends toward each other so that the pins come to the center hole in each plate.** If you have a pair of master-link pliers, use them to grab the two rollers through which the pins of the mas-

4.25 Using Park master-link pliers

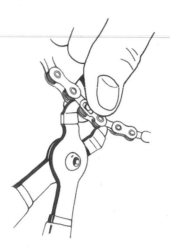

4.26 Wippermann ConneX link—note its orientation with the link's high bump away from the chainring

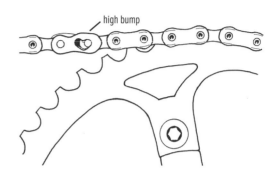

high bump

ter link are inserted (Fig. 4.25). Squeeze the plier handles together while you squeeze the master-link plates toward each other with your fingers. The link will come right apart. Master-link pliers are one of the slickest tools in existence; with them you can easily open SRAM 11-speed and 10-speed PowerLock master links, which are not supposed to be openable.

NOTE: *An old, dirty master link may be hard to open without master-link pliers. To disengage the link plates from the pin grooves while you push the ends toward each other, try squeezing the link plates toward each other with a clothespin or a pair of vise grip pliers set on very low pressure. In desperation, you may have to open the chain somewhere else, reassembling it using a second master link.*

2. **Pull the two halves of the master link apart.**

b. Wippermann ConneX link

The Wippermann link works much the same way as the SRAM, KMC, and Lickton's master links just discussed, but the edges of the link plates are not symmetrical. This asymmetry means that there is a definite orientation for the link, and you don't want to install it upside down.

Orient the ConneX link so that its convex edge is away from the chainring or cog (Fig. 4.26). The link plate is bowl-shaped. If you (incorrectly) have the convex bottom of the bowl toward the cog or chainring, then when it is on an 11-, 12-, or maybe even a 13-tooth cog, the

convex edge will ride up on the spacer between cogs, lifting the rollers out of the tooth valleys and causing the chain to skip under load. The keyhole where you pop the link together forms a heart shape. When the master link is on the top of the cog or chainring, make sure that the heart is right side up (Fig. 4.26).

Remove and install the ConneX link the same way as the SRAM PowerLink in 4-13a, but make sure the high bump of the convex link edge is facing outward from the chain loop (Fig. 4.26). That way, the concave edge can run over the cog spacers on the smallest cogs without lifting the chain.

c. Taya Master Link

Taya's "Sigma Connector" is a no-tools master link available for 1-speed through 11-speed chains.

Connecting

1. **Connect the two ends of the chain together with the master link that has two rivets sticking out of it (Fig. 4.27).**

4.27 Taya Sigma Master Link

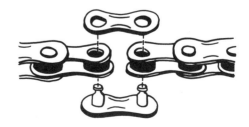

2. **Snap the outer master-link plate over the rivets and into their grooves.** To facilitate hooking each keyhole-shaped hole over its corresponding rivet, flex the plate with the protruding rivets so that the ends of the rivets are closer together.

Disconnecting

1. **Flex the master link so that the pins come closer together.**
2. **Pull the plate with the oval holes off the rivets.**

TROUBLESHOOTING CHAIN PROBLEMS

4-14

CHAIN SUCK

"Chain suck" occurs when the chain does not release from the bottom of the chainring. Instead, it sticks to the ring and gets "sucked" up until it hits the chainstay. Sometimes the chain becomes wedged between the chainstay and the chainring.

A number of things can cause chain suck. To eliminate it, try the simplest methods first.

1. **Clean and lube the chain and see if it improves.** A dry, rusty chain will hold the curved shape of the chainring too long.
2. **Check for stiff links (Fig. 4.13).** Slowly turn the crank backward and watch the chain move through the derailleur jockey wheels. Loosen stiff links by flexing them laterally with your thumbs (Fig. 4.14) or by using a chain tool with back teeth (Fig. 4.22). Set the stiff link over the back teeth closest to the screw handle (Fig. 4.15), and push the pin a fraction of a turn to spread the link.
3. **If chain suck persists, check for bent or torn teeth on the chainring.** Try straightening any broken or torn teeth you find with pliers, and use a file to remove any burrs you find on the teeth.
4. **If the chain still sucks, replace the inner (and perhaps middle) chainring.** The new, unworn rings will release the chain more easily, and some chainrings have teeth that are thinner than others or have a harder, slicker surface on the teeth that releases the chain better.

4-15

SQUEAKING CHAIN

Squeaking is caused by dry or rusted surfaces inside the chain rubbing on each other.

1. **Wipe down the chain (Fig. 4.2) and lubricate it (Fig. 4.1).** A thin and slippery chain lubricant may penetrate and clean up the surfaces enough to rehabilitate a squeaky chain. Avoid wax-based lubricants for this purpose.
2. **Ride it for a half hour or more, and then wipe and lubricate it again and ride it another half hour or more.**
3. **If the squeak does not go away after two rides with fresh lubricant, replace the chain.**

If the initial remedy does not work, the chain is probably too dry and rusted deep inside. Chains often don't heal from this condition. Life is too short and bike riding is too joyful to put up with the sound of a squeaking chain. Replace it.

4-16

SKIPPING CHAIN

There can be a number of causes for a chain to skip and jump as you pedal.

a. Stiff links

1. **Turn the crank backward slowly to see whether a stiff chain link (Fig. 4.13) exists.** A stiff link will be unable to bend properly as it goes through the rear-derailleur jockey wheels. It will jump and move the jockey wheels as it passes through.
2. **Loosen stiff links by flexing them laterally between the index finger and thumb of both hands (Fig. 4.14), or by using a chain tool with back teeth (Fig. 4.22).** Set the stiff link over the back teeth closest to the screw handle (Fig. 4.15), and push the pin a fraction of a turn to spread the link.
3. **Wipe down and lubricate the chain (Figs. 4.1–4.2).**

b. Rusted chain

A rusted chain will often squeak as well as skip. If you watch it move through the rear derailleur, it will look like many links are tight; the links will not bend easily and will cause the jockey wheels to jump back and forth.

1. **Lubricate the chain (Fig. 4.1).**
2. **If this does not fix the problem after a few miles of riding, replace the chain.**

c. Worn-out chain

If the chain is worn out, it will be elongated and will skip because it does not mesh well with the cogs. A new chain will fix the problem if the condition has not persisted long enough to ruin some cogs.

1. **Check for chain elongation as described in 4-6.**
2. **If the chain is elongated beyond the specifications in 4-6, replace it.**
3. **If replacing the chain does not help or actually makes matters worse, see the next section.**

d. Worn cogs

If you just replaced the chain and it is now skipping (despite the derailleurs being in adjustment; see 5-3), probably at least one of the cogs is worn out. If this is the case, the chain will probably skip on the cogs you use most frequently and not on others. However, if it skips only on the smallest cog or two and you have a Wippermann ConneX master link, check to see whether you installed the link upside down (see 4-13b).

1. **Check each cog visually for wear.** If the teeth are hook-shaped, the cog is shot and should be replaced. For cogs up to 21 teeth, Rohloff makes a simple "HG-Check" tool (pictured in Fig. 1.4) that checks for cog wear by putting tension on a length of chain wrapped around the cog. If the last chain roller on the tool snags on the tooth and the last link resists being flipped easily in and out of the tooth pocket while the tool handle is under pressure, or, worse, if the entire measurement chain except the first roller slides easily away from the cog teeth while the handle is under pressure, the cog is worn out.
2. **Replace the offending cogs or the entire cassette or freewheel.** See cog installation in 8-10.
3. **Replace the chain as well, if you have not just done so.** An old chain will wear out new cogs rapidly.

e. Maladjusted rear derailleur

If the rear derailleur is poorly adjusted or bent, it can cause the chain to skip by lining up the chain between gears.

1. **Check that the rear derailleur shifts equally well in both directions and that the chain can be pedaled backward without catching.**
2. **Adjust the rear derailleur by following the procedure described in 5-3.**

f. Sticky shift cable

If the shift cable does not move freely enough to let the derailleur spring over to line up under the cog, the chain will jump off under load. Frayed, rough, dirty, rusted, or worn cables or housings will cause the problem, as will kinked or sharply bent housings. Replacing the shift cables and housings (Chapter 5, 5-7 to 5-15) should eliminate the problem.

g. Loose rear-derailleur jockey wheel(s)

A loose jockey wheel on the rear derailleur can cause the chain to skip by letting it move too far laterally.

1. **Check that the bolts holding the jockey wheel to the cage are tight by using an appropriately sized wrench (usually a 3mm hex key).**
2. **Tighten the jockey-wheel bolts if necessary.** Hold the hex key close to the bend so that you don't have enough leverage to overtighten them. If the jockey-wheel bolts loosen regularly, remove and clean them, and then put Loctite or another threadlock compound on the threads and reinstall them.

h. Bent rear derailleur or rear-derailleur hanger

If the derailleur or derailleur hanger is bent, adjustments won't work. You will probably know when it got bent, too. It was either when you shifted your derailleur into your spokes, when you crashed onto the derailleur, or when you kept pedaling after a plastic bag or a tumbleweed got wrapped into your chain.

1. **Unless you happen to have a derailleur-hanger-alignment tool and know how to use it (Fig. 17.5), take the bike to a shop and have it checked for correct dropout hanger alignment.** The majority of modern derailleur bikes have a replaceable (bolt-

on) right rear dropout and derailleur hanger, which you can purchase and install yourself.

2. **If a straight derailleur hanger does not correct the misalignment, the rear derailleur is bent.** This is generally cause for replacement (see 5-2). With some derailleurs, you can replace the jockey-wheel cage, if that's all that is bent. If you are careful, you can sometimes bend a bent derailleur cage back with your hands. It seldom works well, but it's worth a try if your only other alternative is to replace the entire rear derailleur. Just make sure you don't bend the derailleur hanger in the process.

i. Worn derailleur pivots

If the derailleur pivots are worn, the derailleur will be loose and will move around under the cogs, causing the chain to skip. Replacing the derailleur is the solution.

j. Bent rear-derailleur mounting bolt

If the mounting bolt is bent, the derailleur will not line up straight. To fix it, get a new derailleur or a new bolt and install it following the "upper-pivot overhaul" in 5-32. Observe how the spring-loaded assembly comes apart during disassembly to ease reassembly.

k. Missing or worn chain rollers

You can have a chain that passes the elongation tests mentioned in 4-6 yet skips because one of the cylindrical rollers has broken and fallen off its rivet or is so worn that it is spool-shaped. If you don't happen to check that particular link with the chain-elongation gauge, you'll likely miss broken rollers. The width of the gauge is the same as between the inner plates, so that it won't catch worn-out, spool-shaped rollers either, because it will ride up on the edges of the rollers and not fall down into the center of the narrower waist of the worn roller. You might never know the chain is shot without inspecting every roller.

l. Inverted ConneX master link

If you have a Wippermann ConneX master link upside down (described in 4-13b), the taller link edge will lift the rollers off the cog and will cause the chain to skip. Remove, invert, and reinstall the master link as described in 4-13b.

Never mistake motion for action.
—ERNEST HEMINGWAY

CABLE-ACTUATED SHIFTING SYSTEMS

TOOLS

2mm, 2.5mm, 3mm,
  4mm, 5mm, 6mm
  hex keys
Torx T25 key
Flat-blade and Phillips
  screwdrivers, small
  and medium
Pliers
Indexed-housing cutter
Cable cutter
Grease
Chain oil
Rubbing alcohol

OPTIONAL

Park Tool IR-1 Internal
  Cable Routing Kit
Crochet hook
Vernier caliper

Riding a bike is much more enjoyable when the derailleurs are working well. Feeling the chain respond quickly and positively to shifting commands is sweet. On the other hand, having the chain shift unexpectedly or skip when you pedal hard can ruin your ride.

Cable-actuated derailleurs are simple beasts. When they act up, a few turns of some screws or a cable-tension adjustment will usually get them working again. However, current road bikes generally have brake levers with shift levers integrated into them, and cable-actuating levers do have considerable complexity. The levers on electronic-shift systems are simpler, but the derailleurs are far more complicated; Chapter 6 covers those.

Master this chapter, and you will be able to fix most shifting problems on your bike in seconds, even when you are on the road.

5-1

OPERATING INTEGRATED SHIFT/BRAKE LEVERS

The derailleur adjustments described in this chapter require that you already know how to operate the levers that go with them. However, there may be nuances you're unaware of, so read this section first. Throughout the book, I will also use the term "dual control" to describe road brake levers with an integrated shift mechanism.

SRAM levers are the least obvious to use, so I'll start with them.

a. Rear SRAM DoubleTap (right-hand lever)

To shift to a larger cog (lower gear), push the shift paddle (the lever behind the brake lever in Fig. 5.1A)

5.1A Operating a SRAM DoubleTap front shifter

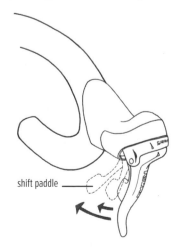

shift paddle

to the left (inward) with your fingers far enough that you feel the second click (if you only go to the first click and release, it will upshift instead of downshift). The most you can move the chain is three cogs with a single push.

To shift to a smaller cog (higher gear), push the shift paddle to the left (inward) with your fingers lightly enough to only hit the first click. Release. It will upshift only one cog at a time.

b. Front SRAM DoubleTap (left-hand lever)

To shift to a larger chainring (higher gear), push the shift paddle (the lever behind the brake lever in Fig. 5.1A) to the right (inward) with your fingers. It takes a firm push to get to the second finger lever click (it will drop back to the inner chainring if you only go to the first click and release).

To shift to a smaller chainring (lower gear), push the shift paddle to the right (inward) to the first click and release.

Trimming in cross-gears

You can trim (feather) the front derailleur so that it does not rub on the chain in some cross-gear combinations. When the chain is on the inner chainring on a 2007 unit and rubs the front derailleur in a cross-gear, give the shift paddle a gentle inward push until you hear a soft click. Year 2007 levers have no trim setting on the big chainring. On 2008 and later 10-speed Red and Rival levers and 2010 and later 10-speed Force levers, the trim setting is on the big chainring. When the chain is on the outer chainring with any of these levers, give the shift paddle a gentle push until you feel the soft click. This may take some practice, as you can easily overdo it and actually perform the shift.

SRAM 22 (11-speed) levers have no trim adjustment; SRAM's 11-speed "Yaw" front derailleur rotates as it swings, and, if it is adjusted properly (5-5f), it should not rub in cross-gears, thus eliminating the need for trim.

c. Rear Campagnolo Ergopower (right-hand lever)

To shift to a larger cog (lower gear), push the finger lever (the shift paddle behind the brake lever; see Fig. 5.1B) to the left (inward) with your fingers. Depending on

5.1B Operating a Campagnolo Ergopower rear lever

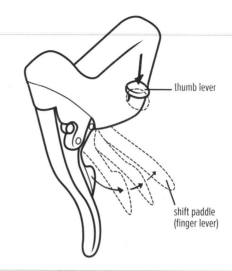

thumb lever

shift paddle
(finger lever)

model and year, you can downshift three to five cogs with a single push.

To shift to a smaller cog (higher gear), push the thumb lever down (yes, with your thumb). A single click upshifts one cog, but depending on model and year, you can move the chain across one to 11 cogs with a single push. High-end Campagnolo Ergopower right levers (Super Record, Record, and Chorus) can shift at least three cogs with a single push on original models; Ultra-Shift models with curvier brake levers and a taller knob on the lever body can upshift all 11 cogs with a single go. Campagnolo lower-end Ergopower right levers (Centaur, Athena, Veloce, Mirage, and Xenon) can only upshift a single cog at a time; you can see the difference because the thumb lever does not run in a long slot with a corresponding long slot in the rubber hood, as is seen on Super Record, Record, and Chorus levers. Instead, there is only a hole in the hood for the thumb lever, indicating that the lever can click only one cog smaller at a time.

d. Front Campagnolo Ergopower (left-hand lever)

To shift to a larger chainring (higher gear), push the finger lever (the shift paddle behind the brake lever) to the right (inward) with your fingers. If the chain does not climb up the outer ring, give another push. If you have a triple chainwheel, you can shift only one chainring at a time.

To shift to a smaller chainring (lower gear), push the thumb lever down.

5.1C Operating a Shimano STI front brake/shift lever

5.1D Operating a Shimano STI front inner lever

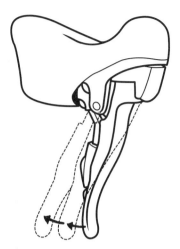

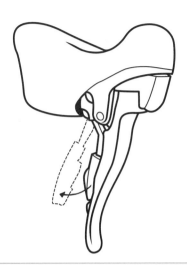

On high-end Ergopower levers with multiple clicks in the thumb lever (and a long slot in the rubber lever hood for it), you can feather (trim) the front derailleur so that it does not rub on the chain in a cross-gear. When the chain is on the inner chainring, give the shift paddle a gentle inward push with your finger to move one click at a time. When the chain is on the outer chainring, give the thumb lever a single-click push at a time; this thumb-lever trim feature is not present on lower-end Ergopower levers.

e. Rear Shimano STI (right-hand lever)

To shift to a larger cog (lower gear), push the brake/shift lever (the larger lever) to the left (inward) with your fingers. The most you can move the chain is three cogs with a single push and two cogs with newer STI levers with shift cables that run under the handlebar tape.

To shift to a smaller cog (higher gear), push the smaller lever behind the brake lever to the left (inward) with your second finger. It will click only one gear at a time.

f. Front Shimano STI (left-hand lever)

To shift to a larger chainring (higher gear), push the brake/shift lever (the larger lever in Fig. 5.1C) to the right (inward) with your fingers. It takes a firm push. If it moves a small click but the chain does not climb up the outer ring, let it return to center then give it another push. If you have a triple, you can shift only one chainring at a time.

To shift to a smaller chainring (lower gear), push the smaller lever (Fig. 5.1D) to the right (inward) with your second finger. If you have a triple crankset, you can shift only one chainring at a time.

Trimming in cross-gears

You can trim (feather) the front derailleur so that it does not rub on the chain in a cross-gear. When the chain is on the outer chainring, give the inner lever a gentle push until you feel the soft click, which moves the derailleur inboard a bit; this latter feature is not present on 10-speed STI levers with under-the-handlebar-tape shift cables. The trim adjustment over the inner chainring is different on 11-speed front derailleurs than on 8-, 9-, or 10-speed ones. On 8-, 9-, or 10-speed systems, when the chain is on the inner chainring, give the brake/shift lever a gentle inward push until you hear a soft click. This moves the front derailleur outboard to increase chain clearance when in small/small gear combinations. On 11-speed systems, by contrast, the front derailleur shifts to the middle position when you shift to the inner chainring, and the trim adjustment is to then provide chain clearance on the small chainring/large cog positions. Therefore, you give the smaller lever another bump to drop the derailleur further inboard as you continue shifting the rear derailleur toward ever-larger cogs. **NOTE:** *There is no trim adjustment with a triple crank.*

Trimming may take some practice, as you can easily overdo it and actually perform the shift.

THE REAR DERAILLEUR

The rear derailleur (Figs. 5.2, 5.3) moves the chain from one rear cog to another, and it also takes up chain slack (such as when the front derailleur is shifted or the bike bounces over a bump). The rear derailleur bolts to a hanger on the frame's rear dropout about which the derailleur can pivot (Fig. 5.4).

Two jockey wheels (the upper, or "guide," pulley [1] and the lower, or "idler," pulley [2]), which live in a guide assembly called a chain cage [3], hold the chain tight and help guide the chain to perform shifts. Depending on brand and model, a rear derailleur has a spring [4] in the (lower) p-knuckle [5] and often another one in the (upper) b-knuckle [6] that pulls the jockey wheels

tight against the chain, creating a desirable amount of chain tension.

The shift cable is affixed to the cable-fixing nut or bolt [7]. Increasing the tension on the rear shift cable (as when you shift to a lower gear) moves the derailleur inward toward the larger cogs. When cable tension is released (i.e., when you shift to a higher gear), a spring [8] between the derailleur's two parallelogram plates [9] pulls the chain back toward the smallest cogs. The two limit screws [10] on the rear derailleur (Fig. 5.3) prevent the derailleur from moving the chain too far to the inside (into the spokes) or to the outside (into the dropout). The limit screws can be on the back of the b-knuckle [6] or on the side of the outer parallelogram plate [9].

5.2 Exploded rear derailleur

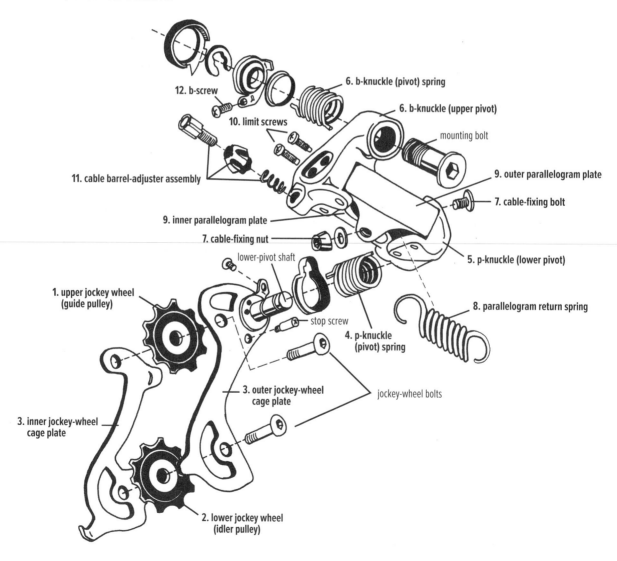

12. b-screw

6. b-knuckle (pivot) spring

10. limit screws

6. b-knuckle (upper pivot)

mounting bolt

11. cable barrel-adjuster assembly

9. outer parallelogram plate

9. inner parallelogram plate

7. cable-fixing bolt

7. cable-fixing nut

lower-pivot shaft

5. p-knuckle (lower pivot)

1. upper jockey wheel (guide pulley)

8. parallelogram return spring

stop screw

4. p-knuckle (pivot) spring

3. outer jockey-wheel cage plate

jockey-wheel bolts

3. inner jockey-wheel cage plate

2. lower jockey wheel (idler pulley)

5.3 Rear-derailleur limit screws and barrel adjuster

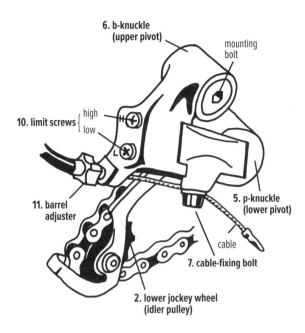

6. b-knuckle (upper pivot)

mounting bolt

10. limit screws { high / low }

11. barrel adjuster

5. p-knuckle (lower pivot)

cable

7. cable-fixing bolt

2. lower jockey wheel (idler pulley)

5.4 Rear-derailleur dropout-derailleur hanger

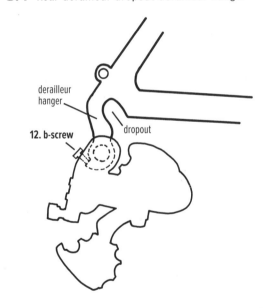

derailleur hanger

dropout

12. b-screw

In addition to limit screws, many rear derailleurs have a barrel adjuster [11] located at the rear of the derailleur, where the cable enters it (Fig. 5.3). The barrel adjuster increases cable tension when it is unscrewed and reduces cable tension when it is screwed in. The barrel adjuster is thus used to fine-tune the shifting mechanism to land the chain precisely on each cog with each corresponding click of the shifter.

Rear derailleurs often have a screw underneath and to the rear (visible in Fig. 5.4). This screw, conventionally called the "b-tension screw" (or "b-screw" [12]), presses against the dropout or a tab attached to the dropout and is largely responsible for controlling the space between the bottom of the cogs and the upper jockey wheel (Figs. 5.5, 5.6). On a 10- or 11-speed Campagnolo rear derailleur, this adjustment is instead made by a screw on the p-knuckle (lower pivot; Fig. 5.9). The other factor that affects the size of this space is chain length.

The chain length, the balance between the springs in the upper and lower pivots, and the b-screw (Fig. 5.4) adjustment determine how closely the derailleur tracks the cogs during its lateral movement and keep the chain from bouncing off the front chainrings when the bike hits bumps.

5-2

REAR-DERAILLEUR INSTALLATION

LEVEL 1. **Apply a small amount of grease to the derailleur's mounting bolt.** Select the appropriate tool for the mounting bolt. This will most likely be a 5mm or 6mm hex key or a Torx T25 (star) tool.

2. **Rotate the derailleur clockwise so that the b-screw or tab on the derailleur ends up behind the tab on the dropout derailleur hanger (Fig. 5.4).**

3. **Thread the bolt a few turns into the hole on the dropout derailleur hanger.** Check to make sure that the tab and/or b-screw is behind the dropout hanger tab and is not getting smashed into the side of the dropout hanger as you tighten the bolt.

4. **Tighten the mounting bolt until the derailleur fits snugly against the hanger.** Consult the Appendix E torque table for appropriate torque.

5. **Route the chain through the jockey wheels and connect it (4-9 through 4-13).**

6. **Install the cables and housings (5-6 through 5-14).**

7. **Pull the cable tight into the groove under the cable-fixing bolt with a pair of pliers and tighten the bolt (Fig. 5.34).**

8. **Follow the adjustment procedure described in the next section.**

ADJUSTMENT OF REAR DERAILLEUR
AND RIGHT-HAND SHIFTER

Perform all of the following derailleur adjustments with the bike in a bike stand or hanging from the ceiling. That way, you can turn the crank and shift gears while you put the derailleur through its paces. After adjusting it off the ground, test the shifting while riding. Derailleurs often perform differently under load than in a bike stand.

Before starting, lubricate or replace the chain (Chapter 4) so that the drivetrain runs smoothly.

a. Limit-screw adjustments

Properly set, these screws (Fig. 5.5) make certain that you will not ruin your frame, wheel, or derailleur by shifting into the spokes or by jamming the chain between the dropout and the smallest cog. It is never pleasant to see your expensive equipment turned into shredded metal. Adjustment requires nothing but a small screwdriver; remember, it's lefty loosey, righty tighty for the limit screws.

b. Low-gear limit-screw adjustment

This screw stops the inward movement of the rear derailleur, preventing it from going into the spokes. This screw is often labeled L, and it is usually the bottom screw (Fig. 5.5). You can check which one it is by shifting to the largest cog and, while maintaining pressure on the shifter or by pulling on the rear-derailleur cable, turning the screw to see if it changes the position of the derailleur.

1. **Shift to the lowest gear.** While turning the crank, shift the front derailleur to the inner chainring on the front and the rear derailleur to the largest cog (Fig. 5.5). Do it gently, in case the limit screw does not stop the derailleur from moving into the spokes. If the cable is not yet tight enough to pull the derailleur up against its inner stop (the low-gear limit screw), pull on the cable with your hand until the derailleur stops moving inboard.

2. **If the derailleur touches the spokes or pushes the chain over the largest cog, tighten the low-gear limit screw until the derailleur does neither.**

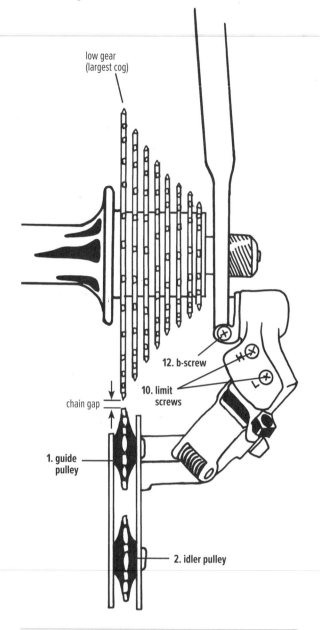

5.5 Low gear

low gear
(largest cog)

12. b-screw

10. limit
screws

chain gap

1. guide
pulley

2. idler pulley

3. **If the derailleur cannot bring the chain onto the largest cog, loosen the screw one-quarter turn.** Repeat this step until the chain shifts easily up to the largest cog but does not touch the spokes or push the chain over the top of the cog.

c. High-gear limit-screw adjustment

This screw limits the outward movement of the rear derailleur. You tighten or loosen this screw until the derailleur shifts the chain to the smallest cog quickly but does not overshift.

How do you determine which limit screw works on the high gear? Often it is labeled H, and it is usually the upper of the two screws (Fig. 5.6). If you're not certain, try both screws. The screw you're looking for is the one that moves the derailleur when you tighten it while the cable tension is released and the chain is on the smallest cog. On most derailleurs, you can also see which screw to adjust by looking between the derailleur's parallelogram side plates. You will see one tab on the back end of each plate. Each is designed to hit a limit screw at one end of the movement. Shift to the smallest cog and notice which screw is touching one of the tabs; that is the high-gear limit screw.

All of these adjustments require you to know how to work your shift levers. For a refresher on shifting Shimano STI, Campagnolo Ergopower, or SRAM DoubleTap, refer back to 5-1.

1. **Shift to the highest gear.** While turning the crank, shift the chain to the large front chainring, and shift the rear derailleur to the smallest rear cog (Fig. 5.6).

2. **If the chain still won't drop without hesitation to the smallest cog, loosen the high-gear limit screw one-quarter turn at a time, continuously repeating the shift.** Loosen the screw until either the chain repeatedly drops quickly and easily or it is clear that the limit screw is not making any further difference and the chain still won't drop.

3. **If the chain still won't drop onto the small cog with the limit screw backed out, or if there is hesitation in the chain's shifting movement, loosen the cable a little to see if it is stopping the derailleur from moving out far enough.** You loosen the cable by either (a) turning one of the two barrel adjusters in the system clockwise—either the barrel adjuster on the back of the rear derailleur (Fig. 5.3) or the barrel adjuster on the threaded boss on the head tube or down tube or along the cable housing (Figs. 5.7, 5.8)—or (b) loosening the cable-fixing bolt, letting a little cable out, and retightening the bolt.

4. **If the derailleur throws the chain into the dropout or tries to go past the smallest cog, tighten the cable or tighten the high-gear limit screw.** See the Pro Tip about this. Turn the barrel adjuster counter-

5.6 High gear

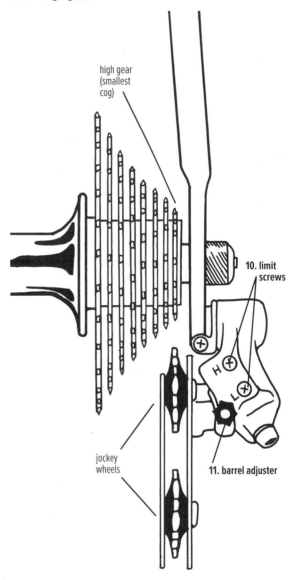

high gear (smallest cog)

10. limit screws

jockey wheels

H

L

11. barrel adjuster

clockwise one-eighth turn or the limit screw clockwise one-quarter turn and redo the shift. Repeat until the derailleur shifts the chain quickly and easily into the highest gear without throwing the chain into the dropout.

NOTE: *Make sure that the washer under the cable-fixing bolt on the rear derailleur is rotated into the correct position, or it may hit the derailleur cage and stop the rear derailleur from getting to the smallest cog. Some derailleurs have a tooth or two on the washer to dovetail into corresponding notches in the derailleur, and a number of positions may seem to fit. Look at the cable groove scored in the washer for a locating hint.*

WHEN THE CABLE TENSION is set correctly so that the derailleur shifts properly on each cog, the high-gear limit screw can be as loose as you like or completely missing, and you won't ever shift the chain into the dropout. This means that you can back out the screw beyond where it needs to be to make sure that it never interferes with the chain dropping onto the smallest cog.

However, note that this also means that whenever you switch wheels, you will need to immediately make sure that the cable tension is accurate so that the derailleur shifts precisely onto each cog. Otherwise, if the cable tension is too loose, which can happen over time or due to switching wheels with a cogset that is set farther inboard, the cable could allow the chain to shift beyond the smallest cog and into the dropout.

d. Cable-tension adjustment on indexed rear shifters

With an indexed shifting system (one that "clicks" into each gear), cable tension determines whether the derailleur moves to the proper gear with each click.

1. **With the chain on the large chainring in the front, shift the rear derailleur to the smallest cog.** While turning the crank, keep clicking the shifter until you are sure it will not let out any more cable.

2. **Downshift one click.** This should pull the cable and move the chain smoothly to the second cog.

3. **If the chain does not climb to the second cog, or if it does so slowly, increase the tension in the cable.** Do this by unscrewing (counterclockwise when viewed from its end where the cable housing enters) either the cable-tension barrel adjuster on the derailleur (Fig. 5.3) or the barrel adjuster on the frame (Fig. 5.7) or along the cable housing (Fig. 5.8). (If you have down-tube shifters, the only barrel adjuster is at the rear derailleur.) If you run out of barrel-adjustment range, retighten (clockwise) both adjusters to within a turn or two from their fully in position, loosen the cable-fixing bolt on the derailleur, and pull some of the slack out of the cable (Fig. 5.34). Tighten the cable-fixing bolt and repeat the adjustment.

4. **If the chain overshifts the second cog or comes close to overshifting, decrease the cable tension by turning one of the barrel adjusters clockwise (i.e., screw it in).** See the following note if you have an inline adjuster. If both barrel adjusters are already fully screwed in, you will need to loosen the cable at the cable-fixing bolt.

NOTE ON INLINE BARREL ADJUSTERS: *Inline barrel adjusters (Fig. 5.8) can be oriented either direction when installed on the cable housing, and the direction you turn them in order to tighten the cable will depend on their orientation. Experiment with the adjuster to see which rotation direction increases or decreases cable tension. Hold both ends when turning it, and the direction that lengthens the adjuster is the direction that will tighten the cable.*

5. **Keep adjusting the cable tension in small increments while shifting back and forth between the two smallest cogs until the chain moves easily in both directions.**

6. **Fine-tune the adjustment.** Shift the rear derailleur back and forth among the smallest six cogs, again checking for precise and quick movement of the chain from cog to cog. Fine-tune the shifting by making small corrections with the cable-tension barrel adjuster.

7. **Shift to the inner ring in the front and to the largest cog in the rear.** Shift up and down one click in the rear, again checking for symmetry and precise chain movement in either direction between the two largest cogs. Fine-tune the barrel adjusters until you get the shifting just right.

8. **Go back through the gears.** With the chain on the big chainring, the rear derailleur should shift easily on all but perhaps the largest one or two cogs in the rear. With the chain on the inner chainring, the rear derailleur should shift easily on all but perhaps the two smallest cogs. Fine-tune while riding by turning the barrel adjusters at the head tube or down tube.

5.7 Barrel adjuster on down-tube shifter boss

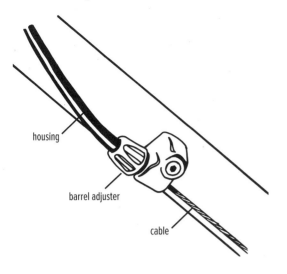

housing

barrel adjuster

cable

5.8 Inline barrel adjuster on shift cable housing

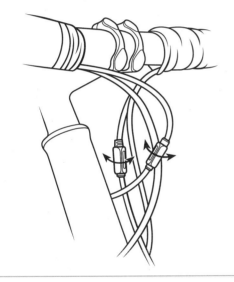

NOTE: *If you can't get the tension to work properly on all cogs, there is likely an incompatibility between the cogs and the shifter, or something is wrong with the cogset. Check 5-42 regarding compatibility between shifters and cogsets to make sure you don't have mismatched cogs, shifter, or derailleur. If the cogs, shifter, and derailleur are supposed to work with each other, then there may be some spacing off within the cogset. If, for instance, it shifts fine in midsize cogs but acts like the cable tension is too high in the large cogs and too low in the small cogs (and your shift cables and housings are new or in good working order; see the Pro Tip on better shifting), you need more spacing somewhere*

within the cogset. Try cutting a circular shim from a beer can that just fits over the freehub body, and slip the shim between a spacer and a cog somewhere in the middle of the cogset. If it improves things, you can play with the number and position of the shims to get it perfect.

ANOTHER NOTE: *If the derailleur is thwarting your adjustment abilities, or it touches the spokes while the chain is on the largest cog, the derailleur hanger on the dropout may be bent. You can straighten it or replace it (17-4).*

e. Cable-tension adjustment on frictional rear shifters

If you do not have indexed shifting, adjustment is complete after you remove the slack in the cable. With proper cable tension, when the chain is on the smallest cog, the derailleur should move as soon as the shift lever does. If there is free play in the lever, tighten the cable by turning the barrel adjuster on the derailleur counterclockwise. If the rear derailleur has no barrel adjuster, loosen the cable-fixing bolt, pull tension on the cable with pliers, and retighten the bolt.

f. Final details: b-screw adjustment

You can get a bit more precision by adjusting the small screw (b-screw) that changes the derailleur's position against the derailleur hanger tab on the right rear drop-out (Fig. 5.4).

View the bike from behind with the chain on the inner chainring and largest cog (Fig. 5.5), and adjust the b-screw so that the upper jockey wheel (the guide pulley) is close to the cog, but not pinching the chain against the cog. Repeat on the smallest cog (Fig. 5.6). You'll know that you've moved the guide pulley in too closely when it starts making noise and even bumping up and down when you turn the crank with the chain on the large cog (i.e., the chain gap shown in Figure 5.5 is narrower than the chain is tall, so that the chain is pinched between the cog and the pulley).

Campagnolo 10- and 11-speed rear derailleurs do not have a b-screw, whereas earlier 8- and 9-speed versions did have one. Instead, you adjust the pulley-to-cog spacing by changing the spring tension in the lower pivot. Set the chain on the large cog. Tighten or loosen the spring-tension adjustment screw under the lower pivot; the screw is up against the jockey-wheel cage plate at the base of the lower derailleur knuckle (Fig. 5.9).

Make the chain gap in Figure 5.5 about 5–7mm, whether it's Campagnolo, Shimano, or SRAM, except make it 6–12mm on SRAM X-Horizon derailleurs (5-3g).

Current Campagnolo 11-speed short-cage rear derailleurs are compatible with 12–29 cogsets, but first-generation ones were not because this lower pivot adjustment could not pull the jockey-wheel cage back far enough. You can upgrade an original Campy 11-speed rear derailleur to work with a 12–29 by replacing the toothed ring the lower pivot adjustment screw drives with one with more teeth. You must remove the jockey wheels and unscrew the bolt holding the jockey-wheel cage onto the lower pivot to do it. Replace the ring, start the bolt, wind the spring by rotating the cage past its stop, tighten the bolt, and replace the jockey wheels.

NOTE: *If, despite your best efforts, you cannot get the rear derailleur to shift well and noiselessly, refer to the chainline discussion under "Troubleshooting Derailleur and Shifter Problems" at the end of this chapter. Also see the troubleshooting chart, Table 5.1, in 5-36.*

5.9 Campagnolo's rear-derailleur lower pivot tension adjustment screw sets the chain gap between the guide pulley and the cogs.

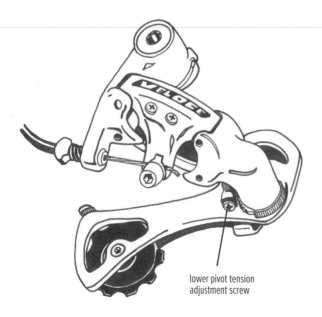

lower pivot tension
adjustment screw

g. SRAM X-Horizon Roller Bearing Clutch rear derailleur adjustment details

Force 1, Rival 1, and Force CX-1 derailleurs are solely for use with a single chainring, and, specifically, to a front chainring without chain guards but with tall, fat/thin/fat alternating teeth. These derailleurs have a roller clutch to control chain bounce and only move laterally. They require a chain at least two and a half links longer than normal; see 4-8, method 3.

To remove the rear wheel, pull the bottom of the jockey-wheel cage forward until you can get the lock button on the cage to engage and leave the chain slack. When you have reinstalled the wheel, pull the cage forward again and the lock button will disengage. Locking the cage is also useful when installing the chain's master link.

Make sure when connecting the cable (5-13) that you route the cable through the little tunnel at the bottom of the curved cable guide on the back of the derailleur. The cable comes through the barrel adjuster, runs in the groove around the back of the curved cable guide, and goes through the little tunnel toward the bolt. The cable needs to wrap over the top of the cable-fixing bolt.

Adjust the derailleur as you would most rear derailleurs, with these exceptions:

- The chain gap, adjusted by the b-screw, should be 6–12mm between the top of the upper jockey wheel and the bottom of the largest cog when in the lowest gear.
- The outer limit screw sets the low-gear limit, and the inboard screw sets the high-gear limit.

THE FRONT DERAILLEUR

The front derailleur (Figs. 5.10, 5.11) moves the chain between the chainrings. The working parts consist of a cage [3, 4], a linkage mechanism, and an arm [5] attached to the shift cable. The front derailleur is attached to the frame, either by a bolt passing through a front-derailleur boss (or tab) attached to the frame's seat tube (Fig. 5.10) or by an integral band clamp surrounding the seat tube (Fig. 5.11). The first type is commonly called a "braze-on" derailleur because the tab that carries it is attached permanently to the frame, and

5.10 Front-derailleur boss (or tab) on seat tube

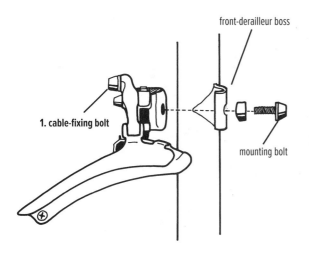

front-derailleur boss

1. cable-fixing bolt

mounting bolt

5.11 Band-clamp front derailleur

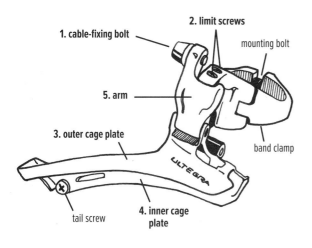

2. limit screws

1. cable-fixing bolt

mounting bolt

5. arm

3. outer cage plate

band clamp

tail screw

4. inner cage plate

in the era of steel frames when these became popular the tab was brazed to the steel seat tube. The second is commonly called a "band-clamp" or "clamp-on" derailleur. In an alternative arrangement, a braze-on-type front derailleur may bolt to a separate wraparound clamp that has a front-derailleur boss shaped like an ear.

When the front shift cable pulls at the cable-fixing bolt [1] (Fig. 5.11), the derailleur swings out until it is stopped by the outer limit screw [2]; the limit screw prevents the outer cage plate [3] from moving so far outward that the chain shifts past the outer chainring. The inner limit screw stops the inner cage plate [4] from moving inward so far that it allows the chain to fall off to the inboard side of the inner chainring.

5-4

FRONT-DERAILLEUR INSTALLATION

LEVEL **NOTE:** For 11-speed cable-actuated front derailleurs, see 5-5f for SRAM and 5-5g for Shimano.

1. **Clamp the front derailleur around the seat tube, or bolt it to the frame boss.**
2. **Adjust the height and rotation as described in 5-5a.** See 5-5f or 5-5g for SRAM and Shimano 11-speed front derailleurs, respectively.
3. **Tighten the mounting bolt (Fig. 5.10 or 5.11).**
4. **Install the chain as in Chapter 4 (4-9 through 4-13).** If the chain is already on the bike, you can leave it on and remove the old derailleur and install the new one without opening the chain. Remove the tail screw ([6], Fig. 5.11) and flex open the front-derailleur cage to get the chain in. Replace the tail screw.
5. **Install cables and housings (5-6 through 5-15).**
6. **Pull the cable tight in its groove on the arm, and tighten the fixing bolt.** For cable clamping on Shimano 11-speed front derailleurs, see 5-5g, step 5.

5-5

ADJUSTMENT OF FRONT DERAILLEUR AND LEFT-HAND SHIFTER

NOTE: *For SRAM 22 Yaw and Shimano 11-speed front derailleurs, skip to sections f and g.*

a. Position adjustments

1. **Position the height of the front derailleur so that, at its closest point, the outer cage passes 1–2mm (1/16 to 1/8 inch) above the outer chainring (Fig. 5.12).**

NOTE: *The lower edge of the derailleur outer cage plate should roughly follow the curve of the chainring, though the tail of the plate will generally be a bit farther above the chainring than will the leading edge, as shown in Figure 5.12. However, if the tail of the cage is way above the chainring when its leading edge is the correct 1–2mm above it, the derailleur is not properly matched to the bike and chainring and may not shift well. This can happen due to trying to mate a front derailleur (especially an older one) designed for large chainrings*

5.12 Proper front-derailleur vertical clearance

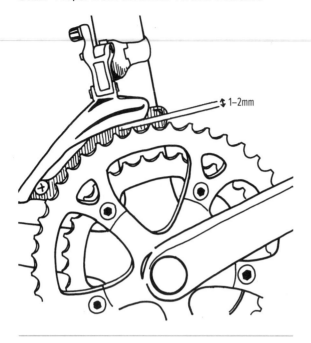

‡ 1–2mm

5.13 Proper rotational alignment of front derailleur on smallest chainring

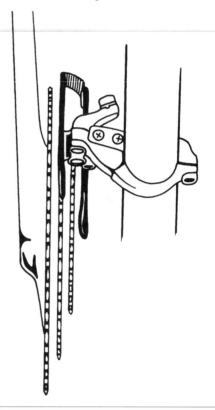

(39–53-tooth chainring pairs or the like) with a "compact" crankset (one that accepts smaller chainrings, like 34–50T). Obviously, getting a front derailleur better mated to the chainring size is ideal, but if it is a braze-on-type front derailleur (Fig. 5.10), you may be able to tip it to approximate the curve of the chainring well enough to work acceptably by placing a wedge-shaped shim between the front-derailleur boss affixed to the frame and the curved mounting face of the front derailleur. You may be able to purchase this shim wedge in a bike shop; otherwise, you can fashion one yourself with a file, a drill, and a little block of aluminum.

2. **Position the outer plate of the derailleur cage parallel to the chainrings or to the chain in the lowest and highest gears when viewed from above.** When on the inner (smallest) chainring and largest cog, the inner cage plate should be parallel with the chainring or the chain (Fig. 5.13). Similarly, check this by shifting to the big chainring and smallest cog and sighting from the top (Fig. 5.14).

b. Limit-screw adjustments

The front derailleur has two limit screws (Fig. 5.11) that stop the derailleur from throwing the chain to the inside or outside of the chainrings. These are sometimes labeled L for low gear (small chainring) and H for high gear

5.14 Proper rotational alignment of front derailleur on largest chainring

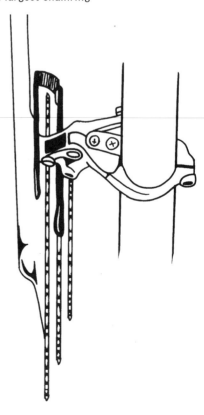

5.15 Front-derailleur limit screws

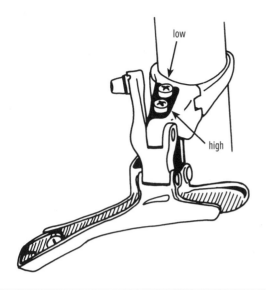

(large chainring) (Fig. 5.15). On most derailleurs, the low-gear screw is closer to the frame.

If in doubt, you can determine which limit screw controls which function by trial and error. Shift the chain to the inner ring, and then tighten one of the limit screws. If turning that screw moves the front derailleur outward, then it is the low-gear limit screw. If turning that screw does not move the front derailleur, then the other screw is the low-gear limit screw.

c. Low-gear limit-screw adjustment

1. **Shift back and forth between chainrings.**
2. **If the chain drops off the inner ring to the inside, tighten the low-gear limit screw (clockwise) one-quarter turn and try shifting again.**
3. **If the chain does not shift easily onto the inner chainring, loosen the low-gear limit screw one-quarter turn and repeat the shift.**

d. High-gear limit-screw adjustment

1. **Shift the chain back and forth between chainrings.**
2. **If the chain jumps over the big chainring, tighten the high-gear limit screw one-quarter turn and repeat the shift.**
3. **If the chain is sluggish going up to the big chainring or does not go up at all, loosen the high-gear limit screw one-quarter turn and try the shift again.**

4. **Pedal hard in your highest gear (big chainring/ smallest cog).** If the chain rubs the front-derailleur outer cage plate, unscrew the high-gear limit screw slightly.

e. Cable-tension adjustment

1. **With the chain on the inner chainring, remove any excess cable slack.** Turn the barrel adjuster on the cable stop or along the cable housing (Figs. 5.7, 5.8) counterclockwise (or loosen the cable-fixing bolt, pull the cable tight with pliers, and tighten the bolt).
2. **Check that the cable is loose enough to allow the chain to shift smoothly and repeatedly to the inner chainring.**
3. **Check that the cable is tight enough that the derailleur starts to move as soon as you move the shifter.** Fine-tune while riding.
4. **Pedal hard in your lowest gear (small chainring/ biggest cog).** If the chain rubs the front-derailleur inner cage plate, loosen the cable tension slightly. You may also need to unscrew the low-gear limit screw slightly if it, rather than the cable, stops the front derailleur's inward movement before it clears the chain.

NOTE: *This method of tension adjustment applies to indexed as well as friction shifters. With indexed front shifting, you may want to fine-tune the cable tension to avoid noise from the chain dragging on the derailleur in some cross-gears or to get more precise shifting.*

ANOTHER NOTE: *Some front derailleurs have a cam screw at the end of the return spring to adjust the spring tension. For quicker shifting to the smaller rings, increase the spring tension by turning the screw clockwise one-quarter or one-half turn.*

NOTE ON SHIFTING TROUBLE: *If you cannot get the front derailleur to shift well, or if it rubs in cross-gears and throws the chain off, refer to the chainline discussion under "Troubleshooting Derailleur and Shifter Problems" at the end of this chapter.*

NOTE ON CHAIN CATCHERS: *If, despite all adjustments, you just can't stop the chain from falling off to the inside, install a "chain catcher" or inner chain stop to nudge the chain back up onto the inner ring whenever it tries to drop off to the inside. Some consist of*

IF YOU ENSURE THAT THE cable tension is set correctly (5-5e) so that the front derailleur stops before dropping the chain off to the inside, then the low-gear limit screw can be as loose as you want (or not even be there at all), and you still won't shift the chain off the inside and into the frame. This means that you can back out the screw beyond where it needs to be to make sure that it never interferes with the chain dropping onto the smallest chainring.

If you do this, however, you must always keep the cable tension accurate or the front derailleur can drop the chain off to the inside as the cable stretches.

a long, bent arm and mount under the mounting bolt of a braze-on front derailleur; K-Edge and SRAM are examples. Others are built onto a band clamp, like the Third Eye Chain Watcher (Fig. 5.54), Deda Dog Fang, K-Edge Clamp-On, or N-Gear Jump Stop.

f. SRAM 22 Yaw (11-speed) front derailleur installation and adjustment

SRAM's front derailleur underwent a major overhaul when the company moved to 11 rear cogs. It not only swings out and up as most derailleurs do, but it also rotates to swing the rear of the front-derailleur cage to match the change in chain angle. This is intended to eliminate chain rub in cross gears without the need to trim the front derailleur. It requires the unique setup procedure below, however.

1. **Tighten the inner limit screw all of the way in.** A new SRAM 22 Yaw front derailleur will come with the inner limit screw already tightened in until the derailleur can't move, but if you're setting up and adjusting a used one, you'll need to do this step so that it will hold the chain on the big chainring during setup.

2. **Set the derailleur height.** With the derailleur cage centered over the big chainring (due to limit screw setting in step 1), slide the derailleur up and down until the tips of the tallest teeth on the outer chainring line up within the laser-scribed line on the inner cage plate (Fig. 5.16), when viewed directly from the side. If the scribed line is hard to see due to it being a used front derailleur scraped by the chain, wipe the inner cage plate clean and darken the alignment line with a marker pen. If you cannot see the line at all, set the outer plate 1–2mm above the chainring.

3. **Line up the scribed marks with the large chainring.** When looking down on the derailleur from above, center the teeth of the big chainring on the marks on the top of the cage on the front and rear crossmembers (Fig. 5.17). If the chain is already on the bike, you'll have to sight through it to see the rear scribed mark as well as the chainring teeth on both ends.

4. **Install the chain.** If it's not already on, install it as in 4-9 through 4-13.

5. **Adjust the inner limit screw.** With the chain on the largest rear cog, unscrew the inner limit screw while turning the crank until the chain drops onto the inner chainring. Keep turning the inner limit screw until the inner cage plate is exactly 0.5mm away from the chain (again, the chain is on the small-big combination).

6. **Double-check derailleur height.** Ensure that the outer cage plate is 1–2mm above the tallest teeth of the outer chainring when the chain is on the inner chainring. If you did step 2 correctly, this step should be irrelevant.

7. **Install the cable (5-7 through 5-15).** Pull the cable tight into its groove and tighten the cable-fixing bolt.

8. **Adjust the outer limit screw.** With the chain on the big chainring and smallest rear cog (Fig. 5.6), push the left shift lever as far as you can. Turn the outer limit screw until the outer cage plate is 1mm from the chain. Release the shift lever.

9. **Adjust the cable tension.** With the chain on the small chainring and largest rear cog (Fig. 5.6) and

without turning the crank, push the left shift lever as far inward as possible and release it. Rotate the crank; the chain should shift to the large chainring. Tweak the cable tension with the barrel adjuster (Fig. 5.7 or 5.8) until it shifts smoothly this way. This method of adjusting cable tension is supposed to ensure SRAM's "Zero-Loss" setting of the high-gear limit screw and cable tension, in which the derailleur goes exactly to the high limit screw on an upshift and doesn't backtrack when the lever is released; adjust the cable tension until you achieve this. Alternatively, adjust the cable tension the same as in section 5-5e.

5.16 SRAM 22 Yaw (11-speed) front derailleur side view: Set derailleur height so that the outer chainring teeth line up with the scribed mark on the inner cage plate.

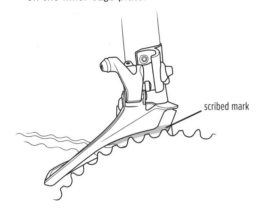

scribed mark

5.17 SRAM 22 Yaw (11-speed) front derailleur top view: Line up the scribed mark with the outer chainring.

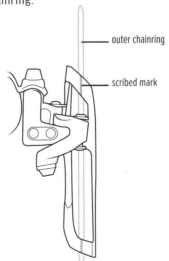

outer chainring

scribed mark

10. Set the chain watcher. Using a 2.5mm hex key, screw the chain watcher mounting bolt into the hole in the center of the derailleur mounting bolt. With the chain on the small chainring and largest rear cog, position the chain watcher so it is as close to the chain as possible without touching it. Tighten its mounting bolt to hold it in place; torque is 0.5–1.0 N-m.

g. Shimano long-arm (11-speed) front derailleur installation and adjustment

With a band-clamp style derailleur, install the derailleur following the procedure in 5-5a. However, follow the first two steps below for the braze-on-style long-arm front derailleur mounted to either a frame mount or to a separate band clamp. With both styles, continue from step 3 below.

1. **Set the derailleur height as in 5-5a, step 1.** Roughly line up the derailleur parallel to the chainrings; you will be readjusting this in a moment.

2. **Adhere the backing plate to the frame.** The cable pulling on this derailleur can flex the frame mount because the cable is lined up close to parallel to its arm, and the long arm provides tremendous leverage. To counteract this, Shimano braces its braze-on-type long-arm front derailleur with an extra support bolt that pushes against the frame, but this support bolt could dent, crack, or bore into the seat tube were it not for the (included) aluminum self-adhesive backing plate under its tip. Depending on the seat tube's shape behind the front derailleur, choose either the flat or curved backing plate. Slip the plate behind the tip of the support bolt (Fig. 5.18). Stick it to the seat tube with the adhesive strip offset behind the bolt tip and the nonadhesive section of the plate directly under the bolt tip as in Figure 5.18.

3. **Set the derailleur's initial rotational position.** Loosen the mounting bolt, rotate the derailleur so that, relative to the face of the chainring, the tail of the outer cage plate is 0.5–1mm further inboard than its leading tip, and tighten the mounting bolt.

4. **Line up the cage parallel to the chainring.** With braze-on derailleurs, use a 2mm hex key to tighten the support bolt (Fig. 5.19) until the outer cage plate

5.18 Installing the reinforcing plate on the seat tube

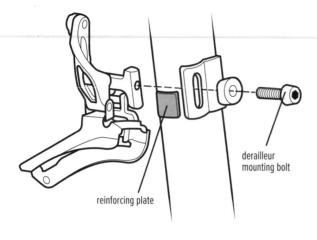

derailleur mounting bolt

reinforcing plate

5.19 Tightening the support bolt

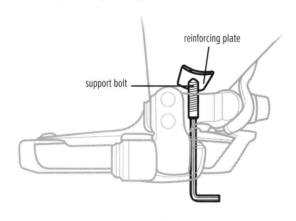

reinforcing plate

support bolt

is parallel with the face of the chainring (Fig. 5.13). With a clamp-on derailleur, loosen the mounting bolt, rotate the derailleur until it is parallel, and tighten the mounting bolt.

5. **Route the cable to the derailleur (5-7 through 5-13).** Don't attach it to the cable-fixing bolt yet.

6. **Ensure that the outer cage plate is flush with the face of the outer chainring.** Turn the inner limit screw (Fig. 5.11) until it is flush; this increases the accuracy of the next steps.

7. **Install Shimano's TL-FD68 tool.** The cable angle and cable-pull ratio is critical with the vertical front-derailleur arm, because if the cable is also vertical or close to it, it will not have any leverage to start the movement of the front derailleur. Depending on where the cable exits toward the derailleur from behind the bottom bracket, Shimano offers two

configurations for the cable attachment under the cable-fixing bolt, one that offers more initial leverage than the other, and this tool determines which to use. Push the tool's plastic top nub into the head of the cable-fixing bolt and the tool's thin pin into the bore of the derailleur-arm pivot pin (Fig. 5.20A).

There is an older, TL-FD90 tool that only works with Dura-Ace 9000 front derailleurs; it installs differently and requires removing the cable-fixing bolt. The TL-FD68 works on Ultegra, 105, and Dura-Ace long-arm front derailleurs (FD-6800, FD-5800, and FD-9000).

8. **Determine which cable-hook configuration is required.** Pull the cable up and into the slit in the plastic nub atop the TL-FD68 (or TL-FD90) tool and see where it lines up relative to the inscribed line on the tool (Fig. 5.20B). If the cable aligns pretty much along the inscribed line, leave the derailleur as is and skip to step 10. If the cable lines up definitively on one side or the other of the line, check that the configuration of the "converter" at the mounting bolt matches the illustration on the tool corresponding to the side of the line your cable is lined up on. If it does, skip to step 10; if it doesn't, continue with step 9.

9. **If needed, switch the converter.** On Ultegra and 105 (FD-6800 and FD-5800) derailleurs, push out the little rectangular converter plate just below the cable-fixing bolt with a 2mm hex key from its recess in the arm. Rotate it 180 degrees, and then push it back into the recess; its protruding pin will now have been moved to the opposite side of the converter plate. On Dura-Ace FD-9000 derailleurs, rotate the notched washer under the cable-fixing bolt so that its notch surrounds the other one of the two protrusions on the derailleur arm.

10. **Pull the cable tight under the cable-fixing bolt.** Make sure it enters and exits under the bolt correctly: On Ultegra and 105 (FD-6800 and FD-5800), the cable will enter on the left side of the converter pin, continue around the right side of the bolt, and come straight up from there; on Dura-Ace FD-9000, no matter which protrusion the converter notch is surrounding, the cable will enter between

5.20A Front and rear of the Shimano TL-FD68 tool

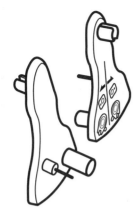

5.20B To use the Shimano TL-FD68, select the converter configuration according to where the cable passes the inscribed line.

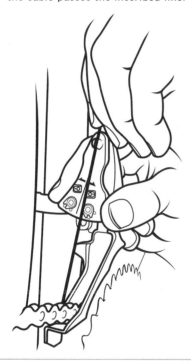

the two protrusions, continue around the right side of the bolt, and come straight up from there.

11. **Tighten the cable-fixing bolt.** Tightening torque is 6–7 N-m.

12. **Set the limit screws and cable tension.** Follow the instructions in 5-5b through 5-5e, with the following exceptions: 1) When checking for chain rub on the inner cage plate in the lowest (small front/big rear) gear combination, make sure that the front derailleur is in its most inward (low trim) position (explained in 5-1f), by activating the shift paddle

three times; adjust cable tension and the inner limit screw to get 0–0.5mm clearance in this combination. 2) Fine-tune the cable tension by shifting to the big/big combination and then giving the shift paddle a soft click to move the derailleur to the high trim position (5-1f). Check that the inner cage plate just barely clears the chain without rub (0–0.5mm clearance), and tighten or loosen the cable accordingly to achieve it. To work correctly, these derailleurs need to have the maximum possible cable tension, and this step ensures that. If, after setting this, the derailleur will not drop all of the way to the inner limit screw (and hence will rub the chain in the lowest gear combination), reduce cable tension just enough for it to barely reach the inner limit screw. 3) Adjust the outer limit screw in high gear (big/small) the same as in 5-5e (0–0.5mm chain clearance).

NOTE: *In some rare cases, the cable exits the frame so far to the right that switching the converter is insufficient to get enough leverage on the derailleur arm to shift without an enormous amount of force from the hand. In that case, you can get good performance by bypassing the converter and wrapping the cable over the top of the cable-fixing bolt from the left side (instead of the Shimano-recommended routings, which have the cable going to the right of the bolt).*

ANOTHER NOTE: *The plastic skid plate on the inner cage plate of Shimano 11-speed front derailleurs is replaceable.*

REPLACING AND LUBRICATING SHIFT CABLES AND HOUSINGS

LEVEL To function properly, derailleurs need to have clean, smooth-running cables (also called "inner wires"). As with replacing a chain, replacing cables is a maintenance operation, not a repair operation. Do not wait until cables break to replace them. Replace any cables that have broken strands, kinks, or fraying between the shifter and the derailleur. You should also replace housings (also called "outer wires") if they are bent, mashed, or just plain gritty, or if the color clashes with your bike (this is really important!).

5.21 Cable-housing types and end caps

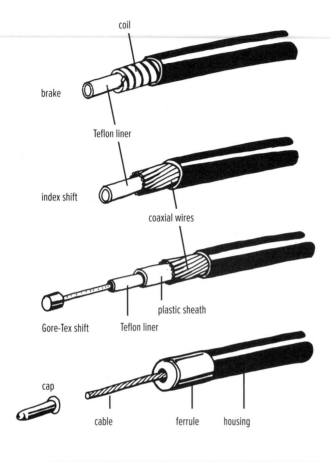

5-6

BUYING CABLES

1. **Buy new cables and housing that are at least as long as the ones you are replacing.**

2. **Make sure that the cables and housing are for indexed shifters.** These cables will stretch minimally, and the housings will not compress in length. Under its external plastic sheath, indexed housing is not made of steel coil like brake housings; it is made of parallel (coaxial) steel strands of thin wire. If you look at the end (Fig. 5.21), you will see numerous wire ends surrounding a central Teflon tube.

3. **Buy two cable-end crimp caps (Fig. 5.21) as well as a cable-housing end (ferrule) for each end of every housing section.** The end caps will prevent fraying, and the ferrules will prevent kinking at the cable entry points, cable stops, shifters, and derailleurs.

PRO TIP — BETTER SHIFTING

SOMETIMES THE BEST THING YOU CAN DO for your bike—especially a bike with a lot of miles on it—is to replace the cables and housings. Drag on shift cables caused by contamination will prevent accurate and consistent shifts.

4. **It is a good idea to buy extra cables, cable caps, and ferrules (Fig. 5.21) to keep on hand in your work area.** They're inexpensive, and if you have a small supply, you will be able to change cables when you need to without making a special trip to the bike shop to get a little cable-end cap.

NOTE: *If your frame has internal cables that run bare inside the frame and you don't have the Park IR-1 magnetic internal-routing tool (Fig. 5.30), buy thin plastic tubing along with the new cables and housings to reduce installation time (5-12). Before you pull out the old cable, slide a long piece of tubing onto the old cable and through the frame as a guide for the new cable. After the new cable is in, you can pull the tube back out. Make sure that if the new internal shift cables cross each other inside the down tube (they will do so if the housings cross in front of the head tube) that they cross a maximum of once. Multiple crosses (i.e., the cables are wound around each other) not only will result in undue friction, but shifting one lever will move both derailleurs.*

5-7

CUTTING HOUSING TO LENGTH

1. **Use a special cutter made for the purpose.** Park, Pedro's, Shimano, SRAM, and Jagwire make good ones (Fig. 1.2). Standard wire cutters (i.e., "side cutters") will not cleanly cut index-shift housing.

2. **Cut the housing to the same lengths as the pieces you are replacing.** If you have no old housings for comparison, cut the new pieces so that they curve smoothly. When you turn the handlebar, the housing should not pull from its cable stop. Allow enough length for the rear derailleur to swing backward (Fig. 5.22) and forward (Fig. 5.23) freely.

5.22 Rear derailleur swinging backward to check housing length

5.23 Rear derailleur swinging forward to check housing length

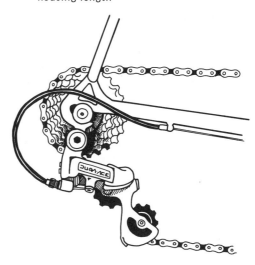

5.24 Opening the end of the inner liner after cutting the housing to length

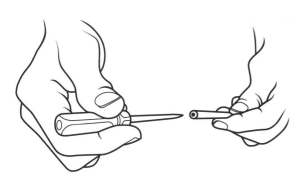

5.25 Crimping the cable end cap

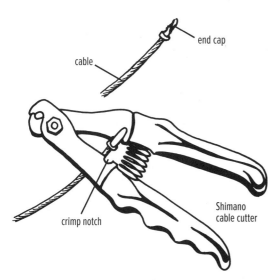

NOTE: *It is critical that you ensure that the housing loop at the rear derailleur is long enough to allow the cable to slide smoothly inside; a tight bend will prevent the derailleur from shifting well to small cogs, because the derailleur's return spring will not be strong enough to overcome the cable friction.*

3. **Open each end of the Teflon liner that has been smashed shut by the cutter.** Use a nail or toothpick (Fig. 5.24).

4. **Place a ferrule over each housing end (Fig. 5.21).** The end at the lever probably will not need one; it depends on the lever, and most will accept only housing without a ferrule on it.

5. **After threading the cable into the housing (see next sections for details), clip the cable 1–2cm past the bolt and crimp a cap onto it to prevent fraying (Fig. 5.25).**

5-8

REPLACING SHIFT CABLE IN SHIMANO STI SHIFT/BRAKE LEVER

1. **Disconnect the cable at the derailleur and snip off the cable end cap (if installed).**

2. **Shift the inner lever to the gear setting that lets the most cable out.** This setting will be the highest-gear position for the rear shift lever (small cog) and the lowest for the front (small chainring).

3. **Push the cable until the cable head emerges far enough from the hole in the lever to grab it.** Pull out the old cable and recycle it.

   a. On original STI levers (on which the shift cable sticks out of the side of the lever and does not run under the handlebar tape), pull the brake lever to reveal the access hole for the shift cable (Fig. 5.26); the hole is on the outboard side of the lever. On Dura-Ace levers with the cable coming out of the side, you must first remove a thin, black plastic cover with a small Phillips screwdriver to get at the access hole.

   b. On 10-speed STI levers where the shift cable goes under the handlebar tape, the cable-access hole is on the lower part of the lever body under the lever hood on the inboard side, toward the front.

   c. On 11-speed STI levers, the cable-access hole is high up on the outboard side of the lever, under the lever hood (Fig. 5.27).

4. **Thread the new cable through the hole and out through the lever (Fig. 5.26) or the lever body (Fig. 5.27).** The recessed pocket into which the cable head seats should be visible through the access hole if you have advanced it far enough by repeatedly shifting the shift paddle; make sure you push the cable through this pocket so its head sets into it. On 11-speed levers (Fig. 5.27), the cable will emerge out of the inboard side of the lever and will be blocked by a tethered plastic cover (under the hood); the cover will get out of the way to let the cable pop out if the hood is off or peeled back far enough. Push downward on the cable as you pull the cable so that it engages the cable hook. On 2009 and later 10-speed STI levers where the shift cable goes under the handlebar tape, the cable will pop up out of its passageway and continue in one of two grooves in a white, low-friction material before entering the cable housing end cap at the exit of the lever body. One groove routes the cable in front of the handlebar, and the other routes it behind the bar, so choose the one you wish. Except on handlebars where the wire goes through tubes inside of

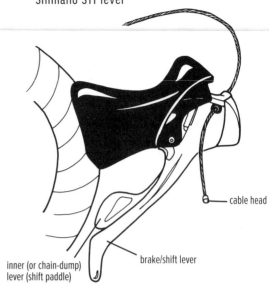

5.26 Threading in a new shift cable into an old-style Shimano STI lever

cable head

inner (or chain-dump) lever (shift paddle)

brake/shift lever

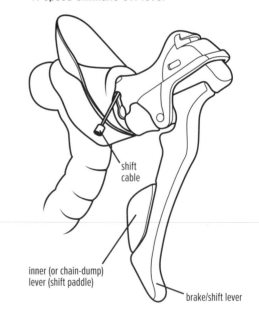

5.27 Threading in a new shift cable into an 11-speed Shimano STI lever

shift cable

inner (or chain-dump) lever (shift paddle)

brake/shift lever

the handlebar, I recommend routing the cable in front of the handlebar. The cable may pop that piece of white low-friction material out of the lever; if so, slide it onto the cable and reinsert it in place.

5. **Route the cable to the derailleur.** With external cables, guide the cable through each housing segment (making sure each housing segment has a ferrule on the end; see Fig. 5.21) and cable guide and cable stop. On 11-speed levers, the housing fer-

rule that is to be inserted into the lever body has a ridge on it to engage the slot in its receptacle on the inboard side of the lever body.

NOTE: *For internal cables, see 5-12.*

Threading in a new shift cable into a high-end 10-speed Campagnolo Ergopower lever

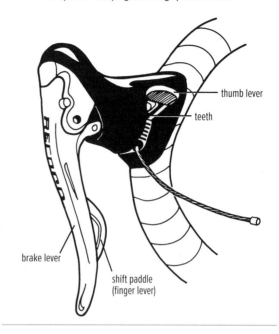

thumb lever

teeth

brake lever

shift paddle (finger lever)

5-9

REPLACING SHIFT CABLE IN CAMPAGNOLO ERGOPOWER LEVER

NOTE: *The Ergopower cable hook (i.e., the countersunk hole into which the cable head seats) is too small to fit a Shimano or other cable head—just another of those maddening parts incompatibilities. It's simplest to just buy a new Campagnolo-compatible shift cable before you begin work. Alternatively, you can file a larger cable head to fit in the Campagnolo lever, but unless you are careful to ensure that the cable head is small enough, expect to push hard with pliers to get the cable back out next time.*

1. **Disconnect the cable at the derailleur and snip off the cable-end cap (if installed).**

2. **Click the thumb lever until it will click no more.**

3. **Push the cable until the cable head emerges from the cable hole near the bottom of the lever body (Fig. 5.28).** It's toward the inboard side—just outboard of the little gear teeth on high-end Ergopower 9- and 10-speed levers visible in Figure 5.28. On QS/Escape levers and 2009 and later Ultra-Shift Ergopower 11- and 10-speed levers (on which the top of the lever body is tall and curves inboard), the access to the cable end is a hole near the base of the lever on the outboard side. Push the cable head out far enough to grab it. Pull out the old cable and recycle it.

4. **Push the new cable in through the hole and up through the lever body until the cable emerges from the housing entry hole at the upper base of the lever body on the outboard side.** On the (taller) Ultra-Shift and Power Shift Ergopower lever bodies, the cable will pop up out of its passageway and continue in one of two grooves in a light-colored, low-friction material before passing through another hole back under the edge of the black lever body before exiting it. One groove

would route the cable in front of the handlebar, and the other would route it behind the bar, so choose the one you wish. Except on handlebars with through-tubes for internal cables that mandate that the cable come from around the outside of the bar, I recommend routing the cable in front of the handlebar. Whichever groove you choose, the cable will not want to go back down and into the hole under the edge of the lever body without some coaxing. First, push the cable out so it extends above the two grooves, and put a little bend in the end of the cable. Pull the cable back in a bit until that bend is just visible. Now push the cable forward again, and push down on the end of the cable with a 2mm hex key or other thin implement to get the cable to pass through the little hole at the end of whichever groove you chose and exit the lever. The cable will emerge into a tunnel for the housing, and at its base there is a brass washer to stop the housing; avoid dislodging this brass washer when you push the cable through. If you dislodge the washer and lose it, pop the washer out of the cable tunnel you're not using (i.e., for cable routing around the back of the handlebar), and use it in the tunnel you are using. Note also that these cable tunnels will fit only 4mm housing; 5mm housing will not fit.

5. **Guide the cable through each housing segment, cable guide, and cable stop to the derailleur.** Except for the end of the housing piece that will insert into a post-2008 (Ultra-Shift or Power Shift) Campagnolo lever (since it doesn't need a ferrule, and one won't fit), place a ferrule on the end of each cable-housing segment (Fig. 5.21). Make sure the housing segment at the lever inserts fully into its hole in the lever body. If the housing won't slip in due to shape of the lever body and/or handlebar where the two meet, loosen the lever clamp bolt to allow it to lift a bit off of the handlebar, push the cable housing into its tunnel, and tighten the lever down. If you insist on installing 5mm housing into an Ultra-Shift or Power Shift lever, you'll have to drill out or cut away the housing tunnel in the lever body; then you'll need a ferrule to support the end of the housing, because the lever tunnel no longer will.

NOTE: *For internal cables, see 5-12.*

5-10

REPLACING SHIFT CABLE IN SRAM DOUBLE TAP LEVER

1. **Disconnect the cable at the derailleur and snip off the cable-end cap (if installed).**
2. **Click the shift paddle with repeated small pushes (Fig. 5.1A) until it won't let out any more cable.**
3. **Push the cable until the cable head emerges from the cable hole at the base of the lever on the inboard side.** Push the cable head out far enough to grab it. Pull out the old cable and recycle it.
4. **Push the new cable in through the hole and up through the lever body until the cable emerges from the housing entry hole at the upper base of the lever body on the outboard side.**
5. **Route the cable to the derailleur.** With external cables, guide the cable through each housing segment (making sure each housing segment has a ferrule on the end; see Fig. 5.21) and cable guide and cable stop.

NOTE: *For internal cables, see 5-12.*

5-11

REPLACING CABLE IN A DOWN-TUBE SHIFT LEVER OR BAR-END SHIFT LEVER (INCLUDING ON AN AERO HANDLEBAR)

1. **Disconnect the cable at the derailleur and snip off the cable-end cap.**
2. **Flip the lever forward to the gear setting that lets out the most cable.** This setting will be the highest gear position for the rear shift lever (small cog) and the lowest for the front (small chainring). This applies to aerobar levers (Figs. 5.43A, B and 5.44A, B) as well as to down-tube shifters (Figs. 5.29 and 5.45).
3. **Push the cable until the cable head pops out of the hole in the shift lever (Fig. 5.29).** Pull out the old cable and recycle it.
4. **Thread the new cable through the hole and out through the other side of the lever (Fig. 5.29).**
5. **Route the cable to the derailleur.** With external cables, guide the cable through each housing segment (making sure each housing segment has a ferrule on the end; see Fig. 5.21) and cable guide and cable stop. With internal cables, see the next section (5-12).

5.29 Threading a new cable into a down-tube shifter

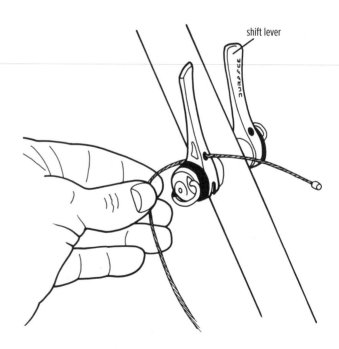

shift lever

5-12

ROUTING INTERNAL CABLES

Generally, internal-cable frames have a removable plastic cover under the bottom-bracket shell that allows access to the shift cables. These frames also usually have at least one cable-entry point with a removable plastic step-down stop or insert for the cable housing. This stop or insert pulls out to give you a large hole to work with, rather than a tiny one that only a cable will fit into.

As I mentioned in the Note in 5-6, if you are replacing a cable, you can clip off the frayed end of the cable and slide a thin plastic tube over it and feed the plastic tube into the frame until it sticks out the other end. You may be able to find this tubing at bike shops, where it will be sold to cover bare stretches of cable to protect a bicycle's finish. Alternatively, electronics stores sell heat-shrink tubing in various sizes and colors, or colored plastic tubing for insulation or wire color-coding; it's often dubbed "spaghetti." Then you can pull the old cables out and slide the new cables in through the tubing. Easy schmeezy; once the new cable is routed through the frame, you can pull out the plastic tube and leave the new cable in.

On the other hand, if you yanked out the old cable before doing this, or if you're working on a new frame without cables already running through it, I highly recommend using Park Tool's IR-1 Internal Cable Routing Kit. This kit works for routing cables, cable housing, electronic wires, and hydraulic hose inside a frame. See 6-6 for routing electronic-shift wires with it.

Without the Park IR-1 tool, you will be stuck with trying to fish cables through with bent wires and tape. Expect frustration.

The Park IR-1 magnet kit (Fig. 5.30) comes with a large magnet and three coated shift cables with a cylindrical magnet attached to the head end. I'll call the "fishing wire" the one that has nothing on the other end. The kit includes two different "lead wires." One is for guiding cables or electronic wires and has a flexible rubber hollow receiver on the end that will stretch over either a Di2 or EPS electronic connector, or over the head of a shift or brake cable. Park's other lead wire

5.30 Park IR-1 magnet kit

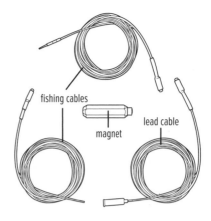

fishing cables
magnet
lead cable

5.31 Cable entry hole, insert, and lead wire

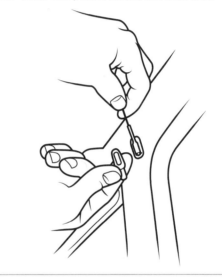

has a conical threaded steel barbed end that can be screwed into the end of cable housing and hydraulic hose, allowing the user to easily pull either of those through a frame. Both lead-wire magnets repel each other and are attracted to the large magnet and the fishing-wire magnet.

Since the cables you will be installing (and the Park cables) are magnetic, there is a chance that you can drag the tip of the cable through the frame with a large magnet on the outside of the frame. It's tough to do this without getting it hung up inside, however.

A far better method is to use the Park IR-1 lead and fishing wires.

1. **Push the magnet end of one of the two Park lead wires down the cable-entry hole at the front of the frame (Fig. 5.31).** You must first pull the plastic cable

5.32 Bottom bracket with cover removed and fishing wire entering opening

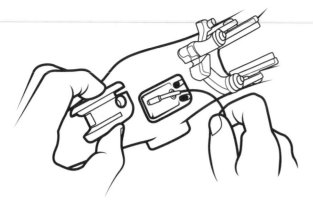

stop or insert out of the frame to have access to the large hole it was filling at the front of the frame.

2. **Push the magnet end of the Park fishing wire up from the large opening at the bottom bracket (Fig. 5.32).** You must first remove the cover over the bottom-bracket cable guides. The two magnets will stick together inside the down tube.

3. **Pull the lead wire out from the hole at the front.** Pull carefully and jiggle it as the magnet end arrives at the hole so that you pull the magnet end of the fishing wire out with it.

4. **Stick the tip of the new cable to the magnet at the end of the fishing wire sticking up out of the frame hole.** The "tip" is the derailleur end of the cable—not the head end. Before connecting it to the magnet, the new cable should already be routed through the shifter, housing, ferrule, and frame cable stop you removed.

5. **Push the new cable into the frame hole.** Allow it to push the Park fishing wire out of the bottom-bracket port. Gently pull on the fishing wire until the tip of the new cable emerges.

6. **Route the new cable through the bottom-bracket cable guide.** Make sure the cable passes under the loop(s) in the guide, if present. If it is a front-derailleur cable, you can now thread it up and out so it emerges behind the base of the seat tube and skip to step 11. If it is a rear-derailleur cable, continue with step 7.

7. **Push the magnet end of one of the Park lead wires down the cable-entry hole at the dropout.**

You must first pull the plastic cable stop or insert out of the chainstay to access the large hole it fills.

8. **If the magnet end of the cable doesn't emerge at the bottom-bracket port, guide it out.** You may use the large magnet to guide it out. Or, you can push the magnet end of the fishing wire into the chainstay until the two magnets stick together, and then carefully pull both magnets out with the fishing wire.

9. **Stick the tip of the new cable to the magnet at the end of the lead wire sticking up out at the bottom bracket.**

10. **Push the new cable into the chainstay.** Allow it to push the Park lead wire out of the hole at the dropout. Gently pull on the fishing wire until the tip of the new cable emerges.

11. **Replace the cover over the bottom-bracket guides and the cable stop(s) or insert(s) at the front of the frame (and at the rear).** You did it!

NOTE: *Make sure that if the new internal shift cables cross each other inside the down tube (they will do so if the housings cross in front of the head tube) that they cross a maximum of once. Multiple crosses (i.e., the cables are wound around each other) not only will result in undue friction, but shifting one lever will move both derailleurs.*

5-13

ATTACHING CABLE TO REAR DERAILLEUR

1. **Shift the rear shifter to the highest gear setting (smallest cog; see 5-1).**

2. **If installed, put the chain on the smallest cog.** The rear derailleur should be aligned under that cog.

3. **Route the cable through each of the frame cable stops and housing segments and the barrel adjuster on the back of the derailleur until you reach the derailleur's cable-fixing bolt.**

NOTE: *For internal cables, see 5-12.*

On a SRAM rear derailleur, make sure that the cable wraps in the groove around the curved cable guide behind the fixing bolt and then routes above the bolt (Fig. 5.33), to the side away from the jockey wheels. On SRAM X-Horizon derailleurs,

5.33 Cable routing to SRAM rear derailleur. (Note cable wraps around curved guide behind cable-fixing bolt and attaches on side of the bolt opposite the jockey wheels.)

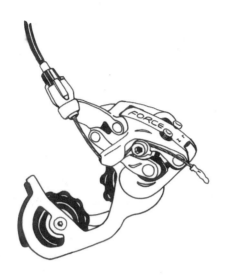

5.34 Attaching rear-derailleur cable

make sure this loop is not too short or kinked

about which side of the bolt the cable goes on, remove the bolt and look for a cable groove on the derailleur or the underside of the bolt's washer.

5. **Tighten the bolt.** On most derailleurs this requires a 4mm or 5mm hex key.

6. **Clip the cable 1–2cm past the bolt and crimp on a cap to prevent fraying (Fig. 5.25).**

5-14

ATTACHING CABLE TO FRONT DERAILLEUR

1. **Click the shifter to the inner chainring position (5-1).** This ensures that the maximum amount of cable is available to the derailleur.

2. **With a hex key, tighten the cable to the cable anchor on the derailleur while pulling the cable taut with pliers (Fig. 5.35).** Be sure that the cable lies in the groove beneath the cable-fixing bolt.

NOTE: *For internal cables, see 5-12. For Shimano 11-speed, see 5-5g.*

3. **Clip the cable 1–2cm past the bolt and crimp on a cap to prevent fraying (Fig. 5.25).**

5.35 Attaching front-derailleur cable

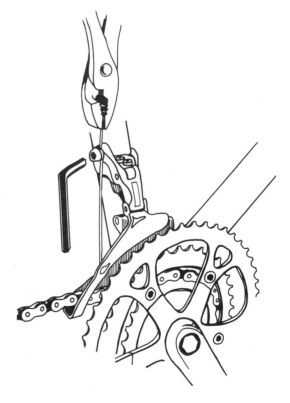

the cable also goes through a little tunnel at the bottom of the curved cable guide on the back of the derailleur (5-3g).

4. **Pull the cable taut and into its groove under the cable-fixing bolt (Fig. 5.34).** On all but SRAM rear derailleurs, the cable usually goes on the side of the bolt toward the jockey wheels. If you're unsure

5-15

FINAL CABLE TOUCHES

A high-quality cable assembly includes the cable-housing end ferrules (Fig. 5.21) throughout and crimped cable caps (Fig. 5.25); cables are clipped about 1–2cm past the cable-fixing bolts.

5-16

CABLE LUBRICATION

New cables and housings with Teflon liners do not need to be lubricated. Old cables can be lubricated with chain oil. White lithium grease and other bike greases can slow cable movement, so use oil or a very light, translucent grease.

1. **Disconnect the cable at the derailleur, and clip off the cable-crimp end.** Be aware that if the cable frays at all when clipped, you may not be able to slide it back in through the housings and may have to replace the cable and perhaps its housings as well. Alternatively, if the frame has slotted cable stops, leave the cable connected and pop the housing ends out of their cable stops. To do so, push inward with your hand on the rear derailleur (or pull outward on the front derailleur to lube a front-derailleur cable on a cyclocross bike) to give the cable some slack.

2. **Coat the areas of the cable that will be inside the cable-housing segments with chain oil.** You won't be able to access the cable sections inside the housing segments taped down to the handlebars.

3. **Squirt oil into each housing section.**

NOTE: *If you have any trouble reinstalling the cable because of fraying, or the housings are dirty and rusty, you should replace both the cable and the housings.*

5-17

REDUCING CABLE FRICTION

In addition to replacing old cables and housings with good-quality cables and lined housings, the following specific steps can improve shifting efficiency:

1. **The most important friction-reducing steps are to route the cable so that it makes smooth bends.** See the Note in 5-7, step 2 regarding the loop at the rear derailleur. Make sure that the front cable-housing sections are long enough that turning the handlebar does not increase the tension on the shift cables.

2. **Choose cables that offer especially low friction.** "Die-drawn" cables, which have been mechanically pulled through a die (a small hole in a piece of hard steel), move with lower friction than standard cables. Die-drawing flattens the outer strands, smoothing the cable surface. Thin cables and Teflon-lined housings with a large inside diameter also reduce friction. Coated cables slide great when new but must be replaced when the coating frays, as it will gum up inside the housings.

3. **If your bike has internal cables, make sure that they aren't crossing each other multiple times inside the down tube.**

THE SHIFTERS

5-18

REPLACING AND INSTALLING INTEGRATED BRAKE/SHIFT LEVERS

Shifters can be replaced as an entire unit and sometimes as separate parts. Brake/shift levers are generally labeled right and left, but if you're in doubt, you can tell which is which because the shift paddle behind the brake lever (Fig. 5.37) should flip to the inside. Here are the steps for you to replace the entire brake/shift lever unit:

1. **Remove the handlebar tape and bar plugs.**

2. **Remove the old brake/shift lever by loosening its mounting bolt with a 5mm hex key or a Torx T25 key and sliding the lever assembly off.** The position of the bolt varies, but it is always on the outside of the lever body, under the lever hood on the outboard or top side. Slip the hex key or Torx T25 key (if the 5mm hex key doesn't fit, your lever probably takes a T25 key) down from the top

between the lever body and the hood (Fig. 5.36), or roll back the hood far enough to get at it from outside of the hood.

3. **Slide the new lever on the bar to where you like it.** With modern handlebar shapes, set the lever height so that the top of the lever body extends the top flat section of the bar level forward or upward. The old-school way is to put a straightedge against the bottom of the bar and slide the lever down until its end touches the straightedge or is slightly above it.

4. **Post-2008 Campagnolo Ultra-Shift and Power-Shift Ergopower levers have a "big hands" insert.**

5.36 The mounting bolt is on the outside of the lever body.

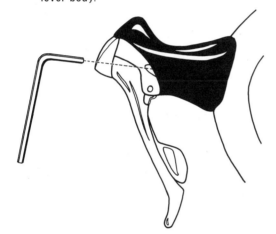

If you want the lever body cocked back so that the reach to the levers is increased, push the insert onto the bottom of the lever base under the hood (Fig. 5.37). They are asymmetrical but not labeled L and R, so line up the hole on the insert with the cable-access hole on the lever body to ensure you have it on the correct lever.

5. **Tighten the mounting bolt.** Again, this will likely take either a 5mm hex key (it's best to have a long one without a ball end that could strip the bolt head) or a Torx T25 key. Put a long straightedge across the top of both levers before they are fully tightened to make sure that they are set at the same height.

6. **Install the cables (5-6 through 5-15; 9-4) and barrel adjusters, if not already in place.**

7. **Wrap the handlebar with tape (12-12).**

5-19

REHABILITATING SHIMANO STI INTEGRAL BRAKE/SHIFT LEVER

LEVEL If you have a jammed Shimano STI shifter, you cannot really go into the mechanism like a watchmaker and replace parts. Shimano doesn't sell the internal parts separately, and opening the

---

5.37 Installing "big hands" insert onto Ultra-Shift Ergopower

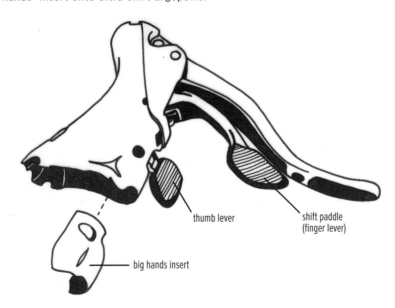

thumb lever

shift paddle (finger lever)

big hands insert

5.38 Exploded Dura-Ace 9-speed STI lever

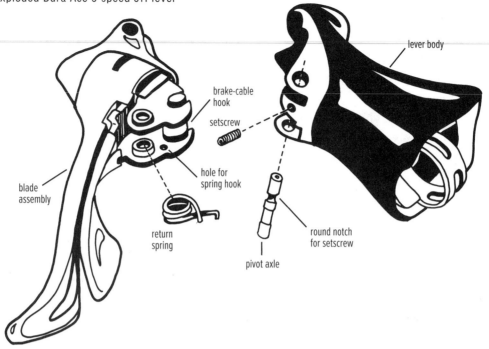

lever body

brake-cable
hook

setscrew

hole for
spring hook

blade
assembly

return
spring

round notch
for setscrew

pivot axle

mechanism voids the warranty. But, with a little lube, you can often rehabilitate a sticky mechanism or an STI lever that does not always engage when you try to shift. Use the thin tube attachment on an aerosol chain lube to flush the guts of the lever from the side by sticking the aerosol tube into the little hole visible when you pull the brake lever; the hole is above the rounded upper section of the main lever on the noncable side. Many riders have used this technique to get years of extra life out of levers that otherwise looked destined for replacement. Don't do this over your carpet.

5-20

OVERHAULING ORIGINAL CAMPAGNOLO ERGOPOWER LEVERS

LEVEL This is a very satisfying task. Whether you want to bring back original shifting performance, or repair a lever body after a crash, or change the number of speeds in the lever, you've come to the right place. As with any mechanism that has lots of precision internal parts, it can be great fun, if you are in the right state of mind, to take an Ergopower (EP) lever entirely apart, clean it up, put it back together, and hear it click cleanly and feel it work smoothly.

Every Record and Chorus Ergopower lever through 2008 and every Centaur, Veloce, Mirage, and below lever through 2006 has two little G-shaped springs (Figs. 5.39, 5.40) that click into teeth in an indexing ratchet, providing the indexing steps as the shift lever behind the brake lever advances the ratchet to pull cable, or as the thumb lever drives the ratchet in the opposite direction to release cable. Since these springs are constantly riding over the gear teeth on every shift in either direction, the G-springs can get worn, flattened, or broken to the point that they will no longer be able to stop the index gear from turning. Get a couple, if you think you need them, and follow along. Your EP levers will be good as new again!

Despite the fact that every little Ergopower part is replaceable, these days only the entire shifter mechanism as a single unit is available for QS/Escape levers (5-21). And when it comes to Ergopower levers with the modern, taller shape and curvier lever blade, that's now also the case with 11-speed Ultra-Shift levers (5-22) and Power-Shift levers (refer to only the first few steps of 5-22 for those).

Refer to the exploded diagrams, Figure 5.39 (8-speed right-hand lever) and Figure 5.40 (9- or 10-speed right-hand lever). You can also magnify all

5.39 Exploded Campagnolo Ergopower lever (8-speed)

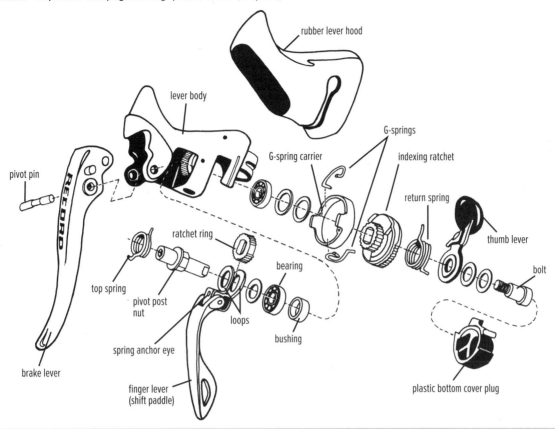

5.40 Exploded Campagnolo Ergopower lever (9-speed) (Note: 10-speed has a different bottom bushing and washer, but is otherwise the same)

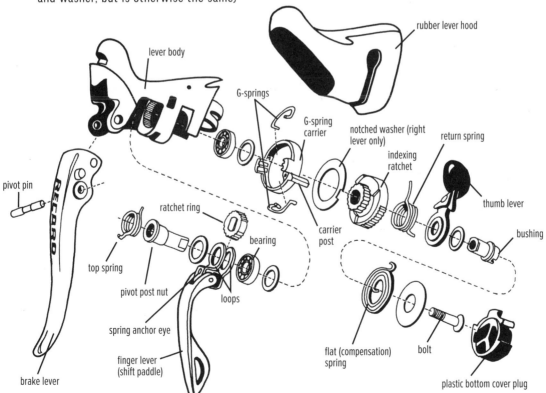

those parts and find their part numbers for your particular year and model of lever on campagnolo.com. Campagnolo North America is now the only complete U.S. source for Ergopower small parts; American distributors still carry a lot of them but often not some critical ones like lever bodies and index gears (which allow changing from 9-speed to 10-speed, for instance).

The following instructions cover overhaul and lubrication, as well as replacing broken or damaged index springs, lever bodies, and other parts. They also cover replacing indexing ratchets to change the number of speeds. When I mention 8-speed levers, I am referring to 1992–1997 Ergopower levers (Fig. 5.39), both right and left. The lever body and the rubber hood come to a point on top on these. The reference to 9- and 10-speed levers applies to left and right 1999–2008 Record and Chorus EP levers and 1999–2006 Centaur and below EP levers whose lever body and rubber hood are rounded on top (Fig. 5.40).

NOTE: *The internals of 1998 9-speed levers are similar to those of an 8-speed lever. On the outside, 1998 levers look like the lever in Figure 5.40, but a Record 1998 lever will be aluminum, whereas from 1999 on, the lever will be carbon fiber.*

Disassembly

1. **Remove the rubber hood.** It's easiest to pull it off the base of the lever, but it will come off over the top as well; squirt rubbing alcohol under it so it will slide off easily. On composite lever bodies, pull off the plastic piece that covers the bottom of the shift mechanism; use pliers if necessary.

2. **Push out the lever-pivot pin by tapping it out with a blunt nail and a hammer.** Support the lever body near the pin so that the edge of the lever does not flex outward as you tap. Holding the lever body flat on a block of wood with a drilled hole that is lined up under the pin does the trick nicely. Pull off the brake lever.

3. **Clamp the lever body onto the end of a handlebar held in a vise.** Place the bar-clamp strap right at the edge of the bar so that the lower part of the lever body is hanging off the end of the bar. You want to be able to get at the lever's mechanism from the

bottom. Hold the bar in the vise so that the lever is upside down. Alternatively, you can also work with the lever loose in your hand.

4. **Shift.**
   a. With a 9- or 10-speed lever, shift to the lowest-gear position with the finger lever (shift paddle) to release tension on the flat coil "compensation" spring at the bottom of the lever; you can see it stick out around the large, flat washer when you get to the low-gear position.
   b. With an 8-speed lever, you do the opposite; shift the thumb lever to the highest-gear position.

5. **Hold the top (pivot post) nut with one hex key while you unscrew and remove the bottom bolt with another hex key.** On a 9- or 10-speed lever, the top nut takes a 5mm hex key, and the lower bolt takes a 3mm hex key (avoid using a 3mm ball-end hex key to start turning this bolt, as you can snap it off—leaving the ball in the bolt head) or a Torx T20 (left lever). On an 8-speed lever, the nut and bolt both accept 4mm hex keys, and the bolt may have a brass washer or two on it, so watch for them.

IMPORTANT NOTE: *The bolt on a right-hand 8-speed lever is left-hand threaded, so it unscrews in a clockwise direction! Also note that 1998 9-speed levers (aluminum brake lever) are configured this way.*

6. **With an 8-speed lever, skip to step 9. With a 9- or 10-speed lever, remove the bottom washer, pop the flat spring out with a thin screwdriver, and take out the thin, play-removing washer, if installed.**

7. **While holding the (9- or 10-speed) assembly together with your thumb, shift to the high-gear position with the thumb lever.**

8. **Holding the thumb lever in place, pull out the bushing in the center of the thumb lever return spring.** Use needle-nose pliers.

9. **Pop out the thumb lever, the spring, the indexing ratchet (also known as the index gear; see Figs. 5.39, 5.40), and, on a 9- or 10-speed right-hand lever, the notched washer.** The two G-shaped index springs, which are the most common replacement item in Ergopower levers, are now

visible. One or two washers that sit against the cartridge bearing underneath may come out with the indexing ratchet. If the bearings are dirty, you can push them out after step 10 and clean, grease, and replace them at that time.

10. **Pop the G-spring carrier and G-springs out.**

11. **Clean and grease all parts.** Unless you're changing the number of gears, you can leave the pivot post nut and finger-lever assembly together and in place (see step 12 for taking it apart).

12. **With an 8-speed or 10-speed lever, unless the top ratchet ring is worn out, skip to step 2 of Reassembly.** To upgrade a 9-speed lever to 10-speed, you need to install a new ratchet ring in the finger-lever assembly, which will come out as an assembly. Flip the lever over so that it is clamped upright on the bar. It is hard to get the top spring out; grab it with thin needle-nose pliers and twist it upward with the other end still hooked into the anchor eye in the loop atop the finger lever. Pull the finger-lever assembly out. Push the center pivot post nut out. Now you can slip the 9-speed ratchet ring out and replace it with a 10-speed ring.

Reassembly

1. **Orient the ratchet ring with the number 9 or 10 or letter L up.** It will be pointed forward when the lever is installed on the bike. Push the pivot nut and washer through the ratchet ring (line up the flats). Hook the top spring's short hook onto the anchor eye atop the upper finger lever loop. Push the assembly back in place, hooking the long spring end into the notch in the lever body with needle-nose pliers. Flip the lever back over to get at the bottom side again.

2. **Put the new (or clean and regreased) G-springs on the underside of the G-spring carrier.** Coat them with grease to hold them in place. Push the G-spring carrier back into place in the lever body.

3. **On a right-hand 9- or 10-speed lever, replace the thin washer on top of the G-spring carrier.** The washer's tab faces down, and the notch (if the washer has it) fits around the vertical post on the G-spring carrier. This washer (and the vertical carrier post) do not exist in left-hand levers or in 8-speed ones, so ignore this step for them.

4. **Drop the (greased) indexing ratchet (index gear) down onto the pivot post nut so that the flats in both parts interlock.** Now is the time to put in a new indexing ratchet if you are changing speeds (for instance, from 8 to 9 or from 9 to 10) or if the ratchet is worn (check the teeth near the cable hole). Make sure that the cable-hook tab on the ratchet butts against the outboard side of the lever body.

5. **Slip the long end of the return spring down into the indexing ratchet.** Feed it out through the ratchet's slot and into the hole in the lever body. The hole goes clear through, out the back, starting at the base of the notch in the body for the thumb lever. Drop the (greased) spring into the ratchet with the short end of the spring sticking up.

6. **Push the small hole in the thumb-lever ring onto the upward-pointing or hooked end of the thumb lever return spring inside the indexing ratchet.** The convex side of the ring should face the spring. Push back on the thumb-lever ring to align it over the ratchet. Make sure the thumb lever is up at the top of its slot in the lever body, so that when you push it back into place it winds the spring tighter, rather than unwinding it.

7. **On 9- or 10-speed levers, while holding the thumb-lever ring down, push the central bushing (with its washer) down through the ring.** Push down and turn the bushing (on a 1998 9-speed lever, use a 5mm hex key; on later 9-speed and all 10-speed levers, use a large screwdriver in the larger set of slots) until the flats on the end of the bushing engage the flats on the end of the pivot post nut protruding into the indexing ratchet. If there were washers on the bushing, make sure you have reinstalled them.

   a. On 1999 and later 9-speed and on all 10-speed levers, the bushing is larger in diameter and has flats that engage over the flats on the pivot post nut, rather than inserting into them. An unfortunate consequence of this larger diameter of the bushing is that it is harder to push the bushing into place without disengaging the return spring from the thumb lever. Keep at it until you get it.

b. On 8-speed levers, insert the bolt, with any washers it had on it, down through the thumb-lever ring until it engages the nut. While holding the nut with a 4mm hex key inserted into the top of the lever, tighten the bolt with another 4mm hex key. Remember that the bolt is reverse-threaded on a right-hand 8-speed lever!

8. **Skip to step 12 for 8-speed levers.**

9. **While holding the bushing in place (it will turn), shift the finger lever all the way to the lowest-gear position.**

10. **Lay the flat spring on top of the thumb-lever ring.** The inner end of the spring hooks into a notch in the end of the bushing, and the outer spring end hooks around either the post on the spring carrier (right-hand lever) or the outboard edge of the lever body (left-hand lever).

11. **Holding the flat spring down, tip the bushing back and forth until you feel the flats in the end of the bushing disengage from the flats on the top bolt.** Use a 5mm hex key on a 1998 9-speed lever and a large screwdriver on 1999 and later 9- and 10-speed levers. Still holding the spring down, turn the bushing about a half-turn to wind the spring (counterclockwise on the right lever; clockwise on the left lever), and jiggle the bushing back and forth with the 5mm hex key or large screwdriver until its flats reengage the top bolt.

NOTE: *For a harder shift feel, leave the flat compensation spring out. Alternative method for steps 10 and 11: Hook the flat spring's inner end into a more advanced bushing slot; the outer end of the spring will be squished up against the back wall of the lever body. While holding the bushing with the hex key or screwdriver (or install the bolt to hold the bushing in place), pull the outer spring end with a hooked awl, a small crochet hook, or a paper clip with a small hook bent into the end and pull it to the post (right-hand lever) or to the notch in the lever body's outboard wall (left-hand lever).*

12. **Holding the flat spring down with your finger, slide the large washer under your finger and on top of the flat spring so that it snaps over the end of the bushing.** The 1999 and later 9- or 10-speed washer has two notches to fit the larger slots in the bushing. Start the bolt while holding the other end of the pivot shaft with a 5mm hex key. Snug the bolt down with a 3mm hex key.

13. **Check the mechanism, taking note of step 11 in the Reassembly section of 5-21.** If it works smoothly, unclamp the lever from the bar, reinstall the brake lever, bottom cap, and rubber hood (engaging the hood's nubs into holes, slots, and protrusions in the lever body). It's usually easier to pull the hood on from the front rather than from the base of the lever body, and lubricating it with rubbing alcohol makes it easier yet. Congratulations! You are done!

5-21

OVERHAULING CAMPAGNOLO QS/ESCAPE ERGOPOWER LEVERS

 LEVEL This mechanism (Fig. 5.41), which appeared on Centaur, Veloce, and other EP levers below them in 2007, eliminates the full swing of the thumb lever and has only a simple hole in the lever body (and in the rubber hood), rather than a long slot, for it to swing through. A hinged pawl clicks into each successive index-gear tooth as a push on the shift lever behind the brake lever rotates the gear to pull cable. The index gear is spring-loaded (and the spring in the derailleur is pulling on the cable), and each time the thumb-lever linkage pops the pawl out of a gear tooth, the springs rotate the index gear in the opposite direction to release cable. The pawl then drops into the next tooth to stop the index gear from rotating any farther than a single shift, and the rear derailleur shifts to the next smaller cog (in the case of the right lever), or the front derailleur shifts to the next smaller chainring (in the case of the left lever).

This mechanism is not as durable as earlier EP levers or later EP Ultra-Shift levers. When worn, it tends to upshift multiple cogs with a single light push of the thumb lever. If this happens, you won't find individual

5.41 Exploded QS/Escape Ergopower lever

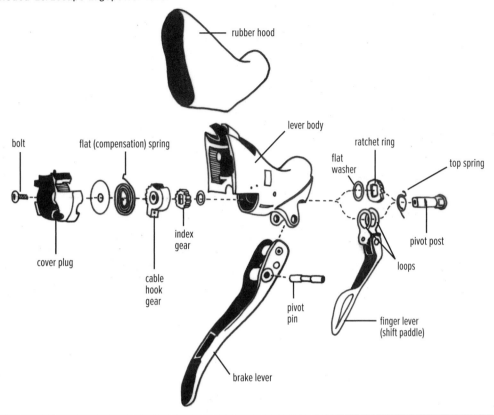

parts; instead, you will have to replace the entire lever body with the shifter assembly already integrated into it.

Disassembly

1. **Remove the rubber hood.** It is easier to pull it off the base of the lever, but it will come off over the top as well, and lubricating inside with rubbing alcohol will make it slide off easily.

2. **Unscrew the handlebar band clamp.** Alternatively, you can leave it on and use it to clamp the lever body to the end of a handlebar fixed in a vise so that the lower part of the lever body overhangs the end of the bar, allowing you to get at the underside of the mechanism.

3. **Push out the brake-lever pivot pin by tapping it out with a blunt nail and a hammer.** Support the lever body near the pin so that the edge of the lever does not flex outward as you tap. Holding the lever body flat on a block of wood with a drilled hole that is lined up under the pin does the trick nicely. Pull off the brake lever, keeping its plastic bushings with it.

4. **Pull off the plastic bottom cover plug.**

5. **With a 2.5mm hex key, remove the bottom bolt.**

6. **Pull off the large, thin washer, flat spring, and plastic cable hook gear.**

7. **Pull off the steel index gear and the tiny, thin washer it conceals.** Turn the lever body upright.

8. **With needle-nose pliers, unhook and pull off the finger-lever return spring (top spring) out through the front of the lever body.**

9. **With a hex key, push the central pivot post nut straight up and out through the top of the finger-lever assembly.**

10. **Pull out the finger-lever assembly and the flat washer under it.** Pull out the ratchet ring from between the loops extending from the top of the finger lever.

11. **Disassembly of the QS/Escape lever is now complete (Fig. 5.41).**

Reassembly

1. **Clean all of the parts and put a thin layer of clean grease on them.**

2. **Replace the greased ratchet ring between the finger lever's extending loops with its engraved 0 up against the top loop (so that it will be pointing forward when on the bike).** Push the greased pivot post nut down through the assembly so that its larger flats engage the flats in the ratchet ring.

3. **With grease, stick the flat washer to the bottom extending loop, around the pivot post nut.**

4. **Push the pivot post nut up so that it does not protrude through the bottom of the finger-lever assembly, and slide the finger lever into its place in the lever body.** Push the pivot post nut down through the assembly and through the brass bushing in the lever body. If the pivot post nut does not come all the way down so that its flange once again contacts the top extending loop, grab its smaller flats extending out of the bottom of the brass bushing with needle-nose pliers. Rotate the post while pulling down until its larger flats line up with the ratchet ring flats and it pulls down until its top flange contacts the top extending loop.

5. **Holding the top spring so that its short, hook end is on the bottom and its long, straight end is on the top, maneuver it in from the top of the lever body.** The long end will sit in the little groove at the top of the wide notch in the lever body for the finger lever. Stick a 4mm or smaller hex key down through the spring coil into the hole in the top of the pivot post nut. With a small hook (a small crochet hook works; I've also used a thin nail by grinding it thinner at the tip and bending it into a hook with needle-nose pliers), pull the hook end of the spring so that it catches on the anchor eye on the top extending loop (the anchor eye is not visible in Fig. 5.41; you can see it in Fig. 5.40). You can also try holding the spring with needle-nose pliers and hooking the short, hook end onto the anchor eye first. Then grab the long end of the spring with the pliers and pull it into its notch in the lever body. Turn the lever over to view it from the bottom.

6. **Replace the tiny, thin washer followed by the index gear onto the end of the pivot post nut.** Have the gear's recessed base away from you and its engraved 0 toward you. Push it in place while pushing the thumb lever to move the pawl riveted into the lever body away and give the index gear clearance to drop into place.

7. **Push the rectangular recess of the thick plastic cable hook gear onto the end of the pivot post nut.** The engraved 0 will be toward you. Turn it and the finger lever as required to engage the recess onto the end of the pivot post nut. Grease the gear teeth.

8. **Push the outboard tip of the flat spring into the small hole in the lever body on the side opposite from the thumb lever.** With needle-nose pliers, wind it an extra turn before pushing its inner end into the slot in the plastic gear.

9. **Grease the flat spring and stick the big flat washer against it.**

10. **Push the bottom screw through the hole in the big flat washer and tighten it into the end of the pivot post nut with a 2.5mm hex key.**

11. **Test the shift mechanism function.** Check that when the finger lever returns, its outboard edge lines up with the outboard edge of the brake lever. This is necessary for the finger lever's engagement teeth to fully disengage from the ratchet ring so that operating the thumb lever properly releases the cable and allows it to advance. If the finger lever does not line up properly, the inboard end of the tiny return spring surrounding the rivet of the finger lever has become disengaged. It needs to hook in diagonally over the outboard notch in the chromed steel flat loop inside the notch at the top of the finger lever. Push the spring tip back into place with a small screwdriver.

12. **Push the plastic bottom cover plug back into the base of the lever body.**

13. **Replace the brake lever, including its plastic bushings.** Line up the holes in the lever and body, and hold the parts together by pushing the pivot pin back in partway by hand. The pin is symmetrical; you can push it in from either side.

14. **Tap the pin in with a hammer.** Ensure that the pin is fully inserted with the same amount protruding on either side.

15. **Install the handlebar band clamp.**

16. **Lube the inside of the rubber hood with rubbing alcohol and slide it back onto the lever body.**

You're done! Cable it up to a derailleur and see how it works.

OVERHAULING CAMPAGNOLO ULTRA-SHIFT ERGOPOWER LEVERS

LEVEL The mechanism in Campagnolo's Ultra-Shift levers introduced in 2009 in Super Record, Record, Chorus, and Centaur (Figs. 5.37, 5.42), whether 10-speed (Centaur) or 11-speed, requires less maintenance and is less prone to failure than the mechanism in prior Ergopower designs, mainly because it has no G-springs to wear out. It is also easier to overhaul, requiring less dexterity when winding the springs.

Campagnolo does not sell small parts for the Ultra-Shift levers separately. If you break the lever body or something inside, you have to replace the entire lever body and shift assembly, which is sold as a complete unit. Brake lever blades and handlebar band clamps are available as separate pieces. The same is true for Power-Shift

levers (see next paragraph). To replace the lever body/shifter assembly or the lever blade, you will only need steps 1–3 in Disassembly and 6–9 in Reassembly.

In 2010, Campagnolo ceased producing 10-speed Ultra-Shift and introduced Power-Shift in Athena (11-speed) and 10-speed Centaur, Veloce, and below. Power-Shift levers are shaped like Ultra-Shift levers, sharing the more curvaceous lever and taller lever body than earlier Ergopower levers. Like QS/Escape levers (5-21), Power-Shift levers can only upshift one cog at a time on the rear, and on the front can only make a complete shift from one chainring to the next, rather than in smaller increments. Just as on QS/Escape levers, eliminating the full swing of the thumb lever results in a simple hole in the lever body (and in the rubber hood) for it, rather than a long slot for the lever's swing. Power-Shift levers also have a directional pivot pin holding the brake lever on, so follow step 2 for removing it.

These instructions apply to the right lever (rear-derailleur control); the left lever is similar, but there are significant differences, including that the thumb-lever return spring is integrated into the cable-hook disc. Running changes to the washers inside Ergopower levers may occur.

5.42 Exploded Ultra-Shift Ergopower lever

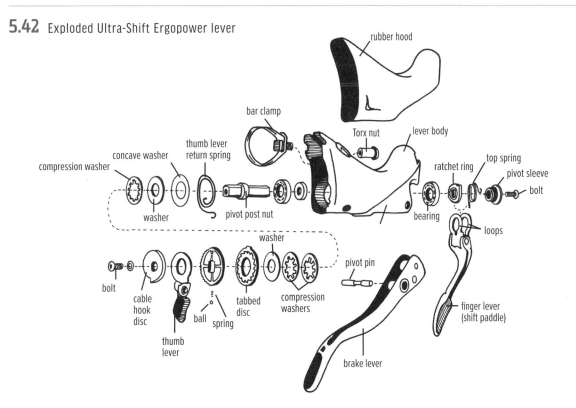

Disassembly

1. **Remove the rubber hood.** It is easier to pull it off the base of the lever, but it will come off over the top as well. Lubricating inside with rubbing alcohol will help it slide off easily.

2. **Push out the brake-lever pivot pin by tapping it out with a blunt nail and a hammer from the inboard side out.** The Ultra-Shift pivot pin has an enlarged head, which goes on the outboard side. Support the lever body near the pin so that the edge of the lever does not flex outward as you tap. Holding the lever body flat on a block of wood with a hole lined up under the pin does the trick nicely. Pull off the brake lever.

3. **With a Torx T25 key, unscrew the band-clamp bolt and remove the handlebar band clamp.** You may instead wish to clamp the lever body to the end of a rigidly fixed handlebar so that the lever is securely held as you work on it. Have the band clamp right at the end of the bar with the lower part of the lever body overhanging so that you can get at the underside of the mechanism.

4. **With a 2.5mm hex key in the bolts at the top and bottom of the lever body, unscrew the bottom bolt and remove it along with its tiny brass washer.**

5. **With needle-nose pliers, pull out the thick plastic disc with the recessed cable entry hole (called "cable hook disc" in Fig. 5.42).**

6. **Unhook the end of the thumb-lever return spring from the spool-shaped pin on the thumb lever.** Use a thin crochet hook or bent paper clip.

7. **Pull the thumb lever straight out to remove it along with the flat ring it's joined to.**

8. **With needle-nose pliers grabbing its raised lips at its center (or by pushing it up from below with a thin screwdriver through the lever-body slot), pull out the big, flat, toothed disc.** Take care not to lose the two tiny springs and two tiny ball bearings that sit in slots on its hidden face (see one ball and spring exploded away in Fig. 5.42; don't let this happen to you!).

9. **With needle-nose pliers or by pushing it up from below with a thin screwdriver through the lever body slot, remove the tabbed disc and washers.**

There will (probably) be two thin, flat washers; a slotted concave compression washer; a mating slotted convex compression washer; a third slotted compression washer concave side up; a thin, flat washer; and a large, thin, concave plastic washer. Some levers differ in the number and orientation of these washers; note their arrangement when removing them.

10. **Pull out the flat thumb-lever return spring.** While not imperative, it's easier to now work on the top of the lever body with it fixed in place; you can clamp it onto the end of a handlebar clamped in a vise.

11. **Unscrew and remove the top bolt with a 2.5mm hex key and, with needle-nose pliers, pull the central pivot post nut down and out of the bottom of the lever body.** The cartridge bearing and spacer above it may come out with the pivot post nut; if not, remove them separately after step 12. You can use the pivot post nut to wiggle the bottom bearing out and then use it to push the upper bearing out.

12. **Unless it fell out when you pulled out the pivot post nut, pull straight out on the finger lever (shift paddle) to remove it.** Hold the parts (shown in Fig. 5.42) between its extending loops with your finger to keep them from falling out.

13. **Pull out the pivot sleeve, finger-lever return spring, and ratchet ring from between the extending loops.** Disassembly of the lever now is complete.

Reassembly

1. **Clean and grease all parts.** If any are worn out or broken, stop now and go buy a new lever body/shifter assembly; the parts are not individually available.

2. **If you removed them, push the cartridge bearings and the spacer between them back in with your finger or thumb.**

3. **Slip the ratchet ring between the loops extending from the finger lever so its raised central lip drops into the hole in the lower loop.** Orient it with its flat side up (stamped with the number 10 or 11 on a right lever) so that it points forward when the lever is installed on the bike. Rotate the ratchet ring

so its tab rests against the flat base between the extending loops.

4. **Install the finger-lever return spring above the ratchet ring.** Hook its short end in the anchor eye adjacent to the upper extending loop.

5. **Push the pivot sleeve through the upper loop and the return spring.**

6. **Push the assembly back in place.** Brace the long end of the return spring against the notch in the lever body.

7. **Insert the pivot post nut up through the bearing and finger-lever assembly while pushing back on the finger lever to line up the ratchet ring.** Compress the top spring (not easy). Check that the flats on the pivot post nut line up with those inside the ratchet ring before pushing the pivot post nut up.

8. **Tighten the top bolt into the pivot post nut.** Ensure that the finger-lever return spring hook end is properly hooked in the eye alongside the top loop riveted to the lever. Turn the lever body over to get at the bottom side again.

9. **Install the thumb-lever return spring.** Insert its central, downturned end into the tiny hole in back adjacent to the bearing. Hook its outboard end around the edge of the thumb lever's slot in the lever body. To do this, you'll probably find it easiest to slip it in through the slot from the side and then angle it so that its hooked end sits up high in back when the downturned end is lined up above the hole. Pull back and push down on the hooked end with needle-nose pliers, pushing the spring down so the downturned end goes into the hole. Holding the spring down to keep the end in the hole, pull the hooked end out to the edge of the lever-body slot with the pliers or with a hook.

10. **Drop the washer stack in over the pivot post nut from the bottom.** Here is the order: large, thin, plastic washer with concave side facing you; thin, large-diameter flat washer; one smaller-diameter flat washer; slotted compression washer with concave side facing you; mating slotted compression washer with convex side facing you; third slotted compression washer with concave side facing you; and other smaller-diameter flat washer. Not all levers have the same stack of washers; yours may not have the third slotted compression washer. Also, the smaller-diameter flat washers can both go on last.

11. **Install the tabbed disc.** The tab must point away from you, and the side with raised teeth around the outer edge must face toward you. As you push the disc down onto the pivot post nut, its tab needs to slide down the square groove running longitudinally in the lever body past the cable access hole.

12. **Ensuring that the tiny springs pushing outward on tiny steel balls are in place in their slots on its face, drop the toothed disc down onto the pivot post nut.** The balls and springs will be on the side facing down toward the tabbed disc below. The teeth around the outer edge of one side of the disc should face out through the thumb-lever slot in the lever body.

13. **Install the thumb lever from the side, through the slot in the lever body.**

14. **Install the thick plastic disc over the end of the pivot post nut.** Ensure that its tab with the recessed hole for the cable head is at the cable access hole in the lever body.

15. **Slip the tiny brass washer onto the bottom bolt and tighten the bolt into the pivot post nut.** Use a 2.5mm hex key.

16. **Loosen the bolt just enough to tip the thumb lever up.** Allow the hooked end of its return spring to pass by the spool-shaped pin on the thumb lever.

17. **With your thin hook, pull the hooked end of the return spring onto the spool-shaped pin on the thumb lever.** You may find it easiest to pull it past the pin, tighten the bolt so it pushes the lever down, and bring the hook back to the pin.

18. **With 2.5mm hex keys in the top and bottom bolts, tighten both bolts.**

19. **Replace the brake lever, including its plastic bushings.** Line up the holes in the lever and body, and hold the parts together by pushing the pivot pin back in from the outboard side as far as you can by hand.

20. **Tap in the pin with a mallet, with the inboard side of the lever body on a block of wood.** You can also

set the end of the pin on a wooden surface and tap the side of the lever body down onto it with the mallet until the pin is fully inserted.

21. **With a Torx T25 wrench in the cap nut, install the handlebar band clamp.**

22. **Lube the inside of the hood with rubbing alcohol and slide it back onto the lever body.**

You did it! Congratulations! Try it and see how it works.

5-23

OVERHAULING BAR-END SHIFTERS

 LEVEL You cannot overhaul Shimano or SRAM bar-end shifter mechanisms, but you can lubricate them; Campagnolo shifters can be rebuilt.

1. **Release the cable at the derailleur.**

2. **Remove the screw holding the shift lever to its mount, and pull the lever off.** At this point, it is possible to remove the lever mount housing (by loosening the expander bolt in its throat with a hex key and releasing the expander plugs and pulling the whole thing out), but there is no need to do so.

3. **Left lever:**

    The Shimano (Fig. 5.43A) and SRAM lever mechanisms cannot be disassembled, but you can rinse or blow grit out of them.

    To get the Campagnolo frictional shifter (Fig. 5.44A) working smoothly, simply clean the parts and put them back together. You generally do not want to grease the mechanism, since that will only demand higher bolt-tightening torque to get the lever to hold its position.

4. **Right lever:**

    If you have a Shimano indexed shifter (Fig. 5.43B), the mechanism cannot be serviced; buy a new lever if the mechanism has failed.

    With SRAM, you can blow out the area around the index gear with compressed air and grease the teeth and the three G-springs, which might rehabilitate it fully.

    With Campagnolo (Fig. 5.44B):

    a. **Pull the top-hat-shaped bushing out of the center of the outboard side of the lever.** Now

5.43A Exploded left lever, Shimano bar-end shifter

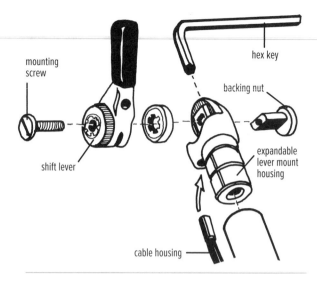

mounting screw
hex key
backing nut
shift lever
expandable lever mount housing
cable housing

5.43B Exploded right lever, Shimano bar-end shifter

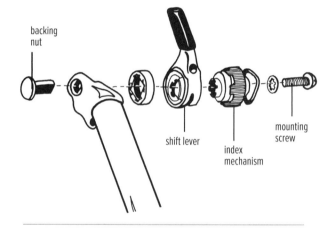

backing nut
shift lever
index mechanism
mounting screw

you can see the index gear and the three G-springs that click into its teeth.

b. **Pull out the plastic G-spring carrier.** To do so, push on the index gear and, with a screwdriver, pry on the edge of the plastic carrier protruding from the lever as well.

c. **Pull or pry the index gear up out from the springs.**

d. **Clean and grease the G-springs and index gear.** You can replace worn springs or change the index gear to change the number of speeds.

e. **Install the index gear.** Line up the arrow on the index gear with the G-spring prong on the side of the G-spring carrier that has a nub on it to engage a notch under the cable hole

5.44A Exploded Campagnolo frictional left-side bar-end shifter

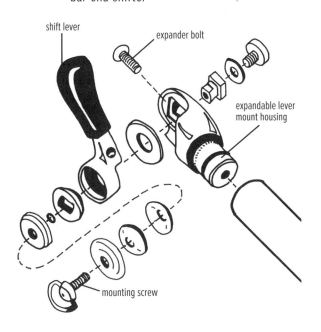

- shift lever
- expander bolt
- expandable lever mount housing
- mounting screw

on the lever. The index gear's arrow marking will point toward the rubber cover on the shift lever. Tip the index gear down to engage two G-springs, fit a screwdriver in between the third spring and the other side of the gear, and pry with it while pushing the gear down into place.

f. **Push the plastic spring carrier back into the lever.** Line its nub up with the notch in the hollow of the lever. Make sure you replace any washers that may have fallen out.

g. **Push the top-hat-shaped bushing back into place.** Line up its two teeth on the underside of its flange with the two notches in the index gear.

h. **Line up the lever in its cable-release position (i.e., flipped down).** Fit the notches inside the bushing's bore over the ledges on the

5.44B Exploded Campagnolo right-side 8-, 9-, or 10-speed indexed bar-end shifter

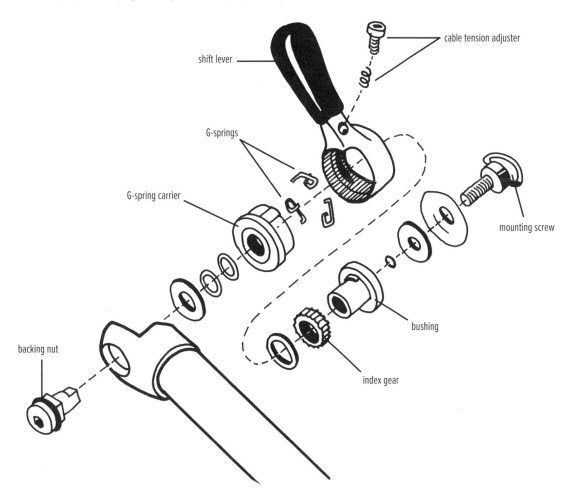

- cable tension adjuster
- shift lever
- G-springs
- G-spring carrier
- mounting screw
- bushing
- index gear
- backing nut

backing nut, once you have pushed it back through the shifter mount housing.

5. **Tighten the mounting screw into the backing nut snugly.** It must be tight enough that the lever does not wobble, but not so tight that it binds.

6. **Install the cable and tighten it at the derailleur.**

5-24

OVERHAULING OR REPLACING DOWN-TUBE SHIFTERS

 LEVEL 1. **Remove the screw holding the shifter to the frame's shifter boss, and pull the shifter off.**

2. **If you have a frictional shifter (Fig. 5.45), all of its pieces come apart.** To get the shifter working smoothly, simply clean the parts, grease them, and put them back together. If you have an indexed Shimano shifter, the mechanism cannot be serviced. Buy a new lever if the mechanism has failed.

3. **Replace the stop piece that fits over the square base of the shifter boss.** With old frictional shifters, the shifter stop boss is a stamped washer with a square hole and a bent tab (Fig. 5.45). With more recent shifters (Fig. 5.29), the stop is a cast piece that has a square stop on it that is to be lined up along the down tube projecting forward.

4. **Put on the brass or plastic washer, slip the shifter on, and install the top washer(s) and the screw.**

5.45 Exploded Campagnolo Nuovo Record frictional lever

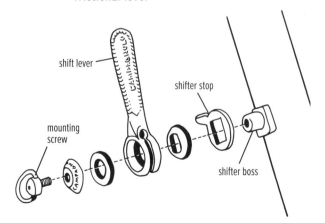

shift lever

shifter stop

mounting screw

shifter boss

Some Shimano left-hand shifters have a return spring in them and must be installed with the lever flipped down in order to fit properly over the stop on the base. It is not until the screw is tightened down that one of these spring-loaded levers will stay in place when rotated counterclockwise to its starting position, pointed forward, parallel with the down tube.

5. **Install the cable at the derailleur.**

CYCLOCROSS SHIFTING CONSIDERATIONS

Cyclocross, in all its dirty glory, is a winter sport often contested in deep mud, snow, ice, and freezing temperatures. Mechanical breakdowns are rampant due to the nature of the sport. You can minimize shifting problems with good maintenance, as well as with good frame and component choice and good planning.

You can reduce shifting problems in general by routing the derailleur cables internally or over the top tube, following the instructions in 5-25 for the front derailleur (whether internal or over the top, the frame must be set up for this cable routing with the correct holes or cable stops) instead of under the bottom bracket where they can get glommed up in the muck or frozen in accumulated snow. The over-top rear shift cable needs to run from the top tube down the right seatstay to the rear derailleur.

Use a short-cage rear derailleur for snappier shifting and a reduced chance of snagging the rear derailleur on course detritus, unless you either insist on using larger cogs than the short derailleur is rated for (see tips in 5-37 to still use a short-cage derailleur) or are running a triple-chainring setup in front (not advisable for 'cross racing; see the Note in 8-15). Tighten the lower knuckle spring tension as in 5-33 so the rear derailleur will keep the chain tight. Do this with short-cage as well as long-cage derailleurs.

You can eliminate front shifting problems with a "1X" (pronounced "one-by"), that is, running only a single front chainring (somewhere in the 38- to 42-tooth range is standard). You have to make sure the chain stays on, though!

Set up a 1X so the chain will stay on with either an X-Sync chainring (whose teeth are tall and alternate fat/thin/fat), or toothless chain-guard rings on each side of the chainring, or with a single outer chain-guard ring and two Third Eye Chain Watchers (Fig. 5.54) or three Deda Dog Fangs clamped around the seat tube, one atop the other (one won't retain the chain as well; if the chain somehow gets under it, you may not be able to get it out without tools and your race may be over).

If you have a single, X-Sync chainring (as with SRAM XX1 or Force CX1, Force 1, or Rival 1), its tall, fat/thin/fat teeth engage the wide/narrow/wide spaces in the chain. If it is paired with a rear derailleur designed for it ("X-Horizon"), forego chain guards. However, without one of these roller-bearing-clutch derailleurs, you may not have sufficient control of chain slack to guarantee chain retention without a guard and inner stop.

If you choose to run a single front chainring, note that if the chain pops off, you won't be able to pedal it back on. Make sure the chain is not too long but is long enough to reach the big chainring without tearing up the rear derailleur—set it up as in 4-8, method 3.

With a front derailleur, run the front cable over the top tube, following the instructions below.

5-25

ROUTING THE FRONT SHIFT CABLE OVER THE TOP TUBE

Unlike front derailleurs for mountain bikes with over-the-top cable routing, road front derailleurs are bottom-pull, not top-pull (i.e., the cable needs to pull from the bottom, not the top). So, if the shift cables go over the top, you need to route the front shift cable down the back of the seat tube and around a little pulley on the back of the seat tube and up to the front derailleur (Fig. 5.46). If your frame does not have a pulley or a threaded hole to accept one, you can buy a pulley that is attached to a band clamp. In a pinch, you can wrap the cable around the chainstay bridge and back up to the front derailleur.

Note that a mountain bike top-pull front derailleur could be an option to avoid a cable pulley, but its cage curvature is designed for smaller mountain bike–size chainrings, so it won't track close enough to bigger road rings to shift well. It also may not match the cable pull of your shifter.

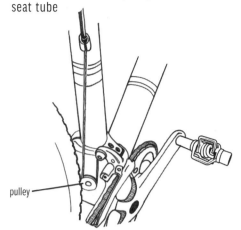

5.46 Over-the-top cyclocross cable routing with front-derailleur cable pulley on the back of the seat tube

pulley

DERAILLEUR MAINTENANCE

5-26

JOCKEY-WHEEL MAINTENANCE

LEVEL With proper attention, the jockey wheels on the rear derailleur will last a long time. They should be wiped off every time you wipe down and lubricate the chain (daily is a good idea). The only other maintenance involved is an overhaul every 1,000–2,000 miles in dirty conditions (more if the bike is pressure-washed); otherwise, an overhaul every time you replace the chain should be frequent enough.

The mounting bolts on jockey wheels also should be checked regularly. If a loose jockey-wheel bolt falls off while you are riding, you'll need to follow the procedure for a damaged rear derailleur in 3-12.

Standard jockey wheels turn on a brass or ceramic bushing. Some high-end models have cartridge bearings. A washer with an inward lip is usually installed on both sides of a standard jockey wheel. Some jockey wheels also have rubber seals around the edges of these washers to keep dirt and grit at bay.

5-27

OVERHAULING STANDARD JOCKEY WHEELS

1. **Remove the jockey wheels by undoing the bolts that hold them to the derailleur (Fig. 5.47A).** The bolts usually take a 3mm hex key.

5.47A Exploded derailleur

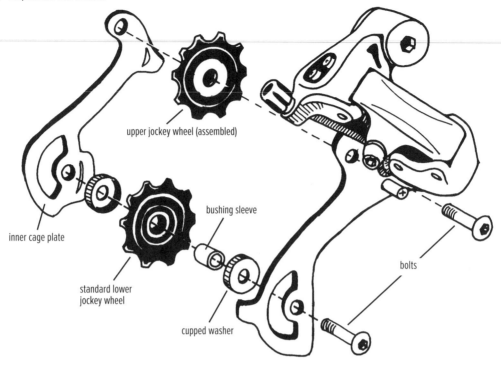

upper jockey wheel (assembled)

inner cage plate

bushing sleeve

bolts

standard lower
jockey wheel

cupped washer

2. **Wipe all parts clean with a rag.** Solvent is usually not necessary but can be used.

3. **If the teeth on the jockey wheels are broken or badly worn, replace the wheels.**

4. **Smear grease over each bolt and bushing and inside each jockey wheel.**

5. **Reassemble the jockey wheels onto the derailleur.** Be sure to orient the inner cage plate properly so that its larger part is at the bottom jockey wheel.

5-28

OVERHAULING CARTRIDGE-BEARING JOCKEY WHEELS

If the cartridge bearings in high-end jockey wheels (Fig. 5.47B) do not turn freely, they can usually be overhauled.

1. **Remove the jockey wheels by undoing the bolts that hold them to the derailleur (Fig. 5.47A).** The bolts usually take a 3mm hex key.

2. **Remove the bearing seals.** With a single-edge razor blade, pry the plastic cover (bearing seal) off one side or—preferably—both sides of the bearing (Fig. 5.48). (Steel covers on bearings cannot

5.47B Exploded jockey wheel with cartridge bearing

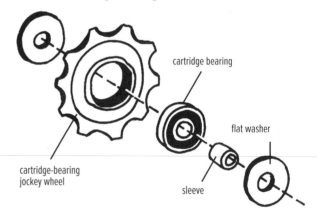

cartridge bearing

flat washer

cartridge-bearing
jockey wheel

sleeve

5.48 Removing a cartridge-bearing seal

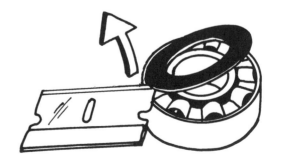

be removed. If such a bearing is not turning freely, replace the entire bearing.)

3. **With a toothbrush and solvent, clean the bearings.** Use citrus-based solvent, and wear gloves and glasses to protect skin and eyes.

4. **Blow the solvent out with compressed air or your tire pump, and allow the parts to dry.**

5. **Squeeze new grease into the bearings and replace the covers.** Use light grease.

6. **Reassemble the jockey wheels onto the derailleur.** Be sure to orient the inner cage plate properly (the larger part of the inner cage plate should be at the bottom jockey wheel).

5-29

UPGRADING JOCKEY WHEELS

You can reduce your bike's drivetrain friction by upgrading the jockey wheels to ones that spin more easily. Simply upgrading from pulleys with bushings (Fig. 5.47A) to cartridge-bearing jockey wheels (Fig. 5.47B) can make a substantial difference. There are many brands available, and they constitute a relatively inexpensive, simple, and quick upgrade.

If you already have cartridge-bearing jockey wheels that are in good shape, you can still upgrade from cartridge bearings containing steel balls to cartridge bearings containing much pricier, harder, smoother, and rounder ceramic balls. The simplest and most advisable way to do a ceramic jockey-wheel-bearing upgrade is to replace the pulleys with complete jockey wheels with ceramic bearings; since the plastic wheel will wear out soon enough anyway, you might as well replace the entire thing. Top-end rear derailleurs already come with ceramic-bearing jockey wheels, and lots of such jockey wheels are available; Friction-Facts.com has comparison tests of frictional drag on different brands.

1. **Remove the jockey wheels by undoing the bolts that hold them to the derailleur (Fig. 5.47A).** The bolts usually take a 3mm hex key.

2. **Reassemble the new jockey wheels onto the derailleur.** Be sure to orient the inner cage plate properly (the larger part of the inner cage plate should be at the bottom jockey wheel).

NOTE: *If you want to just buy a ceramic bearing and upgrade an existing jockey wheel, it can be done with some but not all jockey wheels. You cannot replace a bushing (a metal or ceramic sleeve within the jockey-wheel bore; Fig. 5.47A) with a bearing; you must buy an entire ceramic-bearing jockey wheel.*

Replacing the cartridge bearing

If you have a cartridge-bearing jockey wheel, you must first determine how to get the existing bearing out. Some have no obvious stop against the bearing outside diameter (OD). In that case, find a socket that just fits around the outside of the new bearing (which is the exact same size as the old bearing, right?). Place the new ceramic bearing (of the same size, of course) on a flat surface; set the jockey wheel lined up perfectly on top of it; heat the jockey wheel with a heat gun; set the socket, open end down, on the jockey wheel so it surrounds the bearing; and smack it with a hammer. The old bearing will pop out, and, voilà, the new bearing will have replaced it. Better yet, you can do this in a vise or arbor press.

Some jockey wheels are molded around the bearing. To get the bearing out, you are going to have to push through 1mm or so of plastic that's lapped over the OD of the bearing. Locate two sockets, one whose internal diameter (ID) is just bigger than the bearing's OD, and another whose OD is just smaller than the bearing's OD. With a hand arbor press, a drill press, or a vise, place one socket on one side of the jockey wheel and one on the other, both open toward each other. Apply pressure with the press or vise until the bearing pops out. Clean up the torn plastic edges with a sharp knife and push a new bearing in with the same press or vise, using the old bearing or the smaller socket to do so.

5-30

REAR-DERAILLEUR OVERHAUL

 Except for the jockey wheels and pivots, most rear derailleurs are not designed to be disassembled. If the pivot springs seem to be operating effectively, all you need to do is overhaul the jockey wheels (see previous section) and clean and lubricate the parallelogram and spring, as described next.

REAR-DERAILLEUR WIPE AND LUBE

1. **Clean the derailleur as well as you can with a rag.** Get between the parallelogram plates.
2. **Drip chain lube on both ends of every pivot pin.**
3. **Lube the spring end.** If the derailleur has a clothes-pin-type spring between the plates of the parallelogram (as opposed to a full coil spring running diagonally from one corner of the parallelogram to the other), put a dab of grease where the spring end slides along the underside of the outer parallelogram plate.

REAR-DERAILLEUR UPPER PIVOT OVERHAUL

These steps apply to Shimano and Campagnolo derailleurs.

CAUTION: *Don't undertake this job unless you are installing a new mounting bolt (for instance, a lighter one), or the rear derailleur is so sticky against rotation that a shot of oil won't free it up. The derailleur's strong spring will resist your best intentions at reassembly, and you may have difficulty getting the derailleur back together properly, even with a second set of hands.*

5.49 Rear-derailleur pivots

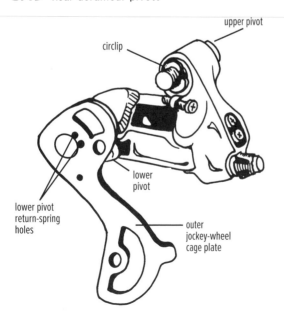

1. **Remove the rear derailleur.** It usually takes a 5mm hex key to unscrew it from the frame and to disconnect the cable.
2. **With a screwdriver, pry the circlip (Fig. 5.49) off the threaded end of the mounting bolt.** Don't lose it; it will tend to fly when it comes off.
3. **Pull the mounting bolt and the upper pivot spring out of the derailleur (Fig. 5.3).**
4. **Clean and dry the parts with or without the use of a solvent.**
5. **Grease liberally, and replace the parts.**
6. **Attach the spring.** Each end of the spring has a hole that it needs to go into. If there are several holes and you don't know which one it was in before, try the middle one. (If the derailleur does not keep tension on the chain well enough, you can later try another hole that increases the spring tension.)
7. **Push it all together and replace the circlip with pliers.** If need be, you can wind the spring before installing the circlip by threading the bolt a few turns onto the derailleur hanger on the dropout. Rotate the derailleur until the flange with the tab or b-screw on it rotates past the stop on the derailleur knuckle, then tighten the mounting bolt until it all pulls together. Then unscrew it off of the dropout while holding the upper pivot assembly together. Now slide in the circlip and snap it in with pliers.

REAR-DERAILLEUR LOWER PIVOT OVERHAUL, JOCKEY-WHEEL CAGE REPLACEMENT, AND SPRING-TENSION ADJUSTMENT

If shifting is sluggish in both directions, the cable is moving freely (5-6 through 5-12), and the cable tension is correct (5-3d), try adjusting the lower pivot spring by turning the b-screw (Fig. 5.4) or Campagnolo's lower-pivot adjustment screw (Fig. 5.9) until the upper jockey wheel is close to the cog but not pinching the chain against it. Perform this adjustment when the chain is on the inner chainring and on the largest cog, as described in 5-3f.

On a Shimano rear derailleur, chain retention and proximity of the guide pulley to the cogs can be further improved by increasing the lower pivot spring tension after disassembling the lower pivot. With the same procedure, you can also replace a bent jockey-wheel cage or increase the capacity of the rear derailleur by substituting a longer cage for a shorter one.

1. **Remove the derailleur from the bike.**

2. **Shimano derailleurs can be divided into two types: those that have a setscrew on the side of the lower pivot, and those that do not.**

   a. If the derailleur has a setscrew (Fig. 5.50), remove it using a 2mm hex key and pull the jockey cage away from the derailleur.

   b. If the derailleur has no setscrew, find and unscrew the tall cage-stop screw on the derailleur cage (Fig. 5.51); it is located near the upper jockey wheel. It is designed to maintain tension on the lower pivot spring and prevent the cage from springing all the way around. Once the stop screw is removed, slowly guide the cage around until the spring tension is relieved. Remove the upper jockey wheel and unscrew the pivot bolt from the back with a 5mm (sometimes 6mm) hex key (Fig. 5.52). Be sure to hold the jockey-wheel cage to keep it from twisting.

3. **Remove the spring.** Determine and mark, if necessary, which hole the spring end is in (Fig. 5.52) before removing it.

4. **Clean and dry the bolt and the spring with a rag.** A solvent may be used if necessary.

5. **Grease all parts liberally.**

6. **Replace the spring ends in their holes in the derailleur body and jockey-wheel cage (Fig. 5.53).** Put the spring in the adjacent hole in the jockey-wheel cage plate if you want to increase its tension. Increasing the lower pivot spring tension pulls the chain tighter; if you have problems with the chain drooping or falling off, or you will be racing cyclocross, increasing the spring tension may solve the problem.

7. **Reassemble the derailleur.**

   a. If the derailleur has a setscrew, push the assembly together, wind the spring, and replace the setscrew (Fig. 5.50).

5.50 Removing and installing lower pivot setscrew from a modern Shimano rear derailleur

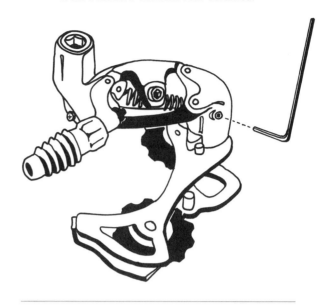

5.51 Removing and replacing the cage-stop screw from an older Shimano rear derailleur

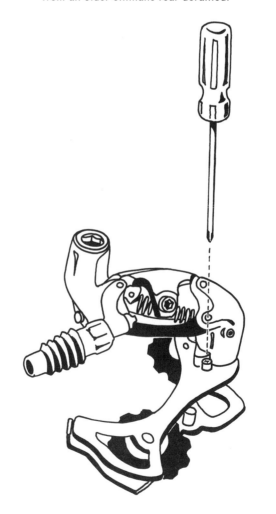

5.52 Removing and replacing the lower pivot center bolt from an older Shimano rear derailleur

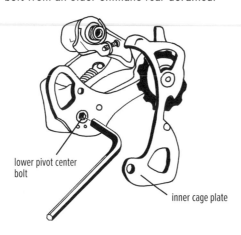

lower pivot center bolt

inner cage plate

5.53 Removing and replacing the lower pivot return spring in one of the two holes in the jockey-wheel inner cage plate

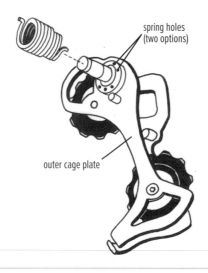

spring holes (two options)

outer cage plate

b. If the derailleur does not take a setscrew, wind the jockey-wheel cage back around, screw it together with the pivot bolt (Fig. 5.52), and replace the stop screw (Fig. 5.51).

5-34

REAR-DERAILLEUR PARALLELOGRAM OVERHAUL

Few derailleurs can be completely disassembled. Those that can (Mavic cable-actuated derailleurs) have removable pins holding them together. The pins have circlips on the ends that can be popped off with a screwdriver.

If you have such a derailleur, disassemble it in a box so that the circlips do not fly away, and note where each part belongs so that you can get it back together again. Clean all parts, grease them, and reassemble.

5-35

REPLACING STOCK REAR-DERAILLEUR BOLTS WITH LIGHTWEIGHT VERSIONS

Lightweight aluminum and titanium derailleur bolts are available as replacement items for some rear derailleurs. Removing and replacing jockey-wheel bolts is simple, as long as you keep all of the jockey-wheel parts together (Fig. 5.47A) and put the inner cage plate back on the way it was. Mounting bolts (Fig. 5.49) are replaced following the instructions outlined earlier in this chapter for overhauling the upper pivot (5-32).

TROUBLESHOOTING DERAILLEUR AND SHIFTER PROBLEMS

Once you have made the adjustments outlined previously in this chapter, the drivetrain should operate quietly and shift smoothly. The drivetrain should stay in gear, even if you turn the crank backward. If you cannot fine-tune the adjustment so that each click with the right shifter results in a clean, quick shift of the rear derailleur, you need to check some of the following possibilities or see Table 5.1. For chain problems of skipping and jumping, see also the "Troubleshooting Chain Problems" section at the end of Chapter 4.

5-36

DERAILLEUR TROUBLESHOOTING

Ensure that the rear derailleur's low-gear limit screw prevents it from going into the spokes before test riding.

Make small adjustments each time before re-checking shifting. Turn cable barrel adjusters one click (or ⅛ turn) each time. Turn limit screws ⅛ turn each time.

If there are multiple options listed for a given problem, perform them in order; if the first fix doesn't work, try the next one.

TABLE 5.1 — TROUBLESHOOTING CABLE-ACTUATED SHIFTING PROBLEMS

REAR DERAILLEUR	
Chain jams between small cog & frame	Tighten cable (5-3d, 5-3e);
	Turn high-gear limit screw clockwise (5-3c).
Rear derailleur touches spokes	Turn low-gear limit screw clockwise (5-3b).
Chain falls between spokes & large cog	Turn low-gear limit screw clockwise (5-3b).
Chain won't go onto large cog	Tighten cable (5-3d, 5-3e);
	Turn low-gear limit screw counterclockwise (5-3b);
	Chain is too short (4-8); replace with a longer one;
	Derailleur cage is too short; replace derailleur or cage (5-33).
Shifting sluggish to a larger cog	Tighten cable (5-3e).
Shifting sluggish to a smaller cog	Loosen cable (5-3e);
	Lubricate or replace cable (5-6 through 5-17);
	Lubricate return spring (5-31).
Shifting sluggish both directions	Replace or lubricate cable (5-6 through 5-17);
	Adjust b-screw (5-3f).
FRONT DERAILLEUR	
Chain falls off to outside	Check derailleur position (5-5a);
	Turn high-gear limit screw clockwise (5-5d).
Chain falls off to inside	Check derailleur position (5-5a);
	Tighten cable (5-5e);
	Turn low-gear limit screw clockwise (5-5c);
	Install inner stop (5-44).
Chain rubs inner cage plate in low gear	Check derailleur position (5-5a);
	Turn low-gear limit screw counterclockwise (5-5c);
	Loosen cable (5-5e).
Chain rubs outer cage plate in high gear	Check derailleur position (5-5a);
	Turn high-gear limit screw counterclockwise (5-5d);
	Tighten cable (5-5e).
Chain rubs a cage plate in cross-gear	Feather (trim) at shifter (5-1b, d, or f);
	Adjust cable tension (5-5e).
Derailleur hits crankarm	Check derailleur position (5-5a);
	Turn high-gear limit screw clockwise (5-5d).
Shifting sluggish to big chainring	Check derailleur position (5-5a);
	Tighten cable (5-5e).
Shifting sluggish to small chainring	Check derailleur position (5-5a);
	Loosen cable (5-5e);
	Turn low-gear limit screw counterclockwise (5-5c);
	Replace or lubricate cable (5-6 through 5-17).

5-37

HOW DO I GET MORE GEAR RANGE?

At some point, everybody gets on a climb they wish they had a lower gear for. If this is a recurring theme for you, maybe it makes sense for you to expand your gear range. In Chapter 11, I discuss compact triple cranks and compact double cranks (a "compact" accepts an inner chainring as small as 34 teeth, rather than the 38- or 39-tooth minimum of standard double cranks).

If you still don't have a low enough gear, then you can address the rear cogs. Rear derailleurs have a specification of maximum tooth-count capacity for the largest cog, as well as a maximum range of teeth over which they can take up chain slack. Modern rear derailleurs for doubles have a lot more range than in the past, even beyond the specified capacity. Current Shimano SS (i.e., short-cage) rear derailleurs have a maximum rear cog size of 28 teeth, a maximum tooth difference on the front chainrings of 16 teeth, and a total capacity of 33 teeth (the sum of the difference in front chainrings, for instance, 50 − 34 = 16, and the total tooth difference over the rear cogs, for instance, 28 − 11 = 17, yielding a total of 33T). The range is similar for SRAM and Campagnolo short-cage rear derailleurs, and all three companies also offer rear derailleurs with a longer jockey-wheel cage and hence a larger maximum rear cog size and a wider total capacity. Shimano GS (long-cage) road rear derailleurs have a maximum rear cog of 32 teeth, a maximum front difference of 16 teeth, and a total capacity of 37 teeth (the sum of the difference in front chainrings, for instance, 50 − 34 = 16, and the total tooth difference over the rear cogs, for instance, 32 − 11 = 21, yielding a total of 37T).

You may want to keep your existing rear derailleur and simply get a larger rear cogset, so what do you do if your cogset is beyond the published range of the rear derailleur? There are three things you can do—one simple and two more complex. The first is to simply tighten the b-screw (5-3f) until the chain runs smoothly on the largest cog. This will rotate the rear derailleur back and move the guide pulley farther from the cogs. I've found on a number of different bikes that by tightening the b-screw (5-3f), 11-speed Shimano SS rear derailleurs will handle an 11–32 cogset, and 11-speed Shimano GS rear derailleurs will handle an 11–36 cogset. You may need a one- or two-link longer chain; check (carefully, in a bike stand!) that the chain is long enough that shifting to the big-big gear combination won't rip apart the rear derailleur.

If it almost works with the b-screw tightened all of the way in but is still noisy on the largest cog, with the guide pulley noticeably bumping up and down over each tooth, you can remove the b-screw and turn it around to get it to rotate the derailleur farther back. You'll need to remove the derailleur to do this. When reinstalling the derailleur, make sure that you twist it back far enough when screwing the mounting bolt into the derailleur hanger that the b-screw is behind the tab on the hanger (5-2, steps 2–4).

Lastly, you can remove the short cage from your derailleur and install a long cage; see 5-33 for instructions.

5-38

STICKY CABLES

Check to see whether the derailleur cables run smoothly through the housing. Sticky cable movement will cause sluggish shifting. Lubricate the cable by smearing it with chain oil or a specific lubricant that came with the shifters (5-16). Do not use grease, which will be too thick to allow free movement. If lubricating the cable with oil does not help, replace the cable and housing (5-7 through 5-15).

5-39

BENT REAR-DERAILLEUR HANGER

A bent hanger will hold the derailleur crooked and bedevil shifting. Instructions for straightening the hanger are in 17-4.

5-40

BENT REAR-DERAILLEUR CAGE

A bent derailleur cage will align the jockey wheels at an angle. Mild bending can be straightened by hand; eyeball the crankset for a vertical reference. A mangled cage can be replaced (5-33).

LOOSE OR WORN-OUT REAR DERAILLEUR

Grab the derailleur and twist it with your fingers to feel for excessive play. Loose pivots, a symptom of a worn-out rear derailleur, will cause the rear derailleur to be loose and floppy. Replace it if it has this problem.

A loose mounting bolt will also mess up shifting by allowing the derailleur to flop around. Tighten the bolt.

DRIVETRAIN COMPATIBILITY ISSUES BETWEEN BRANDS, MODELS, AND SPEEDS

In previous editions of this book, this section was quite long, with lots of history. You can still find it at velopress.com (www.velopress.com/zinnroadsupplement), but given how rare some of the older parts have become, we saved space here by summarizing things more simply below.

Generally, frictional shifters will work with anything, albeit slowly and inefficiently, since the user controls how the derailleur lines up. Indexed (click) shifters, however, require specific derailleurs and cog spacing. Shifting performance will be optimized by using components that were designed to work together as a system—same brand, model, and number of speeds for the shifters, front and rear derailleur, cogs, chain, and, to a lesser extent, the crank. However, some other combinations work.

The rear derailleur's shift-activation ratio—the amount of lateral movement of the rear derailleur divided by the amount of cable pull to generate that amount of lateral movement (i.e., the number of millimeters of lateral displacement of the rear derailleur per millimeter of cable pull)—is built into the derailleur. And the cable pull—the amount the cable moves with each click—is built into the shifter. The cable-pull per shift multiplied by derailleur's shift-activation ratio is equal to the distance the derailleur's jockey wheels move laterally with each shift. To shift properly, this must be equal to the distance from the center of one rear cog to the center of the next (the cog pitch).

Cable pull × Derailleur shift-activation ratio = Cog pitch

Cog pitch is equal to the thickness of a cog (other than the largest or smallest cog) plus the thickness of the spacer separating it from the adjacent cog. Cog pitch decreases as the number of cogs increases. Shift-activation ratios, cable pull, and cog pitches vary not only with the number of rear cogs but also from brand to brand, and even sometimes from model to model within a brand.

Table 5.2 lists cable pull (in millimeters), shift-activation ratio, and cog pitch (in millimeters) based on the number of speeds for the three major component manufacturers.

Complete details on compatibility are on the VeloPress website, but here are a few additional answers to common questions:

- Shimano and SRAM 11-speed cogsets are too wide to fit Shimano/SRAM 10-speed and 9-speed freehubs. The splines are the same, but the 11-speed freehub bodies are 2mm wider.
- Mavic 10-speed freehubs for Shimano/SRAM work with 11-speed cassettes by removing the spacer.
- Shimano 10-speed cogs will fit on Shimano 9-speed freehub bodies. But 9-speed Shimano cogs will not fit Shimano 10-speed aluminum freehub bodies.
- Campagnolo 9-speed freehub bodies have deeper splines that do not fit Campy 8-speed cogs, but its 10- and 11-speed cogsets do fit on the Campy 9-speed freehub bodies. The 5.9mm- to 6.1mm-wide 10-speed chain allows 10 cogs to fit in the same space that used to accept only 9 with the wider 9-speed chain and cogset, and the even narrower 5.4mm-wide 11-speed chain allows 11 cogs to fit in that same space. The thickness of the spacers on either side of each cog has come down to 2.2mm with 11-speed, whereas the width of each chain roller and thickness of each cog have stayed the same. The tooth-to-tooth distance is 4.55mm on Campagnolo 9-speed, 4.15mm on Campagnolo 10-speed, and 3.76mm on Campagnolo 11-speed.
- SRAM, Shimano, and FSA 10-speed cogs fit on all Shimano-compatible 9- and 10-speed freehub bodies other than Shimano 10-speed aluminum bodies (only Shimano 10-speed cogs fit those). All of these 9- and 10-speed cogsets will also fit on Shimano/

TABLE 5.2 — REAR SHIFTING SPECIFICATIONS

BRAND AND NUMBER OF SPEEDS	CABLE PULL	SHIFT RATIO	COG PITCH
Shimano Dura-Ace 7, 8	—	1.9	—
Shimano 6	3.2	1.7	5.5
Shimano 7	2.9	1.7	5
SRAM (1:1) 7 Mountain	4.5	1.1	5
Shimano 8	2.8	1.7	4.8
SRAM (1:1) 8 Mountain	4.3	1.1	4.8
Campagnolo 8	3.5	1.4	5
Shimano 9	2.5	1.7	4.35
SRAM (1:1) 9 Mountain	4	1.1	4.35
Campagnolo 1st-generation 9	3.2	1.4	4.55
Campagnolo 2nd-generation 9	3	1.5	4.55
Shimano 10 Road	2.3	1.7	3.95
Shimano 10 Mountain	3.4	1.2	3.95
SRAM (Exact Actuation) 10 Road/Mountain	3.1	1.3	3.95
Campagnolo 10	2.8	1.5	4.15
Shimano 11 Road	2.7	1.4	3.77
Campagnolo 11	2.6	1.5	3.76
SRAM (Exact Actuation) 11 Road	2.9	1.3	3.79
Shimano 11 Mountain	3.6	1.1	3.9
SRAM (X-Actuation) 11 Mountain	3.5	1.12	3.9

SRAM 11-speed freehubs, but since the freehub body is 2mm wider, you will need a 2mm spacer behind the largest cog.

- In general, 9-speed cranks and chainrings of any brand will work with 10-speed drivetrains from Campagnolo, Shimano, or SRAM. And in general, 10-speed cranks and chainrings will work with 11-speed drivetrains.

- All SRAM Red, Force, Rival, and Apex components are interchangeable, and all SRAM road shifters work with all 10-speed SRAM mountain bike rear derailleurs. For interchangeability among Shimano components or among Campagnolo components, see the website.

- Any 8-speed chain works on any 7- or 8-speed system. All 9-speed chains work with all 9-speed systems. All 10-speed systems require a 10-speed chain, and all 11-speed systems require an 11-speed chain. Experimenting among chain brands is fine.

- Generally, 7- and 8-speed chains are 7.0–7.2mm wide; 9-speed chains are 6.5–6.7mm wide; 10-speed chains are 5.84–6.1mm wide; and 11-speed chains are 5.46–5.74mm wide.

5-43

CHAIN SUCK

Though relatively rare on road bikes, chain suck (where the chain sticks to the chainring and is dragged around until it jams between the chainring and the chainstay) can occur. See 4-14.

5-44

CHAIN FALLS OFF TO THE INSIDE

When shifting from a bigger ring to a smaller one, the derailleur may overshift and throw the chain completely off to the inside. This is more of a problem with stiff shift levers, since they derail the chain so suddenly that it can jump. My recommendation is to always install an inner stop like a Third Eye Chain Watcher (Fig. 5.54), Deda

5.54 Third Eye Chain Watcher

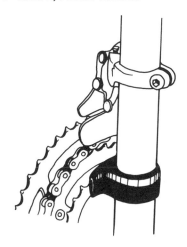

Dog Fang, or N-Gear Jump Stop with any triple crank-set, and it's not a bad idea with a double, too. SRAM, K-Edge, and others make chain stops that mount onto a braze-on front-derailleur, and some bikes come with integrated inner stops that bolt onto the frame.

5-45

CHAIN FALLS ONTO THE GRANNY RING ON A TRIPLE

An inadvertent shift to the smallest chainring on a triple can happen spontaneously when you are riding on the middle chainring and the largest rear cog, and is common on bikes with a 30–39–53 chainring combination. The extra four-tooth difference of the 39–53 outer pair over the outer pair on a 30–42–52 is often more than the system can handle. The derailleur on a 30–39–53 triple sits so much higher relative to the middle chainring than on a 30–42–52 (or on a 30–39–50) that it would have difficulty derailing the chain down to the smallest chainring if the derailleur weren't set to over-shift to the inside, increasing the danger of throwing the chain off to the inside.

Replacing the outer rings with 42–52 chainrings usually fixes the problem. If you don't want to do that, adjust the front derailleur so that its inner cage plate is very close to the chain when it is on the middle ring. Campagnolo Comp Triple cranksets come with 30–40–50 or 30–42–53 combinations, which keep the derailleur a bit closer to the middle chainring. Auto-

dropping to the granny ring is consequently rarely a problem with them.

5-46

CHAINLINE

Chainline is the relative alignment of the front chainrings with the rear cogs; it is the imaginary line connecting the center of the middle chainring with the middle of the cogset (Fig. 5.55). In theory, this line should be parallel with the vertical plane of the bicycle.

5.55 Measuring chainline

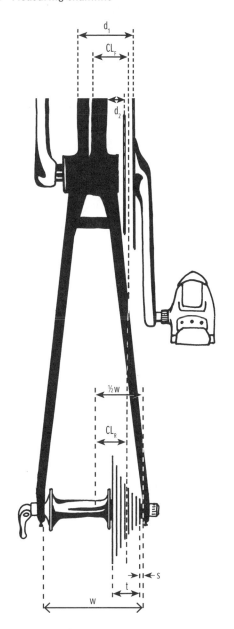

Adjust chainline by moving or replacing the bottom bracket to move the cranks left or right. You can roughly check the chainline by placing a long straight-edge between the two chainrings and back to the rear cogs; it should come out in the center of the rear cogs. (A more precise method is outlined in 5-47.)

If the chain falls off to the inside no matter how much you adjust the derailleur's low-gear limit screw, cable tension, and derailleur position, or you have chain rub, noise, or auto-shift problems in mild cross-gears that are not corrected with derailleur adjustments, a likely culprit is poor chainline (or poor frame alignment).

5-47

PRECISE CHAINLINE MEASUREMENT

You will need a caliper with a vernier scale, a dial, or a digital readout. This measurement only works on a frame with a symmetrical down tube.

The position of the plane centered between the two chainrings relative to the center of the seat tube is often called the chainline, although this is only the front point of the line.

1. **Find the front point of the chainline (CL_F in Fig. 5.55).**

   a. Measure from the left side of the down tube to the outside of the large chainring (d_1 in Fig. 5.55). (Do not measure from the seat tube; this tube is often ovalized at the bottom.)

   b. Measure the distance from the right side of the down tube to the inside of the inner chainring (d_2 in Fig. 5.55).

   c. To find CL_F (the front chainline), add these two measurements, and divide the sum by two.

   $$CL_F = (d_1 + d_2) \div 2$$

2. **Find the rear end point of the chainline (CL_R in Fig. 5.55), which is the distance from the center of the plane of the bicycle to the center of the cogset.**

   a. Measure the thickness of the cog stack, end to end (t in Fig. 5.55).

   b. Measure the space between the face of the smallest cog and the inside face of the drop-out (s in Fig. 5.55).

   c. Measure the length of the axle from dropout to dropout (w in Fig. 5.55); this length is also called the "axle overlock dimension," referring to the distance from end face to end face on either end of the wheel axle.

   d. To find CL_R, subtract one-half of the thickness of the cog stack and the distance from the inside face of the right rear dropout from one-half of the rear axle length.

   $$CL_R = (w \div 2) - (t \div 2) - s$$

3. **If $CL_F = CL_R$ (the rear chainline), the chainline is perfect.** This may not be possible to attain, however, due to considerations of chainstay clearance and prevention of chain rub on large chainrings in cross-gears. CL_F, the rear end point of the chainline, usually comes out between 41mm and 43mm.

   a. Your bike will shift best and run quietest if you get the chainline (CL_F) between 43mm and 45mm. However, this may not be possible on your particular bike because (a) the inner chainring might rub the chainstay; (b) the front derailleur may bottom out on the seat tube before moving inward enough to shift to the inner chainring (this is particularly a problem on bikes with triples and large-diameter seat tubes); or (c) when crossing to the smallest cog from the inner chainring, the chain may rub on the next larger ring (simply avoid those cross-gears).

   b. My general recommendation is to have the chainrings in toward the frame as far as possible without rubbing the frame or bottoming out the front derailleur before it shifts cleanly to the inner chainring.

4. **To improve the chainline, move the chainrings (bikes with traditional bottom brackets only; this is not possible with an integrated-spindle crank).** The chainrings can be moved by using a different

bottom bracket, by exchanging the bottom-bracket spindle with a longer one, or by moving the bottom bracket right or left (bottom bracket installation is covered in Chapter 11).

NOTE: *The chainline can also be off if the frame is out of alignment. If that's the case, it is probably something you cannot fix yourself.*

5. **If improving the chainline does not fix the problem, or if you don't want to mess with the chainline, buy and install an anti-chain-drop device like a Third Eye Chain Watcher (Fig. 5.54), Deda Dog Fang, or N-Gear Jump Stop.** These are inexpensive gizmos that clamp around the seat tube next to the inner chainring. Similarly, SRAM and K-Edge inner stops mount onto a braze-on front-derailleur. Adjust the inner stop's position so that it nudges the chain back on when the chain tries to fall off to the inside.

6

> *All the electronic devices are*
> *powered by white smoke. When*
> *smoke goes out, device is dead.*
>
> —MILAN NIKOLIĆ

TOOLS

2mm, 2.5mm, 3mm,
4mm, 5mm hex keys
7mm, 9mm open-end
wrenches
Flat-blade and Phillips
screwdrivers, small
and medium

OPTIONAL

Park Tool IR-1 Internal
Cable Routing Kit or
Campagnolo equivalent
Shimano TL-EW01 or
TL-EW02 tool

ELECTRONIC SHIFTING SYSTEMS

Even though modern cable-actuated derailleurs shift quickly and precisely with nothing more than the power supplied by your fingers, electronic shifts are faster yet and more powerful.

Electronic derailleurs can also perform some shifts that would be impossible with cable-actuated derailleurs. For example, with simply a touch of a finger, you can upshift an electronic front derailleur from the small chainring to the big chainring while sprinting over the crest of a hill without losing any momentum; that's not possible with a hand-powered shift. With an electronic rear derailleur in a cyclocross race, you can be bouncing down a rough, steep descent you entered at high speed, braking hard for the hairpin at the bottom, and simultaneously shift from high gear all the way to low gear in anticipation of the steep return uphill after the hairpin; that ain't gonna happen with a cable-actuated derailleur!

6-1

OPERATING ELECTRONIC SHIFTERS

Although the operation of the first generation of electronic levers from Campagnolo and Shimano was analogous to what we're accustomed to with cable-actuated shifters integrated into brake levers (5-1), they need not function that way. Unlike a cable shifter that is connected directly to a derailleur, an electronic button is connected to the entire system, and it is the system's programming that mandates what will happen when the shift button is pushed. Apart from limitations such as the number of connection ports on some components, there is no reason why multiple buttons cannot perform the same shift function, which opens the possibility of shifting from many different positions. This goes beyond simply catering to a user's personal preferences; for people with disabilities, it offers options that can allow them to shift as easily as everyone else.

When SRAM introduced their eTap electronic systems, they threw off the confines of having the shifters perform like their cable-actuated brethren. Straight out of the box, both hands control both derailleurs.

Users can now easily reprogram the second generation of Campagnolo's EPS electronic system (the one with an internal battery), including setting it up to work almost identically to SRAM eTap's shifting protocol. With the MyEPS smartphone app, the user can reconfigure which button shifts which derailleur, and how fast the derailleurs perform their shifts.

Shimano Di2 electronic systems can also be programmed to perform shifts differently than how they are set up out of the box. For instance, you can change which button performs which shift of which derailleur, and which derailleurs will be used with which shifters.

With some significant restrictions, Shimano's latest road Di2 components can be mixed and matched with their mountain bike Di2 components, as well as with their previous generations of road Di2 components. For instance, road Di2 derailleurs will work with mountain bike Di2 shifters (on a straight handlebar), with the caveat that both derailleurs must be road Di2 of the same generation. Similarly, if you want to use a much bigger rear cassette on a Di2-equipped road or cyclocross bike than the road rear derailleur can handle, you have to put *both* front and rear mountain bike derailleurs into the system; installing only a mountain bike Di2 rear derailleur will not work. You can also use any Di2 road satellite switches (climbing, sprint, or TT switches) with Di2 road shifters and mountain bike rear derailleurs. But a mountain bike Di2 front derailleur will not handle the 14- or 16-tooth jumps between road-sized chainrings as well as it handles the 10-tooth jumps between the mountain bike chainrings it is designed for.

There are similar limitations to mixing the numbers of speeds or the generations of Shimano components; be aware of this if you are trying to save money by upgrading only part of the Di2 system. For instance, you cannot combine Di2 road 10-speed and 11-speed front and rear derailleurs; they must have the same speed range. (Actually, it may work temporarily to mix, for instance, an Ultegra FD-6770 10-speed front derailleur with an Ultegra RD-6870 11-speed rear derailleur until you update the battery software; after that, it will never work again.) On the other hand, Ultegra ST-6770 Di2 shifters, which were built to be 10-speed but can be reprogrammed with new firmware, will work with a pair of 11-speed front and rear Di2 derailleurs, either Ultegra or Dura-Ace. Similarly, 11-speed shifters, namely ST-6870 (Ultegra Di2) or ST-9070 (Dura-Ace Di2), will work on an 11-speed drivetrain with original Ultegra 6770 10-speed Di2 front and rear derailleurs.

a. Rear Shimano Di2 electronic road shifter (both drop-bar and aerobar types)

Drop-bar shifter

The shift switches on the Di2 electronic lever for drop handlebars (Fig. 6.1A) are oriented along the brake lever blade in approximately the same positions as those shift levers would be on a standard Shimano STI lever (Figs 5.1C, 5.1D). In its standard configuration, the shift paddle upshifts on the right lever and downshifts on the left lever; the brake-lever switch downshifts on the right lever and upshifts on the left lever. So to go to a smaller cog, for example, push the rear, paddle-shaped switch on the right lever. To go to a larger cog, push the longer, forward switch along the edge of the right brake lever.

Shimano also offers satellite shift buttons that plug into the system and offer shift positions in addition to the ones on the brake levers. These are only controllers for the rear derailleur; there are no satellite shifters that operate the front derailleur.

The Di2 "sprint switches" are generally mounted below each brake lever on the inboard side of the handlebar. With the right thumb, push the right sprint button to shift to a smaller rear cog (higher gear). With the left thumb, push the left sprint button to shift to a larger rear cog (lower gear).

The Di2 "climbing switch" is a pod with a pair of buttons that is usually mounted on the top of the bar, just to the right of the stem clamp. In its out-of-the-box configuration, push the big button to go to a higher gear (smaller rear cog) and push the smaller button to downshift to a lower gear (larger rear cog).

Aerobar shifter

Since both shift positions have two buttons each on the Di2 electronic aerobar shifters, to remember which shift button does what on the brake levers as well as on the shifters that ride on the aero extensions (Fig. 6.1B), try this mnemonic device for the standard configuration: The upper buttons are for going uphill and the lower buttons are for going downhill!

In other words, the upper buttons on both sides give you lower gears, and the lower buttons on both sides give you higher gears. For instance, whether your right hand is holding the end of the aerobar or the base

6.1A Shimano Di2 STI right lever

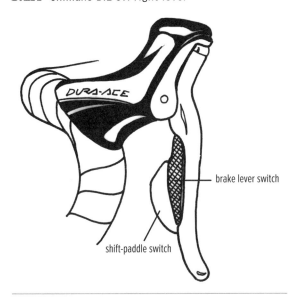

brake lever switch

shift-paddle switch

6.1B Shimano Dura-Ace 7970 Di2 electronic aerobar shifters; SW-R671 shifters shown at ends of aero extensions

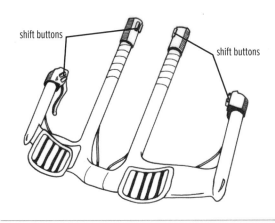

shift buttons

shift buttons

bar, push an upper button to shift to a larger cog (for going uphill) or a lower button to shift to a smaller cog (for going downhill).

If you don't like which button does what, the shift buttons can be reassigned with the eTube diagnostic computer interface.

Shimano also offers single-button Di2 electronic aerobar shifters (model SW-9071). There is only a single button on the top of the SW-9071 unit mounted on each aero extension and each one moves the rear derailleur one way. The standard setup is with the button atop the right aerobar upshifting the rear derailleur (to a smaller cog), and the button atop the left aerobar downshifting the rear derailleur (to a larger cog). Since there are no

front-shift buttons out at the ends of the aerobars, the front derailleur can only be operated from the shifter on the left brake lever.

b. Front Shimano Di2 electronic road shifter (both drop-bar and aerobar types)

Drop-bar shifter

To go to the small chainring, push the rear, shift-paddle switch (Fig. 6.1A) on the left lever. To go to the big chainring, push the longer, more forward button along the edge of the left brake lever. You can reconfigure these commands with the eTube software (PC only).

Don't worry that the front derailleur sometimes moves even though you have not touched the left switches; as you shift through the rear cogs, the Di2 front derailleur automatically moves over at two different gear combinations to avoid chain rub.

Aerobar shifter

Whether at the end of the aerobar or on the brake lever on the base bar, push a left lower button (Fig. 6.1B) to go to the big chainring. Push a left upper button to go to the small chainring. Again, try this mnemonic device: The upper buttons are for going uphill and the lower buttons are for going downhill.

As I mentioned in 6-1a, if there is only a single button on the end of each aero extension, both of them are for shifting the rear derailleur; in that case, the only shift buttons for the front derailleur are on the left brake lever.

As with the rear shifter, if you don't like which button does what, the shift buttons can be reassigned with the eTube diagnostic computer interface.

c. Di2 battery level

To view the battery-level indicator, briefly hold down both shift buttons simultaneously on the right lever; the LED indicator adjacent to the battery symbol on Junction A (Figs. 6.2, 6.3) will light up. The LED will illuminate green, flashing green, red, or flashing red to indicate a full, half-full, quarter-full, or empty battery. If you see red, recharge the battery (6-2a). Shimano also offers an optional wireless computer for Di2, and it has a battery indicator for the shifting system on its screen.

6.2 Shimano Ultegra 6770/6870/Dura-Ace 9070 Di2 Junction A (upper junction box) attached with a band under the stem

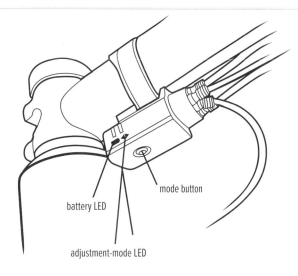

mode button

battery LED

adjustment-mode LED

6.3 Original Shimano Dura-Ace 7970 Di2 Junction A (upper junction box) zip-tied to the front brake cable

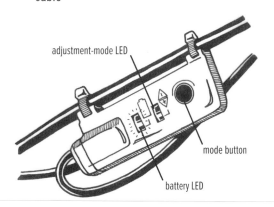

adjustment-mode LED

mode button

battery LED

When the battery runs out of charge, the front derailleur stops working first, then the rear derailleur. They stop at the last gear positions they were in.

If the battery discharges rapidly, you may be due for a firmware update; use the eTube software.

d. Rear Campagnolo EPS electronic shifter (drop-bar, aerobar, and base-bar types)

Drop-bar shifter

The EPS switches (Fig. 6.4A) are the same as on cable-actuated Campagnolo Ergopower (Fig. 5.1B). To shift to a larger cog (lower gear), push the shift paddle (the finger-operated lever behind the brake lever; see Fig.

5.28) to the left (inward) with your finger. Keep pushing it inward for multiple shifts; depending on how long you push the paddle, it will shift only a few cogs or through the entire cog range.

To shift to a smaller cog (higher gear), push the thumb lever down. Hold it down for multiple shifts.

With the MyEPS smartphone app, you can reprogram which button does what; you can even have one level control both derailleurs.

Aerobar shifter

Push the right lever down to shift to a smaller cog (Fig. 6.4B). Pull the right lever up to shift to a larger cog. The lever will return to center after you release it. Hold the lever down or up to shift through multiple cogs up to all 11 cogs.

Base-bar time trial/triathlon shifter

Push the side button on the right brake lever to shift to a smaller cog. Push the button on top of the right brake lever to shift to a larger cog. Holding a button down will shift through multiple cogs up to all 11 cogs.

e. Front Campagnolo EPS electronic shifter (drop-bar, aerobar, and base-bar types)

Drop-bar shifter

To shift to a larger chainring (higher gear), push the shift paddle (finger-operated lever behind the brake lever) to the right (inward) with your finger (Fig. 6.4A). To shift to a smaller chainring (lower gear), push the thumb lever down.

As you shift through the rear cogs, the EPS front derailleur automatically trims its position twice to avoid chain rub.

With the Campagnolo MyEPS smartphone app, you can change this in myriad ways. You can make all shifts come from one lever, for example—the shift paddle can perform downshifts to larger rear cogs, the thumb lever can upshift to smaller rear cogs, and the mode button can shift the front derailleur to whichever front chainring it is not currently on.

You can also program "sprinter" mode: The left shift paddle downshifts to larger rear cogs, the right shift paddle upshifts to smaller rear cogs, the left thumb lever

6.4A Campagnolo EPS electronic Ergopower left lever

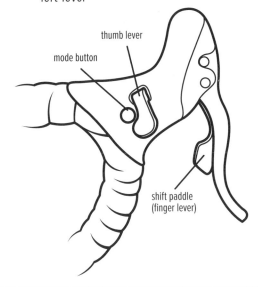

thumb lever

mode button

shift paddle
(finger lever)

6.4B Campagnolo EPS electronic aerobar shifters

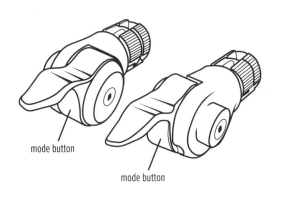

mode button

mode button

shifts to the large front chainring, and the right thumb lever shifts to the small front chainring.

Aerobar shifter

Push the left lever down to shift to the inner chainring (Fig. 6.4B). Pull the left lever up to shift to the big chainring. The lever will return to center after you release it.

Base-bar time trial/triathlon shifter

Push the side button on the left brake lever to shift to the inner chainring. Push the button on top of the left brake lever to shift to the big chainring.

f. EPS battery level

Monitor the battery to make sure the EPS derailleurs will shift for the duration of your ride by pushing and releasing the little mode button on either lever (Figs. 6.4A, B); the LED on the small EPS interface unit on the stem will light up for a few seconds. If the LED glows green, the battery is full; flashing green indicates nearly full; yellow indicates half charge; flashing red indicates low charge, and steady red indicates the need for a charge. Recharging the battery is covered in 6-2a.

g. Front and rear SRAM eTap electronic shifter (both drop-bar and aerobar types)

Like SRAM DoubleTap cable-actuated levers (and unlike Shimano and Campagnolo), eTap has only a single shift switch on each lever. With this single switch, eTap uses paddle-shifting logic like in a racecar: shift up with one hand, and shift down with the other.

Push the right eTap shift paddle (Fig. 6.5A) to shift to a smaller rear cog (higher gear); push the left shift paddle to shift to a larger rear cog (lower gear). Push both shift paddles simultaneously to shift the front derailleur to whichever chainring the chain is currently not sitting on.

Hold down the shifter to shift through multiple rear cogs. When you release the shifter, the rear derailleur stops moving laterally.

6.5A SRAM eTap left lever

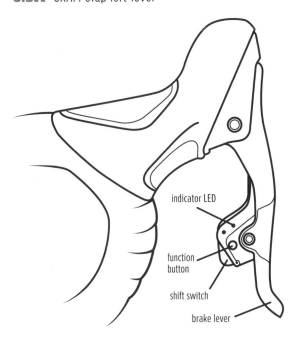

indicator LED

function button

shift switch

brake lever

6.5B SRAM eTap Blips and BlipBox

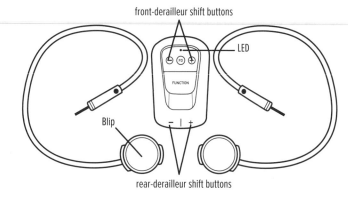

front-derailleur shift buttons

LED

Blip

rear-derailleur shift buttons

SRAM eTap is wireless, but it also offers satellite shift buttons called "Blips" (Fig. 6.5B) that can be plugged into ports on the levers (or to a control box, called a "BlipBox," to run it without the eTap brake levers). The shift logic of the Blips is identical to that of the shifter they are plugged into: pushing a Blip connected to the right shifter shifts to a smaller rear cog (higher gear); pushing a Blip connected to the left shifter shifts to a larger rear cog (lower gear); pushing both Blips simultaneously shifts the front derailleur to whichever chainring the chain is currently not sitting on.

While it normally takes both hands to shift the front derailleur, there is still a fix for someone who wants or needs to use only one hand on the handlebar. By placing a Blip for each shifter side by side on the handlebar, you can perform front and rear shifts with just one hand.

Aerobar shifter

Blips (Fig. 6.5B) are mounted on the ends of the aerobar. Push the right Blip to shift to a smaller rear cog (higher gear); push the left Blip to shift to a larger rear cog (lower gear); push both Blips simultaneously to shift the front derailleur to whichever chainring the chain is currently not sitting on.

Base-bar time trial/triathlon shifter

With tiny aero brake levers mounted on the base bar rather than eTap levers, Blips (Fig. 6.5B) are mounted near the brake levers. Rather than being plugged into eTap levers, they are plugged into a control box (the "BlipBox") that clips into a Garmin computer mount. As with all eTap shifters, push the right Blip to shift to a smaller rear cog (higher gear); push the left Blip to shift to a larger rear cog (lower gear); push both Blips simultaneously to shift the front derailleur to whichever chainring the chain is currently not sitting on. There are also shift buttons for both derailleurs on the BlipBox itself.

Four Blips can be plugged into a BlipBox, two on each side (one pair for the ends of the aerobar and one pair for the base bar). The BlipBox is the shifter, and the Blips are remote switches for that shifter, so the BlipBox will function with zero, two, or four Blips plugged into it.

SRAM's eTap software and firmware are closed; the user cannot reprogram which shifter does what, how many speeds the derailleurs will shift across, or the shifting speed.

h. SRAM eTap battery level

After each shift, an LED indicator light on each component illuminates. On a full charge it glows green, but once the battery has dropped below 25 percent charge remaining, it glows red. Below 15 percent remaining, the LED flashes red.

For the rechargeable batteries on the derailleurs, a green light indicates 15–60 hours riding time remaining. Red means 5–15 hours, and flashing red means you are down to 1–5 hours of function. For recharging instructions, see section 6-2a.

For the non-rechargeable CR2032 watch battery in each shift/brake lever, a glowing green shifter LED indicates 6–24 additional months of riding remaining. A solid red LED indicates 1–6 months remaining, and a flashing red LED indicates approximately 1 month remaining before the battery dies.

To limit battery trickle down, movement sensors in each component automatically put it into sleep mode when you aren't riding. Since movement turns the components on, for extended transportation SRAM recommends that you remove the derailleur batteries and, to keep the battery contacts clean, replace them with the supplied battery port covers. Their red color will remind you to install batteries before riding.

6-2

BATTERY CHARGING AND SHUTOFF

a. Battery charging

Check the battery charge level (6-1c, 6-1f, or 6-1h) at least monthly or every 700km. Charge time for EPS and Di2 batteries is approximately 1.5 hours. For SRAM eTap derailleur batteries, charge time is 45 minutes.

Charge the battery fully before using it the first time. You can charge the battery with any amount of charge left in it. Leaving the battery uncharged for extended periods can damage it, and storing it with at least a half charge in it is preferable. If storing for extended time, recharge it periodically.

Shimano

With a Shimano internal battery, plug the charger into the square port on the side of Junction A (Fig. 6.2). If you have an aftermarket seatpost battery for original Dura-Ace 7970 Di2, you generally will have to remove the seatpost to recharge it (Fig. 6.6). External Di2 batteries (Fig. 6.7) are made to be removed from the bike and charged in a charging pod.

6.6 Installing the Shimano Di2 internal battery into seatpost

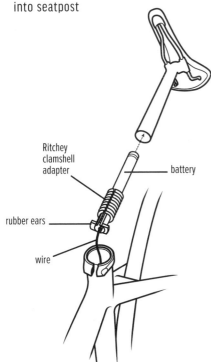

Ritchey clamshell adapter

battery

rubber ears

wire

6.7 Removing the Shimano Di2 external battery: Open the lever, press the release button, and pull the battery out; reverse to install.

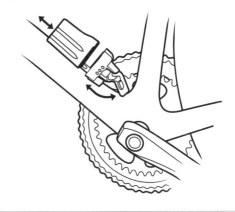

6.8 Installing a Campagnolo EPS V2 seat-tube battery

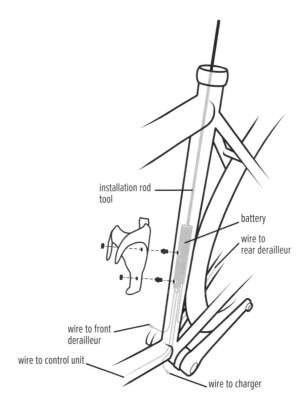

installation rod tool

battery

wire to rear derailleur

wire to front derailleur

wire to control unit

wire to charger

Campagnolo

For EPS charging, locate the charging port. Unscrew the sealing plug from the charging port sticking out of the frame for a V2 internal battery (Fig. 6.8). Pull off the cover over the input/output plug on the more curved end of the external battery (Fig. 6.9). With a V3 internal battery, the charging port is under a rubber cover on the V3 interface attached to the stem. Plug in the charger

6.9 Disconnecting an external Campagnolo EPS battery by inserting the EPS shutoff magnet

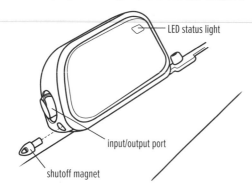

LED status light

input/output port

shutoff magnet

cord, making sure the opposing connectors are properly aligned. With a V2 or V3 battery, the charging connector screws into the charging port.

SRAM

SRAM eTap has two identical rechargeable batteries, one on each derailleur. Flip open the catch atop the front or rear derailleur (Figs. 6.10A, 6.10B) and pull the battery up and out.

If the battery dies on one derailleur, and that's the one you need to get home, you can interchange it with the battery on the other derailleur (or carry a spare battery).

SRAM's recharge cradle obtains power via a micro-USB connector and has four indicator LEDs: The one on the underside next to the power input lights up blue to indicate that it has power, and the three on top indicate increasing battery charge level.

There is also a CR2032 flat watch battery in each handlebar lever that is not rechargeable; replace it within a couple of weeks of when the shift lever's LED first flashes red after performing a shift. These little batteries in the shifters can last two or more years at 15 hours/week of riding time. To replace one, remove the three little screws under the rubber lever hood on the inboard side and open the battery hatch.

b. Battery shutoff/disconnect

Shimano

To disconnect the battery on Shimano Di2, unplug the wire that emerges from the frame at Junction A (Fig. 6.2); you do not have to unplug at the internal battery (Fig. 6.6) or unplug the external battery, if you

have one. To remove an external Di2 battery from its mount, flip open the lever on the back of the mount and press the button on the side of the mount while pulling on the battery (Fig. 6.7). To install the battery, slide it in along its grooves and flip the lever to secure it.

Campagnolo

To disconnect an internal Campagnolo EPS battery, wrap the EPS shutoff magnet strap (a rubber strap with three cylindrical magnets in it) around the seat tube or seatpost (or around the down tube if the battery's in there) where the top of the battery is; that's where its magnetic switch is. The V2 internal EPS battery is bolted to the water bottle bosses, so wrap the magnet strap a couple of inches above the upper bottle boss (very small frames may have the V2 battery upside down in the seat tube to avoid overlap with the seatpost; in this case, wrap the magnet strap around the seat tube a couple of inches below the lower bottle boss). With a V3 battery inside the seatpost, wrap the strap around the post where the top of the battery sits inside the post. If it doesn't shut off the shifting system, move the strap up or down a bit.

To disconnect the external Campagnolo EPS battery, plug in the EPS shutoff magnet (Fig. 6.9), a cylindrical magnet with a plastic key-ring loop on it that plugs into the battery housing on the curved end below the large connector plug.

SRAM

To disconnect a SRAM eTap battery, flip open the catch atop the derailleur and pull the battery up and out (Figs. 6.10A, B).

6.10A Removing a SRAM eTap battery from the front derailleur

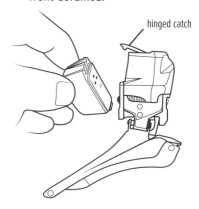

hinged catch

6.10B Removing a SRAM eTap battery from the rear derailleur

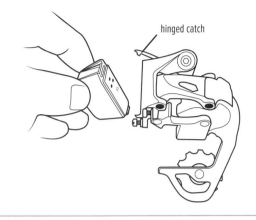

hinged catch

6-3

ADJUSTING ELECTRONIC DERAILLEURS ON THE FLY

You can adjust electronic front and rear derailleurs while you are riding (or while you're stopped). This is useful for fine-tuning and also if you change to a different rear wheel. You simply put the system in adjustment mode and then push (one or more times) the shift switch that moves the derailleur the direction that will eliminate chain rub or improve shifting in that direction.

Campagnolo/Shimano

1. **Push the mode button to go into adjustment mode.** An LED on the monitoring unit attached to the stem will indicate when the system is in adjustment mode. With Shimano Di2, the mode button is on Junction A (Fig. 6.2 on the bottom; Fig. 6.3 on the side); hold the button down until the LED adjacent to the "+/-" symbol on Junction A lights up red. For Campagnolo EPS, hold down the mode button on the appropriate brake/shift lever (Figs. 6.4A, B) until the LED on the EPS interface (attached to the stem) lights up pink (after six seconds).

2. **Push and release the appropriate shift switch.** For instance, if the rear derailleur is slow to shift the chain from a bigger to a smaller cog, push the upshift switch on the right lever once or twice. Each push moves the derailleur a similar amount to a single click of a cable barrel adjuster on a mechanical system (5-3d or 5-5e).

3. **Exit adjustment mode.** Push the mode button again to turn off the LED on the Di2 Junction A or EPS interface.

4. **Shift through the gears.** If shifting is still not ideal, repeat steps 1–4.

SRAM

Hold down the function button on the shifter (Fig. 6.5A, or 6.5B on a time trial or triathlon bike), then execute a shift and let go. Do this on the shifter for the direction you want the derailleur to adjust; for instance, if you want the rear derailleur to move inward slightly more, hold down the function button on the left shifter, and click the shifter once. Each time you do this, it will move the derailleur 0.2mm. There are 13 possible steps, or 2.6mm of total micro adjustment.

INSTALLING ELECTRONIC SHIFTING SYSTEMS

LEVEL Since electronic systems are so different from cable-operated shifting systems, you might have trepidation about installing one of these the first time. However, you may find it to be quicker, if not easier, than installing a cable-shift system. To alleviate consumer worries about durability relative to standard cable systems when converting to electronic shifters, manufacturer warranties tend to be long on electronic parts and wiring harnesses.

6-4

BATTERY INSTALLATION

Make sure that you have the appropriate parts for the bicycle frame. If the frame won't accept internal wires, SRAM eTap or Shimano Di2 with an external battery are your only choices. If you will be using an external Shimano or Campagnolo battery, make sure you have purchased the correct battery mount to fit the mounts on the frame and that there is a nearby hole in the frame sufficiently large to allow all of the wires from the battery to pass through. For Campagnolo EPS, if you will be using the internal battery, the frame must have a hole in the frame down near the bottom bracket for the charging port to stick out.

a. SRAM eTap battery installation

There is an identical battery at the rear of the front and rear derailleurs.

1. **Swing open the catch on top of the derailleur (Figs. 6.10A, B).**
2. **Slide the battery into place, bottom end first.** Ensure that the battery's tab engages the loop on the derailleur body.
3. **Push the top of the battery against the derailleur and swing the catch down to secure it.**

b. Shimano Di2 seatpost battery installation

The Di2 seatpost battery (Fig. 6.6) is cylindrical and very slim. There are a number of ways to secure the battery into the seatpost; here are two, Shimano's method and Ritchey's method.

Except for first-generation Dura-Ace Di2, Shimano components are connected by "E-Tube" wires. The wires are universal, with the same connector at both ends, and they come in a wide array of lengths, every 50mm. They plug into any component in the system interchangeably. Figure your wire lengths (other than the lever wires) to reach from the component, inside the frame, to pop out at the bottom-bracket shell with an inch or two of slack.

Before installing the battery into the seatpost, attach an E-Tube wire of the appropriate length (long enough to extend a few inches past the bottom bracket) to the battery. Using the prongs on the TL-EW02 tool (Fig. 6.15A), pry out the plug in the battery connector. Using the cylindrical end of the TL-EW02 tool, as shown in Figure 6.15A, push the end of the wire into the end of the battery until it clicks.

Shimano

1. **Glue the sleeve into the bottom of the seatpost.** Epoxy or any other glue for metal is fine. Make sure the sleeve (it comes with the battery) is the right size for the seatpost. If you have a PRO-brand Di2 seatpost (PRO is a Shimano subsidiary), this sleeve will already be present in the seatpost.

2. **Insert the battery.** Put the two clamshell adapter pieces around the battery at the wire end and slide it in, wire-end down.
3. **Secure the battery with the snapring.** Sandwich the wave washer between the pair of flat washers, and push the washer sandwich in against the end of the battery. Squeeze the snapring together with snapring pliers (Fig. 1.2), and slide it into the seatpost sleeve so that it snaps into the groove at the end of the sleeve.
4. **Drop the wire from the battery down the seat tube.** You may need to fish it through to the bottom bracket as in 6-6.
5. **Install the seatpost.** Pull out the wire so it sticks out of the bottom-bracket shell a few inches.

Ritchey

1. **Ensure that the adapter is the correct size for the seatpost.**
2. **Put the rubber clamshell Ritchey battery adapter onto the battery (Fig. 6.6).** With the adapter's ears at the wire end, nestle the battery into each half of the adapter. There are grooves and notches that align with corresponding protrusions on the battery.
3. **Push the battery and adapter up into the seatpost.** Keep pushing until the ears are at the end of the seatpost (you will grab these to remove the battery). The adapter's soft ribs will grab the inside of the seatpost and keep the battery secure. If the seatpost is cut at an angle (Ritchey carbon posts are cut this way), don't worry about it; just shove the adapter and battery in as far as you can. It will stay in place just fine.
4. **Drop the wire from the battery down the seat tube.** You may need to fish it through to the bottom bracket as in 6-6.
5. **Install the seatpost.** Pull out the wire so it sticks out of the bottom-bracket shell a few inches.

c. Campagnolo EPS V2 internal battery installation inside seat tube

You will need a special tool to install the long, thin V2 battery inside of the seat tube; the Campagnolo tool is a long, machined rod with an M4 screw thread on one end and a separate extension rod for really long seat tubes (Fig. 6.8). The first time I installed one, I didn't have that tool, so I instead welded an M4 screw to the end of a long steel rod. It has served me well ever since.

The seat tubes on many frames with integrated seat masts are not large enough internally to accept the V2 internal battery; another problem is that the mast may be closed off at the top. In this case, you can mount the V2 battery inside the down tube, but the frame must have an opening between the down tube and head tube large enough to allow slipping the long, thin battery in from the end of the head tube and into the down tube. You will need two different installation tools as well. One is a shift cable with an M4 screw welded to the end. The other is an M4 screw with a hole drilled crosswise through its head that is large enough for a brake cable to easily slide through. This down-tube mounting method is rare enough that I'm skipping it in this book.

If you are upgrading an existing Version 1 Campagnolo EPS system (with an external battery) to a Version 2 (internal) battery, you will need to replace the V1 EPS interface with a V2 EPS interface. This is because the diagnostic function in Version 2 has been moved from the battery to the interface (since indicator lights on the battery would no longer be visible). The V2 interface also allows separate adjustment of the front derailleur over each chainring (V1 only allows one position adjustment), as well as connectivity with the MyEPS smartphone app.

With Campagnolo EPS, you will need wire-entry holes in the frame at least 7mm in diameter to be able to pass the EPS connectors through them. Ideally, the holes should be oval, at least 7mm × 8mm across.

1. **Screw the installation rod into the end of the battery.** Make sure that the battery has an O-ring (supplied) in the groove at that end to silence vibration noise. Note that if you need the extension rod on Campagnolo's installation rod to deal with a super-long seat tube, the threads on the end of the extension rod are left-hand threaded.

2. **Mark the installation depth on the tool rod.** With its wire end down, hold the battery against the seat tube and line up the mounting holes in the battery with the water bottle bosses. Wrap a piece of tape (or mark with a pen) around the installation rod at the top edge of the seat tube. This will allow you to line up the battery holes with the water bottle holes. Also mark on the seat tube where the top of the battery will be; this is where the magnetic shut-off band will go to turn the system off. Alternatively, the battery comes with a paper gauge and a sticker for this. While holding the battery aligned with the bottle bosses, hold the seatpost against the seat tube to the depth it will be inserted, and make sure that the seatpost won't hit the battery. If the seat tube is so short that the battery and seatpost will run into each other, and you cannot shorten the seatpost (because it would not have sufficient insertion into the frame, for example), you will need to install the battery upside down. This will require buying an adapter plate to mount to the bottom (wire end) of the battery that will allow you to screw the installation tool rod into it.

3. **Remove the installation rod.** Unscrew it from the battery and set it aside for the moment.

4. **Wrap the plastic coil around the wire ends.** Two coils to hold wire bundles together are supplied. Use one to wrap around the ends of the shortest wires (charger port and yellow-banded front derailleur) to hold them in place alongside the other wires. Use the other coil to wrap the green-banded wire to the red-banded wire.

5. **Tape the battery against the seat tube.** Turn the battery upside down and tape it so the wire end of the battery is even with the top of the seat tube.

6. **Guide the wires down through the seat tube.** The easiest way is with a magnetic guide tool—a magnet on the end of a shift cable (Fig. 6.12)—but you can easily run these down through the seat tube with a stiff wire from the top or by pulling them down with a wire you run up from the bottom bracket. Guide them out of one end of the bottom-bracket shell.

7. **Untape the battery and screw the installation tool rod into it.**

8. **Holding the end of the rod, slide the battery down into the seat tube.** Stop at the mark on the rod.

9. **Rotate the battery to line up with the bottle bosses.** Looking down into the seat tube, turn the tool so that the hole it's screwed into is offset to

the back, away from the bottle bosses; this ensures that the mounting holes face the bottle bosses.

10. **Screw the longest mounting screw into the battery.** The battery comes with three screws, each of decreasing length. The screws are size M4 and they are thin enough to pass right through the M5 water-bottle holes. Screw the longest screw into the lower bottle mount until the battery is snugged up against the bottle bosses.

11. **Screw in another screw into the battery through the upper bottle mount.** Try the shortest one first. If it won't reach, use the mid-length one.

12. **Tighten the mounting screws with a 9mm open-end wrench.** Be careful; these are small holes threaded into brass, so the tightening torque is low (2 N-m). You can now install a bottle cage onto the protruding studs with a washer and nut inside the cage on each stud. If the stud will hit the bottle, put a (supplied) spacer under the cage. Tighten the nuts to 1.2 N-m.

13. **Wrap the magnetic shutoff band around the seat tube where you marked it in step 2.** This will shut off the system while you complete the installation.

14. **Guide the end of the charging wire out the charging-port hole.** First, remove the spiral wraps from the wires and remove the collar nut from the charger port on the end of the wire. If you can't push the charger port on the end of the wire out through the frame hole with your fingers, you can guide it there: 1) Slide the end of the battery charger's charger wire in through the frame's charger port hole from the outside. 2) Guide it into the bottom-bracket shell. 3) Screw it into the charger port on the end of the battery wire. 4) Pull the charger wire back out through the frame hole until the charger port on the end of the battery wire is sticking perpendicularly out of the frame hole. 5) Without letting the charger port pop back inside of the frame, unscrew the battery charger's wire.

15. **Tighten the collar nut onto the charger port.** There is a small rubber bumper on the charger port inside of the frame, which should keep it tight and silent.

16. **With a screwdriver, install one of the supplied small, hollow plastic screws into the charger port.**

The screw will keep contaminants out of the charger port.

d. Campagnolo EPS V3 seatpost battery installation

1. **Install the battery into the seatpost following the procedure in 6-4b for installing a Di2 battery with a Ritchey ribbed adapter.**

2. **Guide the wires to the bottom-bracket shell.** Do this the same way as for the V2 battery (6-4c).

e. External battery (Shimano or Campagnolo) installation

Regarding Campagnolo, these instructions are for the original water-bottle mount battery (Fig. 6.9). However, with the proper mounting bracket (not included with the battery), the long, thin Campagnolo V2 battery (Fig. 6.8) can also be mounted onto external mounts on the down tube or under the left chainstay. But to use this setup, unless you drill a hole in the frame for the charger port, the wire-entry hole near the battery must be large enough (8mm × 12mm) to not only allow all of the wires from the battery to pass through, but also to allow the fat charging wire to double back and re-emerge from the hole and go to the battery mount where the charging port screws in.

1. **Temporarily install the battery bracket to the frame mounts.** For internal wiring through a hole in the frame between the mounts, partially screw in the bolt at the top of the mount to access the hole in the frame for wires connected to the battery. For bikes without a specific battery mount, bolt it underneath the bottle cage on the down tube (Figs. 6.7, 6.9). Use the long mounting plate for a water-bottle mount; use the short mount for bolt-on mounting under the down tube or left chainstay. With a Shimano Di2 long bottle mount, ensure that there is at least 108mm from the base of the mount to the bottom of the bottle cage; this allows enough room to slide the battery up toward the bottle sufficiently to remove it. A saddle bag with a battery inside requires no battery mount but does require a long battery wire routed through the seatpost and seat tube.

2. **Push the wire(s) into the hole in the frame.** This does not apply to original Di2 external routing. Guide the wire(s) out one end of the bottom-

bracket shell. The easiest way is with a magnetic guide tool—a magnet on the end of a shift cable (Fig. 6.12)—but you can easily run these down through the seat tube with a stiff wire from the top or by pulling them down with a wire you run up from the bottom bracket. Guide them out of one end of the bottom-bracket shell.

3. **Tighten the battery to the mounts.** Use supplied spacers between the mounts and the frame (with Campy, first bolt the EPS mount to the battery).

6-5

INSTALLING ELECTRONIC DERAILLEURS AND SHIFTERS

 Make sure that you have the correct front-derailleur adapter—if one is required—to fit the seat tube.

The motors that drive electronic derailleurs are extremely powerful, and if the front-derailleur upshift button is pushed when your finger is between the derailleur and the large chainring, you will be regretting having had it there, and passersby will hear some colorful language. For your own safety, rotate the cranks whenever shifting, and disconnect the battery whenever installing electronic components (6-2b).

6.11 Shimano Di2 rear derailleur

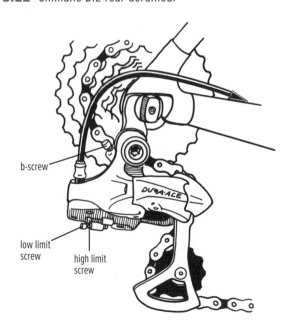

b-screw

low limit screw

high limit screw

1. **Install the rear derailleur (Fig. 6.11) just as in 5-2.**
2. **Loosely mount the front derailleur.** Set it above the height where the top of the chainring will be so that it won't interfere with the crank when you install it.
3. **Install the levers as in 5-18, and install the brake cables as in 9-4.** Di2 levers have a lever-reach adjustment on the back upper part of the lever body, under the rubber hood; turn it clockwise with a screwdriver to decrease lever reach, and vice versa. Adjust reach on SRAM eTap brake levers to one of four possible positions with a 2.5mm hex key under the outboard side of the shifter hood. The shift paddle moves in or out with the brake lever, rather than having its own return spring and reach adjustment as on SRAM DoubleTap levers (Fig. 9.15). Campagnolo's lever reach can't be decreased, but it can be increased by installing the "big hands" insert (Fig. 5.37) under the front edge of the lever body.
4. **Install the Di2 upper junction/EPS interface.**
   a. **SRAM:** Skip this step; eTap is a wireless system.
   b. **Shimano:** Secure Di2 Junction A up near the handlebar. The current modular E-Tube Junction A (Fig. 6.2) connects to the stem with a supplied strap as below. The first-generation Junction A zip-ties to a brake cable (Fig. 6.3).
      » **Wrap the strap over the stem shaft.**
      » **Hook a slot on each strap end over a hook on either side of the Junction A mount.** Orient the mount's short end forward.
      » **Trim the ends of the strap.**
      » **Slide Junction A in from the side until it clicks in.** The wire ports will point forward. To remove Junction A, push the little release tab on the rear of the mount, and slide the junction box out from the side.
   c. **Campagnolo:** Attach the EPS interface under the stem with the long, round cross-section rubber band.
      » **Put a twist in the rubber band and slide it over either end of the stem.**
      » **Hook the band over the tabs on the interface.** The interface will hang under the stem.

6-6
WIRE INSTALLATION

Ensure that you have wires of sufficient length.

Campagnolo wires are attached to the battery and the EPS interface. If they're not long enough, you can purchase Campagnolo extension wires.

Except for first-generation Dura-Ace Di2, Shimano E-Tube wires are universal, with the same connector at both ends. You can get these wires in a wide array of lengths, every 50mm. They plug into any component in the system interchangeably. Figure your wire lengths (other than the lever wires) to reach from the component, inside the frame, to pop out at the bottom-bracket shell with an inch or two of slack.

Di2 considerations

Lever wires for drop bars run about 350mm to connect the levers to Junction A with a little bit of slack. Allow extra slack in the Di2 E-Tube derailleur and battery wires on a coupled travel bike—slack in the wires eases frame disassembly.

If you're using supplementary Shimano Di2 satellite "sprint" switches for drop handlebars, snap them around the handlebar below the brake levers, and plug their wires into an open plug in each lever body.

The Di2 "sprint switches" mount to the inboard side of each curved drop of the handlebar so that the rider can shift the rear derailleur with each thumb while sprinting in the drops of the bars. Generally, the left sprint shifter shifts to a larger cog, and the one for the right hand shifts to a smaller cog, but you can reverse this with Shimano's eTube computer software interface (see 6-8c, "Di2 Diagnostics").

The Di2 satellite (a.k.a. "climbing") switch is a pod with a pair of buttons that mounts on the top of the bar, adjacent to the stem clamp. It allows the rider to shift the rear derailleur while grasping the tops of the bar. You need a five-port Junction A (Fig. 6.2) to plug it into, with the first-generation climbing switch plugged into the right lever. The taller button shifts to a higher gear (smaller cog) and vice versa, although it can be reprogrammed to work the other way around. To update firmware or customize the electronics, download "E-TUBE PROJECT" on e-tubeproject.shimano.com.

Like the satellite switch, current Di2 aerobar shifters (Fig. 6.1B) plug into the upper junction box A (Fig. 6.2), rather than into the levers like Di2 sprint switches. To use them, you need the five-port Junction A, rather than the three-port one.

a. EPS internal routing

Campagnolo and Park Tool each offer a super-useful shift cable with a magnet on the end for guiding the electrical wires through frames and aero handlebars. In addition to EPS wires, you can also use it to guide Di2 wires as well as steel cable housing or hydraulic hose through frames and handlebars. The Campy magnet kit comes with the cable with the magnet on the end (which I'll call the "fishing cable" and apply it to either Campy or Park), a cable with an EPS connector on one end and a magnet on the other end (the "lead cable"), and a separate large cylindrical magnet (Fig. 6.12); the Park IR-1 kit (Fig. 5.30) includes all that plus a cable with a threaded barb on the end opposite the magnet for pulling cable housing or hydraulic brake hose through a frame. Park's lead cable is also a full-length shift cable with a magnet on one end and a rubber sleeve on the end that slides over an EPS or Di2 connector and holds onto it.

6.12 Campagnolo EPS magnet kit for guiding internal wires through the frame

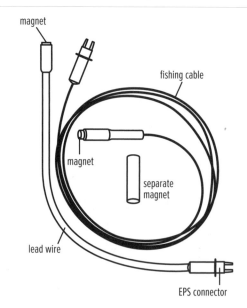

magnet

fishing cable

magnet

separate magnet

lead wire

EPS connector

EPS wires are color-coded: green = rear derailleur, yellow = front derailleur, red = EPS interface, purple = right Ergopower lever, and blue = left Ergopower lever. Arrows or white dots on the male and female ends of the connectors must line up; if need be, mark them with a silver or white paint pen so you can see them easily.

1. **Magnet end first, push the Park or Campagnolo fishing cable in through the hole near the dropout for the rear-derailleur wire.** Shove it through until the magnet appears in the bottom-bracket shell.

2. **Plug the end of the green-banded wire from the battery into the end of the Park or Campagnolo fish-kit lead cable.** This cable will already be in the bottom-bracket shell after 6-4b or 6-4c.

3. **Bring the magnets of the lead cable and the cable from the dropout together.** They should stick together.

4. **Pull the cable out through the hole near the drop-out until the lead cable and the first few inches of the green-banded wire emerge.**

5. **Plug the green-banded connector into the (green-banded) wire emanating from the rear derailleur.**

6. **Magnet end first, push the fishing cable in through the hole in the frame near or on the head tube.** If need be to get around a corner, use the

6.13 Guiding Campagnolo or Park magnetic fishing cable from the head tube into the down tube with external magnet

separate magnet on the outside of the frame to guide the magnet end of the fishing cable from the head tube into the down tube (Fig. 6.13). Push on the cable until the magnet end appears inside the bottom-bracket shell.

7. **Plug the end of the red-banded wire from the battery into the end of the lead cable.** The red-banded wire will already be waiting in the bottom-bracket shell after 6-4b or 6-4c.

8. **Bring the magnets of the lead cable and fishing cable together so they stick.**

9. **Pull the fishing cable out through the hole in or near the head tube until the lead cable and the first few inches of the red-banded wire emerge.**

10. **Connect the red-banded wire from the battery to the red-banded wire from the EPS interface.**

11. **Attach the EPS interface.** This will generally be under the stem. Use the supplied rubber band; double it and slip it over the stem from either end (remove it from the handlebar or steering tube to do so), and hook it on the tabs on either end of the interface. On a time trial/triathlon bike, zip-tie the EPS interface to one of the aerobar extensions.

12. **Magnet end first, push the fishing cable in through the hole in the seat tube near the front derailleur.** Shove it in until the magnet appears in the bottom-bracket shell.

13. **Plug the end of the yellow-banded wire from the battery into the end of the lead cable.** The yellow-banded wire will already be waiting in the bottom-bracket shell after 6-4b or 6-4c.

14. **Bring the magnets of the fish-kit lead cable and fishing cable together so they stick.**

15. **Pull the end of the cable out through the hole in the seat tube until the lead cable and the first few inches of the yellow-banded wire emerge.**

16. **Plug the yellow-banded connector into the (yellow-banded) wire emanating from the front derailleur.**

17. **Peel back the outer bottom edge of the rubber hood on each Ergopower lever.** Slide a thin screwdriver or hex key under it and hold it up to access both tiny screws holding the wire cover onto the lever body.

6.14 Opening the wire cover on Campagnolo EPS Ergopower right lever to access wires

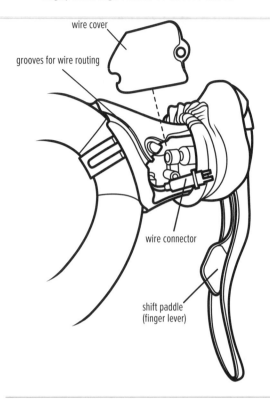

wire cover

grooves for wire routing

wire connector

shift paddle (finger lever)

18. **Remove the wire cover from the outboard side of each lever body (Fig. 6.14).** It takes a small Phillips screwdriver.

19. **Plug in the wire from the EPS interface into each lever.** The purple-banded wire plugs into the purple-banded wire in the right Ergopower, and the blue-banded wire plugs into the blue-banded wire in the left Ergopower.

b. Di2 internal routing

NOTE: *If the bike frame has internal routing holes for shift cables rather than for electrical wires, you will still have to route the electrical wires externally (see 6-6c) if you can't feed them through or if the tube openings into the bottom-bracket shell are not large enough to accommodate the wires and Junction B (Figs. 6.15A–C).*

1. **Route all of the wires into the bottom-bracket shell (Fig. 6.15A).** The wires emanate from 1) Junction A (Figs. 6.2, 6.3) at the handlebar, 2) the front derailleur (Fig. 6.16A), 3) the rear derailleur (Fig. 6.11), and 4) the battery (Figs. 6.6, 6.7).

Thicker, first-generation wires have thin zip-ties pre-attached periodically along their length to prevent the wires from rattling inside the frame, but they also make sliding the flexible wires through the frame more difficult; next-generation modular E-Tube wires have no zip-ties. When pushing a zip-tied wire in through a hole in the frame, push the end in so that the zip-ties around the wire fold back rather than prevent the wire from going through the hole. Do not be afraid to shorten a zip-tie or two to make the task more doable.

The Park Tool IR-1 magnetic cable fish kit (Fig. 5.30) works fantastically well to feed the wires through the frame (you can also use the analogous Campagnolo kit (Fig. 6.12), and tape the Shimano wire ends to the connection on the lead cable meant for a Campy wire end); follow instructions in 6-6a under "EPS internal routing." Otherwise, you can try pushing in a stiff wire from the other end with a hook bent into the end, tape the hook to the wire end, and pull the Di2 wire through.

2. **Pull the ends of the wires out of the bottom-bracket shell and plug all of them into Junction B (Fig. 6.15A)—the lower junction box—until they click.** Use the TL-EW02 or TL-EW01 tool to fully insert them. Slip the end of the wire into the slender plastic TL-EW02 tool (Fig. 6.15A) for Ultegra 6770/6870/Dura-Ace 9070 Di2 and TL-EW01 for

6.15A Connecting Di2 wires at the lower junction box (Junction B) at the bottom bracket

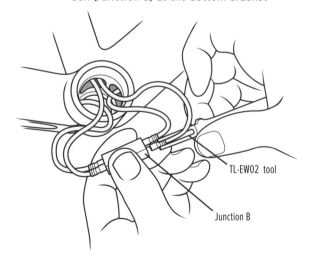

TL-EW02 tool

Junction B

6.15B Shove the lower junction box (Junction B) into the down tube.

6.15C Ensure that the wires do not interfere with the bottom bracket.

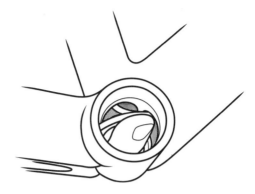

original Dura-Ace 7970 Di2 so that the projection on the connector is aligned with the groove on the narrow end (Ultegra 6770/6870/Dura-Ace 9070 wires and connectors are incompatible with original Dura-Ace 7970 Di2). On first-generation Dura-Ace 7970, slide the supplied shrink tubing over the connection and heat it with a blow dryer to seal water out; subsequent-generation Di2 connections don't require shrink wrap.

3. **Plug in the wires from lower junction box B to the front derailleur, rear derailleur, and battery.** They are labeled FD and RD on first-generation Dura-Ace 7970 Di2 wires, whereas all Di2 wires since then are universal and available in many lengths. Shove Junction B up into the down tube (Fig. 6.15B) so that the only wires visible inside the bottom-bracket shell are the wires going to the front and rear derailleur (Fig. 6.15C), and, in the case of a seatpost battery, a battery wire going up into the seat tube.

c. Di2 external routing (first-generation Dura-Ace 7970 only)

1. **Route all of the wires along the outside of the frame to the bottom bracket.** The wires emanate from 1) Junction A (Fig. 6.3) at the handlebar stem, 2) the front derailleur (Fig. 6.16A), 3) the rear derailleur (Fig. 6.11), and 4) the battery (Fig. 6.7). Route the rear-derailleur wire under the chainstay so the chain can't drop on it.

2. **Provisionally tape the wires to the frame.** Later, cover the wires with the supplied adhesive cover strips (after cleaning the area with alcohol).

3. **Take up any wire slack in the Di2 Junction B looping pegs (under the cover).** You can wind 120mm of slack into it.

4. **Bolt Junction B to the threaded cable-guide hole under the bottom-bracket shell.** Torque is 1.5–2 N-m.

5. **Plug in the wires from lower junction box B to the front derailleur, rear derailleur, and battery.** They are labeled FD and RD on first-generation Dura-Ace 7970 Di2 wires.

d. Connecting Di2 wires to shifters

1. **On Di2, plug the wires from Junction A (Figs. 6.2, 6.3) into the levers.** Peel back the rubber hood, lift the terminal cover hatch, and plug the electrical wire end into the lever terminal with the TL-EW02 or TL-EW01 tool until it snaps in with a click. Use either terminal; the remaining terminal is for a sprinter switch, except on hydraulic Di2 levers with only one terminal.

   If you're using Di2 sprint or satellite switches, plug them in now—plug one sprint switch into each lever; plug a (climbing) satellite shifter into Junction A.

   Cover unused terminals with dummy plugs, again using the TL-EW02 or TL-EW01 tool. Close the hatch and flip the rubber lever hood back down.

   When disconnecting wires, pry connectors off with the other end of the TL-EW02 or TL-EW01 tool, flat side against the component you're disconnecting from.

6-7

COMPLETE COMPONENT INSTALLATION

1. **Install the bottom plastic sleeve and bracket bearings (Chapter 11).** Ensure that both derailleur wires (and battery wire for a seatpost battery) pass above the sleeve (Fig. 6.15C), and the bearing or bearing cup threads do not impinge on the rear-derailleur wire.

2. **Install the crankset (Chapter 11).**

3. **Align the front derailleur (Figs. 6.10A, 6.16B, 6.17) as in 5-4 and 5-5.**

   a. On an eTap system, line up and adjust the front derailleur over the big chainring using the scribed marks on the cage as described in 5-5f for a SRAM Yaw front derailleur (Figs. 5.16, 5.17).

   b. On an EPS front derailleur, do this just as you would a cable-operated one (Figs. 5.12–5.14). The nut takes a 7mm open-end wrench (Fig. 6.17).

   c. On a Di2, rotate the derailleur so that the tail of the outer cage plate is slightly inboard of the tip.

   d. While not necessary when using the band-clamp adapter, with Shimano Di2 front derailleurs mounted to a frame braze-on, adhere the supplied metallic seat tube protector to the seat tube so that the portion of it without foam adhesive sits behind the tip of the front-derailleur support bolt to prevent it from driving straight into the seat tube (Fig. 6.16B). Choose the flat or curved protector depending on seat-tube shape at the support bolt contact point (Fig. 5.18).

   e. With a 2mm hex key, as shown in Figure 5.19, turn the Di2 support bolt (Fig. 6.16A) to align the front-derailleur outer cage plate parallel with the large chainring. If it's already aligned, just tighten the bolt until its tip contacts the protector; this increases the rigidity of the front derailleur during shifting.

   f. Skip to step 7 if the bike has an internal battery and to step 6 with SRAM eTap.

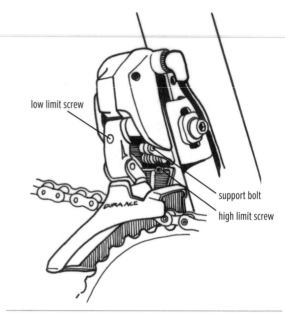

6.16A Shimano Di2 front derailleur

low limit screw

support bolt

high limit screw

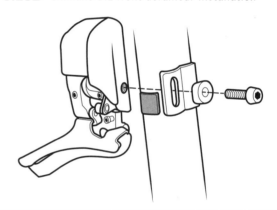

6.16B Shimano Di2 front-derailleur installation

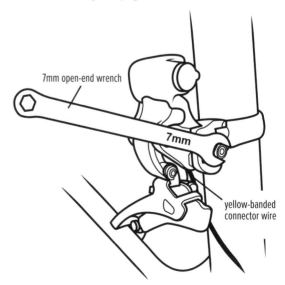

6.17 Installing Campagnolo EPS front derailleur

7mm open-end wrench

7mm

yellow-banded connector wire

4. **If you haven't already done so, install the additional screw(s) in the external battery mount and tighten it down.** The external EPS battery (Fig. 6.9) requires a rubber vibration washer under its tail, whereas the bottle-mount type Di2 battery (Fig. 6.7) requires a zip-tie at its tail (or a third bolt hole in the frame) to keep the battery from bouncing around.

5. **Connect the battery.** Plug in the external Di2 battery (Fig. 6.7), or remove the EPS shutoff magnet from the internal or external battery (Figs. 6.8, 6.9). If the bottom of the seat-tube water bottle hits the external battery on the down tube, Shimano offers an accessory spacer (SM-BA01) that offsets the seat-tube cage upward; its position is adjustable from 32 to 50mm higher than with the cage bolted straight into the seat tube.

6. **Pair the SRAM eTap components.** Since the system is wireless, the components have to recognize each other in order to function. Start at the rear derailleur (the master component) and hold down the function button until its green LED flashes slowly, which indicates that you've opened a pairing session. Hold down the function button on the front derailleur until the LED on the rear derailleur flashes quickly, indicating that they've paired. Repeat with each shifter, or with the BlipBox (Fig. 6.5B) on an aero time trial or triathlon bike that has no eTap levers. You have a 20-second window for each pairing before the rear derailleur LED turns off and the pairing session times out. Once all items have been paired, push the function button on the rear derailleur to go out of pairing mode (after 20 seconds, it will automatically).

7. **Test the system.** Press the lever switches and ensure that the derailleurs move.

8. **Install the chain as in Chapter 4.** With eTap, skip to step 11.

9. **Clean up the wiring.** On EPS, push the connectors and wire slack into the side compartment on the Ergopower lever bodies, route the wire through the grooves (Fig. 6.14), and replace the compartment covers. Wrap the curly plastic wire cover on the exposed wires at the external battery.

   Tape the shifter wires to the handlebar.

6.18 Installing sealing grommet on wire entry into frame

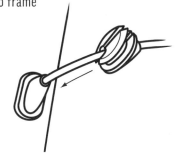

10. **On internally routed wires, slip a rubber grommet around each wire where it protrudes from the frame.** Push the grommet into the hole (Fig. 6.18) until it engages properly and creates a seal. Grommets come in different shapes and sizes depending on frame hole shape and wire size. The standard frame hole size is 8mm × 7mm, with the hole for the battery wires being 14mm × 7.7mm, but Shimano also offers round 7mm grommets. Frame manufacturers using non-standard hole sizes will generally supply grommets with the frame. Hardware stores carry grommets if you have need of an odd size. Make sure there is enough slack in the wire from the upper junction into the frame to allow the handlebars full range.

11. **After reading the following important details, continue to initial system adjustment: 6-8 for Di2, 6-9 for EPS, and 6-10 for eTap.**

Other electrical details

Battery care

Whenever replacing or reinstalling electronic components, disconnect the battery. Wait at least 10 seconds before reconnecting it. The system must individually recognize each component of the electronic shifting system to operate correctly.

As with any battery, do not wet the battery or charger terminals; do not subject the battery to temperatures in excess of 140°F (60°C) or cover the battery or charger when charging; do not allow metal objects or wet objects to connect across the battery or charger terminals; do not use if the battery shows signs of leakage; and do not touch the leaking acid. Use an AC charger only with AC voltages within the range printed on the unit. Do not use the charger with electrical transformers designed for overseas use.

Sealing out the elements

Although EPS, eTap, and Di2 components are designed to be fully waterproof and able to withstand wet-weather riding conditions, avoid submerging in water or solvents and do not subject to high-pressure spray, as in a car wash. Avoid lubricating the rear-derailleur links, as some lubricants can damage the O-rings that seal the electronic components inside. When washing the bike, ensure that all rubber sealing grommets (Fig. 6.18) are installed at wire-entry points and external electrical connections are fully snapped in place or shrink-tubed to properly seal out water.

Each electrical connection on EPS and on 11-speed Di2 has an O-ring seal that requires no heat-shrink tubing over the connection like first-generation Dura-Ace 10-speed Di2 did. Ultegra and 11-speed Dura-Ace Di2 wires are completely interchangeable and can be obtained in different lengths. For Campagnolo, if an EPS wire length is not long enough, install an EPS extension wire with the same connectors.

Secure external electrical wires with tape or zip-ties to prevent snagging them.

On Di2, always connect and disconnect electric wires with the Shimano TL-EW02 (Ultegra 6770/6870/Dura-Ace 9070) or TL-EW01 (Dura-Ace 7970) special tool; make sure the connection snaps together with an audible click to ensure a proper water seal. EPS connectors snap together by hand, but it is recommended to use the Campagnolo UT-CG020EPS tool to disconnect them. Inserting the UT-CG020EPS's attached fork between the tool and the connector and rotating it clockwise pries the male end out; the pin pushes the female end out of the tool.

Crash protection

If an EPS, eTap, or Di2 rear derailleur is hit in a crash, it protects itself by uncoupling the servomotor from the mechanical shaft, and the derailleur moves inward. You may know this has happened if it won't shift properly.

With eTap, the derailleur's spring will pop it back into position, or you can give it a little push if it didn't. You don't have to do anything else; it will start working again.

To re-couple an EPS derailleur, you can repeatedly press the upshift button without pedaling until it hooks up and will again allow shifting to the smallest cog.

To re-couple a Di2 servomotor with its derailleur link, hold down the button on Junction A for five seconds or more.

Alternatively, you can stop and engage the two parts of Campagnolo and Shimano rear derailleurs by pushing inward on the derailleur body with your hand until you hear it click back into place. Unless the derailleur hanger was also bent in the crash, it should now shift properly again.

This uncoupling feature can be used to set the rear derailleur on any cog to ride back home when, for instance, the wire has been cut in a crash, or the battery has been completely discharged.

6-8

ADJUSTING SHIMANO DI2 ELECTRONIC DERAILLEURS AND SHIFTERS

LEVEL Front and rear Di2 derailleurs should be adjusted with the bike in a stand.

a. Di2 rear derailleur

1. **Shift to the fifth largest rear cog.** See 6-1a for shifting instructions.
2. **Hold the mode button down on Junction A (Figs. 6.2, 6.3) until the LED adjacent the "+/-" lights up red.** The system is now in adjustment mode.
3. **Tap the right-hand upshift or downshift switch until the jockey wheel lines up straight under the cog (Figs. 5.5, 5.6).** Listen for chain noise while turning the cranks. Shimano's suggested method (not possible while riding) is to tap the downshift switch until the chain makes noise against the next largest cog. Then tap the upshift switch four times to center the guide pulley under the cog.

NOTE: *In the adjustment mode, each push of a shift switch moves a derailleur less than a single click of a barrel adjuster on a cable-shift system would.*

4. **Push the button on Junction A again until the LED turns off.**

5. **Rotate the cranks and shift through all of the cogs.** Check for silent operation. If not shifting perfectly, repeat steps 1–4.

6. **Shift to the largest cog.**

7. **Tighten the low-gear limit screw.** This is the inboard screw (Fig. 6.11). Tighten it until it touches the link and ensures that the derailleur cannot shift into the spokes. If you tighten it too tightly, it either won't shift to the largest cog or it will continue to run the motor, wearing down the battery rapidly; back off the low-gear limit screw appropriately if that happens.

8. **Shift to the smallest cog.** It will overshift and then come back to center under the cog.

9. **Tighten the high-gear limit screw.** This is the outboard screw (Fig. 6.11). Tighten it until it touches the link, and then back it off (counterclockwise) one full turn. This allows for the overshift-and-return shift method the derailleur utilizes, yet it still ensures that the derailleur cannot shift into the dropout.

10. **Set the b-screw as in 5-3f.** By turning the crank backward in both the small-big (front-rear) and small-small combinations, check that the guide pulley is as close to the cog as possible without bumping up and down as the chain moves.

NOTE: *To be able to shift through the wider gear range of a mountain bike cogset, you can substitute an XTR Di2 mountain bike rear derailleur for either an Ultegra Di2 or Dura-Ace road 11-speed Di2 rear derailleur. See section 6-12 for more details.*

b. Di2 front derailleur

1. **Shift to the inner chainring and largest rear cog.** See 6-1b for shifting instructions.

2. **With a 2mm hex key, turn the front-derailleur low-gear limit screw.** It's on the face of the upper link (Fig. 6.16A); this may be counterintuitive, because the outermost screw sets the inner stop on the derailleur. Adjust the screw to bring the inner cage plate within 0.5mm from the chain. Rotation direction is clockwise to move the cage outward; counterclockwise moves it inward.

3. **Shift to the outer chainring and smallest rear cog.**

4. **With a 2mm hex key, turn the front-derailleur high-gear limit screw.** Again, its location may be counterintuitive, since this screw, which stops the outward movement of the derailleur, is set back behind the linkage. It's just above the front-derailleur cage, below the support bolt you turned in 6-7 step 3d (Fig. 6.16A). Turn the screw to bring the outer cage plate within 0.5–1mm from the chain. As with the low-gear limit screw, rotation direction is clockwise to move the cage outward, and counterclockwise to move it inward; note that this is the opposite direction of how the high-limit screw works on a cable-actuated front derailleur.

5. **Shift to the large chainring and largest cog.**

6. **Hold the mode button down on Junction A (Figs. 6.2, 6.3) until the LED adjacent the "+/-" lights up red.** The system is now in adjustment mode.

7. **Adjust the inner front-derailleur cage plate position to be 0.5mm from the chain.** Push the upshift switch to move the derailleur cage outward, closer to the chain, and the downshift switch (on the shift paddle) to move it further inward (away from the chain). This trim adjustment on Di2 9070 and 6870 is meant to eliminate chain rub in the big-big combination; there is no trim adjustment on the small-small combination, but rub in this combination is less of an issue.

8. **Push the button on Junction A again until the LED turns off.**

The Di2 front derailleur shifts quickly, but in two steps; it pushes hard and fast to derail the chain without pushing so far that it overshifts the chainring, and then it carefully lines up the cage over the chain depending on what cog it is on in the rear.

On either chainring, the front derailleur automatically trims its position twice in either direction when the chain is shifted from one extreme of the cogset to the other. If you set the front-derailleur position and trim adjustment properly, this auto-trim feature will minimize chain rub in cross-gears. With steps 5–8 on Di2 9070 or 6870, you should be able to eliminate chain rub in all gear combinations as long as the chainstays are not shorter than 415mm (Shimano says this is the minimum length to be rub-free in the big-big combination).

c. Di2 diagnostics

1. On the bike.shimano.com website, go to the "INFO" tab and click on "TECH DOCS." Under the road bike section, click on "ULTEGRA Di2" or "DURA-ACE Di2." (Or go to techdocs.shimano.com and choose the "ROAD" section from there.)

2. For first-generation Di2 (Dura-Ace 7970), click on the green "7970" tab and scroll down to find the "SM-EC79" service instructions. To perform the diagnostics, you need the SM-EC79 system checker unit. If you have the unit, you can plug it into each component in the system and push the button on its face. If a green light comes on, that component is fine. The LED indicators on the SM-EC79 system checker illuminate or flash green, red, or orange, and the online instructions guide you through diagnosing problems in the system. You can also use the SM-EC79 system checker to switch the shift commands to change which direction the derailleur moves in response to pressure on a given switch.

3. For Ultegra Di2 6770/Dura-Ace 9070, go to the "INFO" tab on the home page and click on "E-TUBE PROJECT." If you have a Windows computer, you can download the software and plug right into your bike's Di2 system with Shimano's SM-PCE1 PC interface device for Ultegra Di2. You can then diagnose and correct problems in the system on your computer screen.

6-9

ADJUSTING CAMPAGNOLO EPS ELECTRONIC DERAILLEURS AND SHIFTERS

LEVEL Front and rear EPS derailleurs can be adjusted with the bike on a stand or while riding. There is a mode button on the inboard side of each Ergopower lever adjacent the thumb switch (Fig. 6.4A), while it is under each lever at the end of the aero extensions on EPS aerobar shifters (Fig. 6.4B).

NOTE: *The front derailleur seems to have more to say about how the system will behave than does the rear derailleur. Unlike Di2, if you were to remove the front derailleur to run a single chainring (for instance, for cyclocross) with an EPS system, the rear derailleur would not work properly (it would constantly be going into sleep mode). The rear-derailleur adjustment can get screwed up if you start adjusting the front derailleur before the rear derailleur, which confuses it. At that point, the rear derailleur may only shift half of the cogs. If it gets confused, install the shutoff magnet to shut the system down and allow it to reboot itself; then remove the magnet and hold both mode buttons down six seconds or more to move into the master adjustment mode (LED glows blue). You may need to repeat this a number of times.*

a. EPS adjustment mode, initial setup

1. **Shift to the big front ring-biggest rear cog gear combination.** Use the big chainring-second biggest rear cog if it won't go to the biggest cog. See 6-1d and 6-1e for shifting instructions with EPS electronic shifters.

2. **Get into master adjustment mode.** Do this by pressing both mode buttons—one on each lever behind the thumb lever on the inboard side—for at least six seconds until the LED on the EPS interface on the stem glows blue.

b. EPS rear derailleur initial adjustment

1. **Hold down the thumb lever to shift into the second-smallest cog.** Leave the chain on the large chainring.

2. **Tap whichever rear shift lever is necessary for the chain to be quiet and lined up on the second-smallest cog.** In the master adjustment mode (LED glowing blue), each touch of the lever will move the derailleur about the same amount that a single click of a barrel adjuster would on a cable-actuated system.

3. **Press the right mode button to save the adjustment (LED will glow white).**

4. **Push and hold the shift paddle (behind the brake lever) inward to shift into the second-largest cog.** The chain is still on the big chainring.

5. **Tap whichever rear shift lever is necessary for the chain to be quiet and lined up on the second-largest cog.**

6. **Press the right mode button to save the adjustment.** The LED will flash blue and then go back to solid blue. Press the mode button again to exit adjustment mode.

7. **Shift to the largest cog and the inner chainring.**

8. **With a 2mm hex key, tighten the rear-derailleur inner limit screw as in 5-3b and 5-3c until it contacts the tab on the rear derailleur.** This will keep the chain out of the spokes. Then back the screw out a half turn after bringing it into contact with the derailleur tab. This allows for the momentary over-shift used by the electronic system.

9. **As in 5-3f, adjust the lower pivot adjustment screw on the rear derailleur (Fig. 5.9).** Set the distance from the cog to the upper jockey wheel to 5–7mm (when in the lowest gear). The rear derailleur is adjusted.

c. EPS front derailleur initial adjustment

1. **Shift to the inner chainring and the largest cog.**

2. **Go into the master adjustment mode (6-9a, step 2).** The LED will glow blue.

3. **Push and hold whichever front lever is necessary to move the front derailleur's inner cage plate to 0.5mm from the chain.** Holding the left thumb lever down drives the inner front-derailleur cage plate away from the chain; holding the left shift paddle (finger lever) inward drives the inner front-derailleur cage plate toward the chain. There are no limit screws or other front-derailleur adjustments. With the external battery, you're done adjusting the front derailleur. Easy schmeezy. With the V2 internal EPS battery, you can do one more adjustment in the next section (d).

4. **Touch the left mode button to memorize the setting.** The LED will flash blue and then turn off after a few seconds. Tap any mode button again to take it out of standby mode (LED off).

5. **Take the bike for a road test.** Note any gear on which it is not shifting well. On either chainring, the front derailleur automatically trims itself twice in either direction when the chain is shifted from one extreme of the cogset to the other. If you have set the front-derailleur position properly, this auto-trim feature will ensure chain-rub-free riding in any gear combination. If the shifting is not accurate, repeat the adjustments, starting from step a in this section.

d. EPS adjustment mode, fine-tuning or ride setting

This is the mode you use to dial in shifting while riding. Use it if you've gotten a wheel change and the shifting is off. You can also use it if you have chain rub on the front derailleur.

With the V2 internal EPS battery and the EPS interface that goes with it, you can adjust the position of the front derailleur over each chainring individually (you cannot do this with the external EPS battery and V1 interface because the total throw of the derailleur is fixed, so adjusting it over one chainring also sets its travel over the other chainring).

The shifters move the derailleurs differently in this ride-adjustment mode (in which the LED is glowing pink) than in master adjustment mode (6-9a) in which the LED is glowing blue. As we've seen in 6-9b and 6-9c, when the LED is glowing blue, the derailleur moves continuously when you press and hold a shift button. However, when the LED is glowing pink, the derailleur moves only one little step with each press of a shift button (the step is similar in size to the movement caused by turning a barrel adjuster on a cable system by a single click).

1. **Get into fine-tune mode by holding only one mode button down until the LED on the EPS interface glows pink.** This will take about six seconds.

2. **Tap whichever shift button is necessary to move the derailleur you're adjusting in the direction you want.** Do this to eliminate the noise or rub, or to move the rear derailleur toward the cog toward which it is shifting sluggishly. With the front derailleur, provided you have the V2 or V3 internal EPS battery and interface, you can set the position over the large chainring so that it climbs up to it from the inner chainring the way you want without throwing it over, and so that it eliminates chain rub in high gear and/or in cross-gears.

3. **Press either mode button again to memorize the setting.** The LED will flash pink and then go out after a few seconds.

e. EPS diagnostics

It is easy to check battery level as in 6-1f. Expect to get one to three months on a charge. The 12-volt lithium ion battery can be recharged over 500 times without losing significant performance.

Firmware and the device configuration memory can be updated and system diagnosis can be accomplished by a Campagnolo service center or with the MyEPS app (V2 and later only).

6-10

ADJUSTING SRAM ETAP ELECTRONIC DERAILLEURS AND SHIFTERS

a. eTap front derailleur

Don't worry that it takes two hands to shift the front derailleur; the function button can also shift it. Push the function button on the front derailleur itself to shift it during this adjustment. That way, you can shift with one hand while you turn the crank with the other.

1. **Adjust the inner limit screw.** With the chain on the largest rear cog, push the front-derailleur function button while turning the cranks to shift the chain to the small chainring. Turn the inner limit screw (the bottom screw) clockwise until the inner cage plate is exactly 0.5mm away from the chain.

2. **Double-check derailleur height.** Ensure that the outer cage plate is 1–2mm above the tallest teeth of the outer chainring when the chain is on the inner chainring. If you did step 3 of section 6-7 correctly, this step should be irrelevant.

3. **Adjust the outer limit screw.** Shift the chain to the big chainring and smallest rear cog. Turn the outer limit screw (the upper screw) until the front derailleur's outer cage plate is 1mm from the chain.

4. **That's it.** There are no electronic micro adjustments to make with the eTap front derailleur. And don't expect it to reposition itself as the chain moves across the cogs as Shimano and Campagnolo electronic front derailleurs do. The SRAM eTap front derailleur varies how far it moves laterally to perform the shift based on which rear cog the chain is on when you shift. But it always goes back to a single position over each chainring that is supposed to not rub the chain in any cross-gear (if you lined it up properly as in Fig. 5.17 and set the limit screws correctly) and does not change as the rear derailleur shifts.

b. eTap rear derailleur

1. **Center the guide pulley under a cog using electronic micro adjustments.** The micro-adjustment procedure, whether performing initial setup or on-the-fly readjustment, is to hold down the function button (Fig. 6.5A) on the shifter (or BlipBox, Fig. 6.5B, in the case of an aero bike) while activating the shift paddle (or a Blip or a button on the BlipBox). If the pulley needs to go outboard to become centered under the cog, activate the right shifter while pressing its function button. If the pulley needs to go inboard to become centered under the cog, activate the left shifter while pressing its function button.

2. **Adjust the low-gear limit screw.** Shift to the largest cog and, with the guide pulley centered under the cog, turn in the inboard (low-gear) limit screw until it just touches the stop (if it pushes on the stop, it will move the derailleur, so back the screw off until it doesn't push the derailleur, which could strain the servomotor). The procedure is much the same as in 5-3b.

3. **Adjust the high-gear limit screw.** Shift to the smallest cog and, with the guide pulley centered under the cog, turn in the outboard (high-gear) limit screw until it just touches the stop (if it pushes on the stop, it will move the derailleur, so back the screw off until it doesn't). The procedure is much the same as in 5-3c.

4. **Adjust the chain gap.** With the chain on the small chainring and largest cog and using a 2.5mm hex key, adjust the b-screw to set the chain gap to 6mm between the guide pulley and the bottom of the cog (Fig. 5.5) as described in 5-3f.

TROUBLESHOOTING ELECTRONIC SHIFTER PROBLEMS

6-11

ELECTRONIC DERAILLEUR ADJUSTMENT GUIDELINES

Ensure that the rear derailleur's low-gear limit screw prevents it from going into the spokes before test riding.

TABLE 6.1 — TROUBLESHOOTING ELECTRONIC SHIFTING PROBLEMS

REAR DERAILLEUR	
Nothing happens when shift switch is pushed	On Campy EPS, remove battery shutoff magnet (6-2b);
	Charge battery (6-2a);
	Check wire connections;
	Ensure SRAM derailleur battery is ciicked in;
	Check SRAM shifter battery.
Chain jams between small cog & frame	Push rear downshift switch while in adjustment mode (6-3);
	Turn high-gear limit screw clockwise (6-8a, 6-9b, or 6-10b).
Rear derailleur touches spokes	Turn low-gear limit screw clockwise (6-8a, 6-9b, or 6-10b).
Chain falls between spokes & large cog	Turn low-gear limit screw clockwise (6-8a, 6-9b, or 6-10b).
Chain won't go onto large cog	Push rear downshift switch while in adjustment mode (6-3);
	Turn low-gear limit screw counterclockwise (6-8a, 6-9b, or 6-10b);
	Chain is too short (4-8)—replace with a longer one;
	Derailleur cage is too short; replace derailleur or cage (5-33).
Shifting sluggish to a larger cog	Push rear downshift switch while in adjustment mode (6-3).
Shifting sluggish to a smaller cog	Push rear upshift switch while in adjustment mode (6-3);
	Lubricate return spring (5-31).
Shifting sluggish both directions	Adjust b-screw (6-8a, 6-9b, or 6-10b).

FRONT DERAILLEUR	
Nothing happens when shift switch is pushed	On Campy EPS, remove battery shutoff magnet (6-2b);
	Charge battery (6-2a);
	Check wire connections;
	Ensure SRAM derailleur battery is clipped in (Figs. 6.10A, B);
	Check SRAM shifter battery (6-1h).
Chain falls off to outside	Check derailleur position (6-7);
	Turn high-gear limit screw clockwise (6-8b, 6-9c, or 6-10a).
Chain falls off to inside	Check derailleur position (6-7);
	Push front upshift switch while in adjustment mode (6-3);
	Turn low-gear limit screw clockwise (6-8b, 6-9c, or 6-10a);
	Install inner stop (5-44).
Chain rubs inner cage plate in low gear	Check derailleur position (6-7);
	Push front downshift switch while in adjustment mode (6-3);
	Turn low-gear limit screw counterclockwise (6-8b, 6-9c, or 6-10a).
Chain rubs outer cage plate in high gear	Check derailleur position (6-7);
	Push front upshift switch while in adjustment mode (6-3);
	Turn high-gear limit screw counterclockwise (6-8b, 6-9c, or 6-10a).
Chain rubs a cage plate in cross-gear	Push appropriate front shift switch in adjustment mode (6-3).
Derailleur hits crankarm	Check derailleur position (6-7);
	Turn high-gear limit screw clockwise (6-8b, 6-9c, or 6-10a).
Shifting sluggish to big chainring	Check derailleur position (6-7);
	Push front upshift switch while in adjustment mode (6-3).
Shifting sluggish to small chainring	Check derailleur position (6-7);
	Push front downshift switch while in adjustment mode (6-3);
	Turn low-gear limit screw counterclockwise (6-8b, 6-9c, or 6-10a).

Make small adjustments each time before rechecking shifting. Press a shifter switch only once or twice each time while in adjustment mode. Turn a limit screw 1/8 turn each time.

When using Table 6.1, if there are multiple possible solutions listed for a given problem, perform the applicable ones in order; if the first fix doesn't work, try the next one.

6-12

HOW DO I GET MORE GEAR RANGE?

If you have a short-cage electronic rear derailleur and want a wider gear range than it is rated for, there is something you can do beyond tightening in or turning around the b-screw; you can put a longer pulley cage on it! Of course, you can simply install an 11-speed Ultegra Di2 GS rear derailleur instead, but if you have a 10-speed Dura-Ace Di2 7970 rear derailleur, you can interchange its pulley cage with an Ultegra 6700-GS rear derailleur (its pulley cage fits the mechanical Dura-Ace RD-7900 as well). See 5-37 for more information on this specifically, and on expanding gear range in general. See 6-1 for general guidelines on mixing Shimano and Di2 components.

Another option is to use XTR Di2 mountain bike derailleurs with your road Di2 shifters. The XTR rear derailleur will give you the range to shift a huge 11–40 mountain bike cassette. However, you will have to use both XTR derailleurs, because the system won't function with a mix of a mountain bike Di2 rear derailleur and a road Di2 front derailleur. And since MTB front derailleurs are designed for 10-tooth jumps between chainrings, rather than the 16-tooth jumps that road front derailleurs manage with ease, the setup won't shift nearly as well on the front as with the road Di2 front derailleur. So if you make this swap, expect to do some tinkering and tuning to make front shifting acceptable.

All you need in this life is ignorance and confidence, and then success is sure.

—MARK TWAIN

TOOLS

Tire levers
Pump
Patch kit
BBQ grill-cleaning pad
  or brush
Table knife

OPTIONAL

Truing stand
Talcum powder
Tire sealant
Tubular gluing tape
Tubular rim cement
Carogna Remover or
  VM&P Naphtha/
  acetone/white gas
  as glue solvent
Teflon tape
Pliers
Miter clamp
Leather sewing needle
Braided high-test
  fishing line
Thimble
Barge cement (or other
  strong contact cement)

TIRES

A bicycle's tires provide suspension as well as grip and traction for propulsion and steering. The air pressure in the tire is the primary suspension system on a road bike.

Three types of road bike tires are available. "Clinchers" (Fig. 7.1), which are C-shaped in cross-section, are held into a C-shaped rim by a steel or Kevlar bead on each edge of the tire. A separate inner tube inside of the clincher holds the air. "Tubulars" (Fig. 7.2), which are circular in cross-section, have a casing that is wrapped around an inner tube and stitched or glued together. The tire is glued onto a box-section rim that has no vertical rim walls like a clincher rim. The third and most recent type is a tubeless road tire that resembles a clincher tire, but the rim is sealed, and there is no inner tube—merely a valve sealed into the rim (Fig. 7.16). Tubeless tires are generally used with a sealant inside.

This chapter addresses how to fix a flat or replace a tire or tube, and has some general advice on how to choose a tire. Have at it.

CLINCHER TIRES

7-1

REMOVING THE TIRE

 LEVEL This is an easy job once you get the hang of it. See 7-5 for installation.

1. **Remove the wheel (2-2 and 2-11).**
2. **If the tire is not already flat, deflate it.** First remove the valve cap to get to the valve.

   Most road bike tires have Presta valves (also known as Sclaverand or French valves). These valves are thinner (6mm vs. 8mm) than Schrader valves (the kind found on cars) and

7.1 Clincher tire

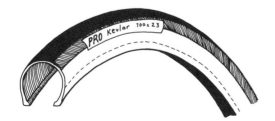

7.2 Tubular tire

7.3 Presta valve

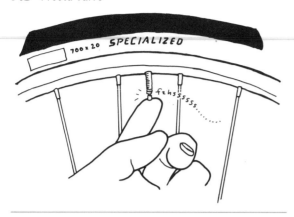

7.4 Schrader valve

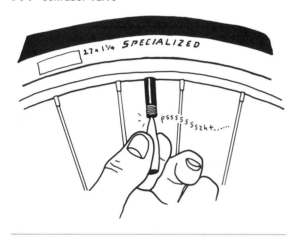

have a small threaded rod with a tiny nut on the end. To let air out, unscrew the little nut a few turns, and push down on the thin rod (Fig. 7.3). To seal, tighten the little nut down again (with your fingers only!); leave it tightened for riding.

To deflate a Schrader valve, push down on the valve pin with something thin enough to fit in that won't break off, like a pen cap or a paper clip (Fig. 7.4).

3. **If you can push the tire bead off the rim with your thumbs without using tire levers, by all means do so.** There is a lesser chance of damaging the tube or the tire if you avoid the use of levers or other tools. First squeeze the tire beads into the center of the rim all of the way around (see the Pro Tip on tire removal and installation) and then push the bead off, starting just to one side or the other of the valve.

4. **If you can't get the tire off with your hands alone, use tire levers.** Insert a lever (with its scoop toward

IF THE RIMS ON YOUR BIKE have a deep section (i.e., with a rim depth greater than 30mm; these are generally carbon rims), the air valves for the tires either will be extremely long or will be fitted with valve extenders, which are thin threaded tubes that screw onto the Presta valve stems.

One type of valve extender is simply a thin tube like a drinking straw with threads inside one end to screw onto the cap threads of the valve (Fig. 7.5A). Due to the wrench flats on a removable Presta valve core's cap threads, these don't work well with Presta valves with removable cores, which are found on most high-quality tubular tires. They also require the valve to be always open, by unscrewing the little nut atop it hard enough that it stays open and does not screw closed on its own. If the valve has a problem such as an imperfect seal or a bent rod, it can leak when the nut is not tightened down.

To deflate tires that have simple straw-type valve extenders installed, you need to insert a thin rod (a spoke is perfect) into the valve extender to release the air.

To install straw-type valve extenders so that they seal properly and allow easy inflation, you need to unscrew the little nut on the Presta valve until it is against the mashed threads at the top of the valve shaft (they are mashed to keep the nut from unscrewing completely). Back the nut firmly into these mashed threads with a pair of pliers so that it stays unscrewed and does not tighten back down against the valve stem while riding, which would prevent air from going in when you pump it. Wrap a turn or two of Teflon pipe thread tape (plumbing tape) around the top threads on the valve stem before screwing on the valve extender; if you do not, air may leak out when you are pumping, and the pressure gauge on your pump will not give an accurate reading of the pressure in the tire. Tighten the valve extender onto the valve with a pair of pliers.

Continues >>

Alternatively, valve extenders for removable-core valves will be threaded at the base with the same thread as on the base of a valve core (Fig. 7.5B), and they will have threads inside the other end to accept a valve core. Deflating and inflating the tire with one of these valve extenders is no different than deflating and inflating any tire with a standard Presta valve. To install one of these valve extenders, unscrew the valve core (counterclockwise) with an adjustable wrench or specific Presta valve-core wrench and remove it. Screw the valve extender into the valve body where the core was; tighten it firmly with pliers or a wrench on its wrench flats. Screw the valve core into the valve extender and tighten it with an adjustable wrench or specific Presta valve-core wrench.

Some valve extenders are straws with a thin knurled knob on top with a shaft running all the way down to a cup that grabs the nut atop the valve (Fig. 7.5C). They actually allow you to tighten or loosen the valve nut with the extender in place, thus behaving like a standard Presta valve. With this type, unless it has a rubber seal at its base, you should also wrap a turn or two of Teflon pipe thread tape around the top threads on the valve stem before screwing on the valve extender; if you do not, air may leak out when you are pumping, and the pressure gauge on the pump will not give an accurate reading of the pressure in the tire. Tighten the valve extender onto the valve stem with a pair of pliers.

Note too that some Vittoria tubulars have a threaded stub where the valve would go, and you thread on a complete valve of the length you need for your specific rim (Fig. 7.5D). It operates like any other Presta valve.

7.5A Straw-type valve extender

7.5B Removable-core valve extender (showing separate valve core)

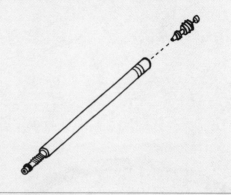

7.5C Topeak/Spinergy valve extender

7.5D Vittoria tube stub with different valve lengths

7.6 Removing clincher tire with levers

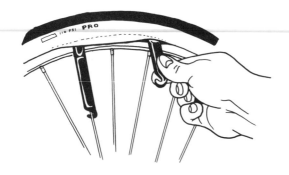

7.7 Pulling out the bead with the third lever

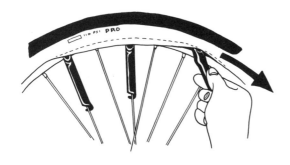

7.8 Removing the inner tube

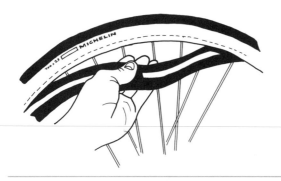

you) between the rim sidewall and the tire (again, close to the valve stem; see the Pro Tip on tire removal and installation) until you catch the edge of the tire bead. Make sure you do not pinch the tube between the lever and the tire.

5. **Pry down on the lever until the tire bead is pulled out over the rim (Fig. 7.6).** If the lever has a hook on the other end, hook it onto the nearest spoke. Otherwise, keep holding it down.

6. **Place the next lever a few inches away, and do the same thing with it (Fig. 7.6).**

REMOVAL OF A CLINCHER OR TUBELESS TIRE is most easily accomplished by starting near the valve stem, and tire installation is similarly best accomplished by finishing at the valve stem. That way, the beads of the deflated tire can fall into the dropped center (valley) of the rim on the opposite side of the wheel, making it effectively a smaller-circumference rim. If you instead try to push the tire bead off (or on) the rim on the side opposite the valve stem, the circumference on which the bead is resting is larger, because the valve stem is forcing the beads to stay up on their seating ledges opposite where you are working. Adhering to this method is particularly critical with a tubeless road tire, because the bead is very tight, it does not stretch at all, and you should install the tire without tire levers to avoid damaging the tire's edges that seal the air inside.

With a standard clincher tire (one with an inner tube), another reason to finish at the valve stem is to avoid ending up with a bit of inner tube trapped under the tire bead where you finished pushing the tire onto the rim. If that happens, it may immediately blow the tire off the rim when you pump it up to pressure (temporarily deafening you; the explosion is very loud) or it may explode later on a ride. In either case, you will have an unpatchable tube with a long rip down its length, possibly endangering your life if the tire blows at high speed on a turn.

You can minimize explosions and flats by finishing the tire installation at the valve stem (Fig. 7.14). First, take the appropriate precaution of having some air in the tube to keep it from twisting and getting under the bead as you push the tire on. Deflate that last bit of air only when it prevents you from getting the final few inches of the tire bead on. Note that the edge of the flat tube can still slip under the bead edge as it pops into the rim at that point. But pushing up on the valve stem after the tire is on (Fig. 7.15) can lift the adjacent sections of inner tube completely into the tire chamber and ensure that none is caught under the edge, waiting to blow the tire off the rim. Always inspect around the tire bead before inflating as in 7-5, step 11.

7. **If needed, place a third lever a few inches farther on, pry it out, and continue sliding this lever around the tire, pulling the bead out as you go (Fig. 7.7).** Some people slide their fingers around under the tire bead, but beware of cutting your fingers on a sharp bead.

NOTE: *There are quick-change tire levers on the market that work differently and more quickly than the separate standard tire levers. If the tire bead is very tight on the rim, however, using separate tire levers may be the only method that works effectively.*

8. **Once the bead is off the rim on one side, pull the tube out (Fig. 7.8).**

9. **If you are patching or replacing the tube, you do not need to remove the other side of the tire from the rim.** If you are replacing the tire, the other bead should come off easily with your fingers. If it does not, use the tire levers as just outlined.

7-2

PATCHING AN INNER TUBE

1. **If the leak location is not obvious, inflate the tube until it is two to three times larger than its deflated size.** Be careful. You can explode it if you put in too much air.

2. **Listen for air, and mark the leak(s).**

3. **If you cannot find the leak by listening, submerge the tube in water or sponge soap suds onto it.** Look for air bubbling out (Fig. 7.9) and mark the spot(s). Make sure you check the valve for leaks, too.

NOTE: *You can only patch small holes. If the hole is bigger than a pencil eraser, a round patch is not likely to work. A slit as long as an inch or so can be repaired with an oval patch.*

7-3

STANDARD PATCHES

Use a patch designed for bicycle tires; it will generally have a gummy bottom layer, usually orange, that sticks out around the edges of the smaller-diameter patch of black rubber.

7.9 Checking for a puncture

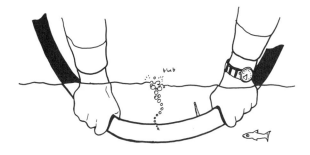

7.10 Applying glue

1. **Dry the tube thoroughly near the puncture and mark the location of the hole with a pen.**

2. **To provide a suitable surface for the patch, clean and then rough up the tube surface within about a 1-inch radius around the hole with a small piece of sandpaper (usually supplied with the patch kit).** Do not touch the sanded area. If the patch kit you are using has a little metal "cheese grater" for the purpose, discard it and replace it with sandpaper. The grater-style rougheners tend to do to a tube what they do to cheese.

3. **Apply glue (patch cement) in a thin, smooth layer all over an area centered on the hole (Fig. 7.10).** Use the end of the glue container or a brush, rather than your finger, to spread the glue around. Cover an area that is bigger than the size of the patch. By the way, the glue is similar to rubber cement, so if the tube in your patch kit has dried out, you can use any rubber cement sold in office supply and hardware stores. If you do this, and use the brush attached to the top of the bottle cap, wipe the brush almost dry before spreading the cement on the tube. You only need a thin layer for the patch.

7.11 Removing cellophane

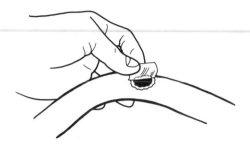

7.12 Installing a rim strip

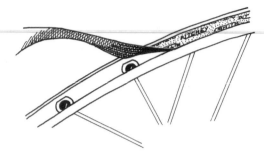

4. **Let the glue dry 10 minutes or so until there are no more shiny, wet spots.**

5. **Peel the patch from its foil backing (but do not remove the cellophane top cover yet).**

6. **Stick the patch over the hole and push it down in place, making sure that all of the gummy edges are stuck down.** With the tube sitting on a hard surface, burnish the patch with the plastic handle of a screwdriver to stick the edges down securely.

7. **Optional: Remove the cellophane top covering, being careful not to peel up the edges of the patch (Fig. 7.11).** Often the cellophane top patch is scored. If you fold the patch, the cellophane will split at the scored cuts, allowing you to peel outward and avoid pulling the newly adhered patch up off the tube. If you can't get the cellophane off without peeling up the patch or you don't mind it being there, just leave it alone. It won't harm the tire.

7-4
GLUELESS PATCHES

There are a number of adhesive-backed patches on the market that do not require cement. Simply clean the area around the hole with an alcohol pad supplied with the patch. Let the alcohol dry, peel the backing, and stick on the patch.

Glueless patches are quick to use and take little room in a seat bag; also, you never open your patch kit to discover that your glue tube is dried up. On the downside, I have not found a glueless patch that sticks nearly as well as the standard type. With a standard patch installed, you can inflate the tube to look for more leaks without having it in the tire. If you do that with a glueless patch, the patch usually lifts enough to start leaking. With glueless patching, you must install the tube in the tire and on the rim before putting air in it. And glueless patches are probably not a permanent fix.

7-5
INSTALLING PATCHED OR NEW TUBE

1. **Feel around the inside of the tire.** Check for anything sticking through that can puncture the tube again. Sliding a rag all the way around the inside of the tire works well. The rag will catch on anything sharp and will save your fingers from being cut by whatever is stuck in the tire.

2. **Replace any tire that has damaged areas (inside or out) where the casing fibers appear to be cut or frayed.**

3. **Examine the rim to be certain that the rim strip is in place and that there are no spokes or anything else sticking up that can puncture the tube.** Replace the rim strip if necessary (Fig. 7.12). Use a stretch-on rim strip or, lacking that, two layers of fiberglass strapping tape. Don't use tape that doesn't have reinforcing fibers in it, since it can stretch or tear into the spoke holes and allow the spoke ends to puncture the tube.

4. **By hand, push one bead of the tire onto the rim.**

5. **Optional: Smear talcum powder around the inside of the tire and on the outside of the tube so the two do not adhere to each other.** Don't inhale this stuff, by the way.

6. **Put just enough air in the tube to give it shape.** Close the valve, if it's a Presta.

7.13 Installing a tire by hand

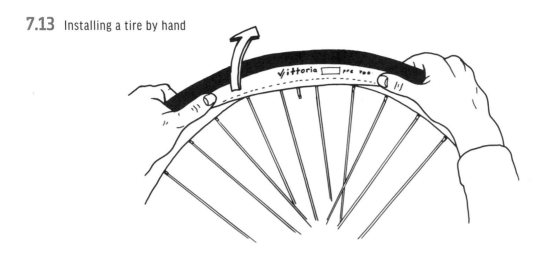

7.14 Finishing at the valve

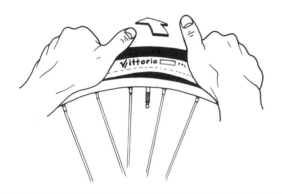

7.15 Seating the tube

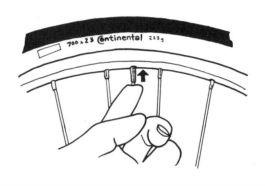

NOTE: *If the rims have a deep section (more than 30mm) and the tube has a standard-length Presta valve, you will need to install a valve extender (Figs. 7.5A–D) so that you can get air into the tire once it is on the rim. See the Pro Tip on valve extenders for instructions on installing the various types of valve extenders.*

7. **Push the valve through the valve hole in the rim.**

8. **Push the tube up inside the tire all the way around.**

9. **Starting at the side opposite the valve stem, push the tire bead onto the rim with your thumbs.** (See the Pro Tip on tire removal and installation for why to start opposite the valve.) Be sure the tube doesn't get pinched between the tire bead and the rim.

10. **Work around the rim in both directions toward the valve with your thumbs, pushing the tire onto the rim (Fig. 7.13).** Finish from both sides at the valve (Fig. 7.14), deflating the tube when it gets hard to push more of the tire onto the rim. (See the Pro Tip on tire removal and installation for why you should

finish at the valve.) Using this method, you can often install a tire without tools. If you cannot, use tire levers, but make sure you don't catch any of the tube under the edge of the bead. Finish the same way, at the valve.

11. **Reseat the valve stem by pushing up on the valve after you have pushed the last bit of bead onto the rim (Fig. 7.15).** You may have to manipulate the tire so that all the tube is tucked under the tire bead.

12. **Go around the rim and inspect for any part of the tube protruding from under the edge of the tire bead.** If you have a fold of the tube under the edge of the bead, it can blow the tire off the rim when you inflate it or while you are riding. It will sound like a gun went off next to you and will leave you with an unpatchable tube.

13. **Pump up the tire.** Generally, 85–110 psi is correct for a good-quality 23–25mm-wide road bike tire (see the Pro Tip on tire pressure), unless you weigh

MANY ROAD RIDERS MAKE THE MISTAKE of using very high air pressure to reduce rolling resistance. The likelihood of hurting yourself in a blowout is high, and using extremely high pressure does not make the tire roll faster; it reduces traction. If the tire cannot absorb small bumps into its surface because it is pumped up too high, the bike will roll slower, even though it may feel fast because it is so stiff and bouncy. Every little bump lifts the bike and rider, providing a backward force on impact. This costs energy compared with absorbing the bump into the tire while the bike and rider continue to roll along smoothly without up-down motion.

Pressures of 140 psi and higher are fast only on the very smooth surfaces found on a velodrome (banked racetrack). On a road surface, anything higher than 120 psi costs you speed.

Regarding rolling resistance, a handmade tire with a casing made of thin, tightly packed threads will generally roll faster than a casing made of fewer thicker, stiffer threads. Thread count is expressed in "threads per inch," or "tpi" for short. So, if you want to roll fast, choose a tire with a high thread count that feels supple when you fold it in your hand, and forget the bomber tire pressures that also endanger you!

more than 200 pounds (in which case you need more air). If you increase the pressure further, you will be pushing the limits of some tires and increasing ride harshness. If you put in less air than recommended, you run the risk of a pinch flat (or "snakebite" flat); see section 3-3a.

7-6

PATCHING TIRE CASING (SIDEWALL)

If the tire casing is cut, get a new tire. No matter what you use to patch the tire casing, the tube will find a way to bulge out of the patched hole, and when it does, it can explode. Imagine coming down a steep hill and blowing the front tire . . . you get the picture.

In an emergency, you can put layers of nonstretchable material between the tube and tire (3-3d, Fig. 3.2). Candidates for this duty include a dollar bill, an empty energy bar wrapper (or two), or a small piece cut from an old tire sidewall.

TUBELESS ROAD TIRES

LEVEL Tubeless tires cannot pinch flat (i.e., the pinching of the inner tube between the tire and the rim when hitting a sharp object, which punctures the tube with two "snake-bite" holes), since there is no tube to pinch. With liquid sealant inside,

tubeless tires also will not lose a significant amount of air to a small puncture. When used in conjunction with tubeless-specific road rims, road tubeless tires also will tend to stay on the rim better than a clincher in the event of sudden pressure loss. You can also run a road tubeless tire at 10–20 psi lower pressure than a clincher and get better traction and more comfort without sacrificing rolling resistance or worrying about pinch flats.

7-7

REMOVING A TUBELESS ROAD TIRE

Tubeless road tires fit very tightly because they must. Since there is no inner tube pressing the tire sidewalls against the rim walls, the tire bead is more critical for retaining the tire. It must not stretch in order to prevent the air pressure from blowing the tire off of the rim. Consequently, road tubeless tires are harder to remove than most clinchers. You will likely need tire levers. To avoid breaking the carbon bead or damaging the rubber edge along it, which would compromise the tire's sealing ability, make sure you use only plastic levers with no sharp edges. Hutchinson Stick'Air tire levers are specifically made for tubeless road tire installation and removal.

If you intend to patch the tire, find the hole before removing the tire from the rim (7-8).

1. **Remove the wheel (2-2 and 2-11).**

2. **If the tire is not already flat, deflate it.** First remove the valve cap (if installed) to get to the valve. To let air out of the valve (it's a standard Presta valve), unscrew the little nut a few turns, and push down on the thin rod (Fig. 7.3). To seal, tighten the little nut down again (with your fingers only!); leave it tightened for riding.

3. **If you can push the tire bead off the rim with your thumbs without using tire levers, by all means do so.** There is less chance of damaging the tire's sealing edges if you unseat the bead by hand. It's easiest if you squeeze the tire beads into the center of the rim all the way around (see the Pro Tip on tire removal and installation) and then start just to one side or the other of the valve to unseat the bead.

4. **If you can't get the tire off with your hands alone, use tire levers.** Insert a tire lever (with its scoop toward you; the pointed end of the tire lever should be on the rim side) between the rim sidewall and the tire until you catch the edge of the tire bead. Again, start near the valve after squeezing the tire beads into the rim valley all the way around to put as much slack into the tire as possible.

5. **Pry down on the lever until the tire bead is pulled over the rim (Fig. 7.6).** If the lever has a hook on the other end, hook it onto the nearest spoke. Otherwise, keep holding it down.

6. **Place the next lever a few inches away, and do the same thing with it (Fig. 7.6).**

7. **If needed, place a third lever a few inches farther on.** Pry the bead over the rim, and continue sliding this lever around the tire, pulling the bead out as you go (Fig. 7.7) until you can get it the rest of the way off with your hands.

8. **Remove the other bead.** Remove by hand if you can.

7-8

PATCHING A TUBELESS ROAD TIRE

You must find the hole before removing the tire from the rim. If the tire has a cut that is more than a few millimeters long, it's best to replace the tire to avoid endangering yourself from a blowout. If a number of casing threads are cut, the tire will also have a bulge and will not ride smoothly.

You can usually eliminate the need to patch the tire by simply deflating the tire and inflating it with a latex-based tire sealant (see 7-9, step 7).

1. **Inflate the tire to no more than 100 psi.**

2. **Listen for where air is coming out, or coat the tire with soap suds or submerge it in water to find bubbles locating the hole.** Make sure you check all the way around, in case there are numerous holes.

3. **Mark the hole(s).**

4. **Remove the tire (7-7).**

5. **Dry the inside of the tire, and rough up and patch the hole(s) from the inside, using the patches in the same way as instructed in 7-3.** Install the tire as in 7-9.

7-9

INSTALLING A TUBELESS ROAD TIRE

Tubeless road tires have carbon-fiber beads to ensure that the beads do not stretch and yet are thin enough to allow both beads to drop into the rim valley together to reduce the mounting circumference of the wheel and make the tire easier to install.

Tubeless-specific road rims have a low ridge on the edge of the tire-mounting ledge to seal against the tire (Fig. 7.16). They also have no holes in the upper rim wall for accessing the spoke nipples.

You can mount a road tubeless tire with sealant on a rim not specifically denoted as a tubeless rim, but doing so runs counter to the tire manufacturers' warnings. Either you would need a rim that has no spoke holes in the upper rim wall, or, in a rim with spoke holes, you would need to use an airtight rim strip like the Stan's

---

7.16 Tubeless road tire with rim cutaway view

NoTubes or Caffélatex rim strip to seal the spoke-nipple access holes. You would also need a valve specifically designed for using a road tubeless tire on a standard road wheel, which will seal the valve hole. One danger of using a road tubeless tire on a rim not designed for it is that the bead may not stay locked on in the case of sudden air loss, allowing the tire to come off the rim. **IMPORTANT SAFETY WARNING:** *NEVER, EVER mount a road clincher tire without an inner tube unless it is specifically branded as a tubeless tire (yes, I know running standard tires tubeless is often done with mountain bike tires and even with cyclocross clinchers, but the pressures are lower). The beads of a standard clincher tire are not sufficiently resistant to stretching and are not precise enough in size to prevent the tire from blowing off the rim. This, compounded with the fact that the tire will have latex sealant inside to get it to hold air in the first place, is extremely dangerous. If the tire does blow off, the rim will be slipping around on the liquid latex in the tire, allowing the rim to slide off of the tire, a situation almost impossible for the rider to control without crashing. Don't do it.*

1. **Install the valve into the rim.** Make sure that the rubber washer on the outside and rubber seal on the inside are properly seated. Put some sealant (like Caffélatex or Stan's NoTubes) around the valve hole before installing the valve; otherwise, wet the rubber base of the valve with soapy water before installing it.

2. **Wet the tire edges.** Use water alone or use soapy water on a sponge.

3. **By hand, push one bead of the tire onto the rim.** Start opposite the valve stem and finish at the valve stem to minimize the rim's mounting circumference (see the Pro Tip on tire removal and installation).

4. **Starting opposite the valve stem (again, see the Pro Tip on tire removal and installation), push the second tire bead onto the rim with your thumbs.**

5. **Work around the rim in both directions toward the valve with your thumbs, pushing the tire onto the rim (Fig. 7.13).** Finish from both sides at the valve (Fig. 7.14). The Pro Tip on tire removal and installation explains why you should finish at the valve. Using this method, you should be able to mount the tire without tools. If not, use tire levers, but make sure you use plastic levers with no sharp edges, finishing at the valve.

6. **Make sure that the beads are well down into the channel of the rim.**

7. **If you're not going to use a sealant, you can now inflate the tire with a high-volume floor pump, air compressor, or gas cartridge to 120 psi maximum.** A small hand pump may not push enough air volume fast enough to get the beads to seat; air will leak out around the edges while you pump until you're blue in the face. You may find that you need the air compressor or cartridge to seat the tire.

Personally, I don't see the point of not using a tire sealant. A tubeless tire has the advantage of eliminating pinch flats, and when used with sealant can also eliminate almost all punctures. The sealant ensures that if you get a puncture, it seals up immediately, and it also can prevent slow leaks around the bead. I therefore recommend using either an aerosol sealant or a liquid sealant to inflate the tire the first time. Method 1, using an aerosol, is to inject sealant through the valve with an aerosol canister like Hutchinson Fast'Air or Vittoria Pit Stop (Fig. 7.21); follow the directions in 7-15a. Method 2 is to pour in a couple of tablespoons of liquid sealant while mounting the tire at the point where there are only a few more inches of the second bead to push on the rim. Another way to get liquid sealant into the tire is to use a Stan's or Caffélatex valve, both of which have removable valve cores, and put the sealant in as in 7-15b. Then inflate the tire with a high-volume floor pump, air compressor, or gas cartridge to 120 psi maximum.

8. **Spin the wheel and ensure that the tire is aligned properly and has no bulges or places where the bead is not fully seated.**

TUBULAR TIRES

LEVEL Tubular tires, or sew-ups, are expensive, slow to install (they must be glued to the rim), hard to repair, and can even be hard to find. So why bother with them?

For one thing, tubular wheelsets (front/rear wheel pairs) are generally lighter than clincher wheelsets, because tubular rims do not require flanges for the tire bead; superlight carbon-fiber tubular rims in particular have made tubulars popular again. Tubular tires by themselves are often lighter than clinchers, too.

Another reason is that most riders find that tubulars ride and corner better than clinchers, particularly for cyclocross. Since the tubular casing is sewn together around the inner tube, tubulars can hold extremely high air pressures, which is why they're almost always used for track racing.

Tubulars can also operate at low air pressures with less risk of a pinch flat, since the rim has no flanges, and the latex tube inside is much tougher against impact than a butyl tube. That's why you'll find most high-end cyclocross bikes equipped with tubulars; reduced air pressure minimizes rolling resistance on rough surfaces and enlarges the tire contact area with the ground for increased traction.

But perhaps the main reason to consider tubulars is their inherent safety. In the event of a blowout, they stay on the rim. Clincher tires, when flat, fall into the rim well, and you may find yourself trying to ride on the slippery metal rim, rather than on rubber.

Tubulars usually deflate more slowly when punctured than clinchers because the air can escape only through the puncture hole. Clinchers let air escape all the way around the rim, and the next thing you know you're skittering around on metal. (Tubeless tires also solve this problem, and they don't require any glue; see 7-7 through 7-9 for more information.)

If the advantages appeal to you—and the disadvantages don't put you off—tubulars are a worthwhile alternative to the standard clincher setup.

7-10
REMOVING A TUBULAR

1. **Remove the wheel (2-2 and 2-11).**
2. **If the tire is not already flat, deflate it.** Tubular tires have Presta valves. To let air out, unscrew the little nut atop the valve stem a few turns, and push down on the thin rod (Fig. 7.3). See Pro Tip on valve

extenders for instructions on dealing with valve extenders of various types.

3. **Push the tire off the rim in one section with your thumbs pushing up against one side.** If you can't get it up any farther than the center of the rim, then work at it the same way from the other side (in the same area) until the entire underside is unstuck there. Avoid using tools, because if you use a tool to pry the tire away from the glue, you will likely tear the base tape at the least and quite likely tear casing cords as well. The tire will always be lumpy in that area after such damage.

If you absolutely cannot get it off by hand in any area of the tire, carefully slide a thin screwdriver blade under it in an area where you have peeled it up as much as possible, working it under without damaging the base tape (alternating from each side if necessary) until the tip extends out from the other side. Then gently roll the screwdriver shaft along the rim, separating the tire as you go.

Stop when you've gone far enough that you can get your hand under the tire.

4. **Peel the tire off the rest of the way around the rim by hand.**

7-11
GLUING A TUBULAR TIRE ONTO A RIM

Gluing tubular tires to the rims properly is critical to continuing the attachment you have with your epidermis. Follow these steps, and your tire will really be secure! Pay particular attention to step 4, because all the rim cement in the world will not keep a tire on if the cement is not adhered to the tire. It is quicker to use tubular gluing tape (7-12), but, save perhaps for some recent developments with adhesives, tape is not as secure as a good glue job.

1. **Stretch the tire on a spare rim before gluing.** To stretch it, install the new, unglued tubular on an unglued rim (Fig. 7.17). Put it on the rim using the method described in step 10. Inflate it, and leave it for at least a few hours, or if possible, overnight.

NOTE: *If the wheel has a deep-section rim (deeper than 30mm) and the tubular has a standard-length*

7.17 Stretching the tubular tire over the rim

Presta valve, you will need to install a valve extender (Figs. 7.5A–C) so that you can get air into the tire once it is on the rim. See the Pro Tip on valve extenders for instructions on installation of the various types of valve extenders. Note too that some Vittoria tubular tires have a threaded stub where the valve would go, and you thread in a complete valve of the length you need for your specific rim (Fig. 7.5D). It operates like any other Presta valve. If you don't have a sufficiently long valve stem, you can use a removable-core valve extender (Fig. 7.5B) between the valve and the stub on the tube; the valve extender would be inverted from the way it works on standard removable-core valves.

2. **Remove the tubular after stretching.** Deflate it first.

3. **Pump the tire (not on the rim) until it turns inside out.** Pump it enough that the base tape faces out to the side so you can access it easily, but do not inflate it further.

4. **Prepare the tire's base tape for glue.** The base tape on most tubulars is cotton and in many cases has a neoprene coating over it to which the rim cement will not bond well. If yours has this slick neoprene coating, scraping the base tape is a way to improve the gluing surface; the tire can roll off if the glue hasn't adhered to the base tape. This step does not apply to most Challenge, Continental, and Tufo tubulars, which usually have no coating over the base tape.

Using the serrations of a table knife or the rough side of a metal file, scrape the base tape back and forth (Fig. 7.18) until its neoprene coating balls up into little sticky hunks. I have also heard of people brushing rubbing alcohol on the base tape to make the surface tacky. I generally discourage the use of solvents on the base tape for fear of solvent penetrating the tape and dissolving the glue holding the tape onto the tire, but rubbing alcohol is probably too mild to be an issue. I have seen many a tire roll right off the base tape because it was not glued well to the tire, even though the tape was well adhered to the rim, so stay away from stronger solvents.

5. **Prepare the rim for glue.** With a new rim, clean off any oil that may be present with alcohol, VM&P Naphtha, or acetone (while wearing rubber gloves and a respirator) and a rag, followed by sandpaper. Roughing up the gluing surface with sandpaper may not help the tire stick to the rim better, but solvent will not remove everything (Teflon, for instance), and sandpaper can remove invisible contaminants that would prevent the glue from sticking to the rim.

With a rim that has been glued before, Carogna Remover softens old glue and tape so they can be more easily scraped and wiped off. Otherwise, scrape or brush off the big lumps with a knife, screwdriver, or barbecue grill brush to get the surface as uniform as you can. You can also strip the entire rim with VM&P Naphtha, acetone, or white gas. Be sure to wear rubber or nitrile gloves when working with these solvents, and remember that the fumes are dense and extremely flammable. Supply plenty of ventilation, and don't work in an enclosed room where an exposed flame may be present, such as in a water heater or furnace.

7.18 Scraping the base tape before gluing

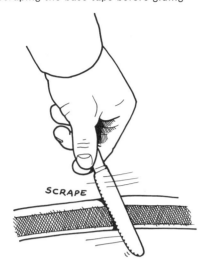

7.19 Applying tubular glue to the rim

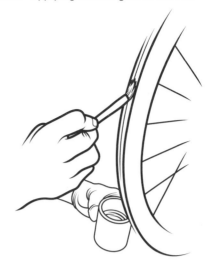

6. **Put a thin layer of glue on the rim (Fig. 7.19), edge to edge, and a thin layer edge to edge on the base tape of the inflated tire.** The best way is to brush glue from a can with an acid brush (hardware stores sell them in the plumbing department); you can even squeeze tubes of glue into a can if you didn't buy it by the can. Otherwise, I recommend squeezing a bead out of the tube onto the rim or tire and then putting a plastic bag over your finger and spreading the glue thinly and uniformly. If you let the layer on the tire get too thick, the base tape of the tubular may become so rigid that it will tear; keep the tire layers especially thin. Let it dry (overnight is best) with the tire deflated so it won't shrink while drying. Repeat two more times, letting the glue dry overnight each time.

7. **Alternatively, for cyclocross, apply CX Tape ("Belgian Tape") on the rim after the second glue layer while it's tacky.** Press the tape into the rim bed. Peel off the paper backing tape, apply a third layer of glue on top of the CX Tape and on the tire, wait 15–30 minutes, and skip to step 9. For more on gluing cyclocross tires, see the Pro Tip on the subject in 7-16.

NOTE ON GLUE TYPE: *I recommend using clear tubular rim cement, ideally Vittoria Mastik'One, especially on a carbon rim, or Continental rim cement, rather than red glues. Red glues can harden up, making the tire base tape rigid and the glue bond weak after a year.*

8. **After the three layers of glue on the rim and tire have dried (overnight), smear or brush another thin layer of glue on the rim.**

9. **If you have a spare tubular wheel with a rim free of glue, first stretch the tire onto this rim as in step 10.** Inflate it, deflate it, and then pull it off. This will make stretching the tire over the glued rim much easier, and you will end up with less glue smeared all over you and the rim.

10. **Mount the (deflated) tire as follows:**
    a. **Stand the wheel up with valve hole at the top.**
    b. **Stretch the tire down onto the rim.** Put the valve stem through the hole, and, leaning over the wheel, grab the tire and stretch outward as you push the base tape into the top of the rim. Keep stretching down on the tire with both hands, using your body weight, as you push the tire down around the rim (Fig. 7.17). I like to lean hard enough on the tire that my feet lift repeatedly off the ground. The farther you can stretch the tire at this point, the easier it will be to get the last bit of tire onto the rim.
    c. **Push the tire all of the way on.** Lifting the rim up to horizontal with the valve side against your belly, roll the last bit of the tire onto the opposite side of the rim. If you can't get the tire to pop over the rim, peel the tire back and start over, pushing down again from the valve stem. Avoid prying a stubborn tire onto

CARBON RIMS CAN BE HARD to glue to. Vittoria Mastik'One seems to be the best available choice, according to research done at the University of Kansas.

Tires tend to snap off carbon rims rather than peel off; they require about the same amount of force to initially dislodge them from the rim, but then it takes less force to remove them than would a well-glued tubular on an aluminum rim. Follow the multiple-layer gluing procedure with clean rims and scraped base tape (7-11). It is worth pulling your tire off a few days after you have glued it on to see how well you did. Then reglue it the same way again, especially if all of the glue pulled off the rim in the process. For cyclocross, see the Pro Tip on gluing 'cross tires at the end of this chapter.

the rim with screwdrivers or other tools, as you will likely tear cords in the base tape and tire casing, leading to a bulge in the tire.

11. **Align the edge of the base tape with the rim.** The sticky tire will resist being pushed from side to side, so you may have to tug hard to center it. You want to see the same amount sticking out from the rim all the way around on both sides around the wheel.

12. **Fix misalignments.** Pump the tire to 100 psi (50 psi for a cyclocross tubular) and spin the wheel. Look for wobbles in the tire. If you find that the tread snakes back and forth as you spin the wheel, deflate the tire and push it over where required. Reinflate and check again, repeating the process until the tire is as straight as you are able to get it. The final process will depend somewhat on how accurately the tubular was made; some brands and models run straighter than others.

13. **Pump the tire up to 120–130 psi (60 psi for a cyclocross tubular) and leave it overnight to bond firmly.** You can get an even better bond by using a woodworker's band (miter) clamp around the entire inflated tire. The miter clamp (see Fig. 1.2) is a piece of nylon webbing with a ratchet-lock buckle on it. Depress the tab on the buckle to let out enough strap to surround the inflated tire and wheel. Pull the end of the strap to tighten the loop around the tire. Use a wrench to tighten the clamp and put extra pressure down on the tire to conform its bottom surface to the rim and bond it tightly. Tomorrow you can release the miter clamp (by using the release [thumb] tab) and ride or race on this wheel.

7-12

TAPING A TUBULAR TIRE ONTO A RIM

Tire gluing is such a hassle that some riders prefer to use tubular rim tape, but most tapes will not hold as well as a superior gluing job. It's not just the bond strength; there are instances of gluing tape delaminating, allowing the tubular to roll off and leaving a layer stuck to the rim and a layer stuck to the tire. You will find opinions both raving about and warning against tubular tape and will have to make your own decision. Don't ride any tire that you can easily push off by hand.

Gluing tapes (Fig. 7.20) like Effetto Mariposa "Carogna," Tufo, and Velox "Jantex" are easy-to-use, double-sided tapes that attach a tubular to a rim without using rim cement. Carogna holds quite well on any tire; Tufo tape adheres Tufo tires well (see Note at end of this section), and Jantex is okay only for road tubulars at high pressure and no aggressive cornering.

CX Tape ("Belgian Tape"; see step 7 of 7-11) is only for use in conjunction with tubular glue; cyclocross mechanics often use that combination to get a stronger bond (see Pro Tip on cyclocross tubular gluing at the end of this chapter).

1. **Take the preparatory steps for proper tire adhesion.** Follow steps 1 through 5 in 7-11, except do not sand the rim; tape needs a smooth surface.

2. **Starting at the valve hole, wrap and stick the tape onto the rim.** The release tape faces up; leave it on! If the tape is wider than the rim, don't center the tape. You'll want to trim it to the same width as the rim, which is best done after the tape is on the rim

7.20 Installing tubular gluing tape; peel back the corners of the backing tape.

but before the tire is mounted (to avoid cutting the tire). Offset the tape to one side so that one edge lines up with the edge of the rim and only one side overlaps the rim; that way you'll only have to trim one side.

3. **Trim the tape's length so the other end also terminates at the valve hole.**

4. **Press the tape down onto the rim.** Use your thumb or, better yet, a piece of dowel rod or a screwdriver handle the same shape as the rim bed.

5. **If the tape is too wide, trim it along the edge of the rim.** A sharp box-cutter knife works well. The sticky stuff will adhere to the knife; I didn't say this would be easy—just that it would be easier than gluing. Excess tape sticking out can pull in contaminants that will unglue the tire, so trimming the tape is important.

6. **Peel back a couple of centimeters of the release tape at each end.** Fold the ends at 90 degrees so they stick out from the sides of the rim, adjacent the valve hole.

7. **Install, inflate, and center the tire.** Follow steps 9 through 12 in 7-11. Centering is easy because the slick release tape allows the tire to move.

8. **Deflate the tire completely.**

9. **Pull the release tape out from under the entire tire.** Be careful not to push the tire laterally while doing so. Make sure you get all of the release tape out; sometimes the thin release tape tears, leaving a strip under the tire.

10. **Inflate the tire.** For a 19–25mm tire, go to 130 psi; go to half that for a cyclocross tire or other tubular 27mm or wider.

11. **Let it sit.** Carogna in particular, as it has a thick, super-sticky side toward the tire, needs time for the thick stuff to flow and evenly fill the voids under the tire; it's best to leave it in a warm place overnight. Tufo tape (with a Tufo tire—see following note) can be ridden immediately.

NOTE ON TUFO TUBULARS AND TAPE: *Tufo tubulars are not stitched together like most tubulars; they are seamless. The benefit to this for rim adhesion is that, unlike most tubulars, they have no bulged ridge along the bottom. Since the radius of curvature of the tire and of the rim bed are the same, the fit is very close, and extra glue is not required to fill voids under the tire. Tufo tape is designed for Tufo tires, and it does not have the ability to fill voids like Carogna, or like rim cement.*

7-13

CHANGING A TUBULAR TIRE ON THE ROAD

If you get a puncture out on the road with a tubular tire, it may be easier to deal with than a flat clincher, provided you can peel the old tire off without too much struggle and too many blood blisters on your thumbs. For your spare, carry an old tubular that has been glued to a rim in the past, or, with a new tire as a spare, either apply glue to the base tape as instructed in 7-11 before packing the tire as a spare, or bring along gluing tape.

Remove the wheel and pull the flat tire off the rim. If you did a good gluing job, this may take some doing. (On the other hand, if the tubular is easy to peel off, you need to improve your gluing technique.) Stretch the spare tire onto the rim as in Figure 7.17; if you have gluing tape, put it down first as in 7-12. Pump it up hard (over 100 psi for a road tire, 50 psi for a 'cross tubular) to get it to stay on the rim for the rest of the ride. Corner carefully going home, as the glue bond is marginal. When you get home, glue a tire securely on the rim before riding that wheel again.

7-14

PATCHING TUBULAR TIRES

In the early 1980s, my racing buddies and I spent countless hours patching tubular tires, often while sitting in the car on the way to distant races. Now that everyone trains on clinchers, nobody seems to patch tubulars anymore. Tubulars arguably are the best tires for racing, being lighter, working on lighter rims, staying on the rim when punctured, and having excellent cornering and descending characteristics. However, even though tubulars are expensive, it makes no sense to risk puncturing during a race due to using a patched tire. Also, tubulars with slow leaks can be easily fixed by injecting liquid sealant into them (7-15).

If you do wish to patch a tubular, here are the steps:

1. **Remove the tire from the rim.**
2. **Pump up the tire to 70 psi (40 psi for a cyclocross tubular), and find the leak by submerging the inflated tire in a bucket of water.** If you're lucky, air will come out through a hole in the tread. In the case of a pinched tube, though, the air may seep out through the casing randomly at the stitches, and be hard to pinpoint. See the next step for help.
3. **For 2 inches on either side of the puncture, peel away the base tape covering the stitching.** If you were unable to precisely locate the hole, try submerging the inflated tire now to watch the bubbles coming out through the stitching. Peel more base tape back if necessary until you are sure that you have exposed the stitching nearest the hole.
4. **Deflate the tire and carefully cut the outer layer of stitching threads for an inch or so on either side of the hole.** Pull the casing open in that spot, and pull enough of the tube out through the hole to find (7-2) and access the hole(s) in it.
5. **Patch the tube in the same manner as outlined in 7-3.** Use the same type of recommended patches.
6. **Push the tube back in place, and sew up the opening in the stitching by hand.** I recommend using a needle made for leather with a triangular cross-section tip and braided high-test fishing line. Stitch one way across the opening, turn the tire around, and double back over the stitches again. For obvi-

ous reasons, be careful not to poke the tube. You may need a thimble to push the needle in and a pair of pliers to pull it out on each stitch.

7. **Inflate the tire to 70 psi or so (40 psi for a cyclocross tubular) and submerge it. Make sure all of the leaks have been patched. Let it dry.**
8. **Deflate the tire and coat the peeled-back section of base tape and the exposed stitching area with contact cement.** Barge cement (a brand originally made for shoes and available in hardware stores) works well. Wait 15 minutes or so for the glue to set, and carefully stick the base tape back down over the stitching. (If the tape stretched when you pulled it loose, it's okay to cut it and overlap the ends.)
9. **Glue the tire onto the rim (7-11).**

7-15

TIRE SEALANTS

Tire sealants can virtually eliminate flat tires caused by tread punctures; they generally do not fix sidewall cuts or holes, pinch flats, or rim-side punctures. Tire sealants used to be made of goo full of chopped fibers that flow to punctures and seal them. These have largely been superseded by thinner sealants that have a liquid latex or similar base. These sealants can come as a liquid in a bottle, like Stan's NoTubes, Caffélatex, Orange Seal, or Schwalbe Doc Blue, or as an aerosol, like Vittoria Pit Stop or Hutchinson Fast'Air (Figs. 7.21–7.22).

NOTE: *You can also purchase tubes with sealant already inside.*

a. Latex sealant foam installation into an inner tube with a Presta valve, a tubular tire, or a tubeless tire

This is a great on-the-road fix for a slow leak.

1. **Deflate the tire, if it is not already deflated.**
2. **Screw or push the end of the nozzle onto the open Presta valve (Figs. 1.1B, 7.3).**
3. **Depress the button on the aerosol sealant can (Fig. 7.21) and hold it down until the full contents are deployed into the tire.** This will inflate the tire as well as fill it with sealant.
4. **Rotate the wheel until the hole is at the bottom, and hold it that way until the sealant has filled the**

7.21 Aerosol latex-based tire sealants

7.22 Injectable tire sealants

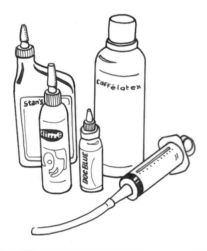

1. **Unscrew the valve core (counterclockwise) with an adjustable wrench or specific Presta valve-core wrench and remove it.** (You can use the Caffélatex with its injector even without a removable valve core, since it won't seal the valve immediately, but you cannot use Stan's sealant this way.)

2. **With the valve at the bottom, jam the tip of a squeeze bottle of liquid latex sealant into the valve stem (or the syringe hose onto the valve stem) and squeeze.** Inject a couple of fluid ounces inside.

3. **Screw the valve core back in (clockwise).**

4. **Pump the tire to full pressure.**

5. **If the tube has a leak, rotate the wheel until the hole is at the bottom, and hold it that way until the sealant has filled the hole and no more air is escaping.** Go ride, or spin the wheel for a while to further spread the sealant in the tube.

c. Maintaining sealant-filled tires

Inflating or deflating

Before opening the valve, have the stem at four o'clock and wait a minute for the sealant to drain away; if you don't, sealant will leak out when the valve is depressed, eventually clogging the valve. Even if you're careful about this, over time the valve may still become clogged with sealant. In that case:

1. **Remove the valve core.**

2. **Push a spoke or similar stiff wire into the valve stem to clear it.**

3. **Pick dried sealant off of the valve core.**

4. **Reinstall the valve core.**

Sealing punctures

1. **If you find the tire has gone flat, pump it up and ride it a bit to see if it seals.**

2. **If you get numerous punctures, you may need to pump and ride repeatedly before the tube seals up.**

3. **Remove embedded nails and other foreign objects.** Rotate the wheel to place the hole at the bottom to seal the hole.

4. **If it won't seal, establish whether it's a sealable hole.**

hole and no more air is escaping. Go for a ride, or spin the wheel for a while to further spread the sealant around in the tube.

Alternatively, the old-school way to seal a slow leak in a tubular or an inner tube, or to add some puncture resistance to either, is to pour a can of evaporated milk into a pump you don't care about and pump it in through the Presta valve. It works quite well for tiny leaks, but if you get a blowout, boy, does it ever stink!

b. Sealing a tubular, an inner tube with a Presta valve (especially with a removable valve core), or a tubeless tire with liquid latex-type tire sealant

The Presta valve must have a removable core, which separate inner tubes rarely have, but which good tubulars and some tubeless valves do have. If the valve core is removable, there will be two wrench flats present on opposite sides of the small valve-cap threads.

Pinch flats, caused by pinching the inner tube between the tire and rim, are nearly impossible to seal because one of the two "snakebite" holes is on the rim side and will not seal because the sealant is thrown to the outside by centrifugal force. You will need to replace the tube.

Sidewall gashes won't seal; the tire needs to be replaced.

CYCLOCROSS TIRES

7-16

TIRE SELECTION

LEVEL **a. Tire type**

Before you can decide on a wheel type (8-14), you need to select the tire type. The choices are clincher, tubeless clincher (or "tubeless-ready," meaning that they require sealant), and tubular cyclocross tires.

Clinchers

Clinchers are the cheapest choice, but they generally roll the slowest and corner the least well. They are easy to mount. For training, an unsponsored rider should use clinchers, because all it takes is an errant piece of sharp metal or a few thorns to reduce an expensive tubular tire to a piece of junk and condemn yourself to yet more hours of tire gluing (7-11). Modern liquid tire sealants can breathe new life into a tubular with small punctures (7-14), but with a clincher, you can also change the inner tube and keep riding the tire for many more miles.

Downsides of clinchers include high weight and rolling resistance, poor grip and poor suspension due to higher pressures required to avoid pinch flats on the sharp edge of a clincher rim (Fig. 7.23), and limited capability of riding when flat. A clincher tire and its tube weigh more than a good tubular, and a clincher wheel is also heavier than a tubular wheel of similar strength and stiffness. Rolling resistance and cornering sluggishness are higher for a clincher, due to its stiffer sidewall cords and thicker tube than a more supple tubular. And clincher rim edges are easily dented.

Tubeless clinchers

Tubeless and "tubeless-ready" clinchers cannot pinch flat, so you can run them at very low pressures to get better traction, a smoother ride, and lower rolling resistance on bumpy sections. If you install them with sealant (7-14), they are essentially impervious to small punctures (you can use sealant in a tubular or clincher, too, by injecting it through the valve, ideally with the valve core removed).

Tubeless and "tubeless-ready" clinchers have the stiff casing of a clincher, but they don't have a tube stiffening and chafing them and can be run at low pressures without fear of pinch flats, so their rolling resistance can approach that of an expensive, supple tubular. They can be as fast to install and remove as a clincher and cost only marginally more. Mounted on a tubeless-specific clincher rim, tubeless-ready clinchers are designed to stay on if ridden when flat, because the "humps" on the rim shelves should "lock" the tire bead on (Fig. 7.16).

The big caveat with tubeless cyclocross tires is that the low tire pressure and intermediate tire diameters used in cyclocross have a tendency to defeat tubeless features that work on road and mountain bike tubeless tires, and "burping" (allowing air to escape suddenly under the bead edge) is often a problem on sharp turns. Also, like any clincher rim, the rim edges are susceptible to denting on rocks, stairs, and curbs. See the Pro Tip on setting up tubeless cyclocross clincher tires.

Tubulars

Tubular tires sit on flatter rim surfaces (Fig. 7.24), are less prone to pinch flats, and can be ridden back to the pit when flat; it's unlikely a flat clincher will stay on the rim long enough for you to do that.

Tubulars, besides being lighter and having lighter rims, lower rolling resistance, and less proclivity to pinch flats, are glued to the rim. Their supple casings keep more tread on the ground than a clincher, unless you were to run the clincher at such low pressure that pinch flats would be a certainty. Tubulars are generally going to be faster than clinchers. But you will pay a high price for tubulars both at the cash register and in time spent gluing (7-11 and the Pro Tip on gluing cyclocross tires).

7.23 Blunt-nose shallow-section clincher rim cross-section

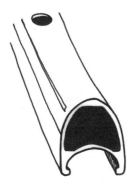

7.24 Deep-section tubular rim cross-section

Some tubulars are tubeless and are sold with a sealant.

Caffélatex sealant (7-15b) or a sealant that came with the tire can be used to fix a slow leak in a cyclocross tubular, or as a prophylactic measure before riding where there are lots of thorns. The downside is that it can solidify inside the tube over the offseason, and rinsing the inner tube and drying it out is not feasible; keep the valve closed at all times to prevent sealant hardening. Sealant removal is relatively easy with a tubeless tubular tire; you can flush it with water via a big syringe, and you can suck water out with a vacuum cleaner necked down to the valve with tape.

b. Tire size

Don't get a tire smaller than 30mm (i.e., 700 × 30). On the other hand, tires wider than 33mm are banned in races sanctioned by the UCI (Union Cycliste Internationale, the governing body of the sport), so a 33mm tire is

---

PRO TIP | GLUING TUBULAR CYCLOCROSS TIRES

GLUING A CYCLOCROSS TIRE is the same as gluing a road tubular (7-11), except there are four important distinctions that make gluing a cyclocross tubular more problematic:

- A cyclocross tire is fatter than a road tire, but most rims used for cyclocross are road rims. Since a tubular tire is round in cross-section (Fig. 7.2), this means that the curvature of a 'cross tire will be lower than the curvature of the concave top surface of the rim, so it will only contact the edges of the rim.

- A cyclocross tire, if inflated as intended, will be ridden at much lower pressure than a road tubular, in order to get better traction, cushion the ride, and reduce rolling resistance on bumpy sections of the course. Tire-gluing studies show that tire adhesion to the rim decreases with decreasing tire pressure (you can see this yourself by putting an unglued tire on a rim and trying to push it off at varying air pressures).

- The continual washing that cyclocross bikes get (2-18), combined with the constant splashing through mud, sand, and water, will tend to dry out the glue joint and infiltrate it with dirt.

- Leverage on a taller tire is higher.

All four may be reasons to use both glue and CX Tape tubular gluing tape (7-11, step 7). The glue-plus-CX-Tape method should hold on any 'cross tire if done correctly, but if not, you can use the following method to shim up the center of the rim to better fit the tire's shape:

1. Peel the base tape off an old tubular and split it down the center with scissors.

2. After applying the second glue layer as in 7-11, step 6, smear glue down one side of the half-width strip of old base tape, and stick it down the center of the rim channel. Apply the third glue layer over it.

3. With a carbon rim, also read the Pro Tip in 7-11 on gluing a tubular to a carbon rim.

YOU HAVE TO BE COMMITTED to some experimentation (ideally before the season) to get tubeless tires to work reliably in cyclocross, especially for a big rider racing with them. The bead needs to fit tightly to not burp air, and there is evidence to suggest that stiffer tire casings are also more likely to burp air than more flexible ones. A common method of addressing burping issues is to add additional layers of rim strip to make the rim bead seat effectively larger in diameter.

You can start with a tubeless-specific road rim, or you can convert a standard rim to tubeless by sealing off the rim's nipple holes. NoTube or Effetto Mariposa tape is the ticket here. You will generally be less likely to burp air if you add an extra layer or two of tape. NoTube's full tubeless conversion kit includes a thick rubber rim strip, and this is perhaps an even better method; like additional tape layers, it also increases the bead seat diameter and effectively reduces the sidewall height (allowing the tire to flex more), but it also seals better around the tire bead.

You can use tubeless-ready clincher tires, and there are some standard clinchers that seem to work well for running without a tube. Use a generous amount of sealant (up to 1/4 cup). Avoid tires that have little vertical sipes along the bead on the outside of the tire, which can prevent making a seal to the rim.

When running any rim and tire tubeless with sealant, you should remove the tire periodically and check for corrosion of the rim; especially remove the valve stem and check there. Rim corrosion can be a problem with some rims and some sealants, and you don't want to let it go until the rim fails.

generally the size to select. Make sure your bike has sufficient mud clearance around the tire; with the tire mounted on a wheel, check under the fork crown, between the chainstays, behind the chainstay bridge, behind the front derailleur and the cable roller on the back of the seat tube, under the brake bridge, and between the seatstays.

A bigger tire gives more cushioning and traction and can be run at a lower pressure, both of which are advantages, but its drawback is extra weight. A tire smaller than 30mm will rattle your teeth out and invite pinch flats.

With clinchers, don't use a road-size inner tube; it will be stretched too thin for durability inside a big cyclo-cross tire. And running sealant inside an undersized inner tube will be less effective; if it punctures, the hole will widen due to the stretched tube. Get a 700 × 28–35 inner tube. While you're at it, make sure the rim strip (Fig. 7.12) is in good shape and can't become dislodged due to water, mud, and being bumped around.

c. Tire tread pattern

With tubulars, choices include a mud tread (Fig. 7.25), an open knobby tread (7.26), a chevron tread (Fig. 7.27), or a file tread (Fig. 7.28). File-tread tires are great in grass, sand, and tacky dirt but have insufficient traction for muddy or loose course conditions, so don't bother buying a pair of these unless you have several pairs of wheels. Obviously, pick the mud tread for muddy courses and the knobby or chevron tread for everything else.

Clinchers and tubeless-ready clinchers offer these tread choices and more, and they're cheaper and easier to change, so experiment; try trading wheels with friends.

d. Tire pressure

'Cross riders coming from a road-riding background tend to use too much tire pressure. Suffice it to say that the best riders in the world are often running 18–28 psi in tubular tires (and you thought 50 psi was super-low!). Traction and rolling resistance improve with lower pres-

7.25 Cyclocross mud tread pattern
(Challenge Limus)

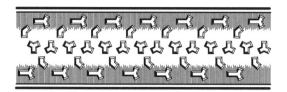

7.26 Cyclocross knobby tread pattern
(Schwalbe Racing Ralph)

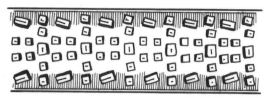

7.27 Cyclocross chevron tread pattern
(Vittoria Cross XG Pro)

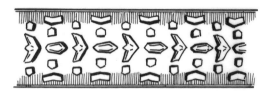

7.28 Cyclocross file tread pattern
(Challenge Grifo XS)

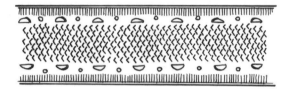

sures, up to a point, so 40 psi should be the maximum for racing, save for on a completely hardpacked, smooth course or if you're a big rider running clincher tires not big enough to evade pinch flats.

If you have tight-fitting, supple tubeless clinchers and don't have burping problems with them (see the previous Pro Tip on this), you can run very low pressures with them, too, but the sharp rim edges are more vulnerable to denting than tubeless rims.

PRO TIP ─ SEALING THE CASING

'CROSS PROS OFTEN COAT their tire sidewalls with Aquaseal urethane adhesive (a McNett product) to seal the casings from mud and allow the mud to wash off more easily.

Also, regular use of 303 Protectant on the tread and the sidewall will extend the tire's life.

*You don't have to invent the wheel,
but you might want to be the
company that invents the rims.*

—M. C. HAMMER

8

WHEELS

TOOLS

Spoke wrench
Grease
2mm, 2.5mm, 5mm,
    10mm hex keys
13mm, 14mm, 15mm
    cone wrenches
Metric open-end or
    adjustable wrenches
Freehub cassette lockring
    remover
Pedro's Vise Whip or
    chain whip

OPTIONAL

Citrus solvent
Truing stand
Freewheel remover
Soft hammer
Rohloff HG-Check
    cog-wear indicator
Fine-tip grease gun

Most road bike wheels are strung together with spokes connecting the hub to the rim. The rim, which usually is made of aluminum but can also be made of carbon-fiber composite, steel, magnesium, or wood, supports the tire and forms the surface on which the brakes are applied. It in turn is supported and aligned by the tension of the spokes. Bearings in the hub, when clean and properly adjusted, allow the wheel to turn freely around the axle.

Composite (i.e., carbon fiber) wheels can use rigid members instead of wire spokes to connect the hub and rim, as is the case with disc wheels or three-, four-, or five-spoke wheels. The spokes in this case are in compression, like the wooden spoked wheels on a Conestoga wagon, not in tension as with a wire-spoke wheel. A composite wheel cannot be trued, although some can be returned to the manufacturer for rim replacement. Some composite wheels use composite spokes in tension, much like a wire-spoke wheel, but the spokes are usually bonded to the rim and hub and hence can't be adjusted to true the wheel. One attempt to bridge the gap between the tension wheel and the wagon wheel is the Mavic R-Sys TraComp (i.e., TRAction/COMPression) wheel. It has thick, rigid carbon spokes backed up against a shell inside the hub to support the rim somewhat in compression, yet the spokes are also in tension and have threaded nipples to allow truing.

Wheels intended for aerodynamic efficiency have either solid sides (disc wheels) or aerodynamically shaped rims and few, blade-shaped spokes. The spokes can be steel, aluminum, titanium, or composite (which generally means carbon fiber).

A cassette freehub or freewheel with cogs attached (Fig. 8.1) allows the rear wheel to spin independently of the chain and pedals while you are coasting, and it engages when force is applied to the pedals. The tires provide suspension as well as grip and traction for propulsion and steering. The air pressure in the tire is the primary suspension system on a road bike.

RIMS AND SPOKES

8-1

CHECKING RIM CONDITION

LEVEL After the tires, your bike's rims are your next line of contact with the road. You never want a wheel to fail on you while you are riding, as the consequences can be severe. To

8.1 The rear wheel

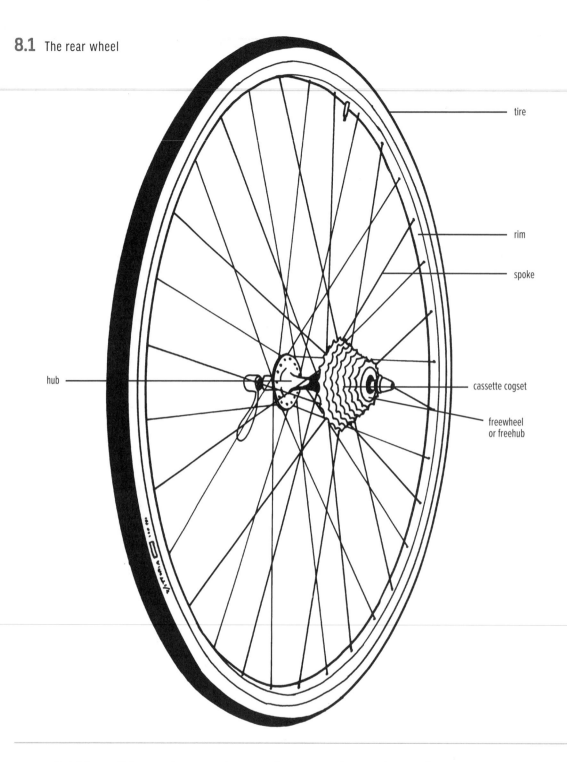

tire

rim

spoke

cassette cogset

freewheel
or freehub

hub

ensure that this won't happen to you, replace any rim that has a significant defect. Chapter 15 describes how to rebuild a wheel with a new rim, should you decide to do it yourself.

Check over the rims for cracks, particularly at the spoke holes, the valve hole, and the seam (opposite the valve hole). If you find a crack, the rim should be replaced immediately.

Inspect the rims for a wear indicator. Modern rims have telltale markings to let you know when brakes have worn the rim sidewalls too thin. On a clincher rim, if the sidewalls become too thin, the tire can push the sidewalls out and fold them open like a limp taco shell, causing braking and tire-retention issues. On a tubular rim, a rim worn thin from braking simply becomes weak enough to collapse.

Rim-wear indicators take different forms. In some cases, they consist of small holes drilled partway into the brake track that, once gone, indicate that too much sidewall material has been worn off. Others have a dark spot underneath the surface that appears once the wear limit has been reached. Mavic rims have a dark hole that appears when the rim is deeply worn; the hole is directly opposite the valve stem hole and is identified by a sticker.

8-2

TRUING A WHEEL

For more information on truing wheels, see Chapter 15 on wheelbuilding (15-4).

You can fix a mild wobble by adjusting the tension on the spokes. An extreme bend in the rim cannot be fixed by spoke truing alone, because the spoke tension on the two sides of the wheel will be so uneven that the wheel will rapidly fall apart. If you have a bent rim that cannot be corrected with spoke truing, you can try banging it into shape to use temporarily to get yourself home (see 3-10 and Fig. 3.11).

1. **Check that there are no broken spokes in the wheel, or any spokes that are so loose that they flop around.** If there is a broken spoke, follow the replacement procedure in 8-3. If there is a single loose spoke, check to see that the rim is not dented or cracked in that area. If the rim is damaged, replace it. If the rim looks okay, mark the loose spoke with a piece of tape and tighten its nipple with a spoke wrench (see step 5 regarding tightening direction) until it feels like it has tension similar to other spokes coming from the same side of the hub (pluck the spoke and feel for tension as well as listen to the tone). Then continue with the truing procedure.

2. **Grab the rim while the wheel is on the bike, and flex it side to side to check the hub-bearing adjustment.** If the bearings are loose, the wheel will clunk side to side. The play in the bearings will have to be eliminated before you true the wheel or the rim will wobble erratically because of the loose hub. Follow the hub-adjustment procedure in 8-6d, steps 29–32.

8.2 Lateral truing if rim scrapes on the left

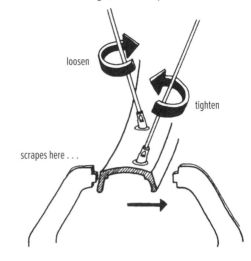

8.3 Lateral truing if rim scrapes on the right

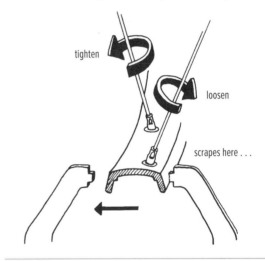

3. **Put the wheel in a truing stand, if you have one.** Otherwise, leave it on the bike and suspend the bike in a bike stand or from the ceiling, or turn it upside down on the handlebar and saddle.

4. **Adjust the truing stand feeler, or hold one of the brake pads so that it scrapes the rim at the biggest wobble.**

5. **Where the rim scrapes, tighten the spoke (or spokes) that come(s) to the rim from the opposite side of the hub, and loosen the spoke(s) that come(s) from the same side of the hub (Figs. 8.2, 8.3).** This process will pull the rim away from the feeler or brake pad. Spokes and nipples have standard right-hand threads: righty tighty, lefty loosey. See the following Note on rotation direction.

NOTE ON DIRECTION TO TURN THE NIPPLES: *Adjusting spokes is like opening or closing an upside-down jar. With the jar right side up, turning the lid to the left (counterclockwise) opens the jar, but you'd turn the lid the opposite (clockwise) direction when you turn the jar upside down (try it and see). Spoke nipples are just like the lid on that upside-down jar. When the nipples are at the bottom of the rim (Fig. 8.4), counterclockwise tightens (shortens the spoke, like closing the lid shortens the jar plus lid), and clockwise loosens (lengthens the spoke, as unscrewing the lid lengthens the jar plus lid until the jar opens). The opposite is true when the nipples to be turned are at the top; then it's like an upright jar and lid. It may take you a few attempts before you catch on, but you will eventually get it. If you temporarily make the wheel worse, simply undo what you have done and start over.*

Note that rotation direction for the nipples on some Shimano wheels is the opposite of this because the nipples screw into the rim. There is also a little sleeve that screws onto the (right-hand) threads on the hub end of the spoke, after the spoke has been slid through its hub hole.

Tighten and loosen about a quarter turn at a time, decreasing the amount you turn the spoke nipples as you move away from the spot where the rim scrapes the hardest. If the wobble gets worse, you are turning the nipples in the wrong direction.

NOTE ON TWISTED SPOKES: *As you turn the nipple on a tight spoke, particularly a thin one or a flat, aero one, it will tend to twist. To avoid this, when you tighten or loosen a spoke, twist it in the direction that corrects the problem, and then twist back half as far. This will be enough to unwind most spokes, but aero ones or superlight ones will still have a twist in them. With superthin round ones, you won't really be able to tell until you go for a ride; the wheel may then "ping" at first as the spokes unwind and settle in. You may need to re-true the wheel once they have settled in. With aero spokes, it is best to hold the spoke from twisting as you turn the nipple. The best tool for steel aero spokes, from DT Swiss, is a red plastic spoke key with a long, conical groove down the spoke side to fit an L-shaped steel tool to keep the spoke from twisting*

8.4 Tightening and loosening spokes

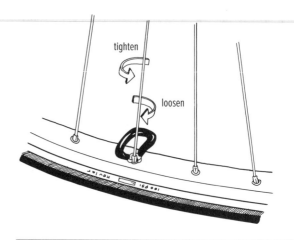

(1-4, Fig. 1.4). The long part of the L-shaped tool is slotted for the aero spoke to keep it from twisting as you turn the nipple, and its conical exterior fits in the groove in the spoke key. For Mavic flat aluminum spokes, use Mavic's plastic ring with slots.

NOTE ON INTERNAL NIPPLES: *Some deep-section wheels have the nipples inside the rim. To adjust the nipples, you must remove the tire and, in the case of a clincher, the rim strip as well. Then you reach down into the spoke holes with the correct-size socket. The simplest tool for this is either a Y-wrench with 5mm and 5.5mm hex sockets and a square-drive socket (for upside-down standard nipples inside the rim) on the ends of its three long arms (1-4, Fig. 1.4) or a specialty wrench for this wheel's particular internal spoke nipples.*

NOTE ON NIPPLES AT THE HUB: *Some wheels have the spoke nipples at the hub, not at the rim. Before you turn them, think carefully about which way tightens and which way loosens. Remember that if you want the spoke tighter (shorter), it is like tightening a jar lid. So if you are looking down the spoke toward the hub, it's like looking at an upside-down jar, whose lid you would thus tighten by turning it counterclockwise.*

6. **As the rim moves into proper alignment, readjust the truing-stand feeler or the brake pad so that it again touches the most out-of-true spot on the wheel.**

7. **Check the wobble first on one side of the wheel and then the other, adjusting spokes accordingly.**

You want to make sure that you don't end up pulling the whole wheel off center by chasing wobbles only on one side. As the wheel gets closer to true, you will need to decrease the amount you turn each spoke; otherwise, you will overcorrect.

8. **Accept a certain amount of wobble if you are truing the wheel in a bike.** The method is not very accurate and is not at all suited for making a wheel absolutely true. If you have access to a wheel-dishing tool, check to make sure that the wheel is centered (15-5, Figs. 15.22, 15.23).

NOTE ON ROUNDNESS: *I have discussed only lateral truing here, but a wheel can also get out of round from banging something hard; it's much less likely to lose radial trueness from spokes loosening than it is to lose lateral trueness. If the wheel is not dented, you may be able to improve its roundness without having so large a tension difference in the spokes from section to section of the wheel that it is unstable and falls apart rapidly. Consult 15-4b, Figures 15.20 and 15.21, to see how to radially true the wheel.*

8-3

REPLACING A BROKEN SPOKE

Go to the bike store and get a new spoke of the same length. Remember, the spokes on the front wheel are usually not the same length as the spokes on the rear wheel. Also, the spokes on the drive (right) side of the rear wheel are almost always shorter than those on the non-drive (left) side. The same goes for spokes on the rotor side of a disc-brake front wheel.

1. **Make sure you are using a replacement spoke of the proper thickness and length.**
2. **Thread the spoke through the spoke hole in the hub flange in the same direction (from the inside out or from the outside in) that the broken one went through.** If the broken spoke is on the drive side of the rear wheel, you will need to remove the cassette cogs or the freewheel to get at the hub flange (8-10 and 8-11).
3. **Weave the new spoke in with the other spokes just as it was before (Fig. 8.5).** It may take some bending to get it in place.

8.5 Weaving a new spoke

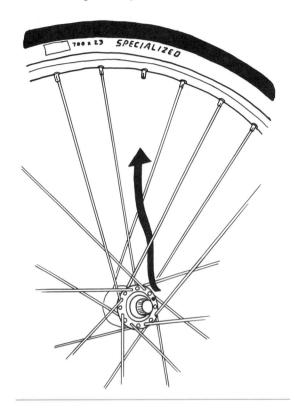

4. **Thread it into the same nipple, if the nipple is in good shape.** Otherwise, use a new nipple; you'll need to remove the tire, tube, and rim strip (or the tubular tire) to install the nipple.
5. **Mark the new spoke with a piece of tape and tighten it about as snugly as the neighboring spokes coming from the same side of the hub.**
6. **Follow the steps for truing a wheel as outlined in 8-2.**

HUBS

8-4

OVERHAULING HUBS

 Hubs should turn smoothly and noise-lessly. If you maintain them regularly, they always will.

All hubs have a hub shell that contains the axle and bearings and is connected to the rim with spokes or, in the case of disc wheels, with sheets of composite material. Beyond that, they diverge into two types:

the loose-bearing cup-and-cone type (Fig. 8.6) and the sealed-bearing (or cartridge-bearing) type (Fig. 8.7).

Cup-and-cone hubs have loose ball bearings that roll along very smooth bearing surfaces called "races" or "cups." An axle runs through the center of the hub. Conical nuts, called "cones" (Fig. 8.6), thread onto the axle. The cones create an inner race for the bearings. In high-quality hubs, the cup-and-cone surfaces that contact the bearings are precisely machined to minimize friction. The operation of the hub depends on the smoothness and lubrication of the cones, ball bearings, and bearing cups. The cones are held in place on the axle by one or more spacers (washers) followed by threaded locknuts that tighten down against the cones and spacers to keep the hub in proper adjustment. In the absence of a locknut, the cone will have a slot with a pinch bolt, which, when tightened, prevents the cone from unscrewing. The rear hub will generally have more spacers than the front, especially on the drive side (Fig. 8.19).

The term "sealed-bearing" hub is a bit of a misnomer, because some cup-and-cone hubs offer better protection against dirt and water than some sealed-bearing hubs. The phrase "cartridge-bearing hub" is more accurate, because the distinguishing feature of these hubs is that the ball bearings, races, and cones are assembled as a complete unit—the cartridge—that is then plugged into a hub shell. Cartridge-bearing front hubs have two bearings, one on each end of the hub shell (Fig. 8.7). A rear hub may have a cartridge bearing on either end of the hub (Fig. 8.18) with loose balls inside of the freehub; it may have multiple cartridge bearings and/or bushings inside the freehub as well as a bearing on either end of the hub (Fig. 8.24); or it may have a cartridge bearing in the left (non-drive) side only and may employ loose hub bearings as well as loose freehub bearings on the drive side.

Cartridge-bearing hubs can have any number of axle-assembly types. Some have a threaded axle with locknuts (Fig. 8.18), similar to a cup-and-cone hub. More common in high-end hubs are aluminum axles (sometimes carbon-fiber or very thin-wall steel axles), often very large in diameter, with correspondingly large bearings. Their end caps usually snap on (Fig. 8.24) or are held with setscrews or circlips. The end caps may also thread into the axle and accept a 5mm hex key.

8.6 Cup-and-cone front hub with standard ball bearings

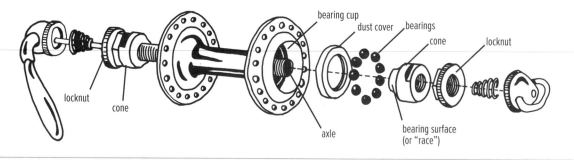

8.7 Front hub with cartridge bearing

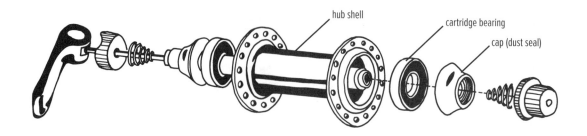

8-5

ALL HUBS, PRELIMINARY

1. **Remove the wheel from the bike (2-2 and 2-11).**
2. **Remove the quick-release skewer or the nuts and washers holding the wheel onto the bike.** With a through-axle hub, flip open the lever, spin the lever to completely unscrew the axle, and pull the axle out; the wheel will now fall out.

8-6

OVERHAULING CUP-AND-CONE HUB (FRONT OR REAR)

To isolate problems, take some time to evaluate the hub's condition before disassembling it. Spin the hub while holding the axle, and turn the axle while holding the hub. Does it turn roughly? Is the axle bent or broken? Wobble the axle side to side. Is the bearing adjustment loose?

a. Disassembly

1. **Set the wheel flat on a table or workbench.** On a rear wheel, work on the left (non-drive) side. Slip a cone wrench of the correct size (usually 13mm, 14mm, or 15mm) onto the wrench flats on one of the cones. If the cone has a slot and a pinch bolt on it (and takes a much bigger cone wrench than those listed here), read the following Note.

NOTE ON CAMPAGNOLO AND FULCRUM: *On current Campagnolo and Fulcrum high-end hubs, you must unscrew the axle into two pieces with 5mm hex keys inserted into either end of the axle. Loosen the setscrew on the large, aluminum, split locknut with a 2.5mm hex key (loosen the setscrew three turns). Unscrew and remove the locknut with your fingers or with a 22mm (or adjustable) wrench. Push the axle end toward the hub to free the slide-on cone. Follow steps 7–27, and then assemble in the reverse order. For a rear hub, see 8-12e for hints on reinstalling the freehub pawls into the hub shell. Adjust the hub by turning the locknut by hand or with a 22mm (or adjustable) wrench until the bearing end play is removed, and tighten the setscrew with a 2.5mm hex key. This adjustment can*

even be performed while the wheel is installed in the frame or fork to get it extremely precise—free-running with no end play.

2. **Put an appropriately sized wrench or adjustable wrench on the locknut on the same side.**
3. **While holding the cone with the cone wrench, loosen the locknut (Fig. 8.8).** This may take considerable force, because the parts are usually tightened against each other securely to maintain the hub's adjustment. Make sure that you are unscrewing the locknut counterclockwise (lefty loosey, righty tighty).
4. **Unscrew the loose locknut off of the axle.** In order to hold the axle in place as you unscrew the locknut, move the cone wrench from the cone on top to the cone on the opposite end of the axle as soon as the locknut loosens. On a rear hub, put another open-end wrench on the opposite locknut.
5. **Slide off any spacers, keeping track of where they came from.** If they will not slide off, the cone will push them off when you unscrew it. Note that some spacers have a small tooth or "key" that corresponds to a lengthwise groove in the axle. Keep these lined up to facilitate removal.
6. **Unscrew the cone from the axle.** Again, you may need to hold the opposite cone with a wrench.
7. **Keep track of the various nuts, spacers, and cones by placing them on your workbench in the**

8.8 Loosening or tightening the locknut

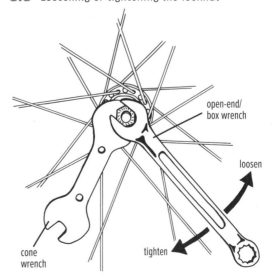

open-end/
box wrench

loosen

tighten

cone
wrench

order they were removed. If that seems too casual, you can slide a twist tie, zip tie, or string through the parts in the correct order and orientation. Either method serves as an easy guide when reassembling the hub.

8. **To catch any bearings that may fall out, put your hand over the end of the hub from which you removed the nuts and spacers and flip the wheel over.** Have a rag underneath the wheel to catch stray bearings.

9. **Pull the axle up and out, being careful not to lose any bearings that fall out of the hub or stick briefly to the axle.** Leave the cone, spacers, and locknut all tightened together on the opposite end of the axle from the one you disassembled. If you are replacing a bent or broken axle, measure the amount of axle sticking out beyond the locknut. Put the cone, spacers, and locknut on the new axle identically.

10. **Remove all of the ball bearings from both sides of the hub.** They may stick to a screwdriver with a coating of grease on the tip, or you can push them down through the center of the hub and out the other side with the screwdriver. Tweezers or a magnetic screwdriver may also be useful for

8.9 Removing the dust caps

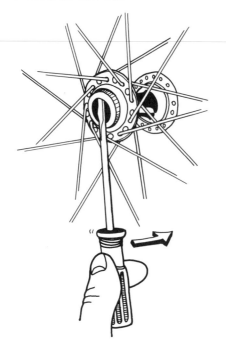

removing bearings. Put the bearings in a cup, a jar lid, or the like. Count the bearings, and make sure you have the same number from each side.

11. **With a screwdriver, gently pop off the seals (i.e., the dust caps) that are pressed into either end of the hub shell (Fig. 8.9).** Be careful not to deform them; leave them in if you can't pop them out without damage. If they are not removed, it is tedious, but not impossible, to clean the dirty grease out of their concave interior with a rag and a thin screwdriver and perhaps some solvent.

b. Cleaning

12. **Clean the hub shell parts.** Wipe out the hub shell with a rag. Remove all dirt and grease from the bearing surfaces. With a screwdriver, push a rag through the hub shell and spin it to clean out the hub-shell axle hole. Wipe off the outer faces of the shell. Finish with a very clean rag on the bearing surfaces, which should be shiny and completely free of dirt or grease. If the hub has been neglected, the grease may have solidified and glazed over so completely that you will need a solvent to remove it. Wear gloves while working with the solvent. If you are working on a rear cassette hub, take this opportunity to lubricate the freehub. (See 8-12 on lubricating freehubs.)

13. **Clean the running parts.** Wipe down the axle, nuts, and cones with a rag. Clean the cones well with a clean rag; strive for spotless. Again, solvent may be required if the grease has solidified. Get any dirt out of the threads on the disassembled axle end to prevent the cone from pushing the dirt into the hub upon reassembly.

14. **Wipe the grease and dirt off the seals.** A rag over the end of a screwdriver is sometimes useful to get inside. Again, glaze-hard grease may have to be removed with a solvent. Keep solvent out of the freehub body.

15. **Clean the bearing balls.** Wipe off the bearings by rubbing all of them together between two rags. This may be sufficient to clean them completely, but small specks of dirt can still adhere to them, so I advise the next step as well.

16. **Polish the bearings.** (If you are overhauling low-quality hubs, you can skip to the next step.) I prefer to wash bearing balls in a plugged sink with an abrasive soap like Lava, rubbing them between my hands as if I were washing my palms. This really gets them shining, unless they are caked with glaze-hard grease. Make sure you have plugged the sink drain! This method has the added advantage of getting my hands clean for the assembly step. It is silly to contaminate your superclean parts with dirty hands. If there is hardened glaze on the bearings, soak them in solvent. If that does not remove it, buy new bearings at the bike shop. Take a few of the old bearings along so that you get the right size.

17. **Dry all bearings and any other wet parts.** Inspect the bearings and bearing surfaces carefully. If any of the bearings have pits or gouges in them, replace all of them. Same goes for the cones. A patina or lack of sheen on balls and cones indicates wear and is cause for replacement. Most bike shops stock replacement cones. If the bearing races (or cups) in the hub shell are pitted, the only thing you can do is buy new hubs. Regular maintenance and proper adjustment can prevent pitted bearing races.

NOTE ON BEARING INSPECTION: *Using new ball bearings when overhauling standard cup-and-cone hubs ensures round, smooth bearings; however, do not avoid performing an overhaul just because you don't have any new ball bearings.*

Inspect the bearings carefully. If there is even the slightest hint of uneven wear or pitting on the balls, cups, or cones, throw the bearings out and complete the overhaul with new bearings.

c. Assembly and lubrication

18. **Press the seals or dust covers in on both ends of the hub shell.**

19. **Smear grease with your clean finger into the bearing race on one end of the hub shell.** I like using light-colored or clear grease so that I can see if it gets dirty, but any bike grease will do. Grease not only lubricates the bearings but also forms a barrier to dirt and water, so use enough grease to cover the balls halfway. Too much grease will slow the hub by packing around the axle.

20. **Stick half of the ball bearings into the grease, making sure you put in the same number of bearings that came out.** Distribute them uniformly around in the bearing race.

21. **Smear some grease on the cone that is still attached to the axle, and slide the axle into the hub shell.** Lift the wheel up a bit (30-degree angle), so that you can push the axle in until the cone slides into position, keeping all the bearings in place. On rear hubs, it is important to replace the axle and cone assembly into the same side of the hub from which it was removed to preserve drive-side cog spacing.

22. **Holding the axle pushed inward with one hand to secure the bearings (Fig. 8.10), turn the wheel over.**

23. **Smear grease into the bearing race that is now facing up.** Lift the wheel and allow the axle to slide down just enough so that it is not sticking up past the bearing race. Make sure no bearings fall out of the bottom. If the race and bearings are properly greased and the axle remains in the hub shell, they are not likely to fall out.

24. **While the top end of the axle is still below the bearing race, place the remaining bearings uniformly around in the grease.** Make sure you have inserted the correct number of bearings.

8.10 Pushing inward on the axle

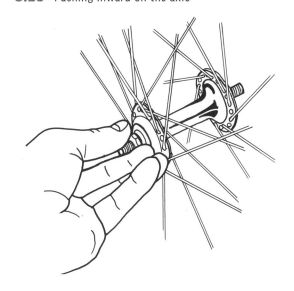

8.11 Seating the bottom cone in the bearings

8.12 Tightening the locknut with another wrench

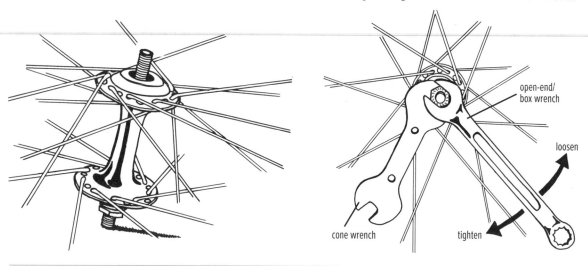

open-end/
box wrench

loosen

cone wrench   tighten

25. **Slide the axle back up into place.** Set the wheel down on the table, so that the wheel rests on the lower axle end, seating the cone into the bearings (Fig. 8.11).

26. **Cover the top cone with a film of grease and then, with your fingers, screw it into place, seating it snugly onto the bearings.**

27. **In correct order, slide on the washer and any spacers.** Properly align any washers that have a little tooth or "key" that fits into a lengthwise groove in the axle.

28. **Use your finger to screw on the locknut.** Note that the two sides of the locknut are not the same. If you are unsure about which way the locknut goes back on, check the orientation of the locknut that is on the opposite end of the axle (this locknut was not removed during this overhaul and is assumed to be in the correct orientation). As a general rule, the rough or concave surface of the locknut faces out so that it can get a good purchase on the dropout.

d. Hub adjustment

29. **Thread the cone onto the axle until it lightly contacts the bearings.** The axle should turn smoothly without any grinding, and there should be a small amount of lateral play. Thread the locknut down until it is snug against the washer(s) and cone.

30. **Tighten the locknut.** Place the cone wrench onto the flats of the hub cone. While holding the cone

steady, tighten the locknut with another wrench (Fig. 8.12). Tighten it about as snugly as you can against the cone and spacers, in order to hold the adjustment. Be sure that you are tightening the locknut and not the cone; you can ruin the hub if you tighten the cone hard against the bearings. On a Campagnolo or Fulcrum hub with a split cone, tighten the cone's pinch bolt now.

31. **Check and adjust play.** If there is too much play in the axle when you are done, or if the bearings are tight, loosen the locknut while holding the cone with the cone wrench. If the hub is too tight, unscrew the cone a bit. If the hub is too loose, screw the cone in a bit. You may need to put a wrench on the opposite-side cone to effectively tighten or loosen the cone you are adjusting.

32. **Repeat steps 29–31 until the hub adjustment feels right.** There should be a slight amount of axle-end play so that the pressure of the quick-release skewer will compress it to perfect adjustment. (A hub held on with nuts—no quick release—should be adjusted with no bearing play.) Tighten the locknut firmly against the cone to hold the adjustment.

NOTE ON HUB ADJUSTMENT: *You may find that tightening the locknut against the cone suddenly turns your "Mona Lisa" perfect hub adjustment into something slightly less beautiful. If it is too tight, back off both cones (with a cone wrench on either side of the hub,* *each on one cone) a fraction of a turn. If too loose, tighten both locknuts a bit. If it's still off, you may have to loosen one side and go back to step 30. It's rare that I get a hub adjustment perfectly dialed in on the first try, so don't be dismayed if you have to tinker with the adjustment a bit before it's right.*

33. **Put the skewer back into the hub.** Make sure that the conical springs have their narrow ends toward the inside (Fig. 8.6).

34. **Install the wheel in the bike and tighten the skewer.** Check that the wheel spins well without any side play at the rim. If it needs readjustment, go back to step 32.

35. **Congratulate yourself on a job well done!** Hub overhaul is a delicate job, and it makes a difference in the longevity and performance of your bike.

8-7

OVERHAULING OR REPLACING HUB CARTRIDGE BEARINGS

Compared with inexpensive hubs with cup-and-cone bearings, cartridge-bearing hubs generally do not need much maintenance, though some are certainly much better sealed than others. If you ride in rain a lot or clean your bike with a high-pressure sprayer, however, you can expect water and dirt to get through any kind of seal. If the ball bearings inside the cartridges get wet or dirty, the cartridge bearings will soon seize up and should be overhauled or replaced.

a. Removing the bearings

If you can easily get at the rubber bearing seal, you may not need to remove the bearings to grease them. See step 1 in 8-7b.

There are many types of cartridge-bearing hubs (Figs. 8.7, 8.13, 8.25, 8.26), and it is outside the scope of this book to explain how to disassemble every one of them, but the instructions below certainly cover the vast majority of them (and see specific details of rear Mavic hub overhaul in 8-12c). It is usually not too hard to figure out how to take apart any hub.

Most cartridge-bearing hubs generally have axle end caps that butt up against the bearings and must be

removed to remove the axle. The axle removal can be initiated by one of the following approaches:

- Pulling or prying off the end caps. You may need an axle-clamp tool (1-4, Fig. 1.4) clamped in a vise to grab the tip of the end cap securely enough to pull it off by pulling up on the wheel.

- Unscrewing the end caps with a 5mm hex key inserted into the central 5mm hex holes in either axle end cap; generally only one cap will come off this way. You may find a receptacle for an 8mm or 10mm hex key in the bore of the axle after one end cap is off; insert that hex key and, with a 5mm hex key still in the opposite end cap or with your fingers on a smooth, conical end cap, unscrew that cap (Fig. 8.24).

- Unscrewing the caps on either axle end with cone wrenches engaged on wrench flats on each cap (Fig. 8.7); generally only one cap will come off this way. You will then tap the axle out from the end whose cap came off; it will drive the bearing on the opposite side out with it. There will sometimes be an enlarged flange with wrench flats behind that bearing; if yours has one, put an open-end wrench on it and a cone wrench on the end cap, and unscrew the end cap to free the bearing.

- On Mavic, unscrewing the axle with a 5mm hex key while holding the other end with either a pin tool in the adjuster ring (front) or a 5mm hex key (front) or 10mm hex key (rear; Fig. 8.24) in the bore of the axle, often after pulling off the end cap on the adjuster end.

- Loosening a tiny setscrew pinching each collar-type end cap around the axle. On a Zipp front hub, one collar is threaded and will unscrew once the setscrew is loosened, and the other (unthreaded) will slide off. Push out the axle bearing and the aluminum bearing cover by pushing on the threaded end of the axle. The rear Zipp hub has a collar with a tiny setscrew; remove the screw and push the axle out of the drive side, and then pull off the freehub body. You can remove the bearings on each end of the freehub body and both ends of the front and rear hub shells with light finger pressure.

- Yanking the rear cassette straight off of the hub by hand (DT Swiss); the freehub body will come off with the cogs and will take off the end cap with it (Fig. 8.26).

- Sliding the end caps off after loosening a setscrew on the side of each cap.
- Loosening three radial setscrews (with a tiny—2mm—hex key) on the axle collar (on the non-drive side on a rear hub), which sometimes must be accessed through a tiny hole in the end of the hub shell (rotate the axle to line up each setscrew under the hole in succession). Once the collar is loose, pull out the end cap, and then pull the entire axle assembly (including the freehub on a rear hub) out from the opposite side.

If you see hex flats inside the axle bore, try sticking a hex key in there (except Zipp; see earlier in this section). If both axle ends have them, put a 5mm hex key in each end and unscrew the end caps. If only one end fits a hex key, the other end must first pull off to reveal a hex hole into which you can insert a second hex key to work against the first to unscrew the axle (Fig. 8.24).

If you see no hex hole in the end, then usually the end cap will pull off; it is usually held on by a rubber O-ring engaging grooves at the end of the axle and inside the end cap, so a good yank will get it off. In practice, this often requires more than just your fingers, however. If you have an axle-clamp tool (1-4, Fig. 1.4), put it in a vise and grab the end of the axle with it to pull the cap off. Without that, you can just grab the outside of the cap with a vise and yank up on the wheel.

Some (generally very old) cartridge-bearing hubs have threaded axles with locknuts that tighten against each other to hold the bearing adjustment, much like a loose-bearing hub (Figs. 8.6, 8.19). You remove the axle just as you would for a loose-bearing hub (8-6).

On many rear cartridge-bearing hubs, once the axle or drive-side axle end cap is out, you can (carefully) pull the freehub body off the hub while turning it counterclockwise. Sometimes the freehub body comes off when you push the axle out. In either case, be ready to catch pawls, springs, or ratchet rings (Figs. 8.24–8.26).

If you were able to pull the axle out (and the freehub as well on a rear hub), you can now drive the bearings out of the hub shell.

Some axles have a shoulder against the inner face of the bearing that allows you to push it out. In this case, smack the end of the axle with a soft hammer (Fig. 8.13) to drive the bearing out (or if the axle end is recessed,

8.13 Tapping out a cartridge bearing using the axle's shoulder

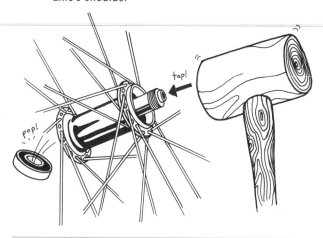

with a rod, hex key, or drift punch just smaller in diameter than the end of the axle). Or you can set the end of the axle down on the workbench and push down on the hub or the spokes emanating from it. If it's too stubborn to just tap it out while you're holding the wheel, first stand the hub up, bearing to be removed down, on two wood blocks or padded vise jaws separated by enough space to allow the bearing to pass between them as it comes out. Unless the fit is very tight, driving the bearing out with the axle usually does not damage a bearing. On a DT Swiss rear hub (Fig. 8.26), you can get the non-drive bearing out this way, but you cannot remove the drive-side bearing without first unscrewing and removing the large threaded ring with inwardly radiating teeth that engage the freehub star ratchets. This requires a special DT Swiss tool held in a very solid vise.

The best way to remove a bearing from a hub whose axle does not have a shoulder designed to push the bearing out is with a specialty tool called a blind-hole bearing puller that slides into the bore of the bearing and then expands to grip it and has a rod with a sliding weight on it that you slide backward against a stop to pull the bearing farther out with each impact. I'm guessing you don't have one, but if you do have one of the right size, by all means use it! Otherwise, with the hub shell standing vertically on two wood blocks or padded vise jaws separated by enough space to allow the bearing to pass between them, slide in a large screwdriver or hex key, and cock it at an angle to hit the backside of the opposite bearing. Smack it with a hammer, alternately

8.14 Tapping out a cartridge bearing with a hex key

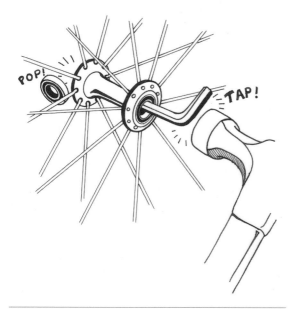

moving its tip around the bearing (like around a clock face) to slowly work it out until the bearing pops out (Fig. 8.14). This is likely to damage the bearing enough to require replacement. Besides the fact that most cartridge bearings are vulnerable to lateral stress, the tip of a screwdriver can damage the seals and retainers.

If the bearing is reusable, you can remove the seals (and even the bearing retainer and the balls themselves if it has a plastic bearing retainer inside); see 11-13 for bearing disassembly instructions.

Driving the bearing(s) out of the freehub body, once it's been removed from the hub, is often simple with a hex key of appropriate diameter, but some freehub bearing systems can only be serviced by the hub maker.

b. Cleaning and greasing the hub cartridge bearings

Once the cartridge bearings are out (and sometimes without removing them at all), you may be able to overhaul them if removing them did not damage them; otherwise you'll need to buy new ones. To overhaul:

1. **Remove the seal.** The most common type of hub bearing is also the simplest to access; it has a rubber-coated metal seal on each side. Gently pop off the rubber-coated bearing seal by sliding a razor blade (Fig. 8.15) or box-cutter knife blade under the edge and prying it up. Avoid tearing or cutting the rubber edges of the seal, but if you

bend the seal, you can easily flatten it out again; it is soft aluminum. If the bearing seals are steel instead, you may find a thin, flat circlip (C-shaped retaining ring) around the outer edge that you can remove; slip a thin blade under one tip of the circlip and work around to pop it out. The bearing cover should come right out with some help from a razor blade. If the bearing has steel seals and no circlip, you probably cannot remove the seals without damaging them (you would remove them by pounding a sharp awl into the edge of the seal and prying up); it's best to buy new bearings.

2. **Clean the bearing.** Squirt citrus-based solvent into the bearing (wear rubber gloves and protective glasses) to wash out the grease, water, and dirt. Scrub with a clean toothbrush.

3. **Dry the bearing with compressed air.**

4. **Pack the bearing with grease and snap the bearing covers back on.** If the bearing does not now spin smoothly, replace it. If you really want to rescue this bearing (for instance, if it's an expensive ceramic bearing whose steel bearing races have rusted), you can disassemble it completely, polish the bearing races and bearings, and bring it close to new. This is possible only if it has a plastic bearing retainer inside; forget it if the bearing retainers are metal (i.e., when the seals are off, you see a bumpy silver metal ring on either side as in Fig. 8.15, rather than a smooth plastic one). See 11-13 for how to disassemble a cartridge bearing.

8.15 Removing the bearing seal

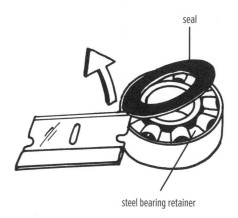

seal

steel bearing retainer

c. Installing the bearings and adjusting the hub

The closest thing the average home mechanic will have to a proper tool for pressing in a cartridge bearing is a quick-release skewer and a socket wrench or the old bearing. The ideal tool is a hub bearing press (Fig. 1.4), which has a central threaded shaft and discs of different sizes to fit various bearing sizes; tightening the wing nut on the end with the proper pair of discs against the bearings drives them in evenly.

NOTE: *Installing a bearing with a hammer can deform the bearing and prevent it from spinning smoothly.*

1. **Grease the bearing and press it into place.** Put a layer of grease around the outside of the new bearing, and place it in proper alignment where it's going to go in. Place a socket whose outside diameter (OD) is just slightly smaller than the bearing's OD against the new bearing, or place the old bearing atop the new bearing (but note that if the bearing seat is deeper than the bearing, the old bearing can get stuck). It's generally better to install one bearing at a time to ensure that it's going in straight, especially because you may have only one socket of the right size anyway. Using large washers or something of the sort to protect the bearing seat on the opposite end, install the skewer (without the springs on it) and tighten it until the bearing is fully pressed in place (Fig. 8.16).

    When you install ceramic cartridge bearings or other high-quality bearings, it is worth thinking ahead of time about maintenance when you are determining the orientation of the bearings. Ceramic bearings for hubs will most likely be "hybrid ceramic" bearings, so even though their ceramic balls cannot rust and are more than twice as hard as steel balls, their races are still steel, may not be stainless, and can rust. Full ceramic bearings are generally not used in hubs, as their brittle outer races can crack when pressed into a tight hole.

    Cartridge bearings have bearing retainers (Fig. 8.15) that separate the balls from each other. This design reduces friction by preventing neighboring balls, whose adjacent sides are turning in opposite directions, from rubbing against each other. The

bearing retainer may be plastic and asymmetrical, so when you remove the bearing seals (with a razor blade slipped under the edge), you'll see the balls on one side and you'll see only the plastic retainer from the other side. Proper maintenance requires cleaning out the bearing and slathering new grease into it, so before you press the bearings in, determine on which side the balls are visible and make sure that side faces outboard. That way, you can pry off the bearing covers and clean and grease them easily. If you were to completely disassemble and overhaul this bearing instead (11-13), you would need to remove it from the hub to get at the other side of the bearing retainer.

If the bearing has a symmetrical retainer (usually steel; Fig. 8.15) concealing the balls on both sides, the best you can do to clean them out is to squirt solvent in followed by compressed air. Disassembling the bearing (11-13) is only possible with a plastic, asymmetrical bearing retainer. Removing a steel bearing retainer (Fig. 8.15) ruins it, and you won't be able to reassemble the bearing.

2. **Reassemble the hub axle, second bearing, and end caps the reverse of the way they came apart.** Install the axle (put a bit of grease on it first) and press in the other bearing, generally using the same method you did with the first bearing. When

8.16 Pressing in the bearing with the skewer and socket

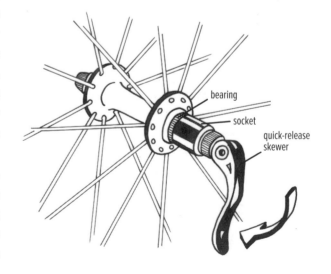

bearing

socket

quick-release skewer

8.17 Pressing the bearing into the freehub

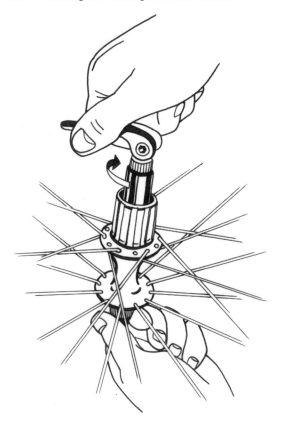

you press in the second bearing with the socket, you'll also need another, bigger socket for the other end of the hub that is large enough to clear the outside of the bearing. Place the bigger socket with its open end over the bearing you just pressed in; this allows the skewer to push against the hub shell, rather than just against the end of the axle. Run the skewer through the axle, both bearings, and the sockets.

If one bearing was originally tightened between a shoulder on the axle and a thread-on end cap, tighten the new bearing onto the axle that way first, and then tap the axle assembly in.

On many hubs, you can also press in the outer bearing in the freehub body with a socket and the skewer once the freehub body is in place on the hub (Fig. 8.17).

3. **Check the bearing adjustment.** Sometimes the bearings will be out of alignment slightly after installation, making the hub noticeably hard to turn. A light tap on either end of the axle with a soft hammer will sometimes free them. If the hub has threaded end caps and you find that tightening them binds up the hub, put Loctite on the threads and tighten them only enough to make the hub adjustment perfect once the wheel is tightened with the skewer into the frame or fork; the threadlock compound will prevent the end caps from loosening.

Many hubs have no adjustment; a tubular spacer between the inner races of the two bearings keeps them at the proper separation, and the end caps push against the other side of the inner bearing races. Other systems have no spacer between the bearings; the bearings just sit in a pocket in each end of the hub, and pressure from the outside has to be adjusted to remove axle end play without side-loading the bearings.

On Mavic hubs, adjust the bearing with a pin tool on the adjuster ring once the hub is tightened into the frame or fork with the quick-release skewer. A little threadlock on the adjuster's threads is a good idea when putting the hub back together.

On White Industries hubs, there are three setscrews pointing radially inward through a sliding collar on one end of the hub. Each setscrew is at 120 degrees from the next, and on some hubs must be accessed through a hole in the hub shell that must be rotated over each setscrew. Loosen the three setscrews with a 2mm hex key and slide the collar inward against the bearing to remove the end play. If the adjustment is still loose, try rotating the collar on the axle end cap first; it may be that the setscrews keep going back into the indentations they made in the collar before and thus prevent changing the adjustment unless you rotate them to a new area of the collar.

8-8

UPGRADING BEARINGS

Following the instructions in 8-6 or 8-7, depending on whether it's a loose-bearing hub or a cartridge-bearing hub, you can replace old bearings with supersmooth ceramic bearings, higher-grade steel bearings, or simply new bearings of the same type. Cartridge bearings

THE MOST BENEFICIAL EFFICIENCY improvements you can make are to reduce air resistance, weight, tire rolling resistance, and bearing friction, in that order.

Reducing wind resistance of the wheels requires purchasing wheels that are more aerodynamic or buying aero spokes and rims and rebuilding them onto the existing hubs (see wheelbuilding instructions in Chapter 15).

Reducing weight also costs money in lighter tires, tubes, rims, spokes, hubs, and/or cogs.

Rolling-resistance improvements that cost you nothing include keeping the tires at proper inflation (see the Pro Tip in 7-5 on tire pressure) and keeping the wheels true.

Reducing bearing friction can cost nothing; the biggest bearing-friction gains you can get may be from overhauling and properly adjusting the hub bearings and cleaning and lubricating the freehub. To reduce bearing friction beyond that requires money to buy better bearings; see 8-8 in this chapter.

are higher-grade if the balls and races are smoother and more uniform. Expect to pay a lot for ceramic bearings and even for high-grade steel bearings. Ceramic bearings should run more smoothly and last longer, since, besides being rounder, smoother, more uniform in size, and harder, the balls cannot rust or be scratched by grit and are less sensitive to lubrication.

Hybrid ceramic bearings are cartridge bearings with ceramic balls and steel races; full-ceramic bearings have ceramic races as well as ceramic balls and are the most expensive. Full-ceramic bearings should be installed at the factory; the ceramic race cannot be pressed in, as it has no flexibility and may crack. Instead, the seat into which it fits must be larger than the bearing OD so that the bearing fits in without pressure, and then it must be glued into place. These sometimes have no bearing seals, as the balls and races are harder than any grit that could get into them.

FREEHUBS, FREEWHEELS, AND COGS

LEVEL Freehubs and freewheels allow the rear wheel to turn freely in the reverse direction of the chain driving direction. Most rely on a series of spring-loaded pawls (Fig. 8.24) that engage internal teeth when pressure is applied to the pedals and disengage from the teeth when the rider is coasting; the springs cause the pawls to click down into the valleys of the teeth as they pass by—that's the clicking sound you hear when coasting.

A freehub is an integral part of the rear hub. The cogs slide onto the longitudinal splines of the freehub body (Fig. 8.18). Gear combinations are changed by removing the cogs from the freehub body and putting on different ones.

A freewheel is a separate unit with the cogs attached to it. The entire freewheel threads onto the drive side of the rear hub (Fig. 8.19). Thread-on freewheels have fallen out of fashion relative to freehubs; changing cogs on freewheels is difficult, and freewheels do not support the hub axle, leaving a long section of axle on the drive end sticking out of the hub unsupported. Freewheels can only be removed and interchanged with a tool that matches the shape at the end of the particular freewheel.

Fixed-gear cogs, as used on track bikes, urban fixies, or some road bike winter-training setups, do not freewheel; the cog drives the chain forward whenever the wheel is rotating. Single-speed freewheels, on the other hand, allow coasting and are found on BMX bikes and other single-speed bikes without coaster brakes. Both fixed gears and single-speed freewheels screw onto a threaded hub.

There are two widths of single-speed gear teeth— thin ones for 3/32-inch-width derailleur chains, and thick ones that accept 1/8-inch-width chains. Toothed drive belts and mating sprockets can also be used on derailleur-free bikes.

With a single-speed set-up, the frame needs to have horizontal dropouts with long slots to allow the wheel to be moved rearward sufficiently to tension the chain or

8.18 Rear freehub with cartridge bearings and cassette cogs

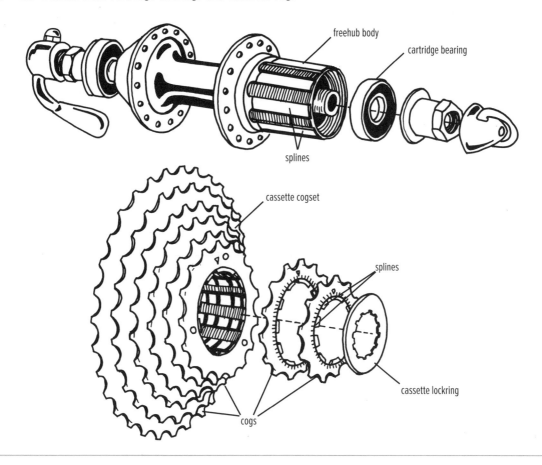

freehub body

cartridge bearing

splines

cassette cogset

splines

cassette lockring

cogs

8.19 Threaded rear hub with standard ball bearings and freewheel

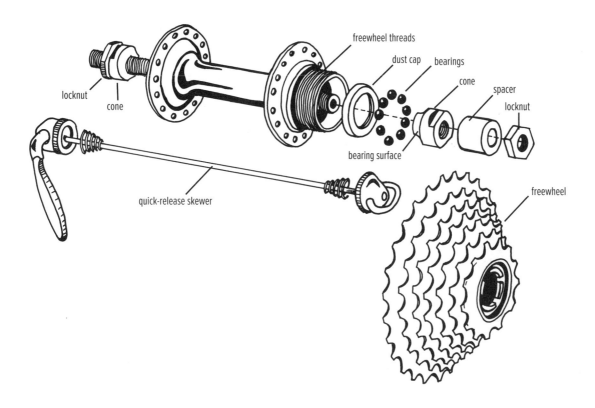

freewheel threads

dust cap

bearings

cone

spacer

locknut

locknut

cone

bearing surface

quick-release skewer

freewheel

belt, or it must have a spring-loaded chain tensioner that pulls the lower section of chain taut; these can thread onto the derailleur hanger for the dropout or be bolted to the chainstay. An alternative design uses an eccentric in the bottom-bracket shell to adjust chain tension.

When put on a standard threaded wheel for a road bike, a fixed-gear cog can unscrew when pedaled backward, so it's best to use a track hub. On a track hub, there is a second set of threads outboard of the standard hub threads. These threads are smaller in diameter and are left-hand threaded. A left-hand-threaded lockring holds the cog on and is tightened whenever the rider pushes backward on the pedals. The cog is removed by unscrewing the lockring in a clockwise direction with a lockring spanner, and then the cog is unscrewed in a counterclockwise direction with a Vise Whip (Fig. 1.2) or a chain whip.

8-9

CLEANING REAR COGS

The quickest, albeit perfunctory, way to clean the rear cogs is to slide a rag back and forth between each pair of cogs while they are on the hub (Fig. 8.20), or, better yet, with Finish Line's thick, twisted-string Gear Floss. The other way—usually unnecessary unless the bike has been neglected—is to remove them (8-10) and wipe them off with a rag or immerse them in solvent.

8.20 Cleaning cogs

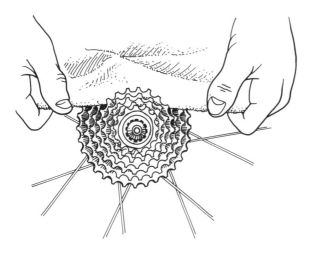

8-10

CHANGING CASSETTE COGS

1. **Make sure you have the tools you need.** Get out a Pedro's Vise Whip or a chain whip, a cassette-lockring remover, a wrench (adjustable or open) to fit the remover, and the cog(s) you want to install. (Some very old freehubs have a threaded smallest cog instead of a lockring. These require two Vise Whips or chain whips and no lockring remover.)

2. **Remove the quick-release skewer.**

3. **Secure the cassette so it cannot spin when you remove the lockring.** Adjust the Vise Whip's jaw-adjustment screw on the end of the handle to fit on any cog you choose, and clamp onto it (Fig. 8.21), or wrap the chain whip around a cog (Fig. 8.22) at least two up from the smallest cog. Wrap in the drive direction (clockwise) so that the cassette is held in place, and keep tension on it so it doesn't fall off.

4. **Remove the lockring.** Insert the splined lockring remover into the lockring. The lockring is the internally splined ring that holds the smallest cog in place. With a wrench on the lockring remover, unscrew the lockring in a counterclockwise direction while using the Vise Whip or chain whip to keep the cassette from turning (Figs. 8.21, 8.22). If the lockring is so tight that the tool pops out without loosening it, install and tighten the skewer, sans springs, through the hub and lockring tool. Once you have broken the lockring free and unscrewed it a fraction of a turn, remove the skewer so you don't snap it; now unscrew the lockring the rest of the way.

5. **Pull the cogs straight off.** Some cassette cogsets are composed of single cogs separated by loose spacers; some cassettes are bolted together with only the smallest one or two cogs and perhaps a spacer or two being loose (Fig. 8.18); some cassettes have groups of two or three cogs attached to splined aluminum carriers along with some loose cogs and spacers; and some SRAM cassettes are machined in a single piece.

6. **Clean the cogs with a rag or a toothbrush.** Use solvent if necessary, observing the usual precautions.

8.21 Using a Pedro's Vise Whip to hold the cassette while unscrewing the lockring

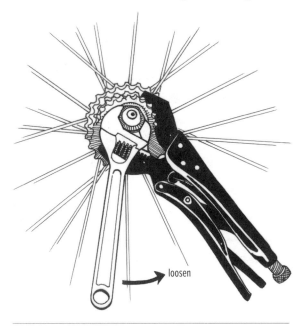

loosen

8.22 Removing a cassette lockring with a chain whip and a locknut remover

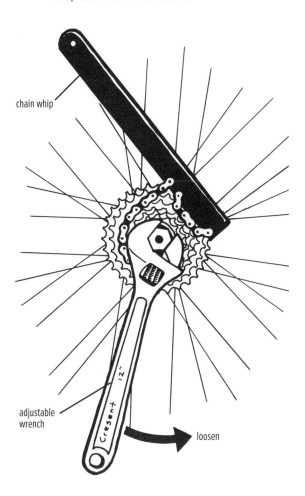

chain whip

adjustable wrench

loosen

7. Inspect the cogs for wear. If the teeth are hook-shaped, they may be worn out and ripe for replacement. Rohloff makes a cog-wear indicator tool; if you have access to one, use it according to its supplied instructions.

8. Replace the cogs. If there was a spacer behind the cassette, make sure you replace it; otherwise, you may find your rear derailleur going into the spokes on shifting to your largest cog.

  a. If you are replacing the entire cassette, just slide the new one on. Usually, you'll find that one spline is wider than the others (Fig. 8.23A), so line them up accordingly. With Campagnolo cogs, read the following Pro Tip.

  b. If you are installing a 9-, 10-, or 11-speed cassette, see the first note under step 9.

  c. If you are replacing individual cogs within the cassette, be certain that they are of the same type and model. For example, not all 16-tooth Shimano cogs are alike. Most cogs have shifting ramps, differentially shaped teeth, and other asymmetries. They differ by model as well as by sizes of the adjacent cogs, so you need to buy one for the exact location and model. Install cogs in decreasing numerical sequence with the numbers facing out.

NOTE: *Some bolt-together cassettes can be disassembled for cleaning and cog interchanging and then reinstalled onto the freehub as separate cogs to facilitate*

8.23A Large spline

future cog changes and cleaning. Note that there are two kinds of bolt-together cassettes: 1) those with three long, thin bolts holding the stack of cogs and spacers together (Fig. 8.18) and 2) those with pairs or groups of cogs bolted or riveted to aluminum spider-shaped carriers that have internal splines to fit on the cassette body. For the type with the three bolts, just unscrew the bolts, take it apart, and put in the replacement cogs. The other type is not to be disassembled from the aluminum spider, and the individual cogs are not to be replaced; you replace each carrier with its attached cogs as a complete assembly.

9. **Install the lockring.** First, ensure that the lockring you are using is the right one for both the freehub and the particular cassette. The diameter of the lockring depends on the size of the first cog, and its thread pitch and diameter depend on the brand of the freehub. With everything back in place, tighten the lockring with the lockring remover and wrench. (If you have the old-type 6- or 7-speed Shimano freehub with the thread-on first cog, tighten that with a Vise Whip or chain whip instead.) Make sure that all of the cogs are seated and can't wobble from side to side, which would indicate that the first or second cog is sitting against the ends of the splines, or that you didn't install all of the spacers. If the cogs are loose after tightening the lockring, loosen the lockring, line up the first and second cogs to make sure they are in place, and tighten the lockring again.

PRO TIP — CORRECT INSTALLATION OF CAMPAGNOLO COGS

CAMPAGNOLO COGS DIFFER from Shimano and SRAM cogs in the shape of their freehub-engagement splines. The splines on Campagnolo cogs are longer (so the grooves in the freehub body are deeper), and instead of a single, wide-splined tooth, Campagnolo uses a narrow, stepped tooth (Figs. 8.23B, C) to align the cogs in the proper orientation on the freehub body. Problem is, some Campagnolo-compatible freehub bodies have either such a long taper to the flanks of their aluminum spline ridges (Zipp, among others; Fig. 8.23B) or such narrow steel spline ridges (Mavic; Fig. 8.23C) that Campy cogs can be slid on in more than one orientation. If a cog is installed incorrectly, all of the pedaling load when the chain is on that cog is carried by a single spline tooth. This puts enormous strain on the cog, often leading to breakage.

Make sure when installing Campagnolo cogs that you slide the stepped tooth on every cog onto the stepped ridge on the freewheel body.

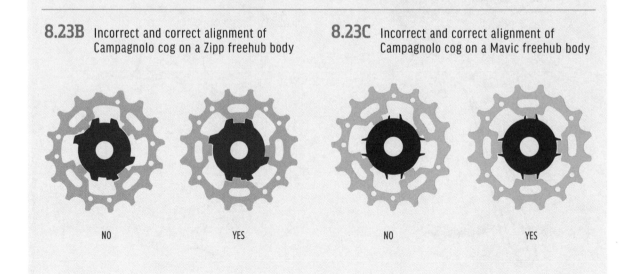

8.23B Incorrect and correct alignment of Campagnolo cog on a Zipp freehub body

8.23C Incorrect and correct alignment of Campagnolo cog on a Mavic freehub body

NO   YES   NO   YES

NOTE ON COMPATIBILITY: *These instructions for remov-ing and replacing cogs apply for 6-, 7-, 8-, 9-, 10-, and 11-speed cassettes. But freehub-body dimensions vary with number of speeds, so make sure you only use a 7-speed cassette on a 7-speed freehub body, and so on. Shimano and SRAM 11-speed freehub bodies are 2mm wider than 9- and 10-speed ones, so putting a 9- or 10-speed cassette on an 11-speed freehub requires putting a 2mm spacer behind the largest cog. Shimano and SRAM 11-speed cassettes are too wide to fit on 9- or 10-speed freehubs. Some 10-speed freehubs can be interchanged for 11-speed ones, but the axle spacing and wheel dish need to be changed as well. Conversely, Campagnolo 9-, 10-, and 11-speed cassettes all fit on the same Campagnolo-compatible freehubs.*

NOTE ON 11-TOOTH COGS: *Although all Shimano 8-speed freehubs are wide enough for a 9- or 10-speed cas-sette, some 8- and 9-speed freehub bodies will not accept 11-tooth cogs (for instance, 1992–1994 Shimano 8-speed freehub bodies will not accept 11-tooth cogs). To accept the small, 11-tooth cog, the freehub splines stop about 2mm before the outer end of the freehub body. You can grind the last 2mm of splines off an old-style 8-speed freehub so that it will accept an 11-tooth cog. The steel is very hard on high-end freehubs, so a grinder, rather than a file, may be needed for this job.*

8-11

CHANGING FREEWHEELS

If the hub has a freewheel (Fig. 8.19) and you want to switch it with another one, follow this procedure. Replacing individual cogs on an existing freewheel is beyond the scope of this book, since it is rarely done these days owing to the preponderance of freehubs and to the lack of availability of spare freewheel parts.

1. **Obtain the appropriate freewheel remover for the freewheel.** Take the wheel to the bike shop to make sure you get the right tool. Round up a big adjustable wrench to fit it.

2. **Remove the quick-release skewer and take off the springs.**

3. **Install the freewheel remover.** Slide the skewer back in from the left side, place the freewheel remover into the end of the freewheel so that the notches or splines engage, and thread the skewer nut back on, tightening it against the freewheel remover to keep it from popping out of its notches.

4. **Unscrew the freewheel.** Put the big adjustable wrench onto the flats of the freewheel remover and loosen it (counterclockwise). It may take con-siderable force to free it, and you may even need to put a large pipe on the end of the wrench for more leverage. Set the tire on the ground for trac-tion as you do it. As soon as the freewheel pops loose, loosen the skewer nut before continuing; otherwise, you may snap the skewer in two.

5. **Loosen the skewer nut a bit.** Unscrew the free-wheel a bit more, and then loosen the skewer nut a bit, and repeat until the freewheel spins off freely and there is no longer any danger of having the freewheel remover pop out of the notches.

6. **Remove the skewer and spin off the freewheel.**

7. **Grease the threads on the hub and the inside of the new freewheel.**

8. **Thread on the new freewheel by hand.** You can snug it down with the freewheel remover and a wrench or with a Vise Whip or chain whip, if you like, but it will tighten itself into place with the first few pedal strokes anyway.

9. **Replace the skewer with the narrow ends of its conical springs facing inward (Fig. 8.19).**

8-12

LUBRICATING FREEHUBS

a. Simple, minor freehub lubrication

Often neglected, freehubs need lubrication and can usually be lubricated for the short term simply by drip-ping chain lube into them. This method is not a long-term fix, however, as the thin lubricant will not protect it for long, and the chain lube can get into the wheel bearings and dilute the grease protecting them.

NOTE: *Do not dunk a freehub in a solvent bath; it will pull in dirt along with the solvent.*

Some freehubs have grease-injection holes on the freehub body, visible after removing the cogs. Remove the cogs to get at the hole, and meticulously clean any dirt out of the hole before injecting the oil or grease. Add oil or very lightweight grease (frequently) to avoid thickening of the grease inside. To be thorough, before injecting the grease, inject diesel fuel or biodegradable chain cleaner into the hole from a squeeze bottle with its thin tip pressed into the hole. Keep adding solvent until the freehub spins without any crunching noises. Rather than using bearing grease in the grease gun, inject a thinner lube into them, such as outboard-motor gear oil or Phil Wood Tenacious Oil by means of an oil squeeze bottle or a fine-tip grease gun (Fig. 1.3). Using heavy oil like this will avoid the problem of grease thickening up inside and sticking the pawls in cold temperatures, preventing engagement of the freehub. It's a dark day if you apply power to the pedal and the freehub slips.

If the freehub has teeth on the faces of the hub shell and freehub (DT Swiss, Hügi, Crank Brothers, and ancient Mavic freehubs have these radial teeth), drip oil into the crease between the freehub and the hub shell as you turn the freehub counterclockwise for a short-term fix. These freehubs come off easily, so you might as well lubricate them well; see 8-12f.

For most loose-bearing freehubs (Shimano), here is the general procedure. The details for lubricating the major types of freehubs follow in separate sections.

1. **Disassemble the hub-axle assembly (8-6a, b).**
2. **Clean the parts.** Wipe clean the inside of the drive-side bearing surface and inspect it. Look for discoloration or wear.
3. **Lubricate the freehub.** With the wheel lying flat and the freehub pointed up toward you, flow chain lube between the bearing surface and the freehub body as you spin the freehub counterclockwise. You will hear the clicking noise of the freehub pawls smooth out as lubricant reaches them. Keep it flowing until old black oil finishes flowing out of the other end of the freehub.
4. **Wipe off the excess lube and continue with the hub overhaul (8-6c, d).**

b. Thorough Shimano freehub lubrication without disassembling the freehub body

1. **Disassemble the hub-axle assembly as described in 8-6a, b.** Remove the cassette as in 8-10.
2. **Remove the freehub body with a 10mm hex key inserted into the internal freehub-fixing bolt.** Pull the bolt out.
3. **Completely flush out the freehub.** Remove the rubber seal (at the back) with a thin screwdriver or pick. With a rubber stopper from a hardware store, close off the bottom of the freehub body. Pour solvent into the outer opening, spinning the mechanism and letting contaminants run out. Repeat until clean. Alternatively, you can soak the entire freehub in solvent and spin occasionally.
4. **Let it dry.** Wipe it off and stand it on a rag to let the solvent drain out the back.
5. **Lubricate.** Squirt outboard gear lube or Phil Wood Tenacious Oil onto the bearings that were revealed when you pulled off the seal and then spin the freehub. Park the body on paper towels and let the excess drain off. With this method, you do not need to remove the freehub body dust seal.

c. Mavic freehub lubrication (or swap)

1. **Remove the axle.** Depending on model, removing the axle usually involves first pulling the non-drive-side dust cap straight off (Fig. 8.24) or unscrewing locknuts from external threads on a steel axle.
2. **Unscrew the parts.** Depending on the model, using two hex keys (two 5mm hex keys, or one 10mm and one 5mm), one in either end, loosen counterclockwise, unscrew, and remove (Fig. 8.24).
3. **Turn the wheel on its side, freehub up.** Put it on a clean surface where you can catch—or at least see—any pawls or pawl springs that fly away.
4. **Rotate the freehub body slowly counterclockwise as you pull up on it, and remove it (Fig. 8.24).**
5. **Clean the pawls, springs, and hub shell.**
6. **Replace the springs and pawls.**
7. **Lubricate it.** Put 10–20 drops of Mavic mineral oil M40122 into the freehub (on the plastic bushing and the ratchet teeth). There is no need to lubricate the rubber seal.
8. **Reinstall the freehub body.** Turn it counterclockwise while holding the pawls down with your fingers.
9. **Replace the axle.** Simple.

8.24 Removing a Mavic Ksyrium axle and freehub in order to lubricate the freehub and pawls

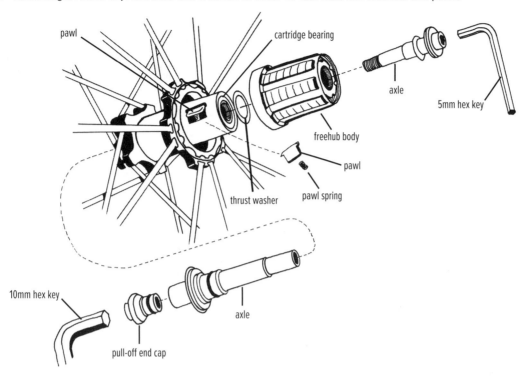

NOTE: *This same procedure can be used to interchange Mavic freehub bodies from, say, a Shimano-compatible type to a Campagnolo-compatible type.*

d. Campagnolo/Fulcrum freehub lubrication

1. **Remove the skewer.**

2. **Unlock the freehub from the hub.** On high-end Campagnolo or Fulcrum rear hubs, insert a 5mm hex key into the drive-side axle end, and put a 17mm open-end or box wrench on the drive-side locknut. While holding the 5mm hex key, unscrew the locknut clockwise (it has left-hand threads, and you need to unscrew it in the opposite direction from what you would expect). Older Campagnolo models instead have a little setscrew on the 17mm slotted locknut that must be loosened with a 2mm hex key in order to unscrew the locknut.

3. **Pull the freehub straight off.** Older Campagnolo models have individual coil springs under each of the three pawls. These can go flying, and they are hard to clean and to insert back into the hub shell. Newer Campagnolo and Fulcrum models have a single, circular wire spring wrapped around all three pawls (it fits in a groove in the freehub body and the flanks of each pawl; Fig. 8.25). With the new style, you can pull the freehub off without parts flying off. When removing older-model freehub bodies, wrap a twist tie around the three pawls as they expose themselves from the hub shell as you pull; if you don't do this, the three pawls and the three springs will fly away.

4. **Clean and grease the three pawls and the radial teeth inside the hub shell.**

5. **Slide the freehub back in while slowly turning it counterclockwise.** If necessary, push inward on each pawl with a pencil tip as you do this, until all three are engaged and the freehub body drops into place. Again, older Campagnolo models require using a twist tie to hold the pawls in place as you push the body in; pull the twist tie off after the pawls enter the hub shell.

6. **While holding the drive-side end of the axle with a 5mm hex key inside its bore, tighten the locknut counterclockwise—it's left-hand threaded!—with a 17mm hex key.** Complete the bearing adjustment as in 8-6d.

8.25 Radial-pawl freehub removal and lubrication

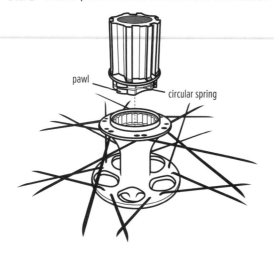

pawl

circular spring

NOTE: *This same procedure can be used to interchange Fulcrum or Campagnolo freehub bodies from, say, a Campagnolo-compatible type to a Shimano- or SRAM-compatible freehub body.*

e. Lubricating freehubs with three radial pawls mounted on cartridge-bearing hubs

The freehub body is essentially the same as the Campagnolo and Fulcrum bodies mentioned earlier, but they exist on myriad cartridge-bearing hubs. The hub shell has radial teeth pointing inwardly on its drive end, and the base of the freehub body slips down inside and has three pawls that flip outward to engage the teeth (Fig. 8.25). Usually, there is a circular spring around all three pawls that flips them outward; it also generally keeps them from flying away when you pull the freehub body off. But it is always possible that the freehub body is the older style with a single tiny coil spring behind each pawl, and these parts definitely can go flying when you pull the freehub body off unless you put a twist tie around them as mentioned in 8-12d.

1. **Access the freehub body by first removing the axle as in 8-7.**
2. **Pull the freehub straight off.** Again, a single, circular wire spring will generally be wrapped around all three pawls (it fits in a groove in the freehub body and the pawls; Fig. 8.25). Nonetheless, exercise care that nothing goes flying.
3. **Clean and lightly grease the pawls and the radial teeth inside the hub shell.**

4. **Slide the freehub back in while slowly turning it counterclockwise.** Push inward on each pawl with a pencil tip as you do this, until all three are engaged and the freehub body drops into place.
5. **Reassemble the hub as in 8-7.**

f. DT Swiss/Hügi freehub lubrication

DT Swiss and older DT and Hügi high-end star-ratchet freehubs pull apart easily for cleaning and lubrication.

1. **Remove the skewer.**
2. **Lay the wheel on its side, cogs up; grasp the cassette and pull up.** The freehub body will come off, bringing the axle end cap with it.
3. **Clean and grease the spring, both ratchets (Fig. 8.26), and the teeth in the freehub body**

8.26 DT Swiss freehub removal and lubrication

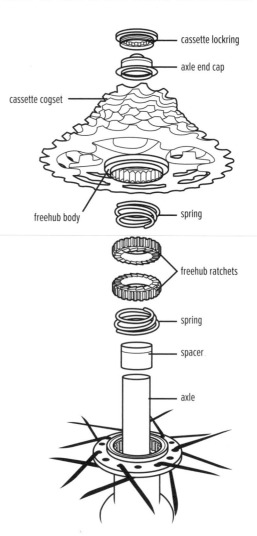

cassette lockring

axle end cap

cassette cogset

freehub body

spring

freehub ratchets

spring

spacer

axle

and hub shell. Do not grease the stepped teeth on the faces of the ratchets, or the freehub may not engage. Those teeth on the two ratchets face each other and must engage instantly when pedaling commences.

4. **Replace the parts.**

5. **Push the freehub and end cap back on, and replace the skewer.** That's it!

8-13

LUBRICATING FREEWHEELS

1. **Clean the parts.** Wipe dirt off the face of the fixed part of the freewheel surrounding the axle (Fig. 8.19).

2. **With the wheel lying flat and the cogs facing upward, drip oil into the crease between the fixed and moving parts of the freewheel as you spin the cogs in a counterclockwise direction.** You will hear the clicking noise inside become smoother as the lubricant flows in. Be sure to keep the flow of lubricant going until the old, dirty oil flows out the backside around the hub flange. Do not dunk a freewheel in a solvent bath; it will pull in dirt along with the solvent.

3. **Wipe off the excess oil.**

CYCLOCROSS WHEELS

8-14

WHEEL SELECTION

In cyclocross, you're carrying your bike a lot, and you're adding weight to it whenever you go through the mud, so you want it as light as possible without sacrificing reliability. This is especially true with the wheels and tires, because you are constantly accelerating them and decelerating them.

Another consideration is steering. It will come as no surprise that a stiff wheel steers and tracks better than a laterally flexible one. But you may not have considered that in deep mud and sand, a tall (deep-section) rim (Fig.

7.24) will cut through and track better than a shallow one (Fig. 7.23).

Mud-shedding ability is also something to consider, and a deep-section, aero-shaped rim will not collect mud the way a shallow, flat rim will.

Trumping all of this will be tire choice. If you're using tubular tires, you can find lighter rims than you can for clinchers or tubeless clinchers, because you don't have the extra rim wall for the clincher to "clinch" into (Fig. 7.23). And tubeless (or "tubeless-ready") tires are designed as a system with tubeless-specific wheels, of which there are few choices. You can mount tubeless tires on standard clincher wheels (see 7-9 and the Pro Tip "Setting Up Tubeless Cyclocross Tires"). With clinchers or tubeless tires, don't count on a flat tire to stay on the rim if you try to ride it back to the pit.

All other things being equal, a shallow-rim wheel will be lighter and more vertically compliant, so in dry conditions, there's your solution. However, 'cross is often muddy, making vertical stiffness a non-issue, and a deep-section wheel will be your friend there. But while a wheel with a shallow aluminum rim can be quite light, that is not the case for a deep aluminum rim—you're better off dragging the mud around and losing steering precision on shallow rims. So the best of both worlds for mud will be a wheelset with deep-section carbon rims that are light, stiff, and strong. If you have rim brakes, you will need carbon-specific brake pads to get decent, nongrabby braking on carbon rims; more on this in Chapter 9.

If you're pursuing cyclocross with the intent of being competitive, you need multiple sets of wheels for training and racing (not to mention two bikes), so it makes sense to have wheelsets for different race conditions, as well as inexpensive clincher wheels for training. If you have rim brakes and approximately the same rim width on all of your wheelsets, or if you are using disc brakes with the rotors on all your wheels spaced the same distance from the axle end, you won't have to readjust the brakes whenever you switch wheels. And if their cassettes are all spaced the same distance from the end of the axle, you won't have to adjust the rear derailleur every time you switch wheels, either.

8-15

COG SELECTION

You might think that you'd want lower gears on a cyclocross bike than on a road bike, but that's not true. The hills are short in 'cross races, and there's usually insufficient traction in mud or even on dry dirt or grass to pedal up the steepest hills; you'll be running up them with your bike. The high gears on a 'cross bike, however, are usually lower than on a road bike. That's because the speeds are lower on dirt, grass, mud, or sand than on pavement, and the steep downhills in 'cross are usually short and technical, emphasizing staying upright rather than powering the pedals.

Your smallest rear cog will depend on the size(s) of your front chainring(s). A fairly standard 'cross double-chainring combination is a 39–46-tooth, although compact (see 11-7 for the difference between "compact" and "standard" double cranks) 36–46 and 34–50 combinations are common. The standard single-chainring choice is somewhere between 38 teeth and 44 teeth. A 46 × 12 high gear is usually plenty for all but the fastest riders, as is a 39 × 11 (see the gear chart in Appendix B). So, with a double, you'd be using a cassette somewhere between 11–25 or 12–25 and 11–28 or 12–28, and with a single chainring, you'd be using those same cassettes or a 1 × 11 system with a 44-tooth chainring and an 11–32 cassette. Many short-cage road rear derailleurs cannot handle a cog larger than 28 teeth, and even that may be a stretch; SRAM 1 × 11 rear derailleurs (5-3g) handle at least a 32T large cog.

NOTE ON TRIPLES: *A triple is not a realistic option on a 'cross racing bike. The extra weight, poorer shifting precision, additional chainrings and derailleur cage in the front to collect mud and grass, and longer derailleurs more vulnerable to being torn off make a triple a net liability. You will never be pedaling in a 'cross race at a slow enough speed to use a granny gear (innermost triple chainring); you will be much faster running.*

It should go without saying that you should use the same number of cogs and compatibility type as your shifter; in other words, don't pair a 9-speed cassette with a 10-speed shifter, or vice versa. And almost as important, be careful about mixing Shimano-compatible and Campagnolo cassettes or shifters. Shimano and SRAM cassettes are interchangeable; they fit the same freehub bodies, and they shift well with either Shimano or SRAM derailleurs. Campagnolo 9- or 10-speed cassettes, however, should only be used with Campagnolo shifters and rear derailleurs dedicated to the same number of speeds. Shimano, SRAM, and Campagnolo 11-speed cassettes all have the same spacing, though, so 11-speed wheel interchangeability is not an issue. Given the great challenges the shifting system already faces in cyclocross, there is no sense in using combinations of questionable compatibility.

What do I say to complaints that my brakes are no good? I'll tell you this: Anyone can stop. But it takes a genius to go fast.

—ENZO FERRARI

9

RIM BRAKES

The most popular brake for road bikes is the dual-pivot sidepull (Fig. 9.1). Its predecessor was the center-pivot sidepull brake (Fig. 9.2), which is also a powerful, lightweight brake. And going way back to the 1960s, dual-pivot center-pull brakes (Fig. 9.3) were the standard. A couple of other dual-pivot center-pull brakes that did not require a cable hanger (cable stop) or a straddle cable (notably Shimano AX, and Campagnolo C-Record and Croce d'Aune Delta brakes) experienced brief popularity in the 1980s.

Center-pull cantilever brakes (simply called cantilever brakes; Fig. 9.4) or sidepull (or "direct-pull") cantilever brakes (generally called V-brakes; Fig. 9.5) are found on cyclocross bikes, hybrid bikes, and many touring bikes and tandems. Both brake types are light and simple and offer good clearance for mud, fenders, and big tires, and both pivot on bosses attached to the frame and fork. V-brakes offer the advantage of not requiring a cable hanger fixed to the frame or fork (as a cantilever brake does), because the cable routes directly to the brake arm. But most road bike brake levers do not pull enough cable to operate a V-brake with-

out some sort of adapter installed to increase cable pull—hence the near-universal use of cantilevers on cyclocross bikes until disc brakes came along. Mini V-brakes work reasonably well with current Shimano road levers, but clearance to the rim is still

9.1 Shimano dual-pivot sidepull brake caliper

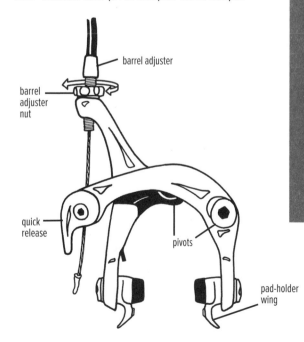

barrel adjuster

barrel adjuster nut

quick release

pivots

pad-holder wing

9.2 Center-pivot sidepull brake caliper

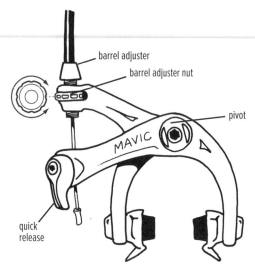

barrel adjuster

barrel adjuster nut

pivot

MAVIC

quick release

9.3 Dual-pivot center-pull brake caliper

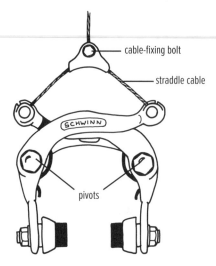

cable-fixing bolt

straddle cable

SCHWINN

pivots

9.4 Center-pull cantilever brake caliper

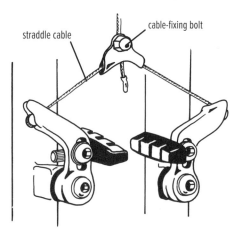

straddle cable

cable-fixing bolt

9.5 Sidepull cantilever brake (V-brake) caliper

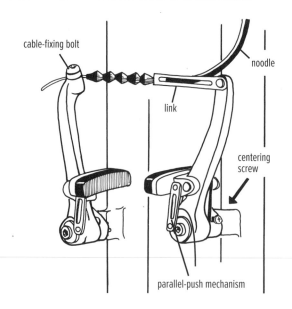

cable-fixing bolt

noodle

link

centering screw

parallel-push mechanism

9.6 SRAM Hydraulic Road Rim (HRR) brake lever and caliper

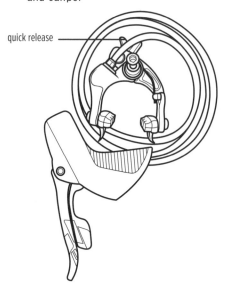

quick release

closer than a cantilever and thus presents more resistance in muddy cyclocross conditions.

Hydraulic rim brakes had brief periods of popularity in the 1990s, and SRAM's Hydraulic Road Rim (HRR) brake (Fig. 9.6), introduced in 2014, incorporates shifters into the hydraulic levers (the lever is the same as on a SRAM hydraulic road disc brake). Road bike models mount in the same bolt hole as standard road bike brakes, but mountain bike versions can be used on bikes with straight handlebars and cantilever bosses on the frame and fork.

RELEASING BRAKES TO REMOVE A WHEEL

Road bike tires are often narrow enough to slip past the brake pads without opening the brakes, especially when mounted on wide rims. With a wide tire or with a brake adjusted for very little clearance between the rim and the pads, the brake will generally need to be opened a bit to get the wheel in or out. The following instructions describe how to open the vast majority of road bike brakes, both old and new. When you put the wheel back in, remember to follow these instructions in reverse so that your brakes will work when you need them.

1. **Opening sidepull brakes:** For most sidepull brakes (Figs. 9.1, 9.2) and SRAM HRR brakes (Fig. 9.6), flip open the quick-release lever on the brake arm. Campagnolo and Mavic dual-pivot sidepull brakes for Campagnolo levers do not have a quick-release mechanism on the brake caliper. Instead, there is a cable-release button on the brake/shift lever. On a Campagnolo Ergopower lever (Fig. 9.7), push the button inward so that it clears the edge of the lever housing and allows the lever to open wider.

2. **Opening cantilevers and ancient dual-pivot center-pull brakes:** For cantilevers (Fig. 9.4) and 1960s and older dual-pivot center-pull brakes (Fig. 9.3), hold the pads against the rim and pull the head of the straddle cable out of the hook at the end of one brake arm.

3. **Opening V-brakes:** For V-brakes (Fig. 9.5), hold the pads against the rim and pull the cable noodle back and up to release it from the brake-arm link.

4. **Opening Shimano AX and Campagnolo Delta (both from the 1980s) dual-pivot center-pull brakes:** On Campagnolo Delta brakes, push the cable-release button on the brake lever (Fig. 9.7), as described in step 1. On Shimano AX brakes, pull the cable-tensioning barrel adjuster up and out.

CABLES AND HOUSINGS

LEVEL Given that cables transfer braking force from the levers to the brakes, proper installation and maintenance are critical to brake performance. Excess friction in the cable system will prevent the brakes from working properly, no matter how well the brakes, calipers, and levers are adjusted. Cables with broken strands should be replaced immediately.

CABLE TENSIONING

As brake pads wear, cable housings compress, and cables stretch, the cable needs to be tightened to remove slack in the system. The barrel adjuster on the brake arm of any road sidepull brake (Fig. 9.8) and on

9.7 Cable-release button on a Campagnolo Ergopower lever

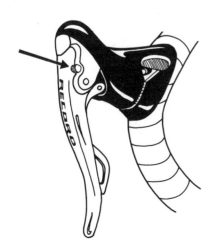

9.8 Turning the barrel adjuster on the brake arm of sidepull brakes

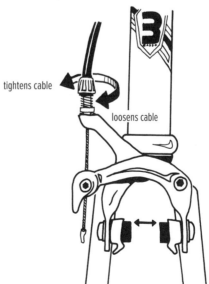

tightens cable

loosens cable

9.9 Headset cable hanger

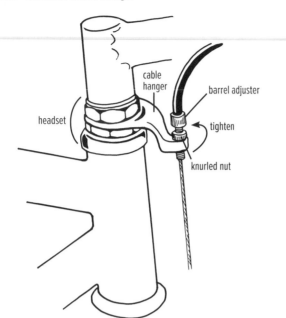

9.10 Pulling the brake cable taut and tightening the cable-fixing bolt

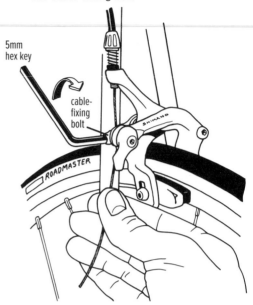

top of Shimano AX and Campagnolo Delta center-pull brakes serves exactly this purpose.

Center-pull brakes, cantilever brakes, and V-brakes (Figs. 9.3–9.5) have no barrel adjuster on the brake lever (or brake/shift lever). Cable tensioning for cantilever brakes and for ancient center-pull brakes requires either turning a barrel adjuster on a cable hanger on the frame or fork (Fig. 9.9) or sometimes on the brake arm (Fig. 9.23), or loosening the cable-fixing bolt on the straddle cable carrier (Figs. 9.3, 9.4, 9.40), pulling the cable taut, and tightening the bolt again. V-brakes generally are set up with a flat handlebar and flat-bar levers, which usually have a barrel adjuster on the lever to adjust cable tension; otherwise, cable tension is increased by loosening the cable-fixing bolt at the brake caliper (Fig. 9.5), pulling the cable taut, and tightening the bolt again.

The cable should be tight enough that the lever cannot be pulled to the bar, yet loose enough that the brakes—assuming they are centered and the wheels are true—are not dragging on the rims.

a. Increasing cable tension

1. **Back out the barrel adjuster to tighten the cable.** Not all of them turn the same way to tighten the cable, so pay attention to which direction tightens the cable and which direction loosens it on yours.

On brakes with a nut on the barrel adjuster (Figs. 9.1, 9.2), turn the nut clockwise when viewed from above to pull the barrel adjuster straight upward in its D-shaped hole in the brake arm. The underside of the adjuster nut usually has notches to hold its adjustment, so holding the pads against the rim with your thumb and fingers will make turning the nut easier; stop when it clicks into a notch, not between notches. On brakes whose barrel adjuster threads into the brake arm (Figs. 9.8, 9.10), turn the barrel adjuster counterclockwise to tighten the cable. For cantilever (Fig. 9.4) or ancient or retro center-pull (Fig. 9.3) brakes, turn the barrel adjuster counterclockwise (Figs. 9.9, 9.23) to tighten the cable, and secure it by tightening the knurled nut.

2. **Increase the cable tension sufficiently that the brake or shift lever does not hit the handlebar when the brake is applied fully.** Be careful not to make the tension so tight that the brake rubs or comes on with very little movement of the lever.

3. **If the barrel adjuster cannot take up enough cable slack to get the brakes as tight as you want, you need to tighten the cable at the brake.** First, screw the barrel adjuster back in most of the way; this step leaves some adjustment in the system for brake setup and cable stretch over time. Loosen

the cable-fixing bolt clamping the cable at the brake (Figs. 9.5, 9.10). Check the cable for wear. If it's badly frayed, replace it (see 9-4). Otherwise, pull the cable tight and retighten the clamping bolt. Tension the cable as needed with the barrel adjuster.

b. Reducing cable tension

1. **Turn the barrel adjuster in to loosen the cable.** Rotation direction depends on the brake. On brakes with a nut on the barrel adjuster (Figs. 9.1, 9.2), turn the nut counterclockwise (when viewed from the top). On brakes whose barrel adjuster threads into the brake arm, turn the barrel adjuster clockwise (Figs. 9.8, 9.10).

   You want some movement of the lever before the pads contact the rim, but not so much that the lever comes back to the handlebar under hard braking. Within that range, it is up to your personal preference. On a cantilever (Fig. 9.4) or ancient or retro center-pull (Fig. 9.3), turn the barrel adjuster on the cable hanger or brake arm clockwise after loosening the knurled nut (Figs. 9.9, 9.23).

2. **Double-check that the cable is tight enough that the lever cannot be squeezed all the way to the handlebar.**

9-3

CABLE MAINTENANCE

1. **If the cable is frayed or kinked or has any broken strands, replace it (9-4).**

2. **If the cable is not sliding well, lubricate it.** Use an oil-based chain lubricant (not a chain wax or other dry lube) or molybdenum disulfide grease. Lithium-based greases and chain waxes can eventually gum up cables and restrict movement.

   To lubricate, open the brake (via the cable quick-release as when you remove a wheel; see 9-1).

   If the bike has slotted cable stops for the rear brake, pull the ends of the rear brake-cable-housing segments out of each stop. On the front brake—and on the rear brake if your bike does not have slotted cable stops—you will have to discon-

nect the cable at the brake, clip off the cable end, and pull out the entire cable.

   Slide the housing up the cable, wipe the cable clean with a rag, rub chain lubricant on the cable section that was inside the housing, and slide the housing back into place. If you have pulled the housing completely off the cable, squirt chain lube through the housing as well.

3. **If the cable still sticks, replace the cable and housing.**

9-4

CABLE REPLACEMENT AND INSTALLATION

1. **Disconnect the cable at the brake caliper, clip off the cable-end cap, and pull out the old cable from the lever.** You will need to pull the lever and then let it back a bit to free the head of the cable from the cable hook in the lever.

NOTE: *When installing a new cable, it is a good idea to replace the housings as well, even if they seem okay. Daily riding in dirty conditions may require cables and housings to be replaced every few months. As with chains and derailleur cables, brake-cable replacement is a maintenance operation, not a repair operation; don't wait until a cable breaks or seizes up to replace it.*

2. **Purchase quality cables and lined housings.** Brake-cable housing is spiral-wrapped to prevent splitting under braking pressure (Fig. 5.21). Plastic-lined housing (i.e., Teflon) reduces friction and is a must.

3. **Cut the housing sections long enough to reach the brakes, and route them so that they do not make any sharp bends.** If you are replacing existing housing, look at the bends before removing the old housings (after unwrapping the handlebar tape to get at them). If the housing bends are smooth and do not bind when the front wheel is swung through its arc, cut the new housings to the same lengths. Otherwise, cut each new segment longer than you think necessary and keep trimming it back until it gives the smoothest path possible for the cable, without the cable tension being affected by steering. Use a cutter specifically designed for cutting housings, or a sharp side-cutter to cut

between two coils, rather than trying to cut across a couple of coils, which will mash them flat.

4. **After cutting, make sure the housing's end faces are flat.** If not, square them off with a file or a clipper.

5. **If the end of the Teflon liner is mashed shut after cutting, open it up.** Use a sharp object like a nail or a toothpick.

6. **Slip a ferrule (a cylindrical cap; see Fig. 5.21) over each housing end for support.** Some brake-arm barrel adjusters and ports in brake-lever bodies function as a ferrule and are too narrow for a ferrule to fit in; they are designed to accept only bare cable housing.

7. **Decide which hand you want to control which brake.** (The U.S. bike standard is that the right hand controls the rear brake, but if you're the only one riding the bike, you can switch it around to match the setup on your motorcycle, for example.) Install the housings into each housing stop, brake lever (or brake/shift lever), and brake caliper.

8. **Turn the adjusting barrel on the brake caliper or cable hanger to within one turn of being screwed all the way in (Fig. 9.8 or Fig. 9.9).**

9. **Insert the cable into the lever, through the lever's cable hook (Fig. 9.11), and out the cable exit hole in the lever body.** Make sure that the cable head is countersunk into its seat in the cable hook. With a cylindrical cable hook, rotate the cylinder so that the more-deeply-countersunk hole is forward. On recent brakes, the cable exits the inboard side of the lever under the edge of the lever hood so that it can be wrapped under the handlebar tape. Many brake levers prior to 1988 or so, and almost all of them prior to 1980, had the cable coming out the top of the lever.

10. **Slide the cable through the housings to the brake caliper.** If your frame has internal cable routing, see the Pro Tip on that. Make sure there is a ferrule on the end of the housing, if one will fit into the barrel adjuster.

NOTE: *Lubrication is not necessary. New cables slide well in new, lined housing, and to avoid attracting dirt it is usually best not to use a lubricant on the cable. Also, some greases can gum up inside the housing. (Down*

9.11 Inserting the cable into the lever, through the cable hook, and out the exit hole

the road when the cable starts to stick, however, you may need to lubricate it; see 9-3.)

11. **Attach the cable to the brake.** (See the section entitled "Brake Calipers" for details on your type of brake caliper.) Pull it taut and tighten the cable-fixing bolt (Figs. 9.3–9.5, 9.10, 9.40).

12. **Adjust cable tension with the barrel adjuster (as in 9-2).** Do this with the wheel in; squeeze the lever hard to seat the cable.

13. **Cut off the cable about an inch past the cable-fixing bolt.** Crimp an end cap on the exposed cable end to prevent fraying (Fig. 5.25). Wrap the handlebar tape (12-12).

14. **Check for free movement.** Once the cable has been properly installed, the lever should snap back quickly when released. If it does not, recheck the cable for kinks and fraying, and check the housing for sharp bends. Release the cable quick-release and hold the pads to the rim while checking the lever for free movement. With the cable still loose, check that the brake pads do not drag on the tire as they return to the neutral position. Make sure the brake arms rotate freely on their pivots and the brake-arm return springs snap the pads away from the rims. If the lever and caliper move freely and spring back strongly, and if there are no obvious binds in the system, check for frayed strands within the housing sections; replace as needed and/or try lubricating the cable as in 9-3.

1. Remove the insert from the cable-exit hole at the rear of the frame (Fig. 9.12). This usually requires a hex key.

2. Push the new brake cable in through the cable hole at the front of the frame.

3. Push the magnet end of the lead wire of the Park IR-1 internal cable-routing tool (Fig. 9.13) into the cable-exit hole at the rear. Endeavor to get those two ends to meet and be held together magnetically.

4. **Push on the cable while pulling on the Park lead wire.** Guide the cable out of the rear hole.

5. **Replace the cable-routing insert into the hole at the rear of the frame.** Slide it onto the cable and into place, and install the screw.

9.12 Removing the insert from the rear brake cable hole

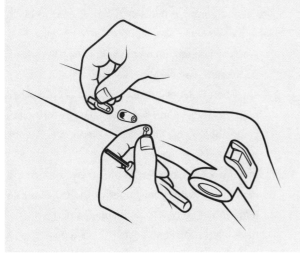

9.13 Park IR-1 Internal Cable Routing Kit

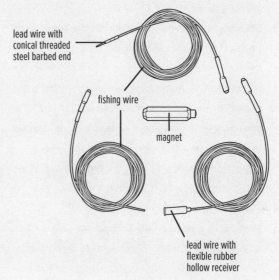

lead wire with conical threaded steel barbed end

fishing wire

magnet

lead wire with flexible rubber hollow receiver

BRAKE LEVERS

 The levers must operate smoothly and be set up so that you can reach them easily while riding.

9-5

LUBRICATION AND SERVICE

1. **Oil all pivot points in the lever.**

2. **Check return-spring function on the lever.** Note that not all levers have springs in them.

3. **Make sure that the lever or lever body is not bent in a way that hinders movement.**

4. **Check for cracks.** If you find any, replace the lever.

5. **Replace torn or cracked lever hoods.**

9-6

REMOVAL, INSTALLATION, AND POSITIONING

 Most current brake levers integrate the brake lever and the shifter in a single unit (Figs. 9.7, 9.14–9.16). Brake/shift levers are generally labeled right and left, but if you're in doubt, you can tell which is which because the shift levers flip to the inside. Here are the steps to replace the entire brake/shift lever unit, or a simple brake lever:

1. **Remove the handlebar tape and bar plugs.**

2. **Remove the old brake/shift lever by loosening its mounting bolt with a 5mm hex key and sliding the lever assembly off.** Some (Campagnolo) levers require a Torx T25 key. The position of

the bolt varies, but it will be on the upper or outboard side of the lever body under the lever hood. Slip the hex or Torx key down from the top between the lever body and the hood (Fig. 9.14) or roll back the hood far enough to get at it from outside of the hood.

Old-style brake levers (Fig. 9.11) have the mounting bolt in the center of the lever body; reach it by pulling the lever and sticking the hex key straight in. Campagnolo and other European levers from the early 1980s and before used a hex nut (accessed with an 8mm socket wrench) rather than a hex nut.

3. **Slide the new lever on the bar.** Some lever bands have an arrow indicating direction of sliding onto the bar, in case they become separated from the lever; use this as a guide for proper orientation. Slide the lever up the bar to where you like it. The current style is to have the levers quite high on the handlebar. Put a long straightedge across the top of both levers after they are tightened to make sure that they are level with each other.

4. **Tighten the mounting bolt.**

5. **Post-2008 Campagnolo Ergopower levers have a "big hands" insert.** If you want to increase the reach to the levers, shove the insert onto the bottom of the lever base under the hood (Fig. 5.37). The inserts are asymmetrical but not labeled per

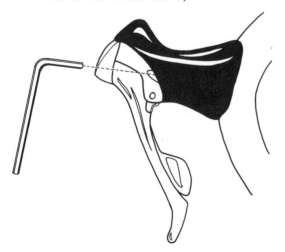

9.14 Tightening a Shimano STI brake/shift lever to the bar with a 5mm hex key

side, so line up the hole on the insert with the cable-access hole on the lever body to ensure you have it on the correct lever.

6. **Install the cables (see 9-4 and 5-6 through 5-17).**

7. **Wrap the handlebar with tape (see 12-12).**

9-7

REACH

Reach adjustment has become standard on dual-control brake/shift levers.

a. SRAM

On the SRAM system, you reduce the reach of the shift lever first, then move the brake lever inward to avoid having them overlap and interfere with each other.

1. **Find the adjustment on the shift lever.** On the outboard side, up near the pivot, you will see either a small screw (labeled "REACH") with a 2.5mm hex hole, or a tiny cam about the diameter of brake-cable housing; the latter will be easier to access by pulling the lever back.

2. **Insert a 2.5mm hex key in the screw, or push the cam inward with your finger or the tip of a pen (Fig. 9.15).** The cam is spring-loaded and will snap back unless you keep pushing inward on it.

3. **Turn the screw or cam counterclockwise until you find its next adjustment position.** With the screw, it's easy; with the cam you turn with your fingers, it's trickier. On some 10-speed levers, it is possible to knock off the circlip retaining this cam, and the cam will go flying and be very hard to find. It takes very little force to turn this cam, as long as it is pushed in. If you're having trouble turning it, push it in more; if you try to turn it when it's not pushed in, you can apply too much force and knock parts off. Once you've turned it a notch (or more), let the cam pop back out against its stop; the shift lever will have pulled in a bit, opening a gap between it and the brake lever.

4. **Peel back the lever hood.**

5. **Tighten the brake lever reach adjustment screw on the back of the lever body (Fig. 9.15).** Turn it

9.15 SRAM reach adjustment (hood removed)

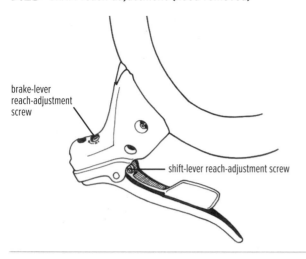

brake-lever
reach-adjustment
screw

shift-lever reach-adjustment screw

9.16 Shimano Dura-Ace 7900 reach adjustment (beneath cover)

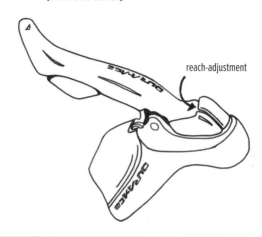

reach-adjustment

clockwise with a 2.5 or 3mm hex key until the brake lever pulls in to barely touch the shift lever.

6. **Readjust the brake-cable tension (9-2) as needed.** (Decreasing lever reach will have pulled the brake pads closer to the rims.)

7. **Repeat steps 1–6 until you get the reach you desire.** Or do the steps in reverse to increase lever reach.

b. Shimano

Shimano has reach adjustment only on levers that have shift cables under the handlebar tape. It's a simple turn of a single screw, but it's under a cover.

11-speed levers

1. **Peel back the lever hood.**
2. **Find the reach adjustment screw.** It is the small slotted screw on the back of the lever body.
3. **With a small screwdriver, tighten the screw (clockwise).** There are many turns of adjustment to this screw; keep turning it until you find the lever reach you like.

10-speed levers (with shift cables under the handlebar tape)

1. **Pull the brake lever.**
2. **Loosen the little Phillips-head screw that is revealed when you pull the lever back.** You do not need to remove this screw all the way; it's a good idea not to, because it is very easy to lose.

3. **Using your fingernail, flip the top of the chrome cover plate forward and remove it (Fig. 9.16).**
4. **Find the reach adjustment screw.** It is the small slotted screw in the hole just outboard of the head of the brake cable.
5. **With a small screwdriver, tighten the screw (clockwise).** There are many turns of adjustment to this screw; keep turning it until you find the lever reach you like.

c. Campagnolo

With the introduction of Campagnolo's taller Ergopower levers, small hands had fewer reach problems, since the new lever shape brought the tip closer to the handlebar. But for some riders with large hands, the reach was too short, so Campagnolo includes a small, plastic "big hands" insert with its levers. It fits between the lever and the handlebar under the bottom edge of the lever body and tips the lever back to increase the distance from the handlebar to the lever blade (Fig. 5.37).

d. Other levers

Even if you don't have adjustable-reach levers, there are a couple of things you can try if you have difficulty reaching the levers.

You can try moving the lever to a different position on the bar. This may bring the lever closer to the bar.

Another option is to buy a bar with a different bend that puts the palm of the hand closer to the lever. There are some bars specifically made to accomplish this.

BRAKE CALIPERS

LEVEL The caliper of a brake is the mechanism that pinches the pads inward against the wheel rim. In most cases, a road bike caliper is a side-pull device that bolts on through a hole in the brake bridge or fork crown (Figs. 9.1, 9.2, 9.17). But "caliper" can also refer to the pair of arms of a cantilever or V-brake that attach to pivot posts welded onto the frame and fork (Figs. 9.4, 9.5). And then there are those old center-pull Weinmann and Universal calipers (Fig. 9.3) from 1960s Raleighs and Peugeots.

9-8

DUAL-PIVOT SIDEPULL BRAKES
(THAT MOUNT WITH A SINGLE CENTER BOLT)

Dual-pivot sidepull brakes mounted on a center bolt (Figs. 9.1, 9.18) have become the industry standard. They are powerful and easy to keep in adjustment.

a. Installation

Stick the center bolt through the hole in the brake bridge or fork crown and tighten it in place with a 5mm hex (or Torx T25) key inserted into the recessed nut (Fig. 9.17). Get a longer recessed brake nut if needed to ensure at least six turns of engagement (this is only an issue on the fork).

b. Cable hookup

Open the quick-release on the caliper (or on the lever on Campagnolo and some Mavic) before you connect the cable. Route the cable housing into the barrel adjuster on the brake arm. On the end of the housing, install a ferrule if one will fit into the barrel adjuster. Push the cable through the housing and the barrel adjuster and under the cable-fixing-bolt washer on the lower brake arm. Pull the cable taut and tighten the bolt with a hex key (Fig. 9.10). Close the quick-release after the cable is connected.

c. Centering

You are trying to achieve an equal amount of space between the pad and the rim on each side. The simplest and quickest way to center these brakes requires no tools. Just grab the brake and twist the entire thing into position (don't mess with the mounting bolt; leave it tight). But do make sure before riding that the recessed nut on the back of the brake bridge or fork is tight (Fig. 9.17).

The centering method built in by Campagnolo, SRAM, and Shimano consists of a setscrew; you can also center SRAM and Mavic calipers with a 12mm or 14mm cone wrench, respectively.

The setscrew is on the side opposite the cable. On Campagnolo, it is on the arm, just above the pad, and it takes a 2mm hex key. On Shimano and SRAM, the

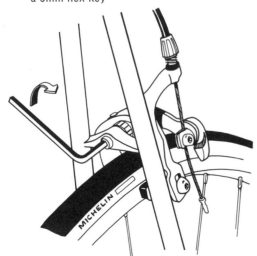

9.17 Tightening a caliper to the brake bridge with a 5mm hex key

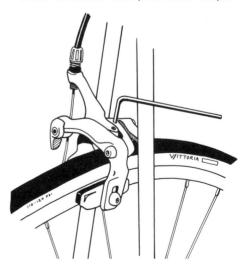

9.18 Turning a setscrew with a 3mm hex key to center a Shimano dual-pivot brake caliper

setscrew is on the upper end of the taller brake arm (Fig. 9.18). As you tighten the screw, the pad on that side moves away from the rim. Loosen the screw, and the other pad (the one on the cable side) moves away from the rim.

SRAM and Mavic dual-pivot sidepull brakes require working a 5mm hex key in the recessed mounting nut while rotating the nut behind the brake caliper with a 12mm or 14mm cone wrench, respectively (similar to center-pivot centering; see Fig. 9.20).

d. Pad adjustment

Loosen the pad-mounting bolt with a hex key (generally 4mm or 5mm). Slide the pad up or down along the slot in the arm to get the pad even with the height of the rim's braking surface. Twist the pad in the vertical plane to have the top edge of the pad follow the curve of the top edge of the rim (Fig. 9.19). While squeezing the brake lever to hold the pad against the rim, tighten the pad-mounting bolt. Make sure the pad does not twist as you tighten (if it does, hold it with your fingers as you cinch the bolt). Also make sure that the pad does not contact the tire, which could quickly wear through the tire sidewall, causing a blowout.

Higher-end brakes also have an orbital adjustment of the pads to align the face of the pad flat or toed in against the rim by means of a concave washer that nests against the convex face of the pad holder. If the brake

9.19 Lining up the pad with the rim

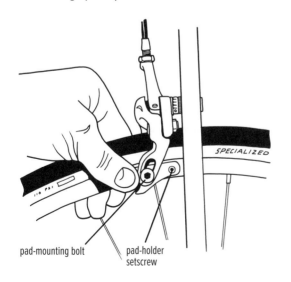

pad-mounting bolt / pad-holder setscrew

squeals or is grabby, toe the pads in a bit so that the forward end of the pad is a little closer to the rim than the rearward end.

e. Spring-tension adjustment

Campagnolo, SRAM, and some Shimano dual-pivot brakes have a setscrew that pushes on the end of the return spring. It is located on the arm, above the cable-side pad. If you tighten this screw (with a 2mm hex key), you will also tighten the spring, thus making the brake both harder to pull and quicker to snap back. See 9-10e and 9-11e for spring-tension adjustment of Shimano direct-mount dual-pivot brakes.

f. Cable-tension adjustment

Follow the instructions in 9-2.

g. Pad replacement

When the pads wear to the point that their grooves are almost gone, replace them. Low-end pads often are molded in one piece with the mounting nut insert or stud, so you just unscrew the old pad and holder assembly and bolt the new one in place.

High-end dual-pivot brakes surround the pad with an aluminum holder that is bolted to the brake arm. The pad can be replaced separately by sliding it from the holder. Some pad holders have a setscrew (Fig. 9.19) that must first be backed out to free the pad. Buy the correct pad according to the year and model of brake and material the braking surface is made out of (aluminum or carbon—see the ProTip on brake pad selection for carbon rims).

Sliding the pad in or out of the holder can be difficult, especially with Campagnolo brakes prior to 2015, as those pads are solely a friction fit (Campagnolo pads now have a setscrew-type pad holder, making pad changes simple). With a friction-fit pad, you may have to yank out the old pad with pliers and slide in the new pad with the aid of a vise or slip-joint pliers, or hold the post in a vise while you push on the pad grooves with a screwdriver.

Be sure to put the proper pad in the proper holder; look at the old one for guidance. Pads often say R or L on the backside and indicate the forward direction;

IT IS ABSOLUTELY CRITICAL to get the right pad if you are using all-carbon rims (with carbon brake tracks). The pad compound for carbon rims is quite different from the compounds made for aluminum rims because normal high-rubber brake pads cannot take the heat of braking on carbon rims and can actually melt. Carbon is an insulator and retains heat well, while aluminum is a conductor and disperses heat well. Carbon rims are also generally lighter than aluminum rims, with less thermal mass to absorb heat.

Carbon rims often come with pads made for them, and approaches vary in their manufacture. Cork pads are also often used for carbon rims due to cork's high coefficient of friction and its resistance to heat.

If you switch back and forth between aluminum rims and carbon rims, switch pads along with the rims. If you do not, braking distance may increase substantially, and aluminum chips that have become embedded in pads used on aluminum rims can damage carbon rims.

Even with specific carbon-rim pads, wear rates are very high on most carbon braking surfaces. After descending a few kilometers of switchbacks, you may notice melted pad material building up on the front of the pads. Make sure that you check the pads frequently and replace them before they get too worn. You would not be the first to have brake failure on a carbon rim.

The buildup on the braking surface of carbon rims is a good thing; when you clean the rims, braking performance may drop, so don't clean the braking surfaces too often. On the other hand, buildup in the pads is not beneficial. Clean the pads and also dig out any chunks of foreign matter from them before you chew up the expensive carbon braking surface.

Campagnolo pads may say DX (right) or SX (left) on the backside.

When you reinstall the pad on the brake arm, make sure that the closed end of the pad holder faces forward. Otherwise, the first time you brake hard, you may see two pieces of rubber fly ahead of you and feel two more hit the backs of your legs. You may not remember anything after that.

9-9

CENTER-PIVOT SIDEPULL BRAKES

Center-pivot sidepull brakes (Fig. 9.2) are still found on lots of bikes because they were the standard from the late 1970s to the early 1990s, and current high-end Campagnolo brake sets use them on the rear to save weight and reduce braking power (most braking is done with the front brake; an overly powerful rear brake can easily lock up the rear wheel). Center-pivot sidepulls work very well and are easy to set up and adjust. Many adjustments are the same as those on dual-pivot brakes.

a. Installation

Follow the instructions in 9-8a.

Some older bikes do not have a countersunk hole in the back of the brake bridge and fork crown. With these, you need a brake with a longer center bolt and a standard nut, which you tighten with a 10mm box wrench.

b. Cable hookup

Follow the instructions in 9-8b, except your brake may require an 8mm box wrench instead of a hex key.

c. Centering

You want an equal amount of space between the pad and the rim on each side. Turn the brake in the direction you need with a cone wrench (usually 13mm or 14mm) slipped onto the flats of the center bolt between the brake and the frame (Fig. 9.20). Hold the brake-mounting nut at the same time, making sure that it is tight when you are finished.

d. Pad adjustment

Follow the instructions in 9-8d.

9.20 Centering a center-pivot sidepull brake with a cone wrench

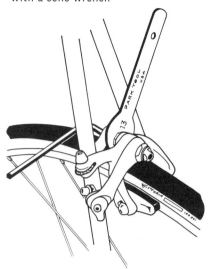

e. Spring-tension adjustment

Some center-pivot sidepull brakes have a spring-tension adjusting screw. And on some Shimano center-pivot brakes, the piece of plastic at each end of the spring can be reversed to tighten or loosen the spring. The hole through which the end of the spring slides is offset in the wafer-shaped plastic piece. Push inward on the end of the spring to free the plastic wafer from the brake-arm tab, flip the wafer over, and push it back in place under the tab. If the hole is to the outside, the spring is looser; if the wafer is flipped so that the hole is toward the inside, the spring is as tight as it is going to get.

f. Cable-tension adjustment

Follow the instructions in 9-2.

g. Pad replacement

Follow the instructions in 9-8g.

9-10

DIRECT-MOUNT DUAL-PIVOT SIDEPULL BRAKES

Direct-mount dual-pivot sidepull road brakes (Figs. 9.21A, 9.21B) mount on a pair of M6 threaded holes above the rim on the fork and seatstays. The casual observer would not tend to notice the difference between a direct-mount and a standard center-bolt dual-pivot sidepull brake (Fig. 9.18).

a. Installation

1. **Screw the mounting bolt of the back arm (the one with the arm sticking up for the barrel adjuster) into the direct-mount hole on the frame or fork (Fig. 9.21A).** Tighten it to 5–7 N-m. Note that the pivot assembly surrounding the bolt should be already adjusted (held by notched nuts) to allow the arm to pivot freely without play. If you have the optional installation tool, you can use that to hold the two parts of the brake together and keep both bolts accessible from the front by the 4mm hex key (Fig. 9.21A); otherwise, mount the back arm before the front one.

2. **Hook the loop on the end of the spring onto the post on the back side of the arm.** On Dura-Ace BR-9010, the spring post is on the front arm (Fig. 9.21B). On Ultegra 6810 and 105 BR-5810, in addition to that spring post, there's also one high up on the back arm with a similar spring configuration to the one in Figure 9.22B. Skip this step if using the optional installation tool.

3. **Screw the mounting bolt of the front arm into the direct-mount hole on the frame or fork (Fig. 9.21A).** Torque is 5–7 N-m.

9.21A Installing Dura-Ace 9010 direct-mount dual-pivot sidepull brake caliper onto direct-mount holes on the fork crown

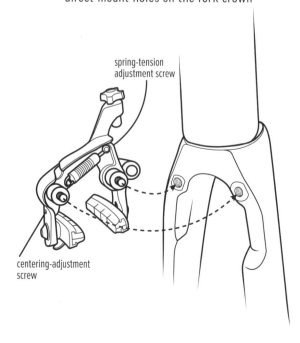

spring-tension adjustment screw

centering-adjustment screw

9.21B Mounting, centering, and adjusting spring tension of Dura-Ace 9010 direct-mount dual-pivot sidepull brake caliper

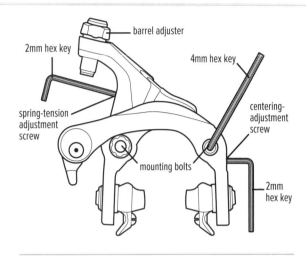

DIRECT-MOUNT REAR SCISSORS BRAKES

Direct-mount rear scissors brakes (Fig. 9.22A, 9.22B) mount on a pair of M6 threaded holes above the rim on the chainstays, behind the bottom bracket.

a. Installation

1. **Screw the mounting bolts into the direct-mount holes on the frame (Fig. 9.22A).** Tighten them to 5–7 N-m. The pivot assembly surrounding each bolt should be pre-adjusted (by means of notched nuts behind and in front of each arm), allowing the arms to pivot freely without play.

2. **If it is disconnected, hook the spring back up.** With thin needle-nose pliers, grab the tiny bent loop on the end of the spring, pull the spring taut, and hook the flat end loop onto the pin (Fig. 9.22B) on the back side of the "Y" arm (the arm with the linkage on it). The fixed end of the spring is attached to the "C" arm (yes, it's shaped like a "C").

b. Cable hookup

Follow the instructions in 9-8b.

c. Centering

To achieve an equal amount of space between the pad and the rim on each side, turn the setscrew on the brake arm opposite the cable with a 2mm hex key (Fig. 9.21B). Tightening it moves the opposite pad toward the rim, and vice versa.

d. Pad adjustment

Follow the instructions in 9-8d.

e. Spring-tension adjustment

On the Dura-Ace BR-9010 brake, tighten or loosen the spring with a 2mm hex key through a notch on the upper back corner of the rear brake arm (Fig. 9.21B), behind the cable and barrel adjuster.

There is no spring-tension adjustment on an Ultegra 6810 or 105 BR-5810 direct-mount dual-pivot sidepull brake (only on the rear scissors version; see 9-11e).

f. Cable-tension adjustment

Follow the instructions in 9-2.

g. Pad replacement

Follow the instructions in 9-8g.

b. Cable hookup

Follow the instructions in 9-8b.

9.22A Mounting and centering Ultegra 6810-R direct-mount rear scissors brake caliper on the chainstays

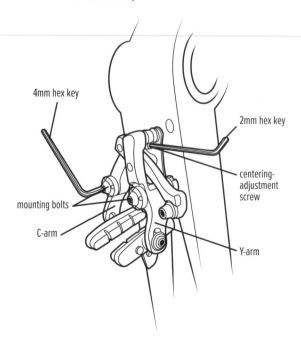

9.22B Back side of Ultegra 6810-R direct-mount rear scissors brake showing spring-tension adjustment

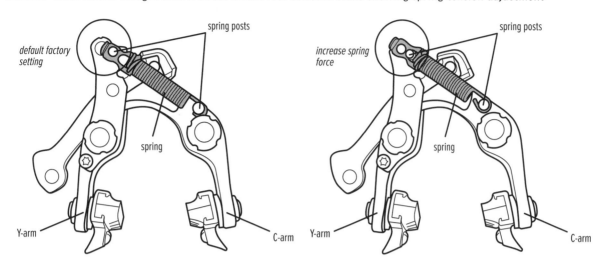

c. Centering

To achieve an equal amount of space between the pad and the rim on each side, turn the centering setscrew on the Y-arm behind the cable (Fig. 9.22A) with a 2mm hex key. Tightening it moves the Y-arm pad toward the rim, and vice versa.

d. Pad adjustment

Follow the instructions in 9-8d.

e. Spring-tension adjustment

On an Ultegra BR-6810-R or 105 BR-5810-R rear scissors brake, there are two spring-tension positions (it comes set on the looser position). To increase the spring tension, use a pair of tiny needle-nose pliers to grab the bent loop on the end of the spring and pull the spring's flat end hole off of the pin it's hooked onto. Hook the second hole over the pin. Center the brake as in step c.

On the Dura-Ace BR-9010-R rear scissors brake, tighten or loosen the spring with a 2mm hex key behind the top of the Y-arm (analogous to adjusting the front brake, Fig. 9.21B).

f. Cable-tension adjustment

Follow the instructions in 9-2. Note that there is no barrel adjuster or cable quick-release on this type of brake caliper. Instead, there is an inline combination barrel adjuster/quick-release; it will be either on the brake cable under the bottom bracket, or it will be up where the cable enters the down tube.

g. Pad replacement

Follow the instructions in 9-8g.

CANTILEVER BRAKES

Whether traditional "cantis" with a straddle cable (Figs. 9.4, 9.23, section 9-12) or sidepull cantilevers (a.k.a. "V-brakes" or "direct-pull cantilevers"; Figs. 9.5, 9.47, section 9-14), cantilever brake calipers consist of two separate brake arms that pivot on posts protruding from the fork legs and rear seatstays. Standard cantilever brakes (Fig. 9.23) have relatively short arms and work well with drop-bar brake levers, but long-arm cantilevers, including V-brakes (Fig. 9.5), require more cable pull than standard road bike levers are able to muster and often have too much braking power with them (see 9-14 for solutions).

Traditional cantilevers allow the user to substantially adjust the mechanical advantage or leverage—the ratio between how much force the hand applies at the lever and the amount of force the brake pads apply at the rim. This is because the user can adjust the angle of the straddle cable as well as the offset of the brake pad from the brake arm, and, on some models, the angle of the arms from vertical as well; see the Pro Tip on cantilever mechanical advantage.

CANTILEVER BRAKE CALIPERS

Found on cyclocross bikes as well as some touring bikes and tandems, standard cantilever brakes ("cantis") offer greater clearance than other rim brakes for fenders, large tires, and mud and wet leaves. Their two separate arms mount onto brake posts integral to the frame and fork and are pulled toward each other when the brake cable pulls up on a transverse cable (a.k.a. "straddle cable") connecting the arms (Figs. 9.23, 9.24).

In order to mount cantilevers, a cable stop is required at the rear on the frame's seatstays or hanging off the seat binder bolt, and another one for the front brake is needed on the fork steering tube above the headset (Fig. 9.9), on the fork crown (Fig. 9.23), or hanging from the stem.

Whether it is on the front or rear, the cable stop often includes a barrel adjuster to pull up cable slack (Figs. 9.9, 9.23), and some cantilever brakes have a barrel adjuster at the caliper (Fig. 9.23).

9.23 TRP EuroX cantilever brake assembly connected via a fork-crown-mounted cable hanger

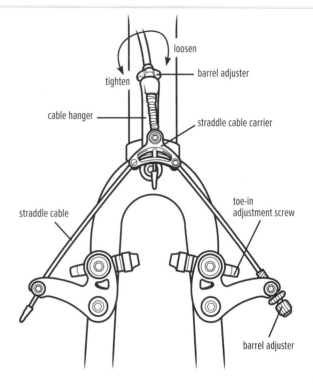

9.24 Cantilever brake assembly with spring-tension adjusting nut in front

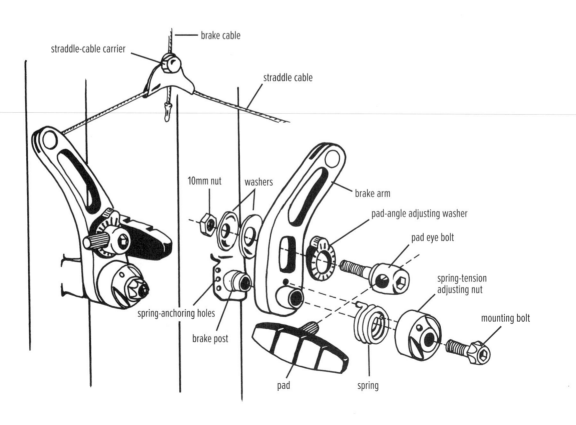

a. Brake arm installation

1. **Grease the brake posts (Fig. 9.24).** Avoid getting grease inside the threads; brake-mounting bolts are treated with threadlock goop to prevent them from vibrating loose. If your brake has an adjustment for brake arm angle (Figs. 9.25A, B), read the Pro Tip on mechanical advantage first.

2. **Make sure you install the brakes with all of the parts in the order in which they were packaged or installed.** The springs often are of different colors and are not interchangeable from left to right.

3. **If the brake has a separate inner sleeve bushing to fit over the cantilever boss, install that first.**

4. **Determine what sort of return system your brakes use.**

   a. If the brake arms have no spring-tension adjustment (Fig. 9.23), or a setscrew on the side of one of the arms for adjusting spring tension (Fig. 9.42), proceed to step 5; the spring in such brakes anchors in a hole in the cantilever boss.

   b. If there is a large nut at each arm for adjusting spring tension, skip to step 7. This spring-tension-adjusting nut is usually in front of the brake arm, surrounding the mounting bolt (Figs. 9.24, 9.25A), but it can also be behind it (Fig. 9.33).

5. **Slip the brake arm onto the boss, inserting the lower end of the spring into the hole in the** cantilever boss. If the boss has three holes as in Figure 9.24, try the center hole first; use a higher hole to make the brake response snappier. Make sure that the top end of the spring is inserted into the corresponding hole in the brake arm as well.

6. **Install and tighten the mounting bolt into the cantilever boss. Skip to step 10.**

7. **If step 4b brings you here, install the spring so that one end inserts into the hole in the brake arm and the other inserts into the hole in the adjusting nut (Fig. 9.24).**

8. **Slide the brake-arm assembly (with any included bushings) onto the cantilever boss.**

9. **Install and tighten the mounting bolt.** Do this while holding the adjusting nut with an open-end wrench so that the pad is touching the rim to facilitate pad adjustment later.

10. **Some brakes are sensitive to the length of the cantilever post on the frame or fork.** There are two common overall post lengths, 21mm and 22mm, measured from the shoulder to the end (Fig. 9.26), and the enlarged section with wrench flats at the base can vary from 4 to 6mm, while the pivot post can range from 16 to 17mm in length.

    If the brake arm binds up when you tighten the bolt, the pivot post may be too short. You may need to switch to a post with a 17mm pivot length. Similarly, if the brake has play and rattles up and

9.25A Wide- and narrow-stance options on an Avid Shorty Ultimate cantilever brake

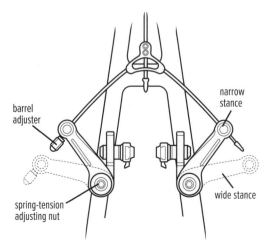

barrel adjuster

narrow stance

wide stance

spring-tension adjusting nut

9.25B Removal and replacement of Torx T10 bolts to change stance on Avid Shorty Ultimate brake

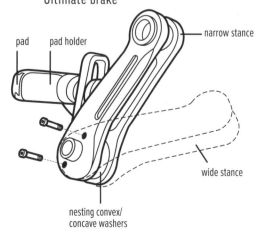

pad   pad holder

narrow stance

wide stance

nesting convex/ concave washers

9.26 Standard cantilever brake pivot post lengths

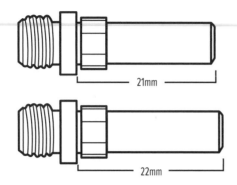

21mm

22mm

down on the pivot, you can get a shorter post or file it shorter.

If the brake spring binds up on itself when the brake is pulled to the rim, then the enlarged section with wrench flats at the base of the post is too short. Replace it with one with longer wrench flats.

Note that while the female threads on all pivot posts are standard M6 × 1 for all brake bolts, the male threads that screw into the permanent threads in the seatstays or fork legs can vary widely. So take your post with you when shopping for a replacement in order to match the male threads.

b. Pad installation and replacement

Older cantilevers had the brake pad integrated with the brake shoe. So if the pad was worn out, you replaced the entire shoe. The shoe could have an unthreaded (Fig. 9.24) or threaded post (Fig. 9.28), depending on brake style.

Modern cantilever brakes have shoes that hold replaceable road pads (Fig. 9.25B). The shoe (a.k.a. "pad holder") usually has a threaded post, but it sometimes has an unthreaded one—sometimes even with a thin, internal bolt that allows pivoting the pad on the end of the post to adjust toe-in (Fig. 9.23).

1. **Remove the old pad, if applicable.** With newer, road-style pads, you'll remove the pad-holder screw (Fig. 9.19) and slide the pad out by hand or by pushing it with a screwdriver. Otherwise, you'll remove the entire shoe, which can have a threaded or unthreaded post.

IF YOUR BRAKE IS ADJUSTABLE for brake-arm angle or stance (Figs. 9.25A, B) and you want to change the arm stance, you must do that before installing it on the brake posts. This consists of disassembling the brake arm from the pad-mounting tab, changing their angle relative to each other, and reassembling.

Why would you want to change the cantilever arm angle? Well, in the wide stance, with the arms sticking out almost horizontally (Fig. 9.25A), leverage is reduced, resulting in the pads moving farther from the rims for a given lever travel, maximizing mud clearance. The narrow stance, by contrast, offers higher power due to increased leverage, and it offers more heel clearance on the rear brake. A full explanation of this is in the Pro Tip on mechanical advantage. In cyclocross, I prefer the narrow stance in the rear to maximize rear braking and to avoid kicking a brake arm into the rear spokes on a botched remount. I like lower power in the front (wide stance) to prevent me from grabbing too much front brake and washing out the front tire, and to maximize mud clearance.

To change the cantilever angle with Avid Shorty Ultimates (Fig. 9.25A), unscrew the twin Torx T10 bolts at each pivot (Fig. 9.25B) to free the separate plates that form each arm. Rotate the arms relative to the pad-mounting tabs, and reengage them at the new angle by inserting the locating pin on each pad-mounting tab into the next hole in the arm plate. Then tighten up the arm assembly again with the Torx T10 bolts.

To change the cantilever angle with FSA K-Force cantilevers, unscrew the collar nut at each pivot to free the separate plates that form each arm. Rotate the arms relative to the pad-mounting tabs, and fix the central shaft at the new angle by engaging the splines of the plastic bushings inserted into each arm plate. Then tighten up the arm assembly again with the collar nut.

After you've changed the stance adjustment, return to step 2 in section 9-12a and continue with installation.

2. **Install the new pad or shoe with integrated pad.** With a road-style pad, slide the new pad in and replace the pad-holder screw (Fig. 9.19).

When installing a brake shoe with a threaded post (Figs. 9.25, 9.28), note the thickness and orientation of the nesting concave and convex washers that come with it. Nest a pair of concave/convex washers on each side of the pad-mounting tab.

Many cantilevers rely on an eyebolt with an enlarged head and a hole through it to accept the pad post (Figs. 9.24, 9.27, 9.33). Some cantilevers have the hole in the eyebolt between two plates forming the brake arm (Fig. 9.23). Lubricate the threads on the eyebolt, slip the post through, and tighten the nut.

3. **If the brake has a spring-tension nut (Figs. 9.24, 9.25A, 9.33), temporarily adjust the spring so that it holds the pad against the rim.** It will make the pad adjustments easier with unthreaded pad posts. If not, you will have to push each arm toward the rim or pull the brake lever as you adjust the pad.

c. Pad adjustment

Pad adjustments are quite easy with some brakes and a real pain with others. Five separate adjustments (a through e in Figs. 9.29–9.31) must be made for each pad:

- Offset distance of the pad from the brake arm (extension of the pad post; distance a in Fig. 9.29)
- Vertical pad height (distance b in Fig. 9.30)
- Pad swing in the vertical plane for mating with the rim's sidewall angle (angle c in Fig. 9.29)
- Pad twist to align the face of the pad with the rim's curvature (angle d in Fig. 9.30)
- Pad swing in the horizontal plane to set toe-in (angle e in Fig. 9.31)

Threaded brake pad posts with nesting concave/convex washers (Figs. 9.25A, B, 9.28) make pad adjustment straightforward. Of brakes with unthreaded posts, cantilevers with a thin central bolt to adjust toe-in (Fig. 9.23) are easy to adjust. So are those that feature a cylindrical brake arm (Fig. 9.27), because the pad is held to the cylinder with a clamp that offers almost full range of motion. Other cantilevers employ a single eyebolt to hold all five pad adjustments (the eyebolt and washers

9.27 Cantilever brake with pad clamps on cylindrical brake arms

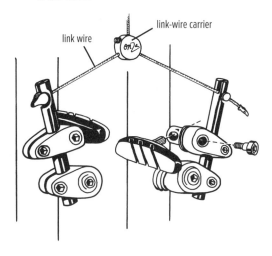

link wire   link-wire carrier

9.28 Threaded brake-pad-post on a one-piece shoe and pad

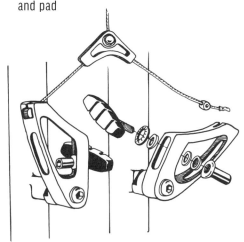

are exploded in Fig. 9.24 and are seen from above in Fig. 9.31). Manual dexterity is required to hold all five adjustments simultaneously while tightening the bolt. Here is the pad-adjustment procedure for all types of cantilevers:

1. **On brakes with unthreaded posts, loosen the pad-clamping bolt and set the pad offset (distance a in Fig. 9.29) by sliding the post in or out of the clamping hole. On brakes with threaded posts, organize the nesting concave/convex washers (Figs. 9.25A, 9.28) to have more washer stack either on the pad side or on the other side.** A good initial position is with the post clamped in the center of its length or the thinner nesting concave/convex washers on the pad side.

9.29 Distance of pad to fixing bolt (a, pad offset) and angle against rim (c, pad swing in vertical plane)

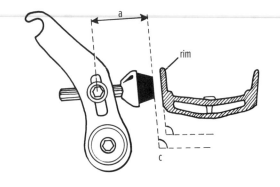

9.30 Up and down (b, vertical pad height) and rotate (d, pad twist)

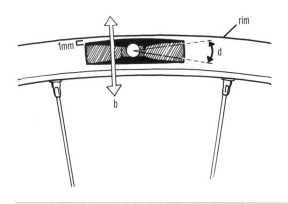

2. **Roughly adjust the vertical pad height (distance b in Fig. 9.30) by sliding the pad-clamping mechanism up or down in the brake-arm slot.** For cantilever brakes with pad clamps surrounding cylindrical brake arms (Fig. 9.27), loosen the bolt clamping the pad holder to the brake arm, and snug the bolt back up once the rough adjustment is reached. With all other types, leave the pad bolt just loose enough to move the pad easily.

3. **Adjust pad swing in the vertical plane (angle c in Fig. 9.29).** You want the face of the pad to meet the rim flat with its top edge 1–2mm below the top of the rim. Fine-tune this adjustment by simultaneously sliding and rotating the pad post up or down.

4. **Adjust the pad twist (angle d in Fig. 9.30).** You want the top edge of the pad parallel to the top of the rim and at least 1mm below it. Make sure that the pad does not contact the tire. With a threaded

pad post, you can pull the brake lever to hold the pad against the rim, move it around to the desired position, and tighten the pad-fixing nut (or bolt). For cantilever brakes with pad clamps on cylindrical brake arms (Fig. 9.27), the pad-securing bolt may now be tightened.

5. **Finally, adjust the pad toe-in (angle e in Fig. 9.31).** The pad should be adjusted either flat to the rim or toed in so that when the forward end (note the arrow in Fig. 9.31) of the pad touches the rim, the rear end of it is 1–2mm away from the rim.

On brakes with a thin central bolt within an unthreaded pad post (Fig. 9.23), adjust toe-in after tightening the large pad-fixing nut by loosening the thin central bolt and rotating the pad to the desired angle; then tighten the bolt.

On cylindrical-arm brakes with two fixing bolts (Fig. 9.27), the toe-in is adjusted by again loosening the bolt that holds the vertical-height adjustment of the pad. Because you have already tightened the other bolt that holds the pad in place, you simply loosen this second bolt and swing the pad horizontally until you arrive at your preferred toe-in setting and retighten the bolt.

Toed in, flat, or toed out—what's the difference? If the pad is toed out, the heel of it will catch the rim and will tend to chatter, making an obnoxious squealing noise. If the brake arms are flimsy or fit loosely on the cantilever boss, or the fork steering tube is overly flexible (see 9-13), the same thing will happen when the pad is flat.

If the brakes work smoothly and powerfully with the pads flat, leave them that way; you'll get the most even pad wear and longest pad life with them set that way. But if the brakes begin to squeal or judder, toe the pads in.

6. **Tighten the pad-fixing nut (or bolt).** That's simple enough with a threaded post or unthreaded posts on brakes as in Figures 9.23 and 9.27. However, with any brake using a single bolt to hold the pad as well as control its rotation, you now have the tricky task of holding all of the adjustments you have made and simultaneously tightening the nut.

9.31 Brake-pad toe-in (e, horizontal pad swing)

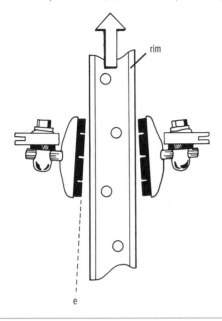

9.32 Curved-face cantilever brake (Ritchey)

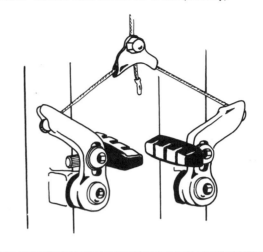

9.33 Ball-joint cantilever brake (Campagnolo)

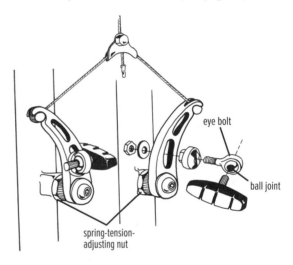

Most eyebolt systems are tightened with a 10mm wrench on the nut on the back of the brake while the front is held with a 5mm hex key (Fig. 9.24). Help from someone else to either hold or tighten is useful here. Probably the trickiest brake to adjust has a big toothed or notched washer between the head of the eyebolt and a flat brake arm (Fig. 9.24). The adjusting washer is thinner on one edge than the other, so rotating it (by means of the tooth or notch) toes the pad in or out; you must hold all of the pad adjustments as you turn this washer, and then keep it and the pad in place as you tighten the nut. It's not an easy job; you may want to slug the guy who came up with this system. Be patient.

Another common type of brake has a convex or concave shape to the slotted brake arm (Fig. 9.32). Cupped washers separate the eyebolt head and nut from the brake arm. The concave or convex surfaces allow the pad to swivel, and tightening the bolt secures everything. Again, you may not get it on the first try. Threaded posts also employ such washers.

NOTE: *Some of these curved-face brakes do not hold their toe-in adjustment well; you may need to sand the brake-arm faces and washers to create more friction between them.*

A rare but simple-to-adjust type of brake has a ball joint at each pad's eyebolt (Fig. 9.33).

d. Straddle-cable adjustment to maximize cantilever performance

The straddle cable should be set in such a way as to provide optimal overall performance. This is not always the adjustment that produces the highest leverage (see Pro Tip on mechanical advantage); sometimes brake feel (i.e., stiffness opposing you when pulling the lever) is improved when leverage is reduced, because you are doing more of the work. I recommend initially setting the straddle cable for relatively high leverage and reducing it from there to improve lever feel and mud clearance.

PRO TIP ┣— CANTILEVER BRAKE MECHANICAL ADVANTAGE

WITH ANY LEVER ARM, the mechanical advantage is highest when the force is applied at right angles to the lever arm. This applies to the straddle cable as well as to the brake arm. Three things determine the leverage of a cantilever brake, which is akin to a bent teeter-totter: 1) the brake lever's mechanical advantage, 2) the ratio of the distance from the pivot (or fulcrum) to the pad face (length FP in Fig. 9.34) to the perpendicular distance from the straddle cable to the pivot (length FC), and 3) the angle of the straddle cable relative to horizontal (angle S).

1. The mechanical advantage of the brake lever is the ratio of the length of the lever (from its pivot to where your finger pulls it) to the distance from the lever pivot to the cable head. You can't do much about a road lever's mechanical advantage, other than pulling closer to the tip. That said, Shimano road levers with shift cables concealed under the handlebar tape (Fig. 9.16) have lower mechanical advantage, and hence higher cable pull, than do most road brake/shift levers.

2. The "cantilever angle" (angle BFP in Fig. 9.34), which is adjustable on Avid Shorty Ultimate (Fig. 9.25) and FSA K-Force cantis, is an important contributor to the mechanical advantage of a cantilever brake. You can see that as the cantilever angle BFP increases, length FC decreases (while FB stays constant), thus reducing the ratio FC/FP and hence the brake's mechanical advantage. Analogously, it becomes progressively harder to lift a kid sitting on the opposite end of a teeter-totter as you move closer to the fulcrum because the ratio FY/FK (fulcrum-to-you distance/fulcrum-to-kid distance) has decreased.

3. The closer the straddle cable is to being horizontal (S = 0), the higher the mechanical advantage, although achieving a horizontal straddle cable that doesn't hit the tire requires tall, "narrow stance" or "low-profile" brake arms (Fig. 9.27) standing up almost vertically (i.e., the cantilever angle BFP is

9.34 Defining angles and lengths on a cantilever brake

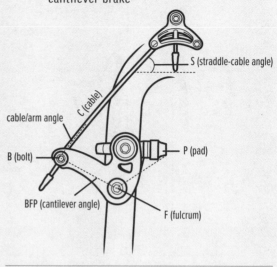

9.35 Straddle-cable angle when open

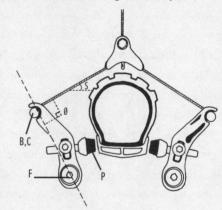

small). And while the leverage is enormous initially, angle S increases rapidly as the brake is applied, reducing its mechanical advantage. A smaller angle S also limits mud clearance and tire size (Fig. 9.35).

The perpendicular distance FC from the pivot (fulcrum) to the straddle cable is at its greatest when equal to the brake arm length FB; that is, when the straddle cable pulls at 90 degrees to the brake arm (Fig. 9.35). But once the pad hits the rim, the actual lever arm is no longer FC but rather PC, from the face of the pad to the cable attachment point on top of the

Continues >>

arm, because the pad face, not the brake post, has become the fulcrum (Fig. 9.36), that is, you are prying the brake post outward rather than pushing the pad further inward.

In other words, you are balancing the straddle-cable's angles relative to the horizontal (S) and relative to the brake arm (FC—and PC once the brake is fully applied) to determine leverage. From Figures 9.34–9.36, you can see that you reduce leverage the longer you make the straddle cable, and vice versa. However, the longer you make the straddle cable, the greater the tire and rim clearance. Understanding that, now you can experiment with what you prefer in your brake setup. Allow at least an inch of straddle-cable clearance over the tire to prevent a tire bulge from engaging the brake.

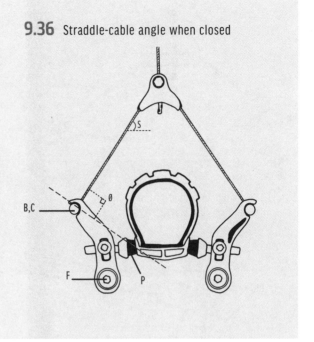

9.36 Straddle-cable angle when closed

The straddle cable (Figs. 9.37–9.39) usually has a metal blob on one end. On an old-school straddle cable (Fig. 9.37), the other end is clamped to one brake arm by the cable-fixing bolt (Fig. 9.42). The blob fits into the slotted brake arm and acts as a quick-release for the brake (Fig. 9.32).

With some cantilevers built since 1988, the cable from the brake lever passes through a round link-wire carrier (Fig. 9.38) and connects directly to one brake arm, and a link wire attached to the carrier hooks to the other arm. Set the cable length from the link-wire carrier to the brake arm approximately the same on both sides (Fig. 9.27). Some post-1993 link-wire carriers have a fixed length of cable housing attached to them (Fig. 9.38) to reach the brake arm. The mechanic has no choice of straddle-cable settings; it is predetermined by the lengths of the link wire and housing segment integrated into the link-wire carrier.

Instead of a cable-fixing bolt on one arm, some brakes have slotted hooks on both arms to accept the blob on the end of a straddle cable or link wire; see Figures 9.23, 9.24, 9.27, 9.28, 9.33, 9.35, and 9.36. In this case, you tighten a small cylindrical clamp on the end of

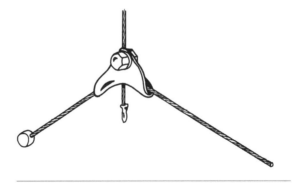

9.37 Straddle-cable carrier, old-school

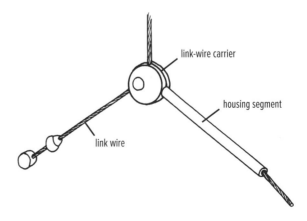

9.38 Straddle cable: link-wire carrier with pass-through housing segment to set straddle cable length

link-wire carrier

housing segment

link wire

9.39 Straddle cable, double-ended

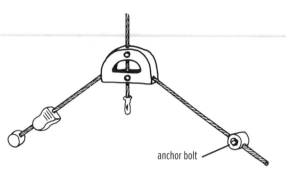

anchor bolt

9.41 An offset straddle-cable stop requires an offset straddle cable carrier

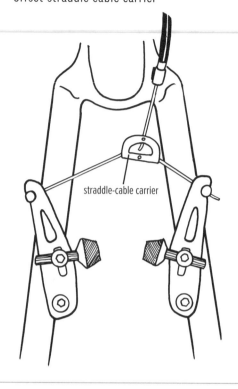

straddle-cable carrier

9.40 Securing cantilever brake cable to straddle-cable carrier

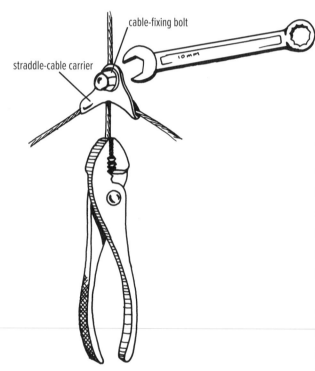

cable-fixing bolt

straddle-cable carrier

9.25A). You adjust cable length simply by loosening the locknut, if included (the brake in Fig. 9.23 has a locknut, and the brake in Fig. 9.25A does not), rotating the barrel adjuster, and retightening the locknut. A barrel adjuster on a cable stop (Figs. 9.9, 9.23) accomplishes the same thing.

To adjust the right-left balance of the brake pads, you can adjust the lateral position of the straddle-cable carrier. Modern straddle-cable carriers (Figs. 9.23, 9.25A), unlike old-school ones (Fig. 9.40), provide friction to fix the carrier's lateral position; the one in Figure 9.25A has a friction bump pressing on the straddle cable, while the ones in Figures 9.23, 9.39, and 9.41 have setscrews pinching the straddle cable. Some brake cables pull asymmetrically as they come around the seat tube or from the headset cable hanger, so the straddle-cable carrier may need to be offset to center the brake pads (Fig. 9.41).

e. Spring-tension adjustment

Spring-tension adjustment centers the brake pads and determines the return force. On brakes with a single setscrew on the side of one brake arm, set the spring

the straddle cable to form a cable end blob (Fig. 9.39); the opposite end may also incorporate a cable barrel adjuster into which the blob on the end of a standard brake cable used as a straddle cable nests (Figs. 9.23, 9.25A).

After you set the straddle cable's length, set its vertical position by loosening the bolt or setscrews that secure it, and slide it along the brake cable. Tighten it in place (Fig. 9.40) so that the brake engages quickly, and the lever cannot be pulled to the bar.

Setting cable length is considerably easier with a cable barrel adjuster on the straddle cable (Figs. 9.23, tension by turning the screw (Fig. 9.42) until the pads hit the rim simultaneously when applied. Higher spring tensions may be achieved by moving the spring to a higher hole on the brake post's platform.

On brakes with a large spring-tension adjusting nut (Figs. 9.24, 9.25A, 9.33) surrounding the mounting bolt, turn the tensioning nuts on both arms to get the combination of return force and centering you desire. You must loosen the mounting bolt while holding the tensioning nut with a wrench. Turn the nut to the desired tension and, while holding it in place with the wrench, tighten the mounting bolt again (Fig. 9.43).

On brakes without an adjustment for it, change spring tension by removing the brake arm and moving the spring to another hole (if present) on the brake post. It is a rough adjustment, and many posts have only one hole. Failing this, you can twist the arm on the post to tighten or loosen the spring a bit.

NOTE: *If the brake arms do not rotate easily on the brake post, there is too much friction. Remove the brake and check that the post is not bent or split, in which case a new one needs to be screwed in or welded on. If not bent, the post is probably too fat to slide freely inside the brake arm, because of paint on it or bulging or mushrooming of the post due to overtightening of the brake-mounting bolt. If it's a replaceable-post type (Fig. 9.26), screw a new one in. Otherwise, carefully sand the circumference of the post to reduce its diameter.*

f. Lubrication and service

Lubricate cables, levers, and/or brake posts whenever braking feels sticky. Cable lubrication and replacement and lever lubrication are covered in 9-3 through 9-5. Lubricate pivots by removing the arms and cleaning and greasing the pivots (9-12a).

9.42 Adjusting return-spring tension with a setscrew

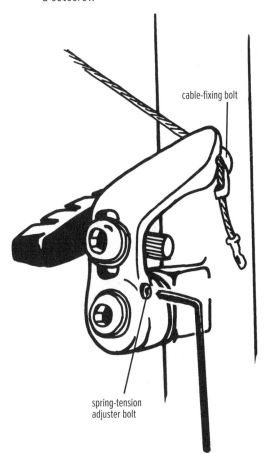

cable-fixing bolt

spring-tension adjuster bolt

9.43 Adjusting return-spring tension with tensioning nut

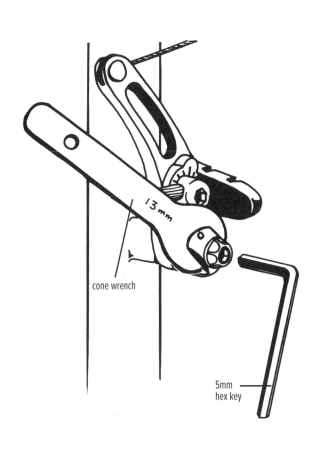

13 mm

cone wrench

5mm hex key

g. Top-mount brake levers

Top-mount or "cross-top" levers (Fig. 9.44) allow the brakes to be accessed from levers mounted on the top, straight section of the handlebar as well as from the normal levers on the drops of the handlebar. The brake cable from the normal lever passes through the top-mount lever and can be pulled from either lever.

1. **Clamp the top-mount lever onto the handlebar.** The lever tip points outward, and lever logos should face up (the mounting bolt is usually underneath).

   Depending on lever type, you will either be mounting the levers just beyond the bulged stem-clamping section, on the narrower 24mm-diameter part of the handlebar (Fig. 9.44), or on the larger 26mm or 31.8mm stem-clamping area.

2. **Orient the levers in a comfortable position for your hands.**

3. **Cut to length and install a section of brake-cable housing from the brake lever, along the front of the bar, to its insertion hole in the top-mount lever clamp (Fig. 9.44).** Reinforce each end of the housing with a ferrule (Fig. 5.21), if possible.

4. **Cut and install a piece of housing to run from the other cable-insertion point on the top-mount** lever (Fig. 9.44) to the brake caliper (front) or frame cable stop (rear). This cable-insertion point will be on the moving part of the lever and usually will have a barrel adjuster on it. Use ferrules (Fig. 5.21) to reinforce the ends.

5. **Install the brake cable, running it right through the top-mount lever to pass from one housing section into the next.** If the old cable is frayed or kinked and won't pass through, buy a new cable.

6. **Connect the cable as described earlier in this chapter for your brake type.** Turn the barrel adjuster on the caliper and/or the top-mount lever to get the right cable tension. You now have another position from which to brake!

9-13

CANTILEVER BRAKE/FORK SHUDDER, BRAKE CHATTER, AND SQUEAL

Squeal and chatter on a front cantilever brake can get to the point where the fork shudders, even when the front brake is only applied lightly. This is due to an overly flexible steering tube on the fork combined with a cable hanger above the headset.

9.44 Top-mount brake lever connected to standard brake lever

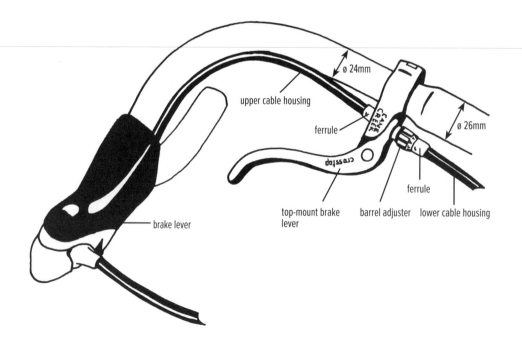

brake lever

upper cable housing

ferrule

ø 24mm

ø 26mm

crosstop

ferrule

top-mount brake lever

barrel adjuster

lower cable housing

Here's the mechanism at work: When the brake pads apply pressure on the rim, the wheel slows and pushes the fork tips back (Fig. 9.45). The fork steering tube and perhaps also the upper part of the fork between the lower headset bearing and the cantilever pivot bosses can flex. If the front brake's cable hanger is above the headset, this will tend to tighten the brake cable, pulling the brake on harder. This will push the fork back yet farther, thus tightening the cable even more and applying the brake yet harder (Fig. 9.45). Eventually, something has to give. Either the bike must come to a stop, conceivably flipping the rider over the front if his or her weight is not back far enough, or the momentum of the bike and rider will force the rim through the brake pads. If the latter happens, the fork will momentarily relax back toward its original shape, and the cable will loosen. But the pressure of continued braking will again slow the wheel, push the fork back, and tighten the cable, which tightens the brake more, which flexes the fork back yet farther, which tightens the brake yet more until something gives again, and the cycle repeats, resulting in a scary fork shudder. Since the bike is essentially applying the brake, it can happen with relatively light finger pull on the brake lever.

What to do?

The least expensive and first thing to try is to adjust the front brake pads with more toe-in (9-12c). This may be enough to fix the problem if the shudder is minor. Some people also have had success with cutting the brake shoes shorter so that there is less pad area to grab. This of course takes away some braking power as well.

A solution that sometimes works with minor shudder and costs nothing except a new straddle cable is to run the front straddle cable very long (combined with more toe-in). As you learned in the previous Pro Tip, this reduces your mechanical advantage and also increases the cable pull. This means that the brake power won't increase as much as the cable tightens due to steering-tube flex.

A more expensive solution is to get a fork with a much stiffer steering tube. This is the method many cyclocross bike manufacturers have chosen: using forks

9.45 Fork shudder in cantilever brake

with steering tubes tapering from 1.5-inch diameter at the fork crown to 1.125-inch at the top (instead of 1.125-inch or 1-inch throughout, as used to be the standards). Unless your frame already accepts a 1.5-inch lower bearing, however, you can't install a tapered-steerer fork, so you are left seeking a stiffer fork with a standard cylindrical steering tube. For a tall bike, you may not be able to find one stiff enough, as the steerer will be so long.

A much less expensive and more effective fix is to use a fork-crown-mounted cable hanger (Figs. 9.23, 9.46), rather than a hanger that surrounds the steering tube above the headset (Figs. 9.9, 9.45). This completely eliminates the steering tube from the equation, and the distance is so short from the fork crown to the pivot bosses that flex there cannot significantly increase brake-cable tension when the brake is applied. Also, two fork legs and a fork crown are stiffer than a single steering tube. The hitch with this solution is that, unless your fork already has a hole through the fork crown, you can't bolt on a hanger as in

Figure 9.46. If you have a steel or aluminum fork crown, you may be able to safely drill a hole through the crown for the hanger bolt, but I cannot recommend this for a carbon fork. Ruckus Components, a carbon-repair shop in Portland, Oregon, will mount a cable hanger to a carbon fork using carbon wrapping. This is the most elegant solution, albeit a pricey one.

Yet another solution is to put a V-brake (Fig. 9.47) on the front. This completely eliminates any flex in the fork from the equation—neither the steerer nor the crown and upper legs affect braking. But, as I discuss in 9-14, even with the shortest V-brake, you will have considerably more brake power and considerably less pad-to-rim clearance and mud clearance over the tire. This setup may be acceptable with a Shimano STI lever with cables that go under the handlebar tape (Fig. 9.16), as these have less leverage and more cable pull than most other road levers, but they still won't offer the clearance of a cantilever. Despite their lower leverage, I think that they should still be set up with the cable tension backed off so far that the brake just comes on as the lever comes to the bar to avoid applying it too hard in a panic situation and flipping over the front. Another solution is a V-brake-

specific drop-bar brake lever, but this only works with a single chainring setup, unless you're willing to use a bar-end or down-tube shift lever for your front derailleur.

Finally, you can use a disc brake (Chapter 10) to eliminate fork shudder. You will need to add a disc-specific fork and disc brake (and new front hub or wheel) to your bike. However, carbon forks for disc brakes that don't have a tapered steering tube are rare, so this solution won't fit your frame if yours accepts a 1⅛-inch or 1-inch steering tube.

If the rear brake is chattering, more pad toe-in or different pads better suited to the rims are generally the answer. Less flexy brake arms also may be required.

9-14

SIDEPULL CANTILEVER (A.K.A. V-BRAKE) CALIPERS

Some hybrid, touring, and tandem bikes have V-brakes, which, like cantilevers, mount on pivot studs attached to the frame and fork.

V-brakes (Figs. 9.5, 9.47) have tall, cantilever-like arms, a horizontal cable-hook link on top of one arm,

9.46 Fork-crown-mounted cable hanger

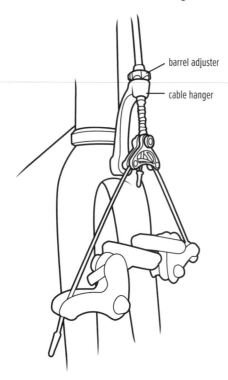

barrel adjuster

cable hanger

9.47 Simple V-brake (a.k.a. sidepull cantilever brake)

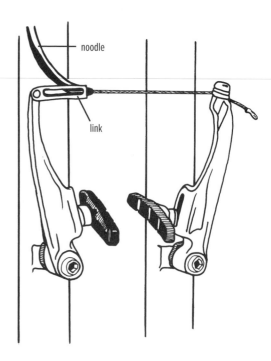

noodle

link

and a cable clamp on the top of the other. A curved aluminum guide pipe, or "noodle," hooks into the horizontal link and takes the cable from the end of the housing and out through the link and then directs the cable toward the cable-fixing bolt on the opposite arm. V-brakes usually have long, thin brake pads with threaded posts. Some V-brakes have "parallel-push" linkages (Fig. 9.5) that move the brake pads horizontally rather than in an arc around the brake post like a cantilever. Simple V-brake designs (Fig. 9.47) mount the pad directly to the arm so that the pads move in a cantilever-like arc.

V-brakes are extremely powerful but require more cable pull than road bike brakes and don't generally work with road bike levers unless 1) you have mini V-brakes with shorter arms coupled to Shimano low-leverage STI levers (these have the shift cables under the handlebar tape); 2) you have V-brake-specific drop-bar levers; or 3) you install a cam unit that replaces the noodle and increases the cable pull.

Unless you're using a cam multiplier or a V-brake lever, I recommend reducing the cable tension to the point that the lever almost reaches the bar when the brake pads hit the rim to avoid pulling them on too hard in a panic stop.

V-brakes set up easily, with a spring pin that goes into the brake pivot hole; just bolt each arm on. Run the cable through the noodle and to the cable-fixing bolt on the opposite arm. Loosen the pad nuts, squeeze the lever, set the pad adjustment, and tighten the pad nuts. For more V-brake setup information, please consult *Zinn & the Art of Mountain Bike Maintenance*.

9-15

SHIMANO AX AND CAMPAGNOLO C-RECORD DELTA AND CROCE D'AUNE DELTA CENTER-PULL CALIPERS

Not many of these brakes were produced, but they were coveted as high-end items, so there are still some around. They attach in the same manner as sidepull calipers, and the pad adjustment and cable-tension adjustment procedure is pretty much the same as well. The major differences have to do with cable connection and centering.

The cable housing stops at a barrel adjuster above the center of the brake. The cable goes straight down through a crosswise hole in the cable-fixing bolt. The Shimano fixing anchor is in a separate triangular yoke that tends to turn as you tighten the bolt.

Centering either of these brakes could not be simpler. Just grab the part sticking straight up (with the cable entry on top) and twist it as needed. Make sure the mounting nut is tight behind the fork or brake bridge.

TROUBLESHOOTING RIM-BRAKE PROBLEMS

9-16

TROUBLESHOOTING BRAKE PROBLEMS

See the following Table 9.1 for possible causes and solutions to common brake problems.

TABLE 9.1 — TROUBLESHOOTING RIM-BRAKE PROBLEMS

SQUEALING

Grease, oil, or fine dust on the rim or pad	Clean the rim with solvent and wipe dry. Use medium-grit sandpaper on dirty pads; avoid using solvents other than rubbing alcohol on pads.
Toe-out of the pads under hard braking so that the heel of the pad does the work	Toe pads in (Fig. 9.31); the front part of the pad should touch first.
Brake arms that are too flimsy for the rider and which chatter or toe-out when the brakes are applied	Remove the pad, put an adjustable wrench on the end of the brake arm, and twist it. If the arms flex too much, get new brakes.
Carbon-fiber or ceramic-coated rims paired with pads not intended for those braking surfaces	Get correct pads.
Loose caliper	Tighten caliper bolts to the proper torque specification.
Squealing and brake chatter on cantilever brakes caused by fork steering tube flexing	Get a stiffer fork or eliminate the steerer flex by using a fork-crown-mounted cable hanger, a mini V-brake, or a disc brake (which also requires a different fork). See 9-13.

LOW POWER

Flexing of brake arms or lever	Install new brakes, but you can try eliminating the other factors first to see whether braking power improves enough.
Stretching of cable	Replace cable and housing (9-4).
Compression of brake housing	Replace cable and housing (9-4).
Squishing of pads	Replace with firmer pads.
Insufficient coefficient of friction between the pads and rim	Common on chrome rims; replace pads or rebuild wheel with aluminum rim.
Oil and grime on the rims and pads	Clean the rims and pads with rubbing alcohol.
Pads incompatible with the rim	Use pads recommended by rim manufacturer; if that fails, try different pads.
Overly long straddle cable on cantilever brake	See 9-12d.
Hydraulic brake system leaking or has air in it	Tighten connections; replace parts as necessary; bleed the system.

TOO MUCH LEVER TRAVEL

Quick-release open	Close quick-release (9-1).
Cable too long	Tighten cable (9-2).
Brake pads worn	Replace pads.
Cantilever cable setup	See 9-12d.
Leaks or air in hydraulic system	Tighten connections; replace parts as necessary; bleed the system.

PAD DRAG

One pad rubs all the way around the rim	Center the caliper for your type of brakes.
Both pads rub all the way around the rim	Loosen the cable (9-2).
Wheel wobbles back and forth against the pads	True the wheel (8-2).
V-brake does not have sufficient clearance	See 9-14 for setup tips.

ANGLED PADS

Pads do not meet flat to the rim	Adjust pad toe-in. If pads cannot be adjusted, remove each pad and twist the end of the arm with an adjustable wrench.
One pad toes in and one toes out	Brake center bolt is bent or the brake hole in the frame or fork is drilled crookedly.

Continues >>

TABLE 9.1 — TROUBLESHOOTING RIM-BRAKE PROBLEMS, CONTINUED

SLOW RETURN	
Bent center bolt or secondary pivot bolt	Replace bolts.
Center bolt nuts too tight	Adjust tightness until there is no play in the caliper while still allowing free movement.
Return spring binding	Make sure the end of the spring is in its plastic friction-reducing piece; for springs without the piece, put a dab of grease between the spring and the spring tab on the arm.
Cable sticking	Lubricate or replace cable (9-3, 9-4), or thaw and dry if frozen.
Cantilever return spring binding	Install longer brake pivot post (9-12a).
LOOSE CALIPER	
Nut(s) holding the caliper together are missing	Replace missing hardware.
Center bolt or nut loose	Tighten the nuts until there is no play in the caliper while allowing free movement. If the brake design uses two nuts, make sure they are both present (the end of the bolt should be covered by a front cap-nut). Hold the back one with one wrench while you tighten the front one against it with another wrench.
PADS WON'T REACH THE RIM	
Short-reach brakes on a long-reach frame	Get a long-reach brake. If the brake nut doesn't fit the small, unrecessed brake hole, you need a brake with a long center bolt and a standard nut and washer. Either buy a new brake with these features, which will be a low-end brake, or find a good old brake. In the 1980s, Campagnolo and others made top-quality brakes with long brake reach. You can also get drop-style center bolts for old Campagnolo short-reach sidepull brakes.
GRINDING NOISE	
Sand or grit on the pads has abraded aluminum rim	Dig aluminum bits out of the pads or replace the pads.
SOFT LEVER OR NO RESISTANCE	
Broken brake cable	Replace cable and housing (9-4).
Loose cable-fixing bolt	Tighten bolt (Fig. 9.10).
Leaks or air in hydraulic system	Tighten connections; replace parts as necessary; bleed the system.
SOFT LEVER THAT PUMPS UP AND GETS FIRMER (HYDRAULIC BRAKES ONLY)	
Leaks or air in hydraulic system	Tighten connections; replace parts as necessary; bleed the system.

10

I couldn't fix your brakes,
so I made your horn louder.

—STEVEN WRIGHT

TOOLS

2mm, 3mm, 4mm, 5mm,
  6mm hex keys
Torx T10, T25 keys
7mm, 8mm box-end
  wrench
7mm, 8mm, 10mm
  open-end wrenches
Isopropyl alcohol
Plastic tire lever
Brake pad spacer
Brake bleed spacer block
Hydraulic hose cutter
Bleed kit (brand-specific
  to the brake)
Park DT-3 rotor-truing
  gauge

OPTIONAL

Drywall-sanding screen
Plastic, grooved blocks
  for clamping hydraulic
  hose
Park DT-3i rotor-truing
  dial indicator
Park DT-2 rotor-truing
  fork(s)
Park PP-1.2 hydraulic
  disc-brake piston press
Park Tool IR-1 Internal
  Cable Routing Kit

DISC BRAKES

LEVEL Disc brakes, both cable-actuated (Fig. 10.1) and hydraulic (Fig. 10.2), are the rage for cyclocross bikes and gravel and endurance road bikes. They provide the best braking in wet conditions of any bicycle brake, and they don't limit clearance for fenders or big tires, or for mud and detritus to pass through.

Disc brakes can offer great stopping and modulation, but installing them correctly is a must. Once properly installed, disc brakes (especially hydraulic ones) generally demand less maintenance than rim brakes, because the tire does not drag dirt and mud into them. There is no need to be intimidated by them; although disc brakes are small and enclosed and therefore somewhat mysterious, they are really quite simple.

Disc brakes have calipers that operate either mechanically or hydraulically. A mechanical disc brake uses a standard brake cable to connect the brake lever to the caliper. A hydraulic disc brake has a master cylinder at the lever filled with hydraulic fluid; a lever-driven piston pumps the fluid through a hose connected to the caliper. Since air is compressible (but fluid is not), a bleed port is provided in the system to remove air bubbles. Inside the caliper, two pistons actuated by the cable or the hydraulic fluid push brake pads against the brake disc (or "rotor"), which is mounted to the wheel hub.

NOTE: *Never squeeze the lever on a hydraulic disc brake without a disc or a spacer between the pads, as you can push a piston in the caliper all of the way out. For bike travel with the wheel out, insert a spacer between the pads—either one that came with the brake or a chunk of corrugated cardboard you cut for the purpose.*

10-1

DISC-BRAKE PAD CHECK AND REPLACEMENT

Disc-brake pads are less easy to see than rim-brake pads, so you should be diligent about checking them for wear. The friction material needs to be at least the thickness of a dime (about 1.2mm, although the brakes will still work until the pad material is 0.5mm thick). Some brakes specify as much as 3–4mm minimum.

Removing some disc-brake pads requires removing the wheel first, while others are "top-loading," offering easy pad change with the wheel installed.

On cotterless types of pads, grab a tab on the pad with your fingers or needle-nose pliers and pull it toward the center of the caliper slot and out (Fig. 10.1). You have to do this from underneath the caliper (with the wheel out), so it's often not easy to get to. Some pads need space to pull toward the other pad and out, one at a time, while others, like the Avid BB7 (shown in Fig. 10.1), will come out together with the spring. With a cable-actuated brake (Fig. 10.1), back out both pad-adjustment knobs first. With a hydraulic brake, you may need to first spread the pads; slip a plastic pad spacer, tire lever, flat-bladed screwdriver, or Park PP-1.2 piston press between the pads and carefully rock the spacer or lever back and forth to push the pistons back into their bores and separate the pads without damaging them.

Cottered pads require that you remove a cotter pin or bolt and then pull the pads out (Fig. 10.2). They're usually easy to get out because they come out of the top of the brake and can often be removed with the wheel in. The cotter pin may be a threaded bolt, a pin with a retaining clip holding it in, or both. Catch the pad spreader spring, too.

Clean the pads with isopropyl alcohol; with a dry, oil-free rag; by rubbing them against each other; or, best yet, by sanding them face down on a drywall-sanding screen so that contaminants fall through the screen. Check for pad wear, scoring, or glazing—anything that could damage the rotor or reduce braking effectiveness. Brakes fade when the pads and rotors get too hot, after which blue discoloration of the rotor and glazing of the pads occur, indicating that resins holding the pad material have broken down and recrystallized on the surface. Glazed pads must be discarded. Always replace pads in pairs.

While not mandatory, you'll reduce the potential for introducing dirt into a hydraulic caliper if, before installing new pads, you clean around the pistons with a Q-tip soaked in the brake fluid used in that brake. Install new pads the way the old ones came out, noting that the left and right pads may differ; it should be obvious if you try to put a pad in the wrong side.

Cotterless pads usually snap back in with a retaining clip (Fig. 10.1), a wire catch, or magnetically. If the pads are not symmetrical and you reverse them, they

10.1 Removing cotterless brake pads

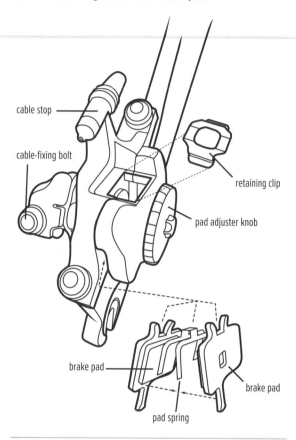

cable stop

cable-fixing bolt

retaining clip

pad adjuster knob

brake pad

brake pad

pad spring

10.2 Removing top-loading cottered brake pads

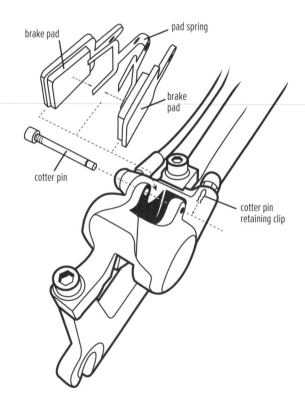

brake pad

pad spring

brake pad

cotter pin

cotter pin retaining clip

may not snap back into place because the piston is often offset from the center of the cutaway for the pad. The spring steel pad spreader between cotterless pads may go in from the bottom after the pads are inserted, or it may go in like a sandwich with the pads (Fig. 10.1). Push the pads deep into the caliper until they click into place.

With cottered pads, the ears on the pads may not line up with the cotter hole if reversed. On cottered pads that have a little butterfly-shaped spring-steel piece that pushes the pads apart (Fig. 10.2), make a sandwich of the new pads and the butterfly spring and push them back in together. Then push in or screw in the cotter pin and replace its circlip, if it has one.

10-2

DISC-BRAKE PAD SELECTION AND BURN-IN

Make sure you buy pads meant for your exact make and model of brake; there are myriad shapes to disc-brake pads, and they're not interchangeable.

When buying pads, you may have a choice of pad compounds, and your choice should be based on the type of riding you do. Metallic pads deal with heat and grit better, whereas resin pads give better initial brake power. Resin pads also wear out faster than metallic pads, especially in wet conditions.

New pads need to be burnished (or "burned in" or "bedded in") with repeated braking before they reach full braking power. If you slam on the brakes when the pads are new, you can damage them so that they won't reach full power, and they may squeal mercilessly to remind you on every ride.

Burn in the pads by braking firmly and evenly without letting the brake get too hot. It's best to do this with one brake at a time, rather than by applying both brakes at the same time. Every manufacturer has a different procedure, but all say that it takes 20 to 40 stops to bed in the pad. I recommend starting out by applying the brake 20 times to bring the speed down from about 10mph to walking speed. Then increase the speed to 15–18mph and brake to walking speed 10 more times. This works the heat cycle evenly over the rotor and reduces the potential for squeal problems. Do not bring the bike to a full stop or lock up the wheel when bedding in the pads.

10-3

PUSHING HYDRAULIC DISC-BRAKE PISTONS BACK IN WHEN PADS RUB

Sometimes hydraulic pistons get pushed so far out that they drag on the rotor or won't let the rotor back in when the wheel is replaced. This can easily happen if the lever is pulled when there is not a rotor or spacer between the pads. You will have to push the pistons back in; on some brakes this is best done with the pads out, while on others it is best done with the pads in.

If the wheel is out and the rotor will not go in, you will first have to push the pads and pistons back by jamming in the plastic pad spacer that is supplied with the brakes (which you should have had in when you pulled the lever when the wheel was out; it would have prevented this from happening). Once you have some space between the pads, you might as well try pushing the pistons back with the pads in. Using a plastic tire lever, a flat-bladed screwdriver, or Park's PP-1.2 piston press, carefully (so you don't gouge the pads) push and twist the lever back and forth until there is enough space between the pads to accept the rotor without rubbing.

On some brakes, you can more successfully get the pistons fully back in place by removing the pads (10-1) and carefully pushing the pistons back in with the box end of a wrench. Then reinstall the pads.

Sometimes, a hydraulic brake in normal usage doesn't retract the pads fully and they rub. This indicates contamination, and you'll have to clean around the piston to get this to stop.

On most hydraulic disc brakes, each piston is pulled back in by an O-ring seal with a square cross-section surrounding the waist of the piston (Fig. 10.3). This "square seal" sits in a groove running around the bore of the piston cylinder; you can see it in cross-section in Figure 10.4. When fluid is forced in behind the piston by squeezing the brake lever, the piston moves outward, and the square seal will start to twist out into the tapered section of the groove shown in Figure 10.4. When the hydraulic pressure is relieved by releasing the lever, the square seal will untwist back to its original configuration, bringing the piston back with it, as long as the seal is not damaged or leaking due to contamination.

10.3 Piston and square seal

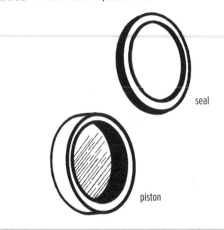

seal

piston

10.4 Caliper cylinder wall, cutaway view

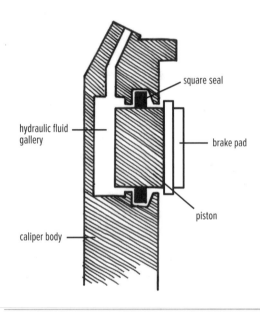

square seal

hydraulic fluid
gallery

brake pad

piston

caliper body

If dirt is present, it can inhibit piston retraction, either by breaking the seal or by creating more friction around the sides of the piston than the square seal can overcome. In this case, simply forcing the pistons back into their bores exacerbates the problem by pushing even more dirt under or against the square seal. So, you want to clean around the piston and lubricate it.

If you have this problem of pad spacing from the rotor being reduced to almost nothing while riding, remove the pads (10-1). While holding one piston in place with a plastic tire lever or a box-end wrench, carefully squeeze the lever to push its mating piston out a bit more to expose more of it for lubrication. Using a cotton swab soaked in hydraulic fluid of the type that's

in your brake, wipe off any grime from around the piston and lubricate it the same way with clean hydraulic fluid.

Carefully prying against the opposite side of the caliper, push the piston back into its bore with the plastic tire lever or box-end wrench. Repeat the procedure to clean and lube the other piston, push it back in as well, and replace the pads.

If you use the wrong implement to push directly on the piston, you can crack it.

Also, if the piston gets out too far and you cannot push it back in far enough against fluid pressure, open the bleed screw slightly while pushing it back, and close it back up immediately to prevent the entry of air.

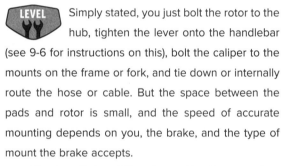

10-4

DISC-BRAKE INSTALLATION AND ADJUSTMENT

LEVEL Simply stated, you just bolt the rotor to the hub, tighten the lever onto the handlebar (see 9-6 for instructions on this), bolt the caliper to the mounts on the frame or fork, and tie down or internally route the hose or cable. But the space between the pads and rotor is small, and the speed of accurate mounting depends on you, the brake, and the type of mount the brake accepts.

After installation, follow the pad bed-in procedure in 10-2 to get full brake performance.

Avoid touching the rotor's braking surface and getting grease or oil on it or the pads. If brake performance drops off, clean the rotor and pads with alcohol. And for obvious reasons, never touch a rotor that's hot after heavy braking.

a. Rotor mounting and removal

Installation

The two rotor-mounting systems are the six-bolt standard (Fig. 10.5) and Shimano's Center Lock (Fig. 10.6), in which a splined aluminum adapter riveted to the steel rotor slips onto the splines of the hub; a single lockring holds it in place. Theoretically, the rotor should be positioned in the same place relative to the axle end with either system, so a wheel with a Center Lock rotor should work fine in a brake set up for a bolt-on rotor of the same diameter. However, if you have a number of wheels you interchange—cyclocross racers will have this concern—the rotors may not all line up in exactly the same place. This is a bummer, since you don't want to have to readjust the caliper every time you switch wheels. If you have this problem, you can get thin shim washers to space a six-bolt rotor away from the hub shell and hence slightly closer to the axle end to get it to more precisely match your other wheel(s). You can also true the rotor by bending it (10-5, Fig. 10.14), and you can move it over slightly one way or another in the process.

10.5 Bolting rotor onto hub

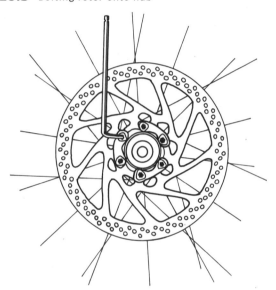

Six-Bolt Rotor

1. **Loosely bolt the rotor to the hub flange (Fig. 10.5).** The logo on the rotor should face outward so that the rotor turns in the proper direction.

2. **Gradually snug the bolts, alternately tightening opposing bolts, rather than adjacent bolts.** A T25 Torx key (like a hex key, but with a star-shaped end) is usually required for this. Torque for rotor bolts ranges from 18 in-lbs (2 N-m) for some manufacturers to 55 in-lbs (6 N-m) for others.

Center Lock Rotor

1. **Slip the rotor splines over the hub splines (Fig. 10.6) with the logo on the rotor facing you.**

2. **Thread on the rotor-securing lockring.** Tighten it with the same splined lockring-remover tool used for rear gear cassettes. If you have a torque wrench that fits the lockring tool, tighten it to 350 in-lbs (40 N-m).

Removal

Six-Bolt Rotor

When removing a bolt-on rotor, you must loosen all of the screws a fraction of a turn before unscrewing any of them fully. On braking, the rotor may rotate relative to the hub a bit and lean against one side of each screw.

10.6 Installing Center Lock splined rotor onto a hub using a cassette lockring tool

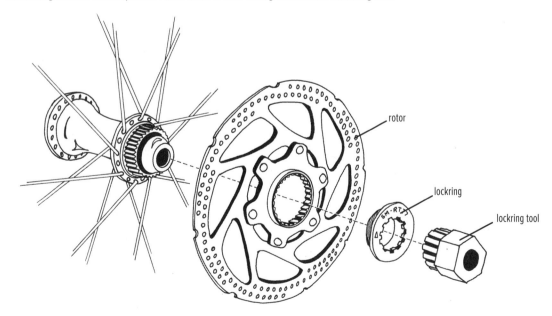

rotor

lockring

lockring tool

If you remove one screw while the others are still tight, the rotor hole's wall will still be pressed against the side of the screw, and the threads on the screw will be damaged. You will then wreck the threads in your hub when you put the damaged screw back in.

Center Lock Rotor

Unscrew the lockring with the splined lockring-removal tool (Fig. 10.6), and pull the rotor off.

NOTE: *Adapters are available to convert Center Lock hubs to accept six-bolt rotors.*

b. Installing a disc-brake caliper

The three types of mounts built into frames and forks are "post mounts" (Fig. 10.7), "International Standard" (IS) mounts (Fig. 10.8), and "flat mounts" (Figs. 10.9, 10.10). IS mounts are drilled transversely (across the frame and toward the wheel) and are not threaded, whereas post mounts are threaded directly into the frame or fork. The beauty of post mounts and flat mounts, whether on the fork (Figs. 10.7, 10.10), frame (Fig. 10.9), or IS adapter bracket (Fig. 10.8), is that the brake can be moved laterally to center it over the rotor.

Flat-mount brake calipers have two threaded holes in their flat base, unlike post-mount calipers, which have ears on either end for the bolts to pass through. Frame flat mounts (Fig. 10.9) are vertical through-holes in the chainstays just forward of the left dropout. Fork flat mounts (Fig. 10.10) are, like post mounts, threaded holes on the back of the left fork leg above the dropout, but, unlike post mounts, they don't have standoffs; they are flush with the back of the fork leg. The bolts thread up into the caliper from the bottom; on the rear, the bolts go through the chainstay right into the caliper (or an adapter for 160mm discs), while on the front, short bolts attach an adapter to the bottom of the caliper, and longer ones bolt the adapter to the fork's flat mounts.

1. **Determine if you need an adapter bracket.** You need an adapter bracket to mount your brake caliper if you are going to:

   a. Mount a post-mount caliper on IS mounts.

   b. Use a larger or smaller rotor than the frame or fork mounts were designed for (on road bikes, the mounts are generally for 140mm rotors, so if you are using a bigger rotor than that, you will need an adapter bracket to

10.7 Mounting a cable-actuated post-mount caliper on fork post mounts (Avid BB7 shown)

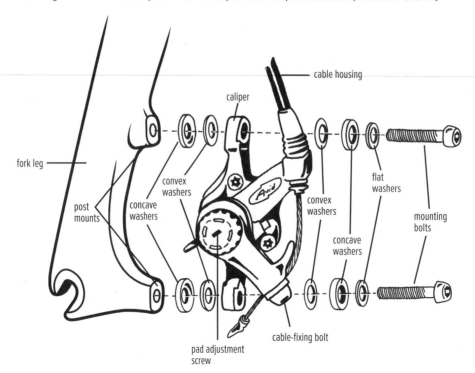

10.8 Mounting a hydraulic post-mount caliper on rear IS mounts

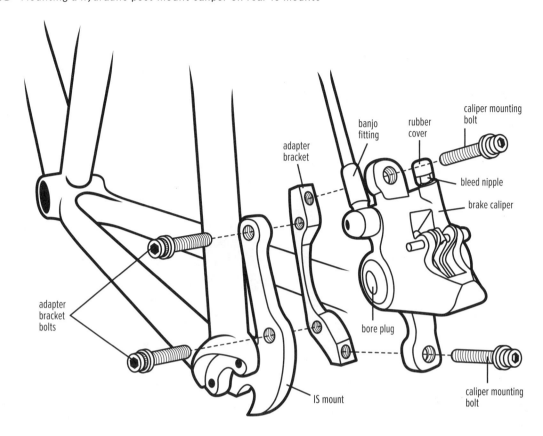

adapter bracket bolts

adapter bracket

banjo fitting

rubber cover

caliper mounting bolt

bleed nipple

brake caliper

bore plug

IS mount

caliper mounting bolt

10.9 Installing a flat-mount caliper on the left chainstay; adapter is used with a 160mm rotor.

10.10 Installing a flat-mount caliper with its adapter bracket onto the fork mounts

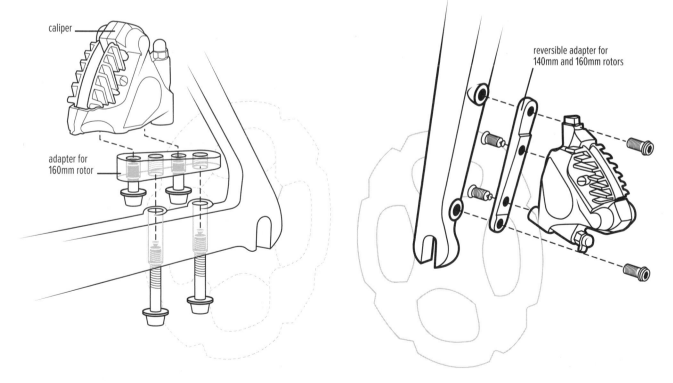

caliper

adapter for 160mm rotor

reversible adapter for 140mm and 160mm rotors

10.11 Adapter for mounting a post-mount caliper on a flat-mount frame

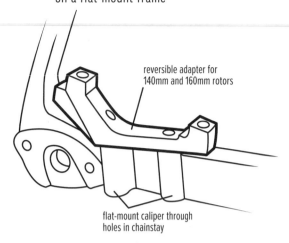

reversible adapter for 140mm and 160mm rotors

flat-mount caliper through holes in chainstay

move the caliper radially outward enough to clear the bigger rotor).

c. Mount a flat-mount caliper on a fork (with flat mounts).

d. Mount a post-mount brake on a flat-mount frame or fork (Fig. 10.11); the adapter is reversible depending on rotor size (140mm or 160mm). The bolts thread up into the adapter bracket from the bottom, through the chainstay flat mounts.

Choose an adapter bracket for your brake model, labeled for the rotor diameter you'll be using.

NOTE: *You cannot mount a flat-mount caliper on post mounts or IS mounts on a frame or fork; no adapter bracket for that exists!*

2. **Install adapter bracket, if applicable.** If you are attaching a post-mount caliper on IS mounts on the frame, tighten the correct adapter bracket to the IS mounts first (Fig. 10.8). Torque is usually 55–70 in-lbs (6–8 N-m).

If it's an adapter bracket to go between a post-mount caliper and post mounts on a fork or frame, stack up the bracket between the brake and post mounts on the fork or frame. Pay attention to the arrow indicating which end is up. Make sure you get bolts that are long enough (but not too long!).

The same goes for an adapter bracket to go between a flat-mount caliper and chainstay flat

mounts; stack up the bracket between the brake and post mounts on the frame, paying attention to the markings indicating which side faces out. Get bolts of the correct length; flat-mount brake calipers have only 13mm of thread in each bolt hole, so make sure there is not more bolt length than that sticking out through the adapter.

For a front flat-mount caliper, firmly bolt the adapter bracket to it, using its supplied bolts. Note the orientation of the bracket; it is labeled for which side goes up based on rotor size.

3. **Loosely bolt the post-mount or flat-mount caliper to the post mounts or flat mounts on the fork (Figs. 10.7, 10.10), frame (Fig. 10.9), or adapter bracket (Figs. 10.8, 10.11).** If the brake has concave and convex washers, keep them in the same order as in Figure 10.7.

NOTE: *If you are installing a hydraulic caliper that is not connected to its lever, skip to section d to cut the hose to length; then skip to 10-6a to fill it with fluid and bleed it; and then come back here and begin with step 3.*

4. **Install the wheel.** The caliper slot will be over the rotor, and the caliper will have some lateral freedom of movement as long as the mounting bolts are still loose.

5. **While squeezing the brake lever, shake and slide the caliper to get it to find its natural position over the rotor, and tighten the mounting bolts.** Alternate between bolts, tightening each a bit at a time to avoid twisting the caliper.

With a cable-actuated brake (Fig. 10.7), first turn the adjuster screw on the wheel side clockwise a few clicks to bring the inboard pad closer to the rotor. Even if there is a knob on the adjustment screw, it is easier on the knuckles to use the Torx T25 key or hex key it requires through the spokes to turn the screw. Then turn the outboard pad adjustment screw (if present; Fig. 10.7) clockwise until it stops (otherwise, pull the cable tight by pulling the lever), which will cause the pads to squeeze the rotor from both sides. Now tighten the mounting bolts to the correct torque setting. Ultimately, you want the spacing to be even on both sides (alternately, you can have a third of the gap between pads and rotor to be on the inboard side and two-thirds of it to be on the outboard side); therefore, after hooking up the cable (9-4 and 10-4c) and tightening the cable anchor bolt to torque, back out the two knobs (or the cable tension and inboard knob) appropriately to achieve this. Be aware that while hydraulic disc-brake pads self-adjust to maintain the same spacing to the rotor as they wear, cable-actuated ones do not. As the pads wear, adjust them with dual adjuster screws (Figs. 10.1, 10.7) farther toward the rotor by turning in (clockwise) both adjuster screws at the caliper, rather than by tightening the cable at the fixing bolt or barrel adjuster; on brakes without an outboard adjuster screw, turn in the inboard adjuster screw and tighten the cable at one of the barrel adjusters.

6. **Spin the wheel to check for brake rub.** If you hear rub, peer through the gap between the rotor and the pads, and, with a white background for contrast, note which pad (or worse, which side of the caliper slot) is rubbing. Loosen the bolts again, and slip a business card or two between the rubbing pad and the rotor.

7. **Repeat steps 4 and 5 until the rotor spins without rub.** If desperate, just loosen the bolts slightly, eyeball the gap, push or tap the caliper as you see fit, and tighten while holding the caliper; expect some frustration.

8. **If the rotor is bent, straighten it.** See 10-5 on rotor truing.

NOTE ON CENTERING BRAKES THAT HAVE ONLY ONE MOVING PAD: *Some disc brakes work by flexing the rotor toward a fixed pad. This applies to some hydraulic disc brakes and to almost all cable-actuated disc brakes. The above procedure will work for installing hydraulic ones, provided you first tighten the fixed-pad adjuster screw about a half turn. Once the caliper is bolted in place, back out the fixed-pad adjuster screw so the pad just barely clears the rotor.*

NOTE ON BENT ROTORS: *A bent rotor will rub or reduce pad adjustment range. See 10-5 on rotor truing.*

c. Hookup of cable-actuated disc brakes

Route the cable housing to the brake following the procedures in 9-4 on cable installation. Tie it down with zip-ties, or run it internally (5-12) on frames with no cable stops. Push the cable through the housing stop on the caliper, and tighten it under the cable anchor bolt.

d. Cutting hydraulic disc-brake hoses to length

When you route the hose to the brake, make it curve smoothly without kinks, without large loops that can catch on things, but not so short that it is tight across spans where it is vulnerable.

If the frame has internal hose routing, use Park's IR-1 internal guide tool (Fig. 5.30), as in 5-12; one of the guide wires has a conical tip you can screw into the end of the hose to pull it through. On a new frame with temporary internal guide tubes, you can slide a cable through the tube and then through the hydraulic hose. Pull the temporary tube out while pushing the hydraulic hose in, with the cable bridging them so the hose can follow.

Or, if the frame has external disc-brake hose guides, use those. Otherwise, secure the hose to the frame or fork with zip-ties, tape, guides that clip or screw into cable guides, or adhesive-backed hose guides.

Don't expect aftermarket brake hoses to be the right length for your bike; you may need to cut them. If one end of the hose has a permanent crimped end on it, don't cut that end. Generally, on an end that you can cut, there will be a brass, olive-shaped ring around the end of the hose (Fig. 10.12); a sleeve nut squeezes down on the "olive" to seal against leaks. The brass olive will need to be replaced after you cut the hose.

10.12 Hose-sealing system: The sleeve nut squeezes the brass olive, which squeezes the hose onto the barbed fitting, which is either threaded or hammered into the hose.

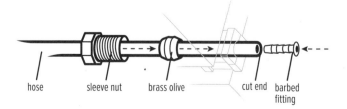

hose   sleeve nut   brass olive   cut end   barbed fitting

10.13A In-line hose connection at caliper

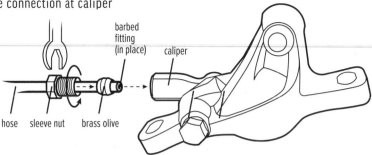

When you cut a hose, point the end up immediately to maintain a dome of fluid at the end, and keep the fluid hole on the lever or caliper pointed up. This may prevent fluid loss and trapped air after you reconnect the hose and the trouble of bleeding the system afterward.

1. **Remove the wheel and the brake pads (10-1).** You don't want to get brake fluid on the rotor or pads.

2. **Disconnect the hose from the fittings.** This is generally best done at the caliper (Fig. 10.13A); do not do it at an end that has a permanent crimped fitting (like at a banjo fitting, Fig. 10.13B).

In general, on a hydraulic road disc brake, it is best to avoid cutting the hose at the lever end, because it is often more of a pain to reconnect it there than at the caliper. In particular, SRAM levers must be disassembled to attach the hose to them; the lever end of the hose terminates in a banjo fitting (like in Fig. 10.13B), and once it's connected, you can't pressure-test the system until you've reassembled the lever (and if it leaks, you have to take everything all apart again). Shimano levers do have an in-line connection (Fig. 10.13C), but it's still more trouble to peel back the lever hood and the tape and turn the nut while it's against the handlebar than it is to get at an in-line fitting at the caliper (Fig. 10.13A).

If you have to install a longer hose, you will have to connect it at both ends, but to simply shorten it, cut it at the caliper if it has no banjo fitting there.

Unscrew the sleeve nut holding the hose. The sleeve nut may be concealed under a plastic or rubber cover; slide it up the hose for access. If the hose attaches at a banjo fitting, don't unscrew the banjo bolt unless you're replacing the entire hose; to shorten it, cut it at the other end.

10.13B Banjo hose connection at caliper

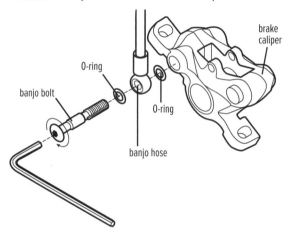

10.13C In-line hose connection at lever

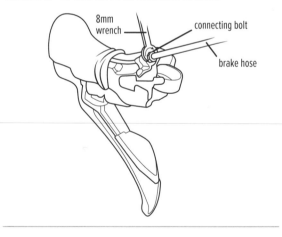

3. **If possible, gently pull the hose straight off.** Pull the hose off and slide the sleeve nut and rubber cover up the hose beyond where you plan to cut. As part of the sealing system, many brakes have a thin barbed nipple extending up inside the hose under the brass olive ring; the olive, compressed by the sleeve nut, tightens around the nipple's barbs (Fig. 10.12). This barbed nipple is generally a separate piece that presses or screws into the end of the hose.

4. **Cut the hose to length with a sharp, perpendicular cut.** Make sure that you will have enough hose length for full rotation of the handlebars. Ideally, use a hydraulic hose cutter (Fig. 1.4). You can also do a good job with a sharp knife while the hose is clamped in a vise, held by a pair of plastic grooved blocks (many brakes come with these blocks and bike shops get so many that they usually have extras). Again, tip the hose end up to prevent loss of fluid and consequent addition of air into the system.

5. **Slide the connection parts onto the hose.** Slip on the rubber hose-nut cover, the hose nut, and the new brass olive. Grease the outside of the olive.

6. **If the brake has a separate barbed fitting, install it.** Tap it into the hose (Shimano) or screw it into the hose with a tiny Torx key (SRAM) while holding the hose securely, ideally in a vise between the grooved blocks mentioned earlier.

7. **Push the hose in place and tighten the sleeve nut (maximum torque: 40 in-lbs/4.5 N-m).** If present, slide the plastic or rubber nut cover back into place over the nut.

8. **Confirm that the hose routing works.** Make sure that the hose does not get yanked when you turn the handlebars. Some brakes allow adjustment of the angle of the banjo at the caliper (or at a satellite master cylinder) by loosening and retightening the banjo bolt to smooth the hose routing.

9. **Skip to 10-6 on bleeding hydraulic disc brakes.** However, if the brake was previously connected, and you think you might have prevented the entry of air into the system with careful cutting and keeping the hose and connection port pointed up, then install the pads and wheel and squeeze the lever. If the lever feels firm, you're done. If it does not, there is air in the line, and you must bleed the air from the system (see 10-6).

e. Lever reach, lever pull, and pad spacing

The reach adjustment, if present, may be similar to those in 9-7.

Lever pull and pad spacing are closely related, as the closer the pads are to the rotor, the less pull it takes to stop. But the pads will rub if too close. The lever on a hydraulic brake may or may not have a lever pull adjustment.

In their out-of-the box configuration, Shimano R785 Di2 hydraulic road levers have some free play before engaging the piston; the lever flips freely outward a few degrees from where it sits at rest. You can eliminate this with the free stroke adjustment screw under the inboard side of the lever hood. Peel back the hood and turn the screw counterclockwise with either a small Phillips-head or thin, flat-blade screwdriver.

Lever pull with cable-actuated brakes can be adjusted by the cable barrel adjuster at the caliper. Pad spacing on cable-actuated disc brakes, which affects lever pull, can also be adjusted with the screw(s) on one or both sides of the caliper. Same goes for hydraulic brakes that have only one moving piston, but only the pad on the wheel side is adjustable.

10-5

TRUING DISC-BRAKE ROTORS

LEVEL Even when you get a caliper perfectly centered, if the rotor is bent, the brake will rub and may even squeal and howl. The spacing between brake pads and rotor is so tight on a bicycle disc brake—around 0.015 inch (0.4mm)—that there is almost no room for any rotor wobble whatsoever. And unlike automotive brake rotors, bicycle rotors are thin and unprotected; they can be bent by rocks thrown up while riding, in a crash, or when packing your wheel in a car or a bike bag. They can also warp due to heat buildup on a long, steep descent. One way or another, the rotors will likely get bent eventually, so you need to be able to straighten them.

For the rider with multiple disc-brake wheels for the same bike, all the rotors must not only be the same diameter, but they must also be vertically true and spaced the same distance from the dropout when installed. Otherwise, the pads will rub when the wheels are switched. Truing is a way to fine-tune the rotor position (beyond installing spacers under the more inboard rotor) so that rotors on two wheels sit identically in the caliper.

If a rotor is really potato-chipped, first remove it from the hub and hammer it flat on an anvil. Then proceed with any of the following methods.

a. Eyeballing rotor in caliper

By eyeballing, you can often do an adequate job to at least minimize brake-pad rub, but be forewarned that this approach requires patience; it can be hard to tell on which pad the rotor is rubbing as the gap is so small. Place a piece of white paper on the floor or the wall, below or level with the caliper, so that you can see the space between the rotor and the pads. Slowly turn the wheel, marking where the disc rubs on each pad with a felt-tipped pen. Carefully bend the disc into alignment with your thumbs at the points you marked, rechecking it constantly by spinning it again through the brake. A rotor bends easily by hand.

b. Using a gauge that grazes the rotor

A more accurate way to judge rotor trueness is to attach a pointer to a truing stand or to the frame or fork in such a way that you can adjust it to graze the rotor. A good solution is the Park DT-3 Rotor Truing Gauge, which bolts to a Park truing stand. It has a microadjustable threaded rod that you can move to just touch the rotor and indicate where its alignment is off.

You can also rig up a pointer to graze the rotor while the wheel is on the bike. The pointer must be mounted securely for this method to work; otherwise, you can make mistakes, thinking the rotor is bent one way when it is actually bent the other.

Bend the rotor with your thumbs away from where the pointer grazes it as you rotate the wheel.

c. Rotor truing with a dial indicator

If you straighten rotors often, the Park DT-3i.2 or other dial indicator (Fig. 10.14) will show the lateral position of the rotor within 0.001 inch (0.025mm), so you can get the rotor as straight as it was when new. The DT-3i.2 dial indicator clamps to a Park truing stand (it attaches to the DT-3 mentioned in method b, which bolts to the truing stand). Set the dial indicator tip against the rotor.

Wherever the needle indicates the greatest deflection in either direction, bend the rotor back, continually rechecking it with the dial indicator.

You can bend the rotor into alignment with your thumbs; better yet, rotor-tuning forks, like Park's DT-2,

10.14 Truing a bent rotor with a dial indicator and rotor-truing forks

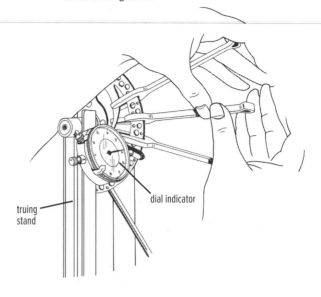

truing stand

dial indicator

provide leverage to precisely bend the rotor. For small bends, one fork is plenty; for bigger bends, if you have three of them, you can stabilize the rotor in position with two forks, one on either side of the bent spot, and bend the rotor to eliminate the warped spot with the third fork (Fig. 10.14).

10-6

HYDRAULIC DISC-BRAKE BLEEDING (OR FILLING)

LEVEL Brakes must be bled whenever they have air in the system. The symptom is a lever that is not firm when pulled and/or becomes more firm with repeated pumping of the lever. Separately, given enough usage in dirty conditions, dirt can get past the seals and contaminate the fluid, so flushing the old fluid out with new fluid will improve performance.

The procedure for filling an empty brake system is the same as for bleeding one. In general, you either move fluid down through the system by filling the reservoir at the top and forcing fluid through to the caliper, or you push fluid up to the lever from a syringe or squeeze bottle at the caliper. Air bubbles float up to the top of the fluid, toward the lever.

With road hydraulic disc brakes in their infancy as of this printing, I'm including general bleeding instructions in addition to specific ones for Shimano and SRAM brakes. You should be able to bleed almost any brake with one of the following methods.

With all brakes:

1. **Remove the wheel.**
2. **Remove the brake pads (10-1).**
3. **Install a spacer block (Fig. 10.15) between the pistons.** If the block won't fit in between the pistons, push them back in their bores with a plastic tire lever or box-end wrench, or with a tool specifically designed for this (like the Park PP-1.2). The spacer allows you to apply hydraulic pressure without the pistons popping out. Hydraulic disc brakes generally come with a bleed block as well as a pad spacer.

IMPORTANT—PAD PROTECTION: *Avoid getting fluid on the pads, which will ruin them. Replace pads contaminated by brake fluid, and clean rotors contaminated by brake fluid with rubbing alcohol.*

IMPORTANT—FLUID TYPE: *Use the recommended brake fluid for your brake. Some systems use mineral oil, and some use DOT (automotive) brake fluid. DO NOT interchange mineral oil and DOT fluid in a brake; doing so will ruin the seals inside.*

Not all mineral oil is the same (viscosity, purity, boiling point, etc. varies), nor is all DOT fluid the same. DOT (which stands for Department of Transportation) has a standardized numbering system. The higher the DOT number, the higher the boiling point, and your brake was designed to operate in a certain temperature range with a DOT fluid for that range. If you were to use DOT 3 or DOT 4 fluid, for instance, in a brake designed for DOT 5.1, you might be without brakes when you need them the most—when they get really hot under heavy braking (see warning below). Again, use the fluid your brake was designed for.

WARNING—BOILING FLUID: *With brakes using DOT fluid, only add fluid from a container that has never been opened before. DOT fluid absorbs water, and the more water it has absorbed, the lower its boiling point. Opening the container for a short time can be enough to bring the boiling point down significantly (if you leave a full glass of DOT fluid out overnight in a humid area, it will overflow the glass by morning!).*

Why is the boiling point of the fluid important? A hydraulic brake works because liquids are essentially noncompressible, so pushing on a piston at one end of a column of liquid (at the lever) can push a piston just as forcefully at the other end of the column of liquid (at the caliper). Gases, on the other hand, are compressible; that's why you have compressed air in your tires. But if hydraulic fluid boils, gas bubbles form in the hydraulic lines, and pulling the lever will only compress the gas; it won't forcefully push the caliper pistons.

"Vapor lock" occurs when the caliper gets so hot that the fluid inside boils. If this happens to you on a ride, let go of the lever so that the caliper can cool in time to brake for the next corner. **With any brake on a steep descent, it is better to intersperse hard braking with no braking to avoid the overheating that continuous braking can produce.**

Once vapor lock occurs, you need to replace the fluid in a DOT-fluid brake with new DOT fluid. With mineral oil, you need only let the brake cool down, since oil doesn't absorb water.

IMPORTANT: *DOT fluid can dissolve paint, so wipe it off wherever it drips on the bike, and rinse the area with isopropyl alcohol quickly. Do not get it on your skin or in your eyes. It is soluble in water as well as in isopropyl alcohol.*

10.15 Installing bleed block into caliper in place of brake pads

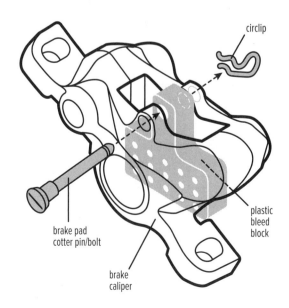

circlip

brake pad cotter pin/bolt

plastic bleed block

brake caliper

10.16 Removing nameplate and bleed screw from Shimano lever

10.17 Bleed funnel attached to Shimano lever; oil stopper plunger at right

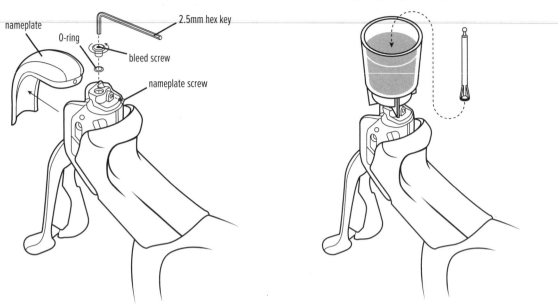

a. Bleeding Shimano brakes

1. **Remove the lever's nameplate.** Peel back the top of the lever hood and remove the nameplate-fixing screw. Hold the lever and pull up the bottom of the nameplate to pull it free (Fig. 10.16).

2. **Turn the handlebar and the lever so the bleed screw is level and the hose trends downward the entire way to the caliper.** You can unbolt the caliper so that it hangs by the hose; if you do so, secure it so that it doesn't move around, which could encourage the bleed hose to fall off.

3. **Remove the bleed screw (Fig. 10.16).**

4. **Install the Shimano bleed funnel.** Thread the funnel's tip into the bleed hole atop the lever (Fig. 10.17).

5. **Connect the syringe to the caliper bleed fitting and open it.** Fill the bleed syringe with Shimano hydraulic mineral oil.

   Some Shimano road disc calipers have standard bleed nipples, while others (specifically, the RS785) have a bleed port hidden underneath, with a bleed screw coming in perpendicularly from the back to open and close the bleed port.

   With a standard bleed nipple (RS785, RS805, or RS505), pull off the rubber cover and put the box end of a 7mm wrench on the nipple. Push the

syringe tube over the bleed nipple, and slide the tube holder over the nipple tip so that the tube will not be disconnected (Fig. 10.18A). Open the bleed nipple ⅛ of a turn.

On a caliper with a hidden bleed port, dig out the rubber cover concealing the bleed port inside the hole underneath the caliper (Fig. 10.18B). Slide the tube holder followed by the adapter sleeve onto the end of the syringe tube, push the tube onto the recessed bleed port, and slide the tube holder up against the underside of the caliper so that the adapter sleeve secures the end of the tube onto the bleed port. Loosen the bleed screw on the back of the caliper ⅛ of a turn with a 3mm hex key to slightly open the bleed port (Fig. 10.18B).

6. **Push the syringe plunger (Figs. 10.18A, B) to push oil into the caliper.** Oil will flow out into the oil funnel at the lever (Fig. 10.17). Keep pushing oil in until no more air bubbles appear in the funnel.

NOTE: *Unlike a bleed method customary on some mountain bike hydraulic brakes, do not depress and release the brake lever repeatedly. Oil without air bubbles may emerge in the funnel as a result, but air bubbles may become trapped inside the brake caliper, and bleeding them out will be more difficult and may require draining out all of the oil and starting over.*

10.18A Bleed syringe pushing oil into the bleed nipple of a Shimano caliper

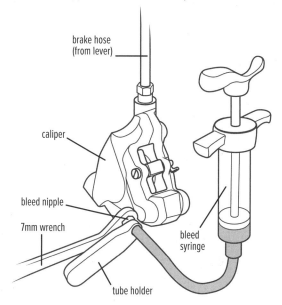

10.18B Bleed syringe, adapter, and tube holder for the recessed bleed port under a Shimano RS785 post-mount caliper

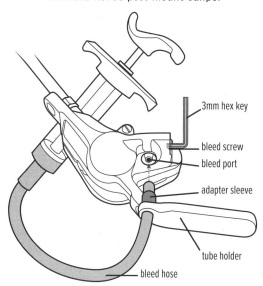

7. **Once there are no more air bubbles in the oil in the funnel, temporarily close the bleed nipple or bleed screw (Figs. 10.18A, B).** Remove the syringe tube; catch drips with a rag.

8. **Attach the catch bag and let oil flow down into it (Fig. 10.18C).** Tie the plastic bag from the bleed kit onto the tube with rubber bands.

   With a standard bleed nipple (RS785, RS805, or RS505; Fig 10.18A), push the tube over the bleed nipple and open the bleed nipple ⅛ of a turn, using the 7mm box wrench that is still on the nipple.

   With the RS785 caliper with a hidden bleed port (Fig. 10.18B), push the tube onto the recessed bleed port. Loosen the bleed screw on the back of the caliper ⅛ of a turn with a 3mm hex key to slightly open the bleed port.

   Oil containing air bubbles will gradually flow via gravity from the caliper bleed nipple/port into the tube and bag. To free more bubbles, gently shake the brake hose, and gently tap the lever body and caliper with the head of a screwdriver. The oil level inside the lever funnel will drop with the flow of oil, so keep adding more oil to the funnel to ensure that the system doesn't ingest any air from that end.

9. **Once no more air bubbles emerge, close the bleed nipple or bleed screw.**

10.18C Catch bag attached to bleed tube on Shimano caliper

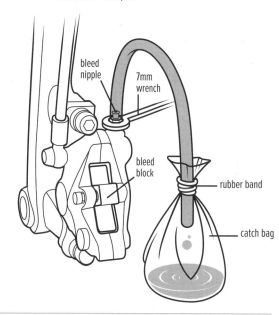

10. **Pull the lever and alternately start and stop the flow of oil.** While pulling the brake lever, open and close the bleed nipple or bleed screw two or three times in rapid succession (for approximately a half second each time) to release any air bubbles that may be in the brake calipers.

    Retighten the bleed nipple or bleed screw. Torque on either type is 4–6 N-m (35–52 in-lbs).

11. **Pull the brake lever repeatedly.** Air trapped in the system can bubble up into the oil funnel (Fig. 10.17). Once bubbles no longer appear, pull the brake lever fully. The lever action should feel stiff.

12. **Loosen the stem clamp bolts, and tilt the handlebar first up and then down by 30 degrees or so.** In each position, pull the lever hard again. If any air bubbles appear, repeat steps 8–11 with the bar in this position until they stop appearing. Retighten the handlebar to torque spec (12-8).

13. **Plug the oil funnel with the larger, O-ring end of the oil stopper plunger.** See Figure 10.17.

14. **Remove the funnel and replace the bleed screw.** Replace the O-ring on the bleed screw if it is damaged. Unscrew the oil funnel while it is still plugged by the oil stopper. Screw in the bleed screw while oil is domed up at the bleed hole.

15. **Clean up.** Wipe away oil on the parts.

16. **Remove the bleed block and install the brake pads and wheel.** Torque on the brake pad retainer bolt (thin screwdriver) is 0.1–0.3 N-m (1–2.5 in-lbs). Squeeze the lever and check brake function.

17. **Zip-tie the lever around the handlebar overnight to check for fluid leaks.** If the zip-tie is still tight around the lever and the handlebar grip the next morning, there are no leaks.

b. Vacuum-bleeding SRAM brakes

LEVEL **NOTE:** *The procedure for bleeding SRAM Hydraulic Road Rim (HRR) brakes is the same as for disc brakes. First, however, close the quick-release lever on the caliper, and turn the barrel adjuster clockwise until it stops. To avoid contaminating the brake pads with brake fluid, remove them (just slide them out of the pad holders).*

There is some air trapped in any DOT fluid, and exerting a vacuum over the fluid, as SRAM's bleed method includes, can draw some of that air out. The SRAM bleed kit has a pair of syringes with screw-on fittings and hose clamps.

1. **If you have set the levers for long reach, check that they are not too far out.** Hold a ruler along the long, straight edge of the hood where it meets the body above and behind the lever pivot. Measure perpendicularly from the ruler to the tip of the lever blade; it should be less than 90mm for bleeding. If not, adjust it inward with a 2.5mm hex key; flip the shift lever back toward the handlebar to get at the adjustment screw under the lever.

2. **Draw fluid up into the two syringes.** Use Avid High-Performance DOT fluid only. Fill one syringe halfway with Avid DOT fluid from a previously unopened container (see warning on boiled fluid earlier in this section), and fill the other syringe ¼ full.

3. **Expel air from the fluid in the syringes.** Hold a rag over the end of the syringe hose, point each syringe up, push the plunger, and expel any bubbles.

4. **Draw dissolved gas bubbles out of the half-full syringe.** Close the clamp on the hose and pull the plunger so tiny bubbles appear in the fluid and get larger the harder you pull (Fig. 10.19). Don't pull hard enough to suck air in past the plunger or past the clamp on the hose. While still pulling the plunger, flick your finger against the syringe to free any bubbles sticking to the sides and bottom, and encourage them to float to the top. Open the clamp and expel collected air as in step 2. Repeat a few times, but don't try to remove all of the bubbles, because there will be no end to it.

10.19 Get more air out of the DOT fluid in the syringe by pulling a vacuum so that bubbles appear; tap syringe to free them.

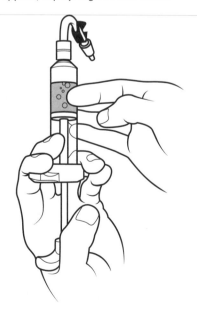

10.20 Pushing and pulling the SRAM bleed syringe plunger at the caliper

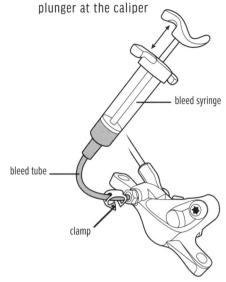

bleed syringe

bleed tube

clamp

10.21 Pushing and pulling the SRAM bleed syringe plunger at the lever

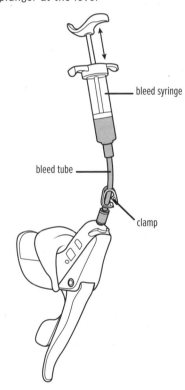

bleed syringe

bleed tube

clamp

5. **Screw the syringes onto the bleed ports.** Peel back the top of the lever hood to access the bleed port. Using a Torx T10 key, remove the bleed screws from the top of the lever body and the side of the caliper. Open the clamps on both syringes and push fluid up to the tip of the hose so there is

no air gap. Screw the half-full, degassed syringe on the caliper (Fig. 10.20); screw the quarter-full, degassed syringe on the lever (Fig. 10.21). Wipe away any fluid that may have dripped before it gets on paint or skin.

6. **Push fluid up from the caliper.** Holding both syringes upright, push the caliper syringe plunger (Fig. 10.20) to force half of its fluid up through the system to the lever syringe, along with any air bubbles. Close the clamp on the lever syringe.

7. **Pull the brake lever to the handlebar and secure it with a strap or rubber band.** This closes the hole connecting the reservoir and the master cylinder.

8. **Suck the air out of the caliper.** Without pulling so hard that you pull air past the plunger seal, pull a vacuum on the caliper syringe's plunger with the syringe upright (Fig. 10.20), drawing air bubbles up through the fluid in the syringe. Push and pull the syringe plunger a few times until bubbles stop coming up the syringe hose.

9. **Holding the brake lever to the bar, remove the band around it.**

10. **Push in the caliper syringe plunger while slowly allowing the brake lever to open.** Close the caliper syringe clamp.

11. **Remove the caliper syringe and replace the caliper bleed screw.** Do not overtighten the tiny T10 bleed screw.

12. **Bleed the lever.** Open the lever syringe clamp, turn the syringe upright, and firmly pull on its plunger (Fig. 10.21). This sucks air out of the lever. Gently push the plunger to pressurize the system without blowing out the bladder in the lever. To free bubbles, pull and release the brake lever, allowing it to snap back 10 times, followed by firmly pulling on the syringe plunger again. Repeat until large bubbles cease to appear in the syringe hose. Lightly push the syringe plunger to pressurize the system, and close the syringe clamp.

13. **Remove the syringe and replace the lever's bleed screw.** Do not overtighten the tiny bleed screw.

14. **Clean DOT fluid off the lever, caliper, and bike.** Wipe it first, then spray it with isopropyl alcohol and wipe again.

15. **Remove the bleed block (Fig. 10.15) and install the brake pads and wheel.** Torque on the brake pad retainer bolt (2.5 mm hex key) is 0.6–0.9 N-m (5–8 in-lbs). Squeeze the lever and check brake function.

16. **Empty the syringes and dispose of the fluid properly.** Do not reuse bled-out fluid. Recycle or dispose of DOT fluid according to local and federal regulations. DOT fluid is toxic; DO NOT pour it down a sewage, septic, or drainage system or onto the ground or into a body of water. To avoid permanently kinking the syringe hoses, store the syringes with the clamps open.

17. **Zip-tie the lever around the handlebar overnight to check for fluid leaks.** If the zip-tie is still tight around the lever and the handlebar grip the next morning, there are no leaks.

c. General instructions for bleeding up from the caliper

1. **Orient the lever so the bleed screw is at the top.** Mount the bike in a stand, turn the handlebar, and rotate the lever on the handlebar such that the lever is the highest point in the system. Turn the lever and/or handlebar until the bleed screw is at the top.

2. **Remove the caliper bleed fitting's rubber cover.** The cover is shown in Figure 10.8.

3. **Push a short section of clear tube onto the tip of a syringe or squeeze bottle.** Push it on so it stays on.

4. **Fill the syringe or squeeze bottle.** Use the recommended fluid for your brake. If it is DOT fluid, take it from a previously unopened container (see warning on boiled fluid earlier in this section).

5. **Push the other end of the tube over the caliper bleed nipple.** First put the box end of a wrench on the bleed nipple if there is room for it; otherwise, grab it with an open-end wrench (Fig. 10.22).

6. **Connect a tube to the master cylinder bleed hole.** The brake's bleed kit should have come with a fluid exit tube and a fitting to hold it into the bleed hole. Plug the proper fitting into the tube. Remove the screw or plug from the master cylinder bleed hole, and push or thread the fitting into it.

7. **Hang a container from the handlebar with wire or a zip-tie.** Direct the other end of the tube into it.

8. **With the caliper bleed fitting closed, squeeze the syringe or fluid bottle repeatedly until any air bubbles in the tube come back into the syringe or bottle.** The syringe or bottle should be pointed straight down; keep it that way throughout the following steps.

9. **Loosen the bleed fitting on the caliper one-quarter turn, and squeeze in fluid from the bottle or syringe for a count of five.** If the tube pops off the caliper bleed nipple, put an "olive"—the barrel-shaped brass hose fitting ring seal (Fig. 10.12)—on the bottle's bleed tube before you stick it on the nipple. Push the olive down over the hose and nipple so it won't pull off.

10. **Pull back slightly on the syringe plunger (or let off for about three seconds on a squeeze bottle until it returns to its natural shape).** This draws air out of the caliper and up into the syringe or bottle.

10.22 Bleeding TRP Parabox hydraulic caliper; note spacer block in place of brake pads.

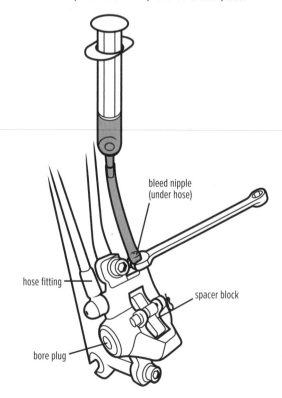

bleed nipple (under hose)

hose fitting

spacer block

bore plug

11. **Push for five seconds, let off for three.** Repeat until no more bubbles come out of the caliper.

12. **Push firmly on the syringe or bottle until clean fluid without bubbles comes out of the tube at the lever.** Tap the syringe or bottle, its tube, and the brake hose, caliper, and master cylinder to break free any stuck bubbles.

13. **While still pushing the syringe or bottle, quickly pull the lever to the handlebar and release.** Look for air coming out of the bleed tube at the lever. Repeat until no more air emerges.

14. **While still pushing the syringe or bottle, close the caliper bleed fitting.** Don't overtighten the fitting—it is small, and you only need to tighten it enough to create a seal!

15. **Remove the tubes from the caliper and lever.**

16. **Replace the lever bleed screw or plug.**

17. **Replace the rubber cover on the caliper bleed fitting.**

18. **Install the pads (10-1) and wheel, and pump the lever.** The lever should feel firm, and it should not come back to the grip. Repeat bleed if it feels spongy.

19. **Clean DOT fluid off the lever and bike.** Wipe it first, then spray it with isopropyl alcohol and wipe again.

20. **Check for fluid leaks by putting a zip-tie overnight around the lever.** If the zip-tie is still tight around the lever and the handlebar grip the next day, the system is completely sealed.

10-7

OVERHAULING DISC BRAKES

LEVEL Regular bleeding and fluid replenishment cleans dirt out of the system and lengthens the time between overhauls of hydraulic brakes. On many disc brakes, overhaul is relatively simple, but you will need an air compressor to get the pistons out of the caliper. There are two kinds of hydraulic calipers: clamshell models whose two pieces bolt together, and single-piece calipers. Buy new seals for the part of the brake you are overhauling before you start. A speck of dirt or hair in a hydraulic disc brake can cause a leak, so work in a clean area with clean methods.

a. Overhauling clamshell hydraulic calipers

1. **Remove the caliper from the bike (Figs. 10.7, 10.8).**
2. **Remove the brake pads (10-1).**
3. **Disconnect the hose.** If it has a banjo fitting (Figs. 10.13B, 10.22), unscrew the hollow banjo bolt holding it on, but don't disconnect the hose from the banjo. Save the two O-rings if you did not get new ones.
4. **Remove the bridge bolts (Fig. 10.23) holding the caliper clamshell halves together.** One of these may be the banjo bolt that you already removed.
5. **Remove the piston(s).** This is best done by blowing compressed air into the fluid-entry hole. If a banjo bolt holds the caliper together, you are blow-

---

10.23 Clamshell hydraulic caliper, exploded view

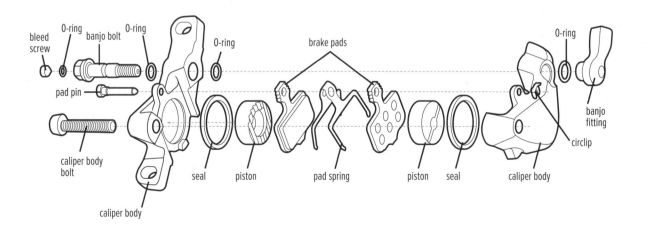

ing into the banjo hole and there is nothing else to plug; do the same on both sides. However, if there are two caliper-body bolts, then blow compressed air into the fluid-transfer hole while plugging either the bleed hole or the fluid-entry hole with your finger, depending on which piston you are removing. Be careful to not get hit with fluid or parts. Wear safety glasses and cover the piston with your hand or hold the piston side of the caliper half face down so that no parts fly away.

6. **Dig the piston seals out of their grooves in the cylinder bores.** Use a fingernail or a toothpick to avoid scratching the bores, or stick an awl straight into the seal without touching the caliper body. You'll find there are relatively few parts inside the caliper (an object that you might assume is much more complicated); generally there is just one seal around each piston (Fig. 10.23).

7. **Clean all parts carefully with isopropyl alcohol.** Inspect the parts. Replace any cracked or scratched parts.

8. **With compressed air, clean the caliper-seal grooves and the bleeder hole.** Wear safety glasses. Check that the seal grooves are completely clean.

9. **Let the parts dry.** Compressed air is humid, which contaminates DOT fluid.

10. **Lubricate the pistons and new seals with brake fluid.**

11. **Put all of the parts back together in the way that you found them.**

12. **Bolt the caliper together to the recommended torque.**

13. **Reinstall the hose.**

14. **Bleed the system (10-6).**

15. **Install and center the caliper (10-4b).**

b. Overhauling one-piece hydraulic calipers

1. **Remove the caliper from the bike.**

2. **Remove the brake pads (10-1).**

3. **Disconnect the hose.** If it has a banjo fitting (Figs. 10.13B, 10.22), unscrew the hollow banjo bolt holding it on, but don't disconnect the hose from the banjo. Save the two O-rings if you did not get new ones.

4. **Remove the bore plug (Fig. 10.22).** You will need a tool specific to the brake caliper.

5. **Push outer piston out.** Reach in with your finger through the open bore to push the piston into the rotor gap so it will fall out.

6. **Blow inner piston out.** Blow with compressed air through the fluid-entry (banjo) hole to push the piston into the rotor gap so it will fall out.

7. **Clean all parts carefully with isopropyl alcohol.** Inspect. Replace any cracked or scratched parts.

8. **With compressed air, blow out the caliper-seal grooves and the bleeder hole.** Wear safety glasses. Check that the seal grooves are completely clean.

9. **Let the parts dry.** Compressed air is humid, which contaminates DOT fluid.

10. **Lubricate the pistons and new seals with brake fluid.**

11. **Install the square seals in the cylinder grooves.**

12. **Install the inner piston.** Slide the piston up into the rotor gap and push it in place with your finger through the open bore.

13. **Install the outer piston.**

14. **Install a new bore plug seal.**

15. **Tighten the bore plug.**

16. **Reinstall the hose.**

17. **Bleed the system (10-6).**

18. **Install and center the caliper (10-4b).**

c. Overhauling cable-actuated disc-brake calipers

Cable-actuated disc brakes (Fig. 10.7) usually push the pistons by means of a number of ball bearings rolling in curved, ramped tracks. They can function well for a long time in dirty conditions, but if they become dirty inside, they won't work as well. Methods to disassemble them vary; you won't find the how-to in their accompanying instruction manual; and it would require too many pages in this book for each one to realistically devote to a task that few readers are likely to undertake. Consult the manufacturer's website for a service manual. It is not particularly complicated to take cable-actuated disc brakes apart and put them back together; it just takes a lot of steps specific to each one.

TABLE 10.1 — TROUBLESHOOTING DISC-BRAKE PROBLEMS

SQUEALING

Squealing disc brake due to oil or winter road de-icer splashed on rotor and pads	Clean rotors with rubbing alcohol and wipe dry. If pads are dirty, sand pads face-down on drywall-sanding screen so contaminants fall away through the screen. Never use solvent other than rubbing alcohol on disc-brake pads.
Squealing or howling due to vibration	Replace or true a bent disc rotor (10-1).
	Try different brake pads.
	Tighten the caliper-mounting bolts to torque spec.
	Stiffen the rear brake mount. To isolate the problem, duct-tape a dowel-rod strut to triangulate the brake-mount area of the seatstay against the chainstay. If the squealing stops, have a frame builder install a permanent reinforcement.
	Try a different fork.

LOW POWER

Flexing of brake arms or lever (cable-actuated brake only)	Install new brakes, but try eliminating other factors first to see whether braking power improves.
Stretching of cable	Replace cable and housing (9-4). Always use compressionless housing.
Compression of brake housing	Replace cable and housing (9-4). Always use compressionless housing.
Insufficient coefficient of friction between pads and disc	Try different pads; replace worn pads. Organic pads offer higher power; metallic ("sintered") pads offer greater durability.
Oil and grime on rotors and pads	Clean rotors and pads with rubbing alcohol. See "Squealing" for pad-cleaning instructions.
Hydraulic system leaking or has air in it	Tighten connections; replace parts as necessary; bleed the system.

TOO MUCH LEVER TRAVEL

Lever quick-release open (Campagnolo)	Close quick-release (9-1).
Cable too long	Tighten cable.
Brake pads worn	Replace pads.
Excessive pad spacing (cable systems only)	Tighten pad adjustment knobs on caliper.
Leaks or air in hydraulic system	Tighten connections; replace parts as necessary; bleed the system.

PAD DRAG

One pad rubs all the way around the rotor	Center the caliper.
Both pads rub all the way around the rotor on a cable system	Loosen the cable or back out the pad adjustment knobs on the caliper.
Both pads rub all the way around the rotor on a hydraulic system	Push pistons back into their bores by prying the pads apart with a plastic lever or by removing the pads and using the head of a box-end wrench to push the pistons back in.
Rotor wobbles back and forth against pads	True the rotor (10-5).

SLOW RETURN

Cable sticking	Lubricate or replace cable (9-3, 9-4), or thaw and dry if frozen.
Hydraulic pistons sticking	Clean around hydraulic pistons and seals. Remove pads. Hold one piston in with a plastic tire lever and squeeze the lever to push the other piston out; clean around it with a Q-Tip soaked in correct brake fluid for the brake system. Push that piston in with the box end of a small box wrench or a tire lever. Clean other piston the same way.

Continues >>

TABLE 10.1 — TROUBLESHOOTING DISC-BRAKE PROBLEMS, CONTINUED

LOOSE CALIPER	
Caliper rattles or clunks	Bolts holding the caliper or caliper adapter are loose or missing. Tighten mounting bolts or replace missing hardware.

GRINDING NOISE	
Sounds like there is sand in the brakes	Clean rotors and pads with rubbing alcohol. See "Squealing" for pad-cleaning instructions.

SOFT LEVER THAT PUMPS UP AND GETS FIRMER (HYDRAULIC BRAKES ONLY)	
Leaks or air in hydraulic system	Tighten connections; replace parts as necessary; bleed the system.

> *When someone tells you something defies description, you can be pretty sure he's going to have a go at it anyway.*
>
> —CLYDE B. ASTER

11

CRANKS AND BOTTOM BRACKETS

The crankset consists of the crankarms, bottom bracket, chainrings, chainring bolts, and crank bolt, incorporating one, two (Fig. 11.1A), or three chainrings (Fig. 11.1B). In modern integrated-spindle cranksets the bottom-bracket spindle is permanently pressed into one of the crankarms. The bottom bracket is traditionally thought of as the spindle and its bearings, but on integrated-spindle cranksets, what is called the bottom bracket has become just the bearings, or the bearing cups and bearings.

The forces applied through the crankset are large, so all parts need to be tight to prevent creaking noises as well as to avoid ruining expensive components by using them when loose. In addition, bottom-bracket bearings need to run smoothly under high loads so that they don't sap your energy.

NOTE: *Cartridge bearings for bottom brackets are specified by their dimensions: their inside diameter (ID), outside diameter (OD), and width. When you need to replace bearings, be sure to get exact replacements.*

CRANKARMS AND CHAINRINGS

11-1

CRANKARM REMOVAL

 LEVEL Most modern cranks are simple to remove with a single hex key (5mm, 8mm, or 10mm). The two-piece (Fig. 11.2), a.k.a. "integrated-spindle," design means that only the left arm needs to come off; the spindle pulls out of the bearings along with the right crankarm.

NOTE: *Some two-piece cranksets (Race Face, for example) have the spindle integrated with the non-drive arm, rather than with the drive-side crankarm.*

ANOTHER NOTE: *"Right," in this chapter and generally throughout this book, refers to the drive side of the bike, and "left" refers to the non-drive side.*

To remove a traditional three-piece crankset (Fig. 11.1), you will need either a thin-wall 14mm (sometimes 15mm or 16mm) socket wrench or a large, 8mm (rarely: 6mm, 7mm, or 10mm) hex key in order to remove the crank bolt. Pre-1980

TOOLS

5mm, 6mm, 7mm, 8mm, 10mm hex keys or socket drivers

Torx T30 key or socket driver

14mm socket wrench

⅜-inch drive ratchet or torque wrench

⅜-inch drive extension

Chainring-nut tool

External-bearing splined bottom-bracket wrench or socket for 24mm- and 25mm-spindle threaded bottom brackets

Shimano TL-FC16 splined left arm cap installation tool or equivalent

Internal bottom-bracket splined socket with large bore for ISIS/Octalink

Snapring pliers

Crank puller with ends for square-taper and splined spindles

Pin spanner (or adjustable pin tool)

Toothed lockring spanner

Adjustable wrench

Pliers

Flat hand file

Single-edge razor blade or box cutter

Grease

Continued on p. 250

249

11.1A Square-taper "three-piece" double crankset

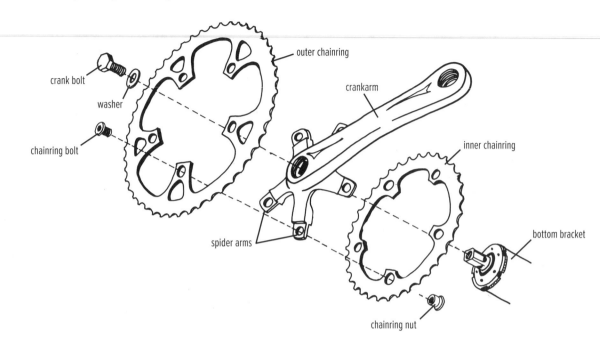

11.1B Third chainring on a triple crank

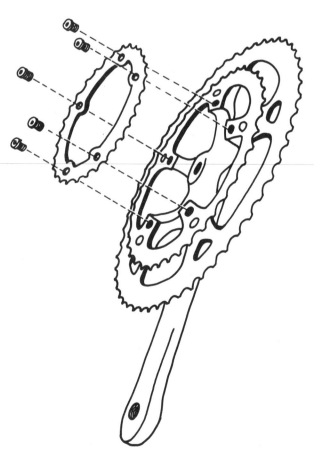

French (TA, Stronglight) cranks have 15mm crank bolts. Campagnolo cranks from the 1980s take a 7mm hex key or a 15mm socket. Early 1980s Shimano Dura-Ace and 600 Dyna-Drive cranks are self-extracting and come off with just a 6mm hex key.

Really old bikes (pre-1970) may have steel cottered cranks, requiring a wrench and a hammer to remove them.

a. Removing integrated-spindle cranks with two pinch bolts on the left arm

LEVEL 1. **Unscrew the bearing preload cap from the left arm completely (Fig. 11.2).** This takes a special splined tool for Shimano (Fig. 11.3); others require a hex key.

2. **Loosen the two pinch bolts holding the arm onto the spindle (Fig. 11.3).** Use a 5mm hex key.

3. **Pull off the left arm.**

4. **Pull the right arm (and attached spindle) straight out.** You may need to tap the end of the spindle with a rubber mallet to get it started. If there is a bearing seal stuck on the spindle, leave it there; when you install the crank, it will go back against the bearing.

11.2 Integrated-spindle (a.k.a "two-piece") crankset (Shimano Hollowtech II shown)

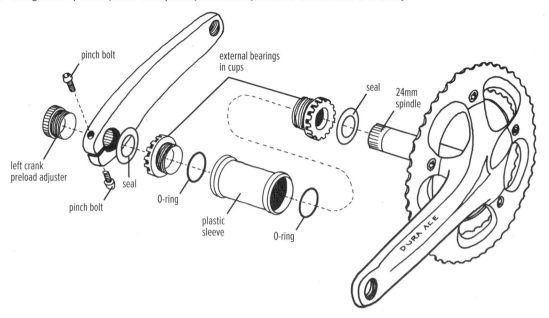

11.3 Removing and installing a left Shimano Hollowtech II crankarm

11.4 Removing and installing the crank bolt

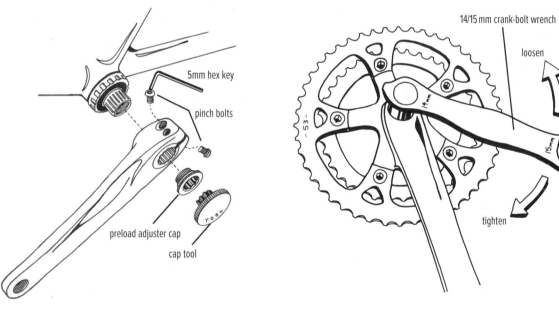

b. Removing integrated-spindle cranks with a single crank bolt other than Campagnolo, Fulcrum, and Specialized

LEVEL

1. **Unscrew the crank bolt as in Figure 11.4, except on the left arm and with a large (8mm or 10mm) hex key.** The arm will come right off. Do not unscrew the cap that surrounds and partially covers the bolt head (it takes a pin tool or a 10mm or larger hex key to get it off); that cap traps the bolt head so that the arm comes off simply by unscrewing the bolt.

2. **Pull the right arm (and attached spindle) straight out.** Tap the end of the spindle with a mallet if it's stuck. If a bearing seal comes off with the arm, you can clean it in place or pull it off and put it back on the bearing.

c. Removing Campagnolo Ultra-Torque (UT), Fulcrum Racing-Torq, and Specialized S-Works integrated-spindle cranks

1. Unscrew the crank bolt with a long hex key. Reach into the spindle from the drive side with a long 10mm hex key (6mm on Specialized) or with a torque or socket wrench with an extension and a 10mm hex driver (6mm on Specialized). Unscrew the bolt and remove it.

NOTE: *On Campagnolo Super Record cranks with a titanium Ultra-Torque spindle, the crank bolt is left-hand threaded.*

2. On Campagnolo and Fulcrum, remove the retaining clip. Find the wire retaining clip around the drive-side bearing cup (Fig. 11.5), and pop both of its ends out of the holes in the cup into which they are inserted. You can pull the retaining clip completely off or leave it on the cup with the ends just off to the sides of the holes. (Campagnolo Ultra-Torque and Fulcrum Racing-Torq cranks have half of the spindle attached to each arm and the bearing pressed onto the spindle, against the arm. The cups are merely receptacles for the bearings, and

the side-to-side position of the crank is maintained by the retaining clip's ends penetrating the holes in the drive-side cup and trapping the drive bearing.)

3. Pull the cranks out. On Campagnolo and Fulcrum, grab the wavy washer out of the left cup (this washer takes up lateral slack in the system).

d. Removing Campagnolo Power Torque (PT) cranks

Unless you resort to the unorthodox method (see Note in step 6) this job requires special tools, namely, the Park CBP-5 adapters and CBP-3 puller, or the equivalent.

1. Unscrew the crank bolt as in Figure 11.4, except with a 14mm hex key. The arm will not come off.

2. Remove the washer. If the washer did not come off with the crank bolt, get it out of the crankarm hole. If you don't, you won't be able to pull the crank off and may wreck the tool and the crank trying.

3. Pad the crank. If it's a carbon Power Torque crank, install one of the cardboard curved pads under the head of the crank. If it's an aluminum crank, slip the molded plastic cup pad shown in Figure 11.6 under the head of the crank. (The head of the aluminum PT crank has a curved edge terminating in a ridge around the end, and the feet of the CBP-3 bearing puller that you will use to pull the crank off cannot grab it well without marking it, hence the plastic molded cup pad to protect the crank finish. The carbon PT crank, by contrast, has a flat back face that mates well with the bearing puller's fingertips, so a cardboard pad is sufficient to protect it.)

4. Insert the extension plug. The plug will push on the end of the spindle (as long as you removed the washer that was under the crank bolt) when the bearing puller's push rod pushes on it.

5. Install the CBP-3 bearing puller. Hook the puller's fingers under the pad surrounding the head of the crank (there are little recesses for the fingertips under the edges of the molded plastic cup pad), and tighten the two side knobs to remove play from the puller's fingers so they can't slip off (Fig. 11.6).

6. Pull the crank off. Tighten the push rod until the crankarm comes off.

11.5 Removing and installing a Campagnolo/Fulcrum bottom-bracket bearing retaining clip with pliers

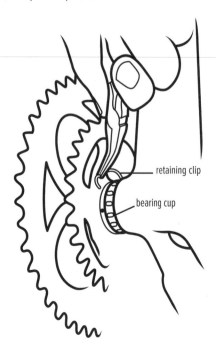

retaining clip

bearing cup

11.6 Positioning a Park CBP-3 puller to pull off a Campagnolo Power Torque left crankarm

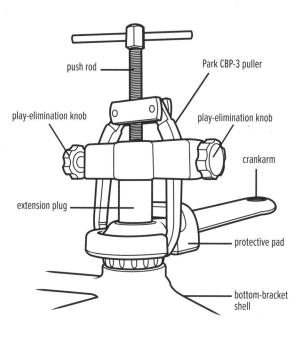

- push rod
- Park CBP-3 puller
- play-elimination knob
- play-elimination knob
- crankarm
- extension plug
- protective pad
- bottom-bracket shell

NOTE: *In a pinch, to remove a Power Torque crank without the right tools, in place of steps 2–6, simply unscrew the left bottom-bracket cup with a C-shaped, splined wrench (Fig. 11.19). The left arm will come off. Continue with step 7.*

7. **Remove the retaining clip.** See 11-1c and Figure 11.5.

8. **Yank out the drive arm.** Pull the spindle out by pulling on the drive crank. If it's stubborn, tap the end of the spindle with a soft hammer. Catch the wavy washer.

e. Removing three-piece cranks (square-taper, Shimano Octalink, and ISIS)

LEVEL 1. **Remove the dust cap covering the crank bolt if present.** This requires either a 5mm hex key, a two-pin dust cap tool, or a screwdriver.

2. **Remove the crank bolt with the appropriate wrench (Fig. 11.4).** Unless the crankarm comes off when you unscrew the bolt, make sure that you extract the washer (Fig. 11.1A) with the bolt. If you leave it in, you will not be able to pull the crank off.

NOTE: *Some cranks that accept a hex key in the crank bolt are self-extracting and don't require a crank puller*

(Fig. 11.7). The crank bolt is held down by a retaining ring threaded into the crank; as the bolt is unscrewed, its lip pushes on the ring and pushes the crank off.

ANOTHER NOTE: *"Square taper," "Octalink," and "ISIS" are three different bottom-bracket and crankarm interface standards. Square-taper bottom-bracket spindles are square on the end (Fig. 11.1A) and fit into a square hole in the crankarm. The spindle ends are tapered (at a 2-degree angle) to tighten into the crank as the arm is pushed into the spindle. ISIS (Fig. 11.25) and Shimano Octalink (Fig. 11.24) are both oversized hollow spindles (a.k.a. "pipe spindles") with longitudinal splines on the ends. The spline patterns look similar, but the measurements are different and cannot be interchanged.*

3. **Unscrew the crank puller's center push bolt (Fig. 11.7) so that its tip is flush with the face of the tool.** Make sure the flat end of the push bolt is the right size for the bottom bracket; the push bolt end is much smaller for a square-taper spindle than for an ISIS or Shimano Octalink splined spindle.

4. **Thread the crank puller into the hole in the crankarm.** Be sure that you thread it in (by hand) as far as it can go; otherwise, you will not engage sufficient crank threads when you tighten the push bolt, and you will damage the threads. Future crank

11.7 Using a crank puller (two types shown); be sure to thread in the crank puller completely.

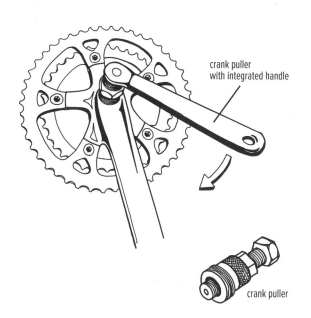

- crank puller with integrated handle
- crank puller

removal depends on those threads being in good condition.

5. **Tighten the push bolt clockwise (Fig. 11.7) until the crankarm pulls off of the spindle.** Use a socket wrench or the included handle.

6. **Unscrew the puller from the crankarm.**

f. Removing old steel cottered cranks

A steel cottered crank is secured to the spindle by means of a tapered, wedge-shaped cotter bolt with a nut on the end. The cotter bolt runs transverse to the crankarm and wedges into a notch in the spindle. To remove these cranks, first remove the nut from the bolt, and then smack the cotter pin on the threaded end with a mallet to knock it loose.

11-2

CRANKARM INSTALLATION

a. Installing integrated-spindle cranks with two pinch bolts on the left arm

 1. **Grease the spindle tip and the bore of each bearing (Fig. 11.2).**

2. **Grease the dust shields.** There will be a rubber-coated dust shield to cover the bearing on each side. Install the shield on each side, and grease it.

3. **Push the spindle (attached to the right crankarm) in through the bearings from the drive side.**

4. **Slide the left arm onto the end of the spindle.** Check that the crank is at 180 degrees from the right arm.

5. **Gently tighten the left-side preload adjuster cap (Fig. 11.3).** Use the special plastic splined cap tool for Shimano or a hex key for FSA, Easton, and others. Torque is not high—3.5–6.2 in-lbs (0.4–0.7 N-m)—just enough to pull the right and left cranks over against the bottom-bracket cups, so do not overtighten.

6. **Tighten the two opposing (greased) pinch bolts (Fig. 11.3).** Using a 5mm hex key, alternately tighten each bolt one-quarter turn at a time. Torque is 88.5–133 in-lbs (10–15 N-m).

7. **Recheck the torque after one ride as the crank may settle in and the bolt will need retightening.**

b. Installing integrated-spindle cranks with a single crank bolt other than Campagnolo UT and PT, Fulcrum, and Specialized (but including Campagnolo Over-Torque)

 1. **Grease the spindle tip and the bore of each bearing.**

2. **Grease the dust shields.** There will be a rubber-coated dust shield to cover the bearing on each side. Install the shield on each side, and grease it.

3. **Push the spindle (which is attached to the right crankarm) in through the bearings from the drive side.**

4. **Slide the left arm onto the end of the spindle.** If there is a wavy washer, put it onto the spindle before the left arm. Check that the crank is at 180 degrees from the right arm.

5. **Tighten the crank bolt with an 8mm or 10mm hex key.** Torque for this bolt is high; see Appendix E. Campagnolo Over-Torque (OT) cranks instead require a special Campagnolo pin tool to tighten the threaded collar that holds the left crankarm on. Before you put the arm on, slide Campagnolo's expanding preload adjuster sleeve for OT on the spindle splines. Once the left arm is tightened in place, turn the preload ring with your fingers counterclockwise to remove side play (it's threaded on the inner ring that engages the spindle's splines, so it expands as you unscrew it) and then lock it in place with its setscrew.

6. **Recheck the torque after one ride as the crank may settle in and the bolt will need retightening.** Periodically check the torque from then on.

c. Installing Campagnolo Ultra-Torque (UT), Fulcrum Racing-Torq, Campagnolo Power Torque (PT), and Specialized S-Works integrated-spindle cranks

1. **Prepare the clip.** On Campagnolo and Fulcrum, set the wire retaining clip (Fig. 11.5) so that its ends are not in the holes in the drive-side cup but are just adjacent to them.

2. **Install Specialized washers.** Apply a film of grease at the base of each spindle where the Specialized spindle meets the crankarm. Install the bearing spacers on the arms; the cone-shaped side faces

the crankarm, and the stepped side faces the bearing. Install the wavy washer on the left arm.

3. **Push the drive-side crank into the right cup.**

4. **Install the wavy washer.** On Campagnolo Power Torque, slide the bearing seal, then the wavy washer, and then the rubber cup-shaped dust cover on the end of the spindle, which should now be sticking out of the left bearing. On Campagnolo Ultra-Torque and Fulcrum Racing-Torq, place the wavy washer into the left cup (Fig. 11.8).

5. **Install the left arm.** Push the left arm onto the spindle on Power Torque, and push the left bearing into the cup on Ultra-Torque and Racing-Torq. Clock the arms so that the teeth at the ends of the spindle stub attached to each crankarm engage with the cranks at 180 degrees from each other.

6. **Install the (greased) crank bolt with washer.** Except on Power Torque, put it in from the drive side.

NOTE: *On Campagnolo Super Record cranks with a titanium Ultra-Torque spindle, the crank bolt is left-hand threaded.*

If you have a long 10mm hex bit for a socket wrench (6mm on Specialized), use a torque wrench on it and tighten it to 372 in-lbs (42 N-m), or 300 in-lbs (34 N-m) on Specialized. If you only have a long hex key, tighten it as hard as you can. Only use a hex key with a flat end for this; a hex key with a ball end will not engage the bolt well enough for such high torque. On Power Torque, put the washer and bolt into the left arm and tighten to torque with a 14mm hex key.

7. **On Campagnolo and Fulcrum cranks, push in the retaining clip (Fig. 11.5).** Pop its ends into the holes in the right cup, thus securing the crankset in the proper lateral position.

d. Installing three-piece cranks (square-taper, ISIS, and Octalink)

LEVEL 1. **Slide crankarm onto the bottom-bracket spindle.** With square-taper spindles, clean off all grease from both parts. Grease may allow the soft aluminum crank to slide too far onto the spindle and deform the square hole in the crank. With an ISIS or Shimano Octalink splined spindle, however, do grease the parts. With ISIS and Octalink cranks you must be careful to line up the crank splines with those on the spindle before tightening the crank bolt.

11.8 Campagnolo Ultra-Torque crankset assembly

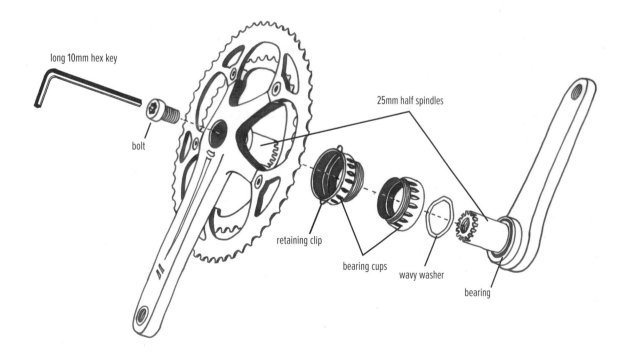

long 10mm hex key

bolt

25mm half spindles

retaining clip

bearing cups

wavy washer

bearing

2. **Install the crank bolt.** Apply grease to the threads and tighten (Fig. 11.4). Apply titanium-specific anti-seize compound for titanium spindles and for titanium crank bolts. If you have aluminum or titanium crank bolts, first tighten the cranks with the greased steel bolt to the specified torque, then replace the steel bolt with the lightweight bolt and tighten it to spec.

NOTE: *Here is where a torque wrench comes in handy; tighten the bolt to about 300–435 in-lbs (32–49 N-m), and as high as 522 in-lbs (59 N-m) for some steel oversized bolts in ISIS spindles (see the torque table, Appendix E). If you're not using a torque wrench, make sure the bolt is really tight, but don't muscle it until your veins pop.*

3. **Replace the dust cap.**

4. **Check the front-derailleur adjustment (see 5-5).** Removing and reinstalling the right crankarm could affect chainring position and hence shifting.

5. **Recheck the torque after one ride as the crank may settle in and the bolt will need retightening.** Periodically check the torque from then on.

e. Installing old steel cottered cranks

LEVEL To reinstall a steel cottered crankarm, you need to buy a new pair of cotter bolts at a bike shop because after removal, the old ones will be deformed and will not secure the crankarms properly. Installation consists simply of sliding the crankarm back into place on the spindle, inserting the new (greased) cotter, and tightening the nut to pull the cotter tightly into the notch in the spindle.

11-3

CHAINRINGS

Get into the habit of checking the chainrings regularly. They do wear out and need to be replaced. It's hard to say how often, so include chainrings as part of your regular maintenance checklist. Always check the chainring teeth for wear when you replace the chain. Check the chainring bolts periodically for tightness and the chainrings themselves for trueness by watching them as they spin past the front derailleur.

11.9 Worn teeth

1. **Wipe the chainring clean and inspect each tooth.** The teeth should be straight and uniform in size and shape. If the teeth are worn into a hook shape (Fig. 11.9), the chainring needs to be replaced. The chain should be replaced as well (4-7), because this tooth shape effectively changes the spacing between teeth and accelerates wear on the chain, and it indicates that the chain was already worn in order to cause the hook shape in the first place.

CAUTION: *Don't be deceived by the erratic tooth shapes (some tall, some short) and bulged spots below the teeth on modern chainrings; the shapes are designed to facilitate shifting (Fig. 11.10). Shifting ramps on the inboard side, meant to speed chain movement between the rings, often look like cracks on inexpensive chainrings, where they are pressed into the ring rather than being a separate steel piece riveted on like the one shown in Figure 11.10.*

NOTE: *Another wear evaluation method is to lift the chain from the top of the chainring; the greater the wear of either part, the farther the chain separates. If it lifts more than one tooth, the chain and perhaps the chainring as well need to be replaced.*

2. **Remove minor gouges in the chainrings with a file.**

3. **Straighten bent teeth carefully.** If an individual tooth is bent, try bending it back carefully with a pair of pliers or a Crescent wrench (Fig. 11.11) and smooth it with a file. If it breaks off, take the hint and buy a new chainring.

4. **Check chain release.** While turning the crank slowly, watch where the chain exits the bottom of the chainring. See if any of the teeth are reluctant to let go of the chain. If the chain gets pulled up a bit as it leaves the bottom of the chainring, it can get sucked up between the chainring and the chainstay. Locate any offending teeth and see if

11.10 Chainring shifting ramps and asymmetrical teeth

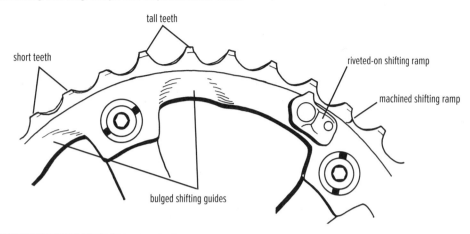

tall teeth

short teeth

riveted-on shifting ramp

machined shifting ramp

bulged shifting guides

you can correct the problem. If the teeth are really chewed up or cannot be improved with pliers and/ or a file, the chainring should be replaced.

NOTE ON CHAINRING POSITION: *Never rotate the chainring position relative to the spider arms of the crank, because the shifting ramps (Fig. 11.10) will not be in the proper places to function correctly in picking up the chain as you shift. The outer chainring will generally have a protruding pin to locate it behind the crankarm, and each of the other two rings will have a radially inward-pointed tooth or a small engraved triangle to mark where it is designed to line up behind the crankarm.*

11-4

CHAINRING BOLTS

Check that the bolts are tight, and tighten them (Fig. 11.12) by turning them clockwise. In the past, this job always required a 5mm hex key, but now it often requires a star-shaped Torx T30 key. As you turn the bolt, its nut may also turn. If so, hold the nut with a two-pronged chainring-nut tool (Fig. 11.12), a 6mm hex key, or a wide screwdriver, depending on what the nut requires.

Some lightweight aluminum chainring bolts cannot take much tightening and will snap off easily. Be careful not to overtorque them (see Appendix E).

If you're having trouble with chainring bolts loosening, put some threadlock compound on the threads. If you're having trouble with the chainring bolts creaking

11.11 Straightening warped chainrings

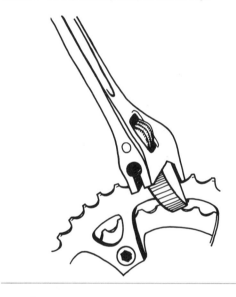

11.12 Removing and installing chainring bolts

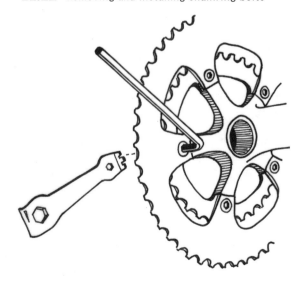

when pedaling (or you're trying to eliminate that as a possible source of creaking while pedaling), try grease on the threads and heads; you can do this right over dry threadlock compound.

11-5

WARPED CHAINRINGS

Looking down from above, turn the crank slowly to check whether the chainrings wobble back and forth relative to the plane of the front derailleur.

If they do wobble, first tighten the chainring bolt(s), and then push the crankarms back and forth to make sure there is no play in the crank or bottom bracket. If there is play with a loose-bearing bottom bracket, also adjust the bottom bracket (11-11, step 15). With an integrated-spindle crank, you may be missing a spacer or a wavy washer. A small amount of chainring lateral flex is normal when you pedal hard, but excessive wobbling will compromise shifting. Small, localized bends can be straightened with an adjustable wrench (Fig. 11.11). If a ring is really bent, replace it.

11-6

BENT CRANKARM SPIDERS

If you installed a new chainring and are still seeing serious back-and-forth wobble, chances are good that the spider arms on the crank are bent. If the crank is new, this is a warranty item, so return it to your bike shop.

If you insist on fixing it, you can find which spider arm is bent by tracking the movement of the chainring relative to the front-derailleur outer cage plate. Wrap the spider arm with a rag, grab it with an L-shaped adjustable wrench or pipe wrench, and give it a little tweak in the appropriate direction by pushing on the end of the wrench handle. Obviously you can't do this with a carbon crank, because carbon won't bend. And you can only do it on a crank with a standard double-chainring spider; the arms on a triple or compact double crank neither will allow the wrench to fit nor will bend easily, due to the short spider-arm length on a compact and the threaded standoffs for the granny gear on the spider arms of a triple crank.

11-7

CHAINRING REPLACEMENT

Most road cranks have either four or five spider arms, and if you're replacing a chainring, you need to know more details about the crank to ensure that the new chainrings will fit.

A standard double crank has a bolt circle diameter (BCD) of 130mm—the circle formed by the chainring bolts has a 130mm diameter. The smallest chainring you can mount on it has 38 teeth.

A standard Campagnolo double crank has a 135mm BCD, and the smallest chainring you can mount on it has 39 teeth. A compact double crank has a bolt circle diameter of 110mm (the spider arms are 10mm shorter than on a standard double crank), and you can mount a 33-tooth chainring on it. A Campagnolo compact double crank also has a 110mm BCD, but one bolt hole is offset so it only accepts Campagnolo compact chainrings.

A standard road triple crank has a BCD of 130mm for the outer two chainrings and 74mm for the inner ("granny") chainring, which can be as small as 24 teeth.

You want to get chainrings in pairs (or in threes, in the case of a triple) to ensure that their chain ramps are appropriately positioned for optimal shifting; you want the correct tooth to be there to pick up the chain as it moves from one ring to the next. Chainrings nowadays often have not only their own size (number of teeth) stamped on them, but the number of teeth on the adjacent chainring they're meant to work with as well (Fig. 11.13). Also look for a stamp on the chainring that indicates the number of rear cogs it should be used with (for example, if you have an 11-speed cogset, be sure to get 11-speed-compatible chainrings).

11.13 Chainring tooth-number and speed-number stamps on a 10-speed chainring pair

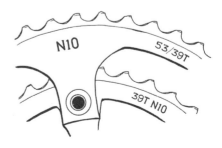

a. Double chainrings

Replacing either of the chainrings on a double (Fig. 11.12) or the two largest chainrings on a triple is easy.

1. **Unscrew the chainring bolts.** They will take either a 5mm hex key or a Torx T30 key (Fig. 11.12). You may need to hold the nut on the backside with either a chainring-nut tool, a 6mm hex key, or a thin, wide screwdriver.

2. **Install the new chainrings.** Lubricate the bolts and the little recesses that accept them in the chainring faces, and tighten them (Fig. 11.12). The outer chainring has a protruding pin meant to keep the chain from falling between it and the crankarm. Make sure this pin lines up behind the crankarm and faces away from the bicycle. The middle ring (and the inner ring on a triple) has a small chainring-orientation bump protruding radially inward or an engraved triangle that is also to line up with the crankarm. And both (or all three) chainrings have recesses for the heads of the chainring bolts and nuts, so make sure that these recesses receive those parts and are not facing inward toward the spider. If the chainrings are rotated relative to the crank or inverted, the shift ramps will not work as designed.

NOTE: If you're positioning a Rotor chainring, the ring is not round, and how you clock the ring changes your pedaling. There will be at least three positioning options and sometimes five: three on one side of the chainring and two on the other that are effectively half steps between the other three positions. These holes will be marked with a number of dots above them. Start by locating the middle of the three marked holes (the one with two dots) aligned with the crankarm. Install the chainring bolts as described here. Down the road you can change the position if you wish to experiment further.

   a. Be careful not to overtighten lightweight aluminum chainring bolts.

   b. If you're having trouble with chainring bolts loosening up, you can put some threadlock compound on the threads instead of grease, but grease will do a better job of eliminating creaking noises. You can grease right over the dried threadlock compound that comes on many new chainring bolts.

NOTE: Whenever you change the size of the outer chainring, you must reposition the front derailleur for proper chainring clearance, as described in 5-5.

b. Replacing the inner chainring on a triple

1. **Pull off the crankarm (11-1).** If you don't remove the crankarm, you will have to remove the two outer chainrings to get at the inner chainring, which is more work.

2. **Remove the bolts holding the chainring on.** Use a 5mm hex key (or, for some, a Torx T30 key). With the exception of Dura-Ace (and some Ultegra) triples (Fig. 11.1B), the bolts are threaded directly into the crankarm. On Dura-Ace (and some Ultegra) triples, the inner ring is attached to a special middle chainring with long tabs extending radially inward.

3. **Install the new chainring and then lubricate and tighten the bolts.** The ring will have a small chainring-orientation bump protruding radially inward or an engraved triangle; make sure it lines up under the crankarm. Some inner chainrings have two of these indexing bumps to clock the chainring differently depending on the size of the adjacent chainring. The chainring combination will be stamped near the bump. So if the outer chainring is a 50-tooth and you're installing an inner 36-tooth ring, position the bump labeled "36-50" behind the crankarm, not the one labeled "36-52."

Make sure that the heads of the chainring bolts and nuts are recessed into the countersunk holes (i.e., make sure the holes are not facing inward toward the spider). Some lightweight aluminum chainring bolts cannot take much tightening and will snap off easily; don't overtorque them.

If you're having trouble with chainring bolts loosening, you can put threadlock compound on the threads instead of grease, but grease will do a better job of eliminating creaking noises. You can grease right over the dried threadlock compound that comes on many new chainring bolts.

NOTE: Some cranks have a removable spider or chainrings attached together as a set. At the more expensive end, first-generation FSA carbon ISIS road cranks and Sibex titanium cranks use the same splined

11.14 Removing and installing a chainring spider on a first-generation FSA carbon ISIS road crank or on a Shimano 1996–2002 XTR or 1997–2003 XT crank

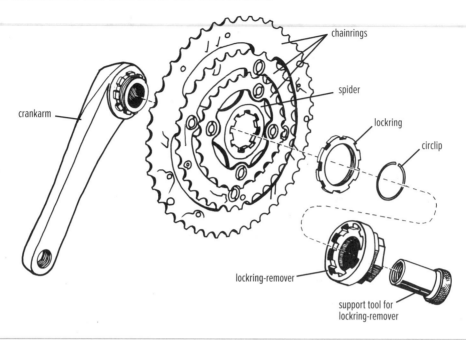

slip-on spider system as the Shimano 1996–2002 XTR and 1997–2003 XT cranks shown in Figure 11.14. After removing a circlip (by prying it off with a screwdriver), a special lockring-remover tool loosens the chainring-spider-securing lockring; a female-threaded tool that goes on the crank bolt holds the lockring-remover tool in place (Fig. 11.14). Once the spider is off, you can interchange chainrings within the set or simply pop on a whole new set.

Inexpensive cranks sometimes have chainrings riveted to the crank or riveted to each other and bolted to the crank as a unit. If these chainrings are damaged, you may have to replace the entire crankset.

4. **Replace the crankarm (11-2, Figs. 11.3, 11.4).**

PRO TIP ── ADVANTAGES OF A COMPACT-DRIVE DOUBLE CRANK

IF YOU WANT LOWER GEARS FOR CLIMBING but do not want a triple crank (which is heavier than a double and spaces the feet farther apart), consider a compact-drive double crank. A compact-drive road crank has a 110mm bolt-circle diameter, considerably less than the standard 130mm or the 135mm of Campagnolo, which limit the minimum inner chainring size to 38 teeth (or 39T). The "compact crank" allows the use of a 34-tooth inner chainring.

You can switch from, for example, a 39–53 chainring setup to a 34–50 setup (lowering the front derailleur accordingly). You generally will need to replace the bottom bracket as well, but the new crank may work with the existing one, and you may wish to use the rest of your components as is (possibly shortening the chain by one link). For junior-racing gearing, you can use 33–41 compact chainrings.

The tighter curvature of the compact chainrings, and the larger 16-tooth jump (50 minus 34) versus the 14-tooth jump (53 minus 39) that you had before means that the double front derailleur you have been using may not shift crisply. Current front derailleurs are compatible with both compact and standard chainring sizes, but if yours is pre-2005 or so, it may only be intended for standard chainrings (i.e., 52T or 53T outer chainring). Its tail may be too high above the chainring for optimal shifting. In this case, you may need to install a new front derailleur as well.

BOTTOM BRACKETS

Traditionally, bottom brackets other than on the most inexpensive of bicycles thread into the frame's bottom-bracket shell. Originally, these were cottered cranks (cranks held on by a cotterpin) and then were square-taper bottom brackets (Fig. 11.20). In both cases, the crankarms were separate from the bottom-bracket spindle (three-piece cranks). External-bearing bottom brackets for two-piece (integrated-spindle) cranks (Fig. 11.2) then took over.

Now, however, the trend is toward frames with unthreaded bottom-bracket shells into which the bottom-bracket bearings are pressed. These are available in a number of different widths and diameters.

Bottom-bracket bearings tend to be fairly well sealed toward the outside but not toward the inside, and water can come in around the seatpost. So, if your frame's bottom-bracket shell does not have a drain hole in it and you ride it in the rain, I recommend drilling one to let water out. Drill it as close to straight under the bottom when the complete bike is standing upright, but place it so that it will not be covered by the screw-on derailleur cable guide.

Threadless bottom brackets

Eliminating bottom-bracket threads reduces machining and assembly time in bike factories. It allows the use of full-carbon bottom-bracket shells straight out of the frame mold without any machining afterward, so threadless bottom brackets with press-in bearings have become popular with the frame manufacturers. However, there is so far no single standard for width or diameter of the frame's bottom-bracket shell. Apart from some early designs with small bearings for three-piece cranks, threadless bottom brackets take integrated-spindle cranks, but that spindle can be either 24mm or 30mm (and even bigger in isolated cases). Table 11.1 lists the key features of the current threadless bottom-bracket types; details on each system follow.

TABLE 11.1 — THREADLESS BOTTOM-BRACKET SPECIFICATIONS

BOTTOM-BRACKET TYPE	SPINDLE DIAMETER	BB SHELL WIDTH	BB SHELL INSIDE DIAMETER	BEARING SIZE	THREAD-TOGETHER OPTION?	NOTES
BB386EVO	30mm	86.5mm	46mm ID	46mm OD, 30mm ID	Yes	Crank also fits BB30 & PF30 w/ 9mm spacers. Also fits BBright w/ drive-side spacer.
BB30	30mm	68mm	42mm ID	42mm OD, 30mm ID	Yes*	Thread-together system uses outboard bearing.
PF30	30mm	68mm	46mm ID	42mm OD, 30mm ID	Yes	—
BB30A (Cannondale)	30mm	73mm asymmetrical	42mm ID	42mm OD, 30mm ID	—	Only fits BB30A-specific crank.
PF30A (Cannondale)	30mm	73mm asymmetrical	46mm ID	42mm OD, 30mm ID	—	Only fits BB30A-specific crank.
OSBB carbon (Specialized)	30mm	61 or 62mm	42mm ID	42mm OD, 30mm ID	—	—
OSBB aluminum (Specialized)	30mm	68mm	42mm ID	42mm OD, 30mm ID	—	—
BB86/PF24 (Shimano)	24mm	86.5mm	41mm ID	37mm OD, 24mm ID	Yes	BB386 crank fits with aftermarket BB.
BB90 (Trek)	24mm	90mm	37mm ID bearing seat	37mm OD, 24mm ID	—	Campagnolo UT, Fulcrum RT cranks fit with washers.
BBright PF (Cervelo)	30mm	79mm asymmetrical	46mm ID	42mm OD, 30mm ID	Yes*	Accepts PF30 and BB386 bearing sleeves.
BBright CA (Cervelo)	30mm	79mm asymmetrical	42mm ID	42mm OD, 30mm ID	—	Accepts BB30 press-in bearings.

* Convert to 24mm spindle.

11.15 BB386 crank assembly

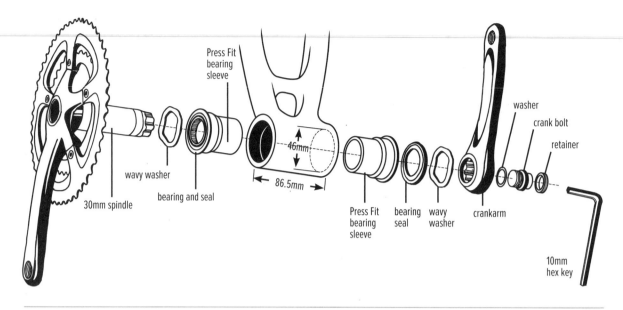

Press Fit bearing sleeve

wavy washer

bearing and seal

30mm spindle

46mm

86.5mm

Press Fit bearing sleeve

bearing seal

wavy washer

crankarm

washer

crank bolt

retainer

10mm hex key

BB386

Let's begin here, because BB386 cranks, called by their creator FSA "BB386EVO" (Fig. 11.15), are currently the most versatile; they fit in more bike frame types than any other. Like BB30 cranks, BB386 cranks have 30mm-diameter bottom-bracket spindles and pass through 30mm inside diameter (ID) cartridge bearings. They will also fit in BB30 and PF30 bottom brackets (with the addition of 9mm spacers on either side) as well as BBRight bottom brackets (with the addition of a drive-side spacer). They have longer spindles than BB30 and PF30 cranks, however, which allows them to fit threaded bottom-bracket shells, since the crankarms are far enough apart to allow external bearings to fit between them and a threaded bottom-bracket shell (Fig. 11.16B). Bottom-bracket face to bottom-bracket face distance is 90mm, the same as threaded external-bearing systems.

A BB386 bottom-bracket shell is 86.5mm wide (Fig. 11.16A); this, combined with the "3" for the 30mm spindle, explains what "386" denotes. The bearings are housed in plastic or aluminum press-fit sleeves that press into the 46mm ID shell from either end and overlap inside. Adapters are available to allow the use of any 24mm or 25mm integrated-spindle crank into a BB386 shell. Some bottom-bracket companies like Wheels Manufacturing and Praxis Works make bottom brackets

11.16A BB386 bottom-bracket shell cutaway with PF30 bearings; if this shell were 68mm wide, it would be a PF30 bottom bracket

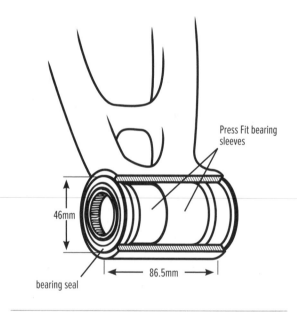

Press Fit bearing sleeves

46mm

bearing seal

86.5mm

for BB386 frames whose cups thread together inside of the bottom-bracket shell; this makes them less likely to creak or come loose than press-in cups.

Other BB386 variations

Campagnolo Over-Torque, Race Face Cinch, Clavicula, and Rotor cranks are similar in length to BB386EVO, but not exactly. They are completely compatible with BB386,

11.16B ISO threaded (a.k.a. "English threaded" or "BSA") bottom-bracket shell cutaway with BB386 external bearings; threads are 1.37 in. x 24 TPI.

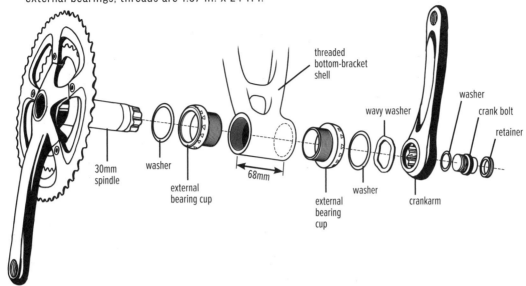

11.16C BB30 bottom-bracket shell cutaway with BB386 cranks and spacers; the shell is 68mm wide with a 42mm inside diameter.

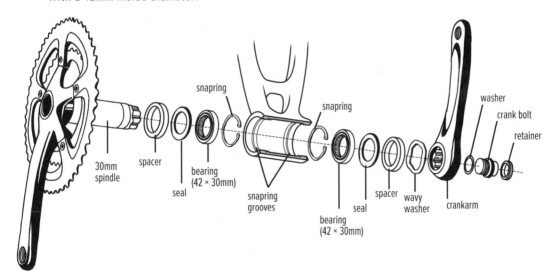

but it often takes some experimentation with spacers to fit them without play on a BB386 bottom bracket.

BB30 and PF30

Both BB30 and Press Fit 30 (a.k.a. "PF30") bottom brackets fit BB30 cranks, which, like BB386 cranks, have 30mm-diameter bottom bracket spindles, but they are shorter than BB386 spindles. BB386 and BB30 cranks also differ in the consequent shaping of the crankarms due to the spindle-length difference; BB30

cranks generally are splayed out more and have more ankle clearance.

On BB30 road bikes, the bearings press into a 68mm-wide bottom bracket shell (Figs. 11.16C, 11.17) with a 42mm inside diameter (ID). The bearings are prevented from going in farther by snaprings seated in grooves in the inner diameter of the shell (Fig. 11.16C). To fit the longer spindle of a BB386 crank into a BB30 bottom bracket shell, you only need a pair of thick spacers (Fig. 11.16C).

11.17 BB30 crank assembly; note the absence of the two thick spacers in Figure 11.16C that are required to adapt a BB386 crank to fit a BB30 bottom-bracket shell.

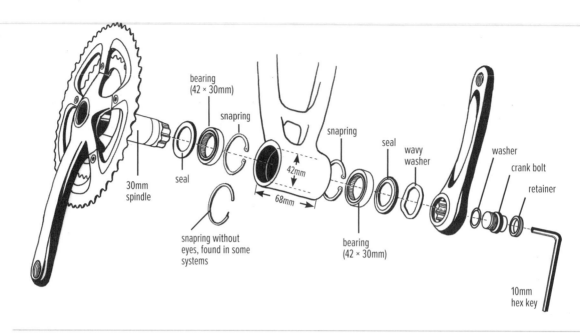

PF30 bearings are the same size as BB30 bearings (42mm OD, 30mm ID), but they are housed inside plastic or aluminum sleeves that are 46mm in diameter that press straight into the bottom-bracket shell without any need for snaprings. The bottom bracket is the same as BB386 (Fig. 11.16A) except that the length of the sleeves may be shorter on PF30; it's PF30 when it's in a 68mm-wide PF30 bottom-bracket shell, and it's BB386 when it's in an 86.5mm-wide BB386 shell like the one in Fig. 11.16A. Some bottom-bracket companies like Wheels Manufacturing and Praxis Works make bottom brackets for BB30 and PF30 frames whose cups thread together inside of the bottom-bracket shell; this makes them less likely to creak or loosen than press-in cups.

Other BB30 variations

BB30A and PF30A are both Cannondale standards. They are the same as BB30 and PF30 but with the non-drive side of the frame 5mm wider (so the shell is asymmetrical and 73mm wide). They only fit a BB30A-specific crank (made by FSA for Cannondale or by Cannondale). Some bottom-bracket companies like Kogel and Wheels Manufacturing make conversion bottom brackets to fit Shimano or SRAM GXP cranks on BB30A and PF30A frames.

Strange as it may sound, Specialized's OSBB (over-sized bottom bracket) "standard" has different dimensions depending on what material the frame is made of! OSBB on carbon frames is simply a narrower version of PF30 (61 or 62mm shell width, rather than 68mm). On aluminum frames, Specialized OSBB is identical to BB30.

PF24 (BB86)

BB86 and BB92 are often called the "Shimano press-fit system" standards for road and mountain bikes, respectively, even though other crank manufacturers make bottom brackets for this standard as well. Shimano refers to it simply as "Press Fit."

In an attempt at more clarity and consistency with the morass of bottom-bracket standards, I'll call it "Press Fit 24" or PF24, since the spindle diameter is 24mm and it has the same type of plastic cups (adapters) as PF30. PF24 bottom brackets accept standard integrated-spindle cranks for external-bearing bottom brackets (Fig. 11.2), which have a 24 × 90 road integrated crank spindle (24 × 90 refers to its 24mm diameter and the 90mm distance from the outside face of one bearing to the outside face of the other).

A PF24 (BB86) road bottom-bracket shell is 86.5mm wide and has a 41mm ID. The PF24 (BB86) system is

11.18 SRAM GXP crank in the BB90 shell of a Trek Madone carbon frame

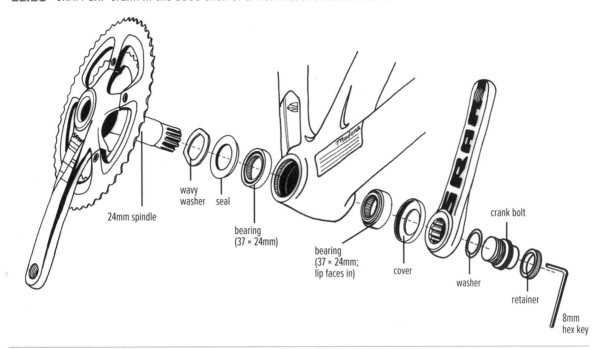

24mm spindle
wavy washer
seal
bearing (37 × 24mm)
bearing (37 × 24mm; lip faces in)
cover
crank bolt
washer
retainer
8mm hex key

similar to BB386 and PF30 (Fig. 11.16A); the bearings are incorporated into plastic or aluminum sleeves that press into the bottom-bracket shell. On PF24, however, the bearings have a 37mm OD, and the sleeves have a 41mm OD. Each sleeve's shoulder is 1.75mm wide, creating a 90mm overall width for road BB86 (86.5mm + 1.75mm + 1.75mm), exactly the same as the Trek system below or as a threaded external-bearing system.

You cannot install a BB30 or BB386 crank into a PF24 (BB86) bottom-bracket shell, but, with an aftermarket bottom bracket, you can fit a BB386EVO (or equivalent, like Campagnolo Over-Torque, Race Face Cinch, etc.) into one. Some aftermarket bottom-bracket companies like Wheels Manufacturing make bottom brackets for PF24 (BB86) frames whose cups thread together inside of the bottom-bracket shell; this makes them less likely to creak or loosen than press-in cups.

BB90 (Trek system)

BB90 and BB95 are Trek's slip-fit bearing systems for road and mountain bikes, respectively. The road bottom-bracket shell is 90mm wide and has a 37mm-ID bearing seat molded directly into the frame (Fig. 11.18). The 37mm OD × 24mm ID bearing is the same as you would find inside the threaded cup of external-bearing

cups for 24mm spindles, and it is compatible with any standard external-bearing/integrated 24mm-spindle crankset (Fig. 11.2).

Trek and aftermarket bottom-bracket companies like Kogel and Wheels Manufacturing supply bearing sets for all 24mm integrated-spindle cranksets, and the bearings slip into place with finger pressure alone. Campagnolo Ultra-Torque and Fulcrum Racing-Torq cranksets fit right in without cups or retaining clips, but you have to insert a flat washer into each side under the bearings, a wavy washer under the left bearing, and fit covers on both bearings.

You cannot install a BB30 or BB386 crank into a BB90 bottom-bracket shell.

BBright

Cervelo's BBright system is designed for 30mm-spindle cranks and is neither as wide as BB386 nor as narrow as BB30. The bottom-bracket shell is 79mm wide and is not centered on the down tube; the added width is on the non-drive side. Two frame types with two different shell diameters are available: BBright PF, which has a 46mm ID and fits PF30 or BB386 press-fit bearing sleeves, and direct-fit BBright CA (only on California-made Cervelo frames), which has a 42mm ID and fits

BB30 press-in bearings. BBright frames fit BB386 cranks with an additional spacer on the drive side; they can also be adapted to 25- or 24mm-spindle cranks.

BB94

Wilier Triestina's discontinued BB94 system has a 94mm-wide bottom-bracket shell into which a Campagnolo Ultra-Torque or Fulcrum Racing-Torq crankset fits directly without cups, wavy washers, or a retaining clip. Using composite spacers from Wilier Triestina and 37mm OD × 24mm ID bearings like the Trek system, you can install Shimano, SRAM, FSA, and other 24mm-spindle cranks on it.

Look ZED

Look's ZED system consists of a one-piece carbon crank/bottom-bracket spindle. The shell has a 65mm inside diameter and is 68mm wide.

Early press-in

Some old bottom brackets with a square-taper spindle do not thread into the bottom-bracket shell. One type, found on old Fisher, Klein, and Fat Chance frames, uses cartridge bearings held into an unthreaded bottom-bracket shell by snaprings in machined grooves, much like BB30 but with a much smaller-diameter shell and a spindle with square-taper ends. Early Merlin titanium frames used a similar system except that the bearings are pressed and glued into place with Loctite, without a snapring.

Threaded bottom brackets

Until the recent renaissance of threadless bottom brackets, the bottom-bracket type on most bike models simply threads into the frame's bottom-bracket shell and accepts the crankarms (Figs. 11.19, 11.20). Simple enough, but not all threaded bottom-bracket shells are the same.

Almost all current threaded bottom-bracket shells in bicycle frames have ISO (a.k.a. "English," "BSA," or "BSC") standard threads; these have a 1.37-inch diameter and a thread pitch of 24 threads per inch and, on a road frame, are 68mm wide. The diameter and thread pitch are usually engraved on the bottom-bracket cups.

11.19 Installing and removing threaded external-bearing cups; threads are ISO, a.k.a. "English" or "BSA."

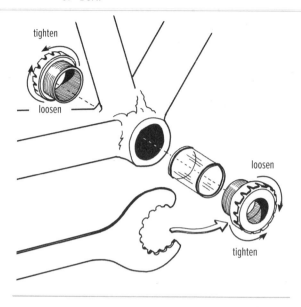

11.20 Cup-and-cone bottom-bracket assembly

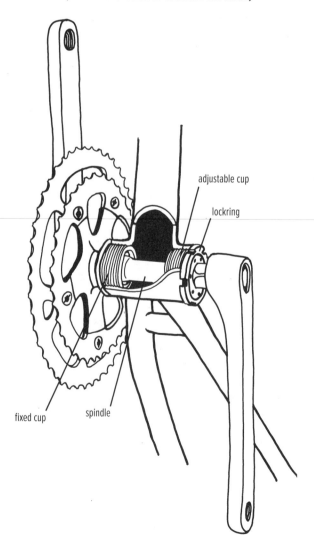

If you are replacing a bottom bracket, make sure that the new cups have the same threads. It is important to remember that the drive side (right side) of an English/ISO standard bottom bracket has left-hand threads. Turning counterclockwise tightens the drive-side cup (Fig. 11.19), while turning clockwise tightens the left cup into its standard right-hand threads (Fig. 11.19).

Other bottom-bracket threads you may run across are Italian (with a 36mm diameter; note that both cups have right-hand threads), French, and Swiss (both come in 35mm diameter, but use different thread directions). The latter two thread patterns are rare, although French threading was common until the early 1980s.

Currently the most common threaded crank system on high-end bikes is an external-bearing bottom bracket with an integrated-spindle crank (Fig. 11.2). The bearings are external to the bottom-bracket shell, have a 37mm OD with a 24mm (usually) ID, and are housed in threaded cups (Fig. 11.19). The spindle and crankarm on at least one side are a single piece; most systems use a 24mm spindle permanently pressed into the drive crankarm (Fig. 11.2), although Race Face X-Type left crankarms are integrated with their spindles. Campagnolo Ultra-Torque and Fulcrum Racing-Torq cranks have a 25mm-diameter split spindle; half of it is integrated into the left arm and half into the right arm (Fig. 11.8). The spindle meets in the middle, with the crank bolt pulling the teeth on each end together.

Two-piece threaded cranksets

With the bearing cups of integrated-spindle, external-bearing cranksets (Fig. 11.2) being external to the bottom-bracket shell (Fig. 11.19), the bearings and spindle can be far larger (and hence stiffer) than prior designs (whose bearings are contained within the bottom-bracket shell). The external bearings can also be right up against the crankarms (Fig. 11.3), adding more stiffness. These are called "two-piece" cranks to mean two crankarms with an integrated bottom-bracket spindle.

For integrated-spindle bottom brackets to work properly, the threads on both sides of the frame's bottom-bracket shell must be aligned, and the end faces of the shell must be parallel. If you are installing an expensive bottom bracket and have any doubts about the frame, it is a good idea to have the bottom-bracket shell tapped (threaded) and faced (ends cut parallel) by a qualified shop possessing the proper tools. These tools are pictured in Figure 1.4. This procedure will improve durability and freedom of movement and will reduce the likelihood of creaking while pedaling.

Three-piece threaded cranksets

For three-piece cranks (i.e., two crankarms and a bottom-bracket unit), the most common type of bottom bracket is probably the cartridge type (Fig. 11.21 or 11.25); it has cups that accept the splined removal tool shown in Figure 11.25. These bottom brackets can have a square-taper spindle (Fig. 11.21), an ISIS splined spindle (Fig. 11.25), or a Shimano Octalink splined spindle (Fig. 11.24).

While most Octalink bottom brackets are the cartridge type (Fig. 11.21 or 11.25) with an Octalink spindle, the first generation of Shimano Dura-Ace Octalink bottom brackets had four sets of loose, adjustable, and overhaulable bearings: two sets of tiny ball bearings and two sets of needle bearings (Fig. 11.24). Once Octalink cartridge bottom brackets came out, Octalink split into version 1 and version 2, which vary in the depth of mounting grooves on the spindle.

To counter Shimano's patented Octalink designs, which offer increased stiffness and lower weight than the square-taper designs shown in Figures 11.21–11.23, a number of manufacturers banded together in the late 1990s to create the ISIS standard for bottom brackets. Like Octalink, ISIS has a larger-diameter splined spindle (Fig. 11.25), but it features longer and deeper splines than Octalink.

The most common bottom bracket prior to the 1990s was the square-taper "cup-and-cone" style with loose ball bearings (Figs. 11.20, 11.22). As discussed earlier regarding two-piece threaded cranksets, the bottom-bracket shell for cup-and-cone bottom brackets must be threaded concentrically, and the faces of the shell must be "faced" parallel to each other and perpendicular to the spindle. Otherwise, the bearings will drag and wear excessively.

Another older bottom-bracket type has cartridge bearings secured by an adjustable cup and lockring at either end (Fig. 11.23).

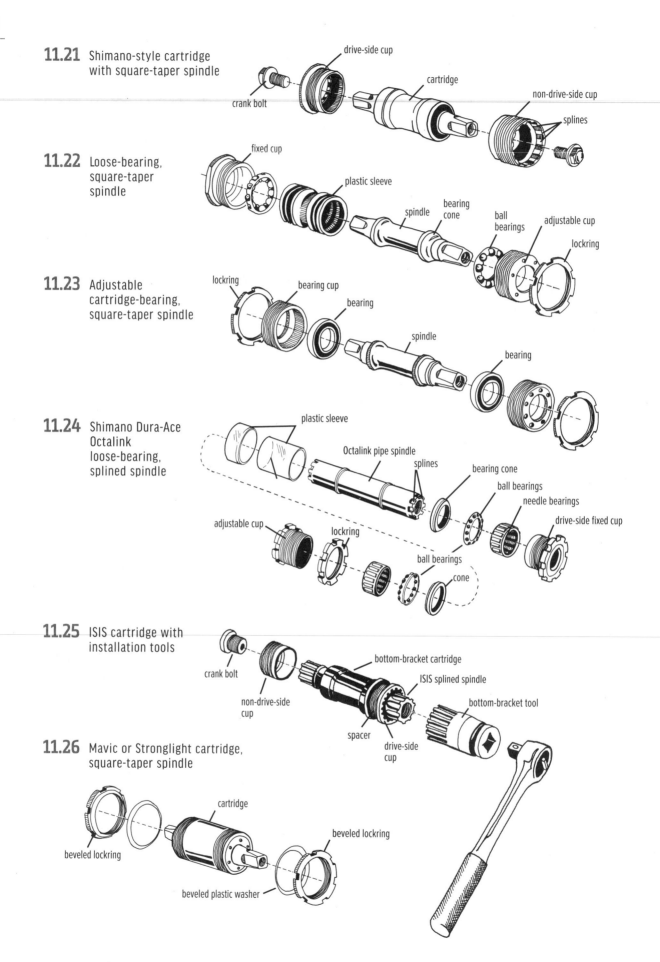

11.21 Shimano-style cartridge with square-taper spindle

drive-side cup
cartridge
non-drive-side cup
splines
crank bolt

11.22 Loose-bearing, square-taper spindle

fixed cup
plastic sleeve
spindle
bearing cone
ball bearings
adjustable cup
lockring

11.23 Adjustable cartridge-bearing, square-taper spindle

lockring
bearing cup
bearing
spindle
bearing

11.24 Shimano Dura-Ace Octalink loose-bearing, splined spindle

plastic sleeve
Octalink pipe spindle
splines
bearing cone
ball bearings
needle bearings
drive-side fixed cup
adjustable cup
lockring
ball bearings
cone

11.25 ISIS cartridge with installation tools

bottom-bracket cartridge
ISIS splined spindle
crank bolt
non-drive-side cup
bottom-bracket tool
spacer
drive-side cup

11.26 Mavic or Stronglight cartridge, square-taper spindle

beveled lockring
cartridge
beveled lockring
beveled plastic washer

Another type made first by Mavic and then by Stronglight includes a cartridge that is externally threaded on each end (Fig. 11.26). It slips into the bottom-bracket shell and is held in place by tapered lockrings threaded onto the cartridge. The lockrings have a convex 45-degree taper to bind against the bottom-bracket shell, which is machined with a matching 45-degree concave taper on its ends.

BOTTOM-BRACKET INSTALLATION

Threaded bottom brackets

The most important item in the installation of a separate (non-integrated-spindle) threaded bottom bracket is to make sure that the axle length in the bottom bracket is correct. If it's incorrect, the chainrings will not line up with the rear cogs (i.e., the chainline will be off; see 5-46 and 5-47). No amount of fiddling with the derailleurs will get such a bike to shift properly. Get a bottom bracket specifically recommended for the crankset, and double-check that it has the proper threading for the frame. Before installing a new bottom bracket of a brand and model different from the crank, see Figure 5.55 and read the chainline sections (5-46 and 5-47).

With one-piece cartridge-style bottom brackets (Figs. 11.21, 11.25, 11.26), absolute precision of the threads and faces of the bottom-bracket shell is not critical. However, for integrated-spindle cranksets (Figs. 11.2, 11.8) and cup-and-cone bottom brackets (Figs. 11.20, 11.22, 11.23) to spin freely and not wear rapidly, the threads inside both ends of the frame's bottom-bracket shell must be lined up with each other, and the end faces of the shell must be parallel. If you have any doubts about the frame and are installing an expensive bottom bracket, it is a good idea to have the bottom-bracket shell tapped (threaded) and faced (ends cut parallel) by a qualified shop possessing the proper tools. Doing so will reduce binding and the likelihood of creaking with integrated-spindle cranksets and will improve adjustment and freedom of movement with loose-bearing bottom brackets.

Always grease the threads of the bottom-bracket cups and the threads in the frame when installing bottom brackets (or use an antiseize compound on them).

THREADED EXTERNAL-BEARING/ INTEGRATED-SPINDLE BOTTOM-BRACKET INSTALLATION AND REMOVAL

LEVEL 🔧🔧

1. **Grease the threads.**

2. **Start the cups by hand.** Turn the right (drive-side) cup counterclockwise and the left cup clockwise with ISO (English-threaded) frames. Turn them both clockwise with Italian threads.

NOTE: *A removable plastic sleeve (Fig. 11.19) keeps contamination away from the inner side of the bearings. Keep the sleeve on the right cup when installing.*

ANOTHER NOTE: *Even if the bottom-bracket cups came with spacers, do not install either spacer on a double crank unless the bottom-bracket instructions specifically call for them.*

3. **Tighten the cups.** Use the splined tool designed for the purpose (Fig. 11.19). Sockets as well as a crow's-foot splined outboard bottom-bracket cup tool from RWC (OBBCT2; Fig. 1.4) are available to allow use of a torque wrench; turn the OBBCT2 at 90 degrees to the torque wrench so that the torque reading is accurate. You need a bigger socket for BB386 threaded external-bearing bottom-bracket cups. Torque is high (310–442 in-lbs, or 35–50 N-m); if you don't have a torque wrench, yank on the tool arm pretty hard.

4. **Install the spindle and crankarms.** Follow the instructions in 11-2a, b, or c, depending on type.

Bottom-bracket removal is obviously done by reversing the rotation directions in step 2. Use a standard wrench, not a torque wrench, to unscrew the bolts (it's not a good idea to use a torque wrench to unscrew bolts, as it can stretch the spring and thus throw off its subsequent torque readings).

PRESS-IN BOTTOM-BRACKET INSTALLATION

LEVEL 🔧🔧 There is no room for error with a press-in bearing system in a bottom bracket, be it BB386, PF30, BB30, PF24 (BB86), BB90, BB94, or some other system. The bearings must line up straight with each other and must fit tightly enough that they won't rattle or move around, yet loosely enough that they won't become deformed when pressing them in. Proper preparation of the bottom-bracket shell must be done at the factory; you cannot expect to correct a malformed shell.

Threadless bottom-bracket installation is similar to threadless headset installation (12-22).

1. **Clean and grease the inside ends of the bottom-bracket shell.**

2. **Grease the outside of the bearings or Press Fit bearing sleeves (Fig. 11.27).**

3. **Press in the bearings and install the spindle and crankarms.** Follow one of the following procedures, depending on the bottom-bracket shell and crank type. But see step 3 below if you are installing a replacement bottom bracket that threads together.

Bicycle parts companies like Wheels Manufacturing and Praxis Works make bottom brackets for some press-fit bottom-bracket shells that thread together where they meet inside of the bottom-bracket shell (Fig. 11.30); this tends to minimize the chances for creaking and loosening. Creaking can be an unpleasant irritant while riding with press-fit bottom brackets if the fit is not perfect between the shell and the bearings or the press-in cups. With one of these thread-together bottom brackets, step 3 above instead becomes:

3. **Thread the cups together.** Push the cups in from either side until they meet, following indications on the parts for drive- and non-drive-side. Include any recommended spacers between the left cup and the bottom-bracket shell. Using a knurled installation wrench for external bearings on each cup (Fig. 11.19), screw them tightly together.

a. Press Fit bearings housed in plastic or aluminum cups: PF30, BB386, PF24 (BB86)

1. **Press in the bearing cups.** The bearing cups (or "adapters") are specific to the bottom-bracket shell and the crankset. Each cup has a bearing inside and slips into or over the other cup (Fig. 11.27). Using a headset press, press in one cup at a time (Fig. 11.28). Ideally, use drifts against the bearings

that fit their ID and outer face, in which case you can press in both cups simultaneously; otherwise, push them in one at a time because the flat faces of the headset press will contact the ends of the cups and can't guarantee alignment. Ensure that both bearing cups are seated fully. Grease and install the manufacturer-supplied dust shields over them (Fig. 11.27). See the following Note if you are installing a replacement bottom bracket that threads together.

a. While these instructions apply to Campagnolo Over-Torque cranks (30mm spindle), Campagnolo Ultra-Torque, Fulcrum Racing-Torq, and Campagnolo Power Torque cranksets have 25mm spindles and can't be installed in threadless bottom brackets until the correct Campagnolo adapter cups have been installed into the ends of the shell.

11.27 Press Fit bottom bracket: PF30, BB386, or PF24 (BB86)

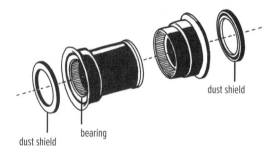

dust shield

bearing

dust shield

11.28 Pressing a bearing into a threadless bottom-bracket shell with a headset press

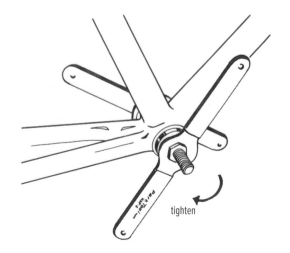

tighten

11.29 Pressing Campagnolo Ultra-Torque adapter cups into a threadless bottom-bracket shell

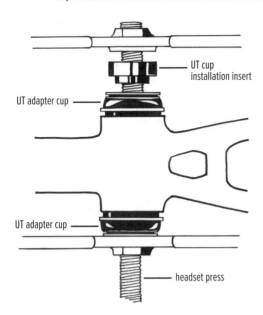

UT cup installation insert

UT adapter cup

UT adapter cup

headset press

11.30 Tightening aftermarket threaded bottom-bracket cups into a threadless bottom-bracket shell (BB86/PF24 shown here)

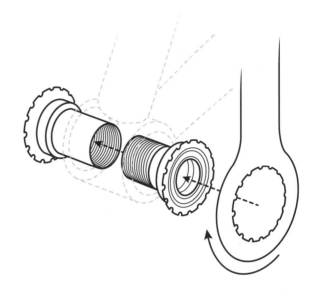

Push them in by hand until the O-ring disappears from view and resistance increases. To press them in the rest of the way, you can use the Campagnolo shop tool designed for the job or the correct bearing drifts (like from Park's BB30/BB86 bearing installer; Fig. 1.4) and a headset press. Run the headset press or the Campy press through the cups and bearing drifts and carefully turn the handle until the cups just meet the bottom-bracket shell (Fig. 11.29). Without these tools, have a shop install the cups, or use a headset press without inserts, and do it one at a time to increase your odds of getting them in straight. Make sure you don't crush the walls of the cups in the process. Once the cups are pressed in, install the cranks as in 11-2c.

NOTE: *For a press-fit bottom-bracket shell with aftermarket threaded bottom-bracket cups that thread together where they meet inside of the shell, step 1 instead becomes:*

1. *Thread the cups together. Push the cups in from either side until they meet, following indications on the parts for drive- and non-drive-side. Include any recommended spacers between the left cup and the bottom-bracket shell. Using*

a knurled installation wrench for external bearings on each cup, screw them tightly together (Fig. 11.30).

2. **Install the spindle and dust shields, and tighten the crankarms.** Follow the instructions in 11-2a, b, or c, depending on type.

b. BB90: Trek frame with a 90mm-wide bottom-bracket shell

1. **Push the bearings into the frame by hand (Fig. 11.31).** Trek has a plastic bearing-installation tool for its BB90 frames, but you usually don't need it.

   a. For Shimano, Race Face X-Type, Easton (and others), place the Trek-supplied seals over the bearings (except FSA, where the seal is already in the bearing). Install the crankarms as in 11-2a.

   b. For SRAM/Truvativ GXP, make sure to put the bearing with the smaller bore (22mm, rather than 24mm) in the left side with the inner bearing ring's lip facing inward. Place the flat, rubbery bearing seal against the right bearing, and slip the wavy washer onto the crank spindle. Install the crankarms as in 11-2b.

   c. For Campagnolo and Fulcrum cranks, install the bearing as in Figure 11.31, only with Power

11.31 Pressing in bearing by hand in a Trek Madone BB90 bottom-bracket shell

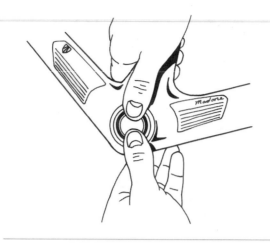

11.32 Installing a Campagnolo Ultra-Torque crankset in a Trek Madone BB90; bearing seal seat and thin washers not shown.

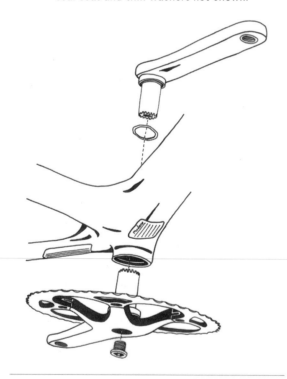

on either side with UT and Racing-Torq, and on the right side only with PT. Install the crankarms (Fig. 11.32) as in 11-2c.

c. BB30: All types

1. **Clear the bottom-bracket shell of any metal chips or other detritus.**

2. **Grease the contact surfaces.** Apply a thin layer of grease to the snapring grooves and surfaces outboard of them, and to the snaprings themselves.

3. **Insert the snaprings.** If the snapring has a hole (eye) on either end, push the tips of the snapring pliers into the holes, squeeze the handles to reduce the snapring's diameter, and install it into the groove inside one end of the bottom-bracket shell (Fig. 11.33). If the snapring does not have holes, push the ring into the shell to compress it, push its square-cut end to the groove so that it drops into it, and work around, pushing the rest of it in place. Check that the snapring is fully engaged in the groove all of the way around. Repeat for the other snapring in the other end. (To remove a snapring with eyes on the ends, compress it with snapring pliers and pull it out. To remove a snapring without eyes, slip a screwdriver blade under the pointed end of the snapring, push it inward, beyond the groove, and keep pushing on it, working from that end toward the square-cut end until the snapring is free.)

4. **Press in the bearings.** Beg, borrow, or buy bearing drifts for BB30 bearings (Wheels Manufacturing, FSA, Park, and Cannondale sell such drifts). With a headset press (or a bearing press like the one from Wheels) and depending on the bearing drifts you have, press in both bearings simultaneously or one bearing at a time (Fig. 11.28). Obviously, you are pressing them in until they stop at the snapring; this should be right when the bearing face is flush with the face of the shell. Work carefully, feeling the increasing pressure from the press's handles, and keep checking for symmetry and flush alignment with the face of the shell. Without a headset (or bearing) press, if you're careful and have BB30 bearing drifts, you should be able to press the bearings in straight using a large bench vise.

Torque cranks and only on the left (non-drive) side. It is a 6805 bearing, 37 × 25 × 7mm. With a Campagnolo Ultra-Torque or Fulcrum Racing-Torq crankset, first Loctite the Trek-supplied seals for the bearing seats into the shell on either side (right side only with PT) and let sit for 24 hours. The next day, place the Trek-supplied thin washer into the seat

5. **Grease the back side of each bearing.** This protects it from water trapped in the shell.

6. **Place the rubber-coated aluminum dust shields against the bearings.** Their grooves face inward, toward the bearings.

7. **Push the spindle in from the drive side.** First, lightly grease the spindle bearing areas, splined end, and threads. You may need a rubber mallet to push in the spindle, since BB30 spindles are designed to be a light interference fit.

8. **If supplied, put a wavy washer on the left end of the spindle.** It goes between the bearing shield and the crankarm and takes out lateral play.

9. **Push on the left crankarm and tighten the crank bolt.** Ideally, use a torque wrench (8mm or 10mm hex driver) to torque spec (torque is often less than with external-bearing cranks; see Appendix E for torque specs).

10. **Check the wavy washer.** It should be compressed somewhat, but not flattened. If it is not slightly compressed, remove the crank, add as many spacers against the bearing shields as are required to fill the space so that the wavy washer will be slightly flattened when the crankarm is on, and reinstall the crankarm.

11. **If the bearings fit loosely or creak, they need Loctite.** Remove them (11-14b), smear a thin layer of Loctite 609 retaining compound where they sit inside the bottom-bracket shell, and reinstall them.

NOTE ON CAMPAGNOLO BB30: *"Over-Torque" refers to Campagnolo's 30mm-spindle carbon cranks. They are the company's lightest and stiffest, but they are not labeled as Record, Super Record, etc.; they are stand-alone products dubbed Comp One and Comp Ultra. Over-Torque cranks have longer spindles like BB386 and will fit BB30, PF30, BB386, and BSA (English-threaded); in addition, Wheels Manufacturing makes direct-fit bearings to adapt them to BB86 (PF24). Over-Torque won't fit BB90 (Trek), and currently there are no bottom brackets to fit them to Italian-threaded frames. Over-Torque crank installation requires a special pin tool to tighten the lockring that holds the left arm on. The bearing preload adjuster is a small threaded ring that engages the spindle's splines between the left arm*

11.33 Installing a snapring in a BB30 shell

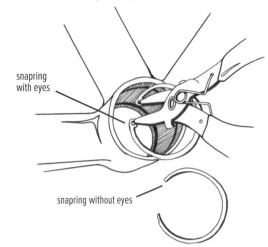

snapring with eyes

snapring without eyes

and bottom bracket; once the left arm is tightened in place, turn the preload ring with your fingers to remove side play and then lock it in place with its setscrew.

Campagnolo Ultra-Torque, Fulcrum Racing-Torq, and Campagnolo Power Torque cranksets, on the other hand, can't be installed in BB30 or other threadless bottom brackets until the correct Campagnolo adapter cups have been installed into the ends of the shell. Push them in by hand until the O-ring disappears from view and resistance increases. To press them the rest of the way in, you can use the Campagnolo shop tool designed for the job or the correct bearing drifts (such as the ones from Park's BB30/BB86 bearing installer, Fig. 1.4) and a headset press. Run the headset press or the Campy press through the cups and bearing drifts, and carefully turn the handle until the cups just meet the BB shell (Fig. 11.29). Without these tools, have a shop install the cups, or use a headset press without inserts, and install them one at a time to increase your odds of getting them in straight. Make sure you don't crush the walls of the cups in the process. Once the cups are pressed in, install the cranks as in 11-2c.

d. BB94: Wilier Triestina frame with a 94mm-wide bottom-bracket shell

1. For Shimano, SRAM/Truvativ GXP, FSA MegaExo, Race Face X-Type, Easton (and others), push the bearings in by hand and place the Wilier-supplied seals over them. Install the cranks as in 11-2a or 11-2b.

2. With a Campagnolo Ultra-Torque or Fulcrum Racing-Torq crankset, install the arms at 180 degrees from each other by hand into the bottom-bracket shell. Tighten the cranks as in 11-2c. For Power Torque, push the left (non-drive) side bearing in by hand and install the right (drive) arm as in 11-2c.

e. Square-taper press-in

Small-diameter, unthreaded bottom-bracket shells with snapring grooves were popular in the 1980s. A snapring at each end of the shell retains the bearings, and the spindle has a square taper. If the shell is bored to the correct diameter, you can probably seat the cartridge bearings by hand. If not, you can press them in with a vise or a headset press (Fig. 11.28), using a large socket or other flat-ended cylindrical object as a drift pushing against the bearing, as long as it is just slightly smaller in diameter than the OD of the bearing. The inboard bearing stops are usually shoulders or snaprings on either end of the axle, rather than lips inside the shell.

If the bearings will go in by hand, install one snapring with snapring pliers into the groove in one end of the shell. Push the entire assembly of axle and two bearings in from the other side of the bottom-bracket shell. Install the other snapring, and you're done.

If you can't press them in by hand, press one bearing in as far as the snapring groove by using one vise jaw or headset-press face against the shell face, the other against first the bearing, until it reaches the shell, and then against the drift to push the bearing into the bottom-bracket shell's bore. Install the snapring. Install the spindle and push the other bearing in the same way until you can install the other snapring.

11-10

THREADED CARTRIDGE BOTTOM-BRACKET INSTALLATION

These instructions apply to square-taper or ISIS cartridge bottom brackets like those shown in Figures 11.21 and 11.25, as well as to ones with this type of cartridge body and a Shimano Octalink splined spindle (Fig. 11.24). Most cartridges fit a Shimano splined

socket (similar to Fig. 11.25), but Campagnolo cartridges require a different socket.

NOTE: *Press-in aluminum sleeves with internal threads are available to convert a BB30 or PF30 unthreaded bottom-bracket shell to a threaded one.*

1. Thread the left cup (the one without a lip on the end) in three to four turns (clockwise) by hand.

2. Slide the cartridge into the bottom-bracket shell, paying particular attention to the "right" and "left" markings on the cartridge. The cup with the raised lip is the drive-side cup (the cup shown on the left in Fig. 11.21 and the cup on the right in Fig. 11.25). The drive-side cup is left-hand threaded on an English-threaded bottom bracket and right-hand threaded on an Italian-threaded one.

3. Tighten the drive-side cup. By using the splined cup socket (again, Campagnolo bottom brackets require a socket with slightly different splines than all of the other brands) with either an open-end wrench or a socket or torque wrench on it, tighten the drive-side cup into the drive side of the bottom-bracket shell until the lip seats against the face of the shell (as in Fig. 11.34, except on the drive side instead of the non-drive side as illustrated). Note that a splined bottom-bracket socket meant for ISIS and Octalink cartridge bottom brackets

---

11.34 Tightening or loosening the non-drive-side bottom-bracket cup

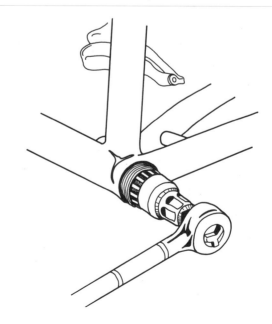

(Fig. 11.25) will also fit square-taper cartridge bottom brackets (except Campagnolo), but the reverse is not true; the bore of the square-taper bottom-bracket socket will be too small to swallow the ISIS or Octalink splined spindle end. Recommended torque is high: 442–620 in-lbs (50–70 N-m); see Appendix E.

NOTE: *On most bikes, this drive-side cup will tighten counterclockwise. On Italian-threaded bikes, it will tighten clockwise.*

4. **With the same tool, screw in the left (non-drive) cup clockwise until it tightens against the cartridge (Fig. 11.34).** The torque required is the same as for the right cup (see Appendix E). There is no adjustment of the bearings to be done; the bottom bracket is now ready for crank installation (11-2d).

11-11

CUP-AND-CONE BOTTOM-BRACKET INSTALLATION

Cup-and-cone (or "loose-ball") bottom brackets (Figs. 11.20, 11.22, 11.24) use ball bearings that ride between cone-shaped bearing surfaces on the axle and cup-shaped races in the threaded cups. One cup, called the fixed cup (the cup on the left in Figs. 11.20 and 11.22, and on the right in Fig. 11.24), has a lip on it and fits on the drive side (right side) of the bike. The other, called the adjustable cup (the right cup in Figs. 11.20 and 11.22, and on the left in Fig. 11.24), has a lockring that threads onto the cup and against the face of the bottom-bracket shell. The individual ball bearings are usually held together by a retaining cage, which varies in shape depending on bottom bracket. Some folks prefer to do without the retainer; it works fine either way.

In order for cup-and-cone bottom brackets to turn smoothly, the bearing surfaces of the cups must be parallel. Because the cups thread into the bottom-bracket shell, the threads on both sides of the shell must align, and the end faces of the shell must be parallel. If you have any doubts about the frame, it is a good idea to have the bottom-bracket shell tapped (threaded) and faced (ends cut parallel) by a qualified shop possessing the proper tools.

1. **Unless you have a fixed-cup tool, have a shop install the fixed cup for you.** The shop tool ensures that the cup goes in straight and very tight. The tool pictured in Figure 11.35 can be used in a pinch, but it can let the cup go in crooked and will slip off before you get it really tight. The fixed cup must be tight (see Appendix E for torque) so that it does not vibrate loose. Remember that English-threaded fixed cups are tightened counterclockwise.

2. **Lubricate the cups.** Wipe the inside surface of both cups with a clean rag, and put a layer of clean grease on the bearing surfaces. Apply enough so that the balls will be half covered; more than that will be wasted and will attract dirt.

3. **Wipe the axle (also called a spindle) with a clean rag.**

4. **Figure out which end of the bottom-bracket axle is the drive side.** The drive side may be marked with an R; if not, you can tell by choosing the side with the longer end (when measured from the bearing surface). If there is writing on the axle, it will usually read right side up for a rider sitting on the bike. If there is no marking and no length difference, the axle orientation is irrelevant.

5. **Slide one set of bearings onto the drive-side end of the axle (Fig. 11.36).** If you're using a retainer cage, make sure you put it on right. The balls, rather than the retainer cage, should rest against the bearing surfaces. Because there are two types of retainers with opposite designs, you need to be careful to avoid binding as well as smashing the retainers. If you're still confused, there is one easy test: If it's in right, it will turn smoothly; if it's in wrong, it won't. If you have loose ball bearings with no retainer cage, stick them into the greased cup. Most setups rely on nine balls; you can confirm that you are using the correct number by inserting and removing the axle and checking to make sure that they are evenly distributed in the grease with no extra gap for more balls.

6. **Slide the axle into the bottom bracket so that it pushes the bearings into the fixed cup (Fig. 11.36).** You can stick your pinkie in from the other side to stabilize the end of the axle as you slide it in.

7. **Insert the protective plastic sleeve (shown in Figs. 11.22 and 11.24) into the shell against the inside edge of the fixed cup.** The sleeve keeps dirt and rust from falling from the frame tubes into the bearings; if you don't have one, get one.

8. **Now turn your attention to the other cup.** Place the bearing set into the greased adjustable cup. If you are using a bearing retainer, make sure it is properly oriented. If you are using loose balls, press them lightly into the grease so that they stay in place.

9. **Without the lockring, slide the adjustable cup over the axle and tighten it clockwise by hand into the shell, being certain that it is going in straight.** Screw the cup in as far as you can by hand—ideally, all the way until the bearings seat between the axle and cup.

10. **Locate the appropriate tool for tightening the adjustable cup.** Most cups have two holes that accept the ends, or pins, of an adjustable cup wrench called a "pin spanner" (Fig. 1.2). The other common type of adjustable cup has two flats for a wrench; on this type, you may use an adjustable wrench.

11. **Carefully tighten the adjustable cup against the bearings, taking great care not to overtighten.** Turn the axle periodically with your fingers to ensure that it moves freely. If it binds up, you have

gone too far; back off a bit. The danger of over-tightening is that the bearings can force dents into the bearing surfaces of the cups, and the bearings will never turn smoothly again.

12. **Screw the lockring onto the adjustable cup.**

13. **Tighten the lockring against the face of the bottom-bracket shell with the lockring spanner while holding the adjustable cup in place with a pin spanner (Fig. 11.37).** Lockrings come in different shapes, and so do lockring spanners; make sure yours mate with each other.

14. **As you snug the lockring against the bottom-bracket shell, check the axle periodically.** The lockring can pull the cup out of the shell minutely and loosen the adjustment. The axle should turn smoothly without free play in the bearings. I recommend installing and tightening the drive-side crankarm onto the drive end of the axle (Fig. 11.4) at this time so that you can check for free play by wiggling the end of the crank; it will give you a better feel for any play in the system.

15. **Adjust the cup so that the axle play is just barely eliminated.** While holding the cup in place, tighten the lockring as much as you can (Fig. 11.37) so that the bottom bracket does not come out of adjustment while riding (recommended torque is in Appendix E; tightening it as much as you can is

11.35 Tightening or loosening the drive-side fixed cup

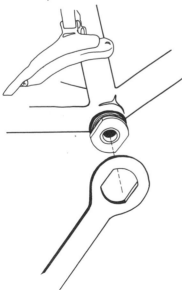

11.36 Placing the axle and drive-side bearings in the bottom-bracket shell

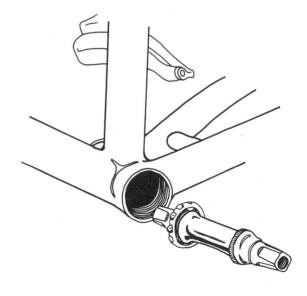

11.37 Tightening the lockring: Hold the adjustable cup in place with a pin spanner.

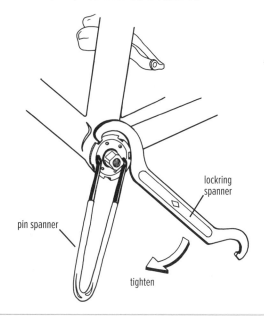

pin spanner

lockring spanner

tighten

about right). You may have to loosen the lockring, adjust the cup, and tighten the lockring a time or two until you get the ideal adjustment.

11-12

INSTALLING OTHER TYPES OF BOTTOM BRACKETS

The four bottom-bracket types just described probably represent about 95 percent of the road bikes in circulation. There are, however, a few variations worth mentioning.

a. Cartridge-bearing bottom brackets with adjustable cups

Cartridge-bearing bottom brackets with adjustable cups (Fig. 11.23) are reasonably easy to install. They have an adjustable cup at each end. With this type, you simply install the drive-side cup and lockring, slide the cartridge bearing in (if it is not already pressed into the cup), slip the axle in, and then install the other bearing, cup, and lockring. Tighten each lockring while holding the adjustable cup in place with a pin spanner (Fig. 11.37). Adjust for free play as in 11-11, steps 11–15.

The advantage of having two adjustable cups is that you can center the cartridge by moving it side to

side in the bottom-bracket shell. If the chainrings end up too close or too far away from the frame (see chainline discussion in 5-46 and 5-47), you can shift the position of the spindle.

Sometimes cartridge-bearing bottom brackets bind up a bit during adjustment and installation. A light tap on each end of the axle with a rubber mallet usually seats them.

b. Stronglight (or Mavic) cartridge-bearing bottom brackets

Stronglight (or Mavic) cartridge-bearing bottom brackets (Fig. 11.26) require each end of the bottom-bracket shell to be chamfered at an angle to seat the angled lockrings. You need to have the chamfer cut by a shop equipped with the correct cutting tool. Once the shell has been cut, simply slip the cartridge into the shell, slide on one of the angled plastic rings from either end (pictured in Fig. 11.26), and screw on a lockring, angled side inward, from either side. Holding the cartridge with a pin spanner, tighten the lockrings on each side (Fig. 11.37). The beauty of these bottom brackets is that they work independently of the shell threads, so they can be installed in shells with ruined threads or nonstandard threads. Mavic stopped producing them in 1995, but Stronglight resumed production for a while.

OVERHAULING THE BOTTOM BRACKET

A bottom-bracket overhaul consists of cleaning or replacing the bearings, cleaning the axle and bearing surfaces, and regreasing the bearings. Both crankarms must be removed first (11-1).

11-13

OVERHAULING INTEGRATED-SPINDLE BOTTOM BRACKETS

 LEVEL **a. Simple bearing scrub and lube from one side**

1. **Remove the crankarms as in 11-1a, b, c, or d.**
2. **With Campagnolo or Fulcrum, skip to step 3. With all others, remove the rubber bearing seal or dust shield covering the bearing.** In many cases,

11.38 Prying the bearing cover up off of the external bearing with a box cutter blade

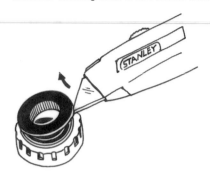

11.39 Prying the bearing seal out of a bearing with a box cutter blade

the drive-side cover seal stays stuck on the crank spindle and pulls off with the right crank (with Race Face X-Type cranks, the left bearing cover seal comes off with the left crank). To get this cover seal off to reveal the bearing, slip a blade under the edge and pry it up (Fig. 11.38), possibly working around the outside with a thin screwdriver for FSA and Shimano cover seals that extend inside the bearing bore.

3. **Remove the seal.** Now that the bearing is visible, you'll see that it has a circular seal between its inner and outer rings that covers the balls inside. With a razor blade or knife blade, get under the edge of it and pry it off (Fig. 11.39).

4. **Clean the bearings.** With a clean, dry cloth, or with solvent and a clean toothbrush, scrub the bearing to clean the dirty grease off the ball bearings, which usually will be concealed under a bearing retainer. You can blow the bearing out with compressed air; wear safety glasses. Repeat until clean. You can't get at the other side of the bearing without remov-

ing it from the cup (see 11-14a), but do your best to flush the bearing and dry it out afterward.

5. **Repack the bearing with clean grease and replace the bearing seal and the bottom-bracket cover seal.**

6. **Reinstall the crankarms as in 11-2a, b, or c.** If the bearings still do not turn well after this or are gritty, you'll need to overhaul (see 11-13b) or replace the bearings; see 11-14, or simply replace the entire cup and bearing (11-8).

b. Disassembling and overhauling a cartridge bearing

LEVEL This only works with high-end bearings with plastic bearing retainers separating the balls. You would ruin a steel bearing retainer trying to get the balls out. If you remove the bearing seal as in Figure 11.39 and see a shiny steel flat ring underneath it with a bump over each ball bearing, you cannot reassemble it if you take it apart, so don't even start. And with inexpensive bearings, even if you have a plastic bearing retainer, the amount of time and effort it takes does not make much sense economically. But I do think it makes sense to overhaul expensive hybrid ceramic bearings.

1. **Remove the bearing seal.** Do this while the bearing is still in the frame to determine if it can be disassembled. Pry off the cover seal (Fig. 11.38); then get underneath the seal with a box cutter or razor blade and pry it up (Fig. 11.39). Don't cut the rubber edges, but if you bend it a bit, don't be concerned—you can easily straighten it back out; it's soft aluminum.

2. **Wipe the grease away and inspect what's inside.** If you're looking at a shiny steel bearing retainer that has a bulge at each ball, you can't disassemble it without ruining it. You'll need to either replace the bearing or do the simple clean and grease in section 11-13a. Go on to step 3 if it's a plastic bearing retainer; one side will be smooth, concealing the balls; the other side will have prongs coming up between each ball, so that each ball is clearly visible. With a threaded external bearing cup, read 11-14a, step 8, first. If the spine of the plastic bearing retainer is facing you, you may be able to overhaul

the bearing without removing it from its external cup (which is an involved process; see 11-14a). If this is the case, skip to step 5.

3. **Separate the cartridge bearing from the bottom bracket as in 11-14a, b, c, d, or e.**

4. **Pry off the other bearing seal as in step 1.**

5. **Remove the bearing retainer.** With its spine facing you, pry it up with an awl at each ball.

6. **Push all of the balls together on one side of the bearing.** You can slide them around in their tracks with the awl.

7. **Pull out the inner bearing race.** Loop your finger through the inner race opposite the collected balls and pull it over until it touches the outer race. The balls should now fall out, freeing the race.

8. **Clean the balls and races.** Wipe the races clean with a rag and polish them with a polishing or rubbing compound. Clean the balls and the bearing retainer by wiping them with a clean rag and then by washing them by hand with soap in a (plugged!) sink. Dry everything thoroughly.

9. **Group all of the bearings in the track together inside the outer bearing race.** Each ball should be touching its neighbors. Use grease to hold them in place.

10. **Install the inner bearing race.** Push it in on the opposite side from the balls.

11. **Space the balls evenly around the bearing between the races.** You can slide them around in their tracks with the awl.

12. **Carefully push in the bearing retainer.** You may find that you will have to snip the ears off of the last prong of the bearing retainer to go in, as the balls can no longer move laterally once there is a prong between each pair of balls all of the way around except the last pair.

13. **Pack the bearing with grease from both sides.** For ceramic bearings, it's preferable to use grease designed specifically for them.

14. **Install the bearing covers with your fingers.**

15. **If required, press the cartridge bearings back into the cups or onto the spindle.**

16. **Take pride in a complicated job well done.** The bearing is good as new again!

11-14

REPLACING THE BEARINGS IN INTEGRATED-SPINDLE BOTTOM BRACKETS

You can replace the bearings with the same type you had before to bring the bottom bracket back to the way it was. Or you can make it spin better than ever by replacing the cartridge bearings with upgraded steel or ceramic bearings. Ceramic bearings are expensive, but they give the performance advantage demanded by top pro cyclists. Ceramic balls are lighter, smoother, 2.5 times rounder, 2.5 times harder, and 50 percent stiffer than steel balls and are less affected by heat.

If you're replacing only the bearings and not the entire cup assembly with bearings in place, first select the bearings you will need from WheelsMfg .com, EnduroForkSeals.com, CeramicSpeed.com, BocaBearings.com, or your bike shop; make sure they are the correct ones for the crankset.

a. Interchanging the bearings inside external-bearing cups

LEVEL In most cases, replacing or upgrading the bearings means replacing the cups with new ones that contain new bearings, unless you have access to the specialty tool required to pull the bearings out of external-bearing cups. SRAM, FSA, and others offer cups with ceramic bearings inside. Instructions for installing the new cups are in 11-8; removal is just the opposite, after removing the cranks as in 11-1a. But if you want to get the tool and upgrade the bearings inside external-bearing cups, the instructions are below.

You need a special tool to remove bearings from external-bearing cups without marring the cups and to install new ones ensuring complete bearing insertion and proper alignment. Enduro has a nice one made by Sonny's Bike Tools (Fig. 11.40), and Phil Wood also makes such a tool. The key to the external bottom-bracket bearing puller is a two-piece collet with a band around it. The instructions here are for the Enduro tool; the Phil Wood tool is slightly different. Bearing installation into the cups is the same for Shimano Hollowtech II, FSA MegaExo, and Race Face X-Type bottom brackets; an extra step is required for SRAM/Truvativ GXP cups.

1. **Remove the crankarms (11-1b).**

2. **Unscrew the cups from the frame with the proper tool (11-8, Fig. 11.19).** Remember, the drive-side cup will be left-hand threaded unless it is an Italian-threaded frame.

3. **Remove the bearing cover seals.** These are the seals covering the face of the outboard cup, not the seals on the cartridge bearings themselves. To remove a cover seal, slip a razor blade under the edge and pry it up (Fig. 11.38), possibly working around the outside with a thin screwdriver blade for FSA and Shimano cover seals that extend inside the bearing bore.

4. **Drop the collet into the bearing from the inboard side.**

5. **Expand the collet.** Push the rounded-nose cylindrical "collet expander" into the collet to spread it inside the bearing bore. The lips of the collet will catch the back of the inner bearing race. Many bearing cups have an internal shelf that would prevent you from being able to push the bearing out if you did not have this collet.

6. **Put the bearing cup's outboard end face down into the tool's cup holder.**

7. **Push the bearing out by applying pressure on the collet expander.** With the Enduro tool, you accomplish this by running a big bolt through it and tightening it. This often takes considerable force. And especially in the case of Truvativ/SRAM GXP, the bearing finally comes free with a loud pop.

8. **Determine bearing orientation.** When you install cartridge bearings, it is worth thinking ahead about maintenance. Even though ceramic balls cannot rust and are more than twice as hard as steel balls, the races are made of steel and can rust. Ceramic cartridge bearings, like all cartridge bearings, have bearing retainers that separate the balls, but the ones in ceramic bearings are always plastic and can be removed to disassemble, clean, polish, and repack the bearing as in 11-13. The retainers reduce friction by preventing neighboring balls, whose adjacent sides are turning in opposite directions, from rubbing against each other. A plastic bearing retainer will be asymmetrical, so when you remove the bearing seals (with a razor blade slipped under the edge), you'll see the balls with the peaks of the plastic retainer between them on one side and you'll only see the spine of the plastic retainer from the other side. The simplest maintenance, which is sufficient if done frequently enough (SRAM recommends a service interval of 100 hours for the ceramic bearings in its Red crankset), is to simply pry off the bearing cover, wipe the dirty grease as much as possible out of the bearing, repack it with new grease, and replace the bearing cover (11-13a). You can do this most effectively from the side of the bearing where you see the balls (rather than the spine of the bearing retainer), so if this will be your method of maintaining your expensive bearing, before you press the bearings in, ensure that the spine of the bearing retainer faces inboard. On the other hand, if you orient the bearing so that the bearing retainer's spine faces outward, you can completely disassemble the bearing without removing it from the cup; you'll be able to remove the retainer, balls, and inner race in order to do a complete bearing overhaul as in 11-13b without having to break out this Enduro or Phil Wood tool. This is a good thing, because each time you press a bearing in and out of the cup, you may stretch it slightly or scrape off some aluminum from its bore, so the press fit within the cup may become slightly loose. Plan ahead for the maintenance you intend to perform with the bearing orientation you choose. If the new bearing has a symmetrical retainer (i.e., a shiny steel one with bumps concealing the balls on both sides), so be it; you'll have to just do your best to clean and grease the bearings or replace them when the time comes.

9. **Place the new bearing on an insert that fits snugly in the bearing's ID.** One end of the insert is 24mm in diameter, and the other is 25mm. Stock bearings for FSA and Shimano have a 25mm ID; a thin plastic shim integral with the outer bearing cover brings the bore down to the 24mm diameter of the spindle. If you are simply replacing the bearings and reusing the stock seals, get 25mm ID bearings. Otherwise, Enduro's Shimano/FSA replacement kit

11.40 Pressing the new bearing into the external-bearing cup with an Enduro tool

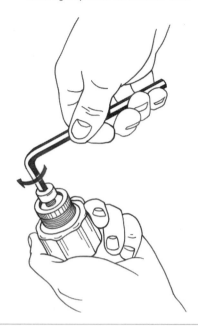

instead uses a 24mm ID bearing and a thin outer silicone cover seal instead of the stock one.

10. **Place the insert and bearing inside the bore of the tool's cup holder, facing upward.**

11. **Place the bearing cup over it and put a support ring atop it.**

12. **Run the bolt through and tighten it.** Tighten with an 8mm hex wrench until it hits a dead stop (Fig. 11.40).

13. **Check that the bearing is in as far as the old one was.** If not, flip the bearing cup over, with the insert still inside of the bearing bore, and put it back into the tool's cup holder with the cup's threaded section down inside. Reinstall the bolt and tighten it until it stops. This second pressing step is always necessary with Truvativ and SRAM GXP cranks (24mm ID bearings), because the replacement bearing (a standard size, 7mm thick) is 1mm narrower than the (proprietary) bearing employed by Truvativ.

14. **On a non-drive-side Truvativ/SRAM GXP bearing, install its sleeve.** GXP cranks require an 11.5mm-wide, 1mm-thick sleeve that stops the non-drive shoulder of the GXP spindle and establishes the side-to-side position of the crankset. Use the special GXP insert in the tool to install the sleeve.

15. **Install the cups as in 11-8 and the crankarms as in 11-1.** Spin them and smile.

b. Interchanging bearings on BB30

LEVEL With a BB30 crank (Fig. 11.17), you'll need a special bearing puller. Enduro has a removal/installation tool that works similarly to the tool described in 11-13a. As in 11-13a, you insert the collet and collet expander into the bearing and drive it out by tightening the tool's center bolt.

The following instructions are for Park Tool's BBT-39 T-shaped BB30 bearing puller (actually, it's a pusher-outer) (Fig. 1.4).

1. **Angle the BBT-39's T-end in through one bearing and push it straight in against the other bearing (Fig. 11.41).** Make sure it does not hit the snapring's eyes (Figs. 11.17, 11.33).

2. **Center the BBT-39 shaft with the dummy bearing insert in the near side.**

3. **Smack the handle with a hammer.** The bearing on the far side will pop out.

If you don't have the correct tool, you'll probably want to have a shop do this to avoid mauling the inside of the bottom-bracket shell, but if you're careful, you can reach in against the back side of the opposite bearing (not against the snapring!) with a rod or big hex wrench that you tap with a hammer. Work a little bit on each side of the bearing, moving the end of the rod or wrench from side to side and around the bearing as you tap it to slowly walk the bearing out.

Install the new bearings as in 11-9c.

11.41 Removing a BB30 bearing with a Park BBT-39 tool

c. Interchanging Press Fit bottom-bracket bearings

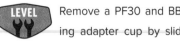

LEVEL Remove a PF30 and BB386 plastic bearing adapter cup by sliding in a "rocket" headset cup remover tool so it expands behind the bearing. Smack the bearing out just like you would a headset cup (see Figs. 12.37, 12.38; 12-20, step 4).

Remove PF24 (BB86) cups in like manner, but you'll need a smaller-diameter cup-remover rocket. The diameter of a headset rocket is usually 1 inch (25.4mm), which is of course larger than the ID of a PF24 bearing, as it's intended for a 24mm spindle. The Park BBT-90.3 rocket tool is specifically designed for removing PF24 (BB86) bearing adapters.

You will mangle the plastic adapters if you try to remove the bearings from them. Interchanging bearings means buying and installing a new bottom bracket, whether it's PF24 (BB86), PF30, or BB386. The molded plastic adapters are inexpensive.

Install the new bearings as in 11-9a.

d. Interchanging bearings on BB90 or BB94

LEVEL With a BB90 (Fig. 11.18) or BB94 system, you can pull the bearings out with your finger. If need be, walk it out a bit at a time by placing a big hex key against it from the opposite side and gently tapping it with a hammer as you move the hex key tip around from side to side on the backside of the bearing.

Put the new bearings in as described in 11-9b.

e. Interchanging Campagnolo Ultra-Torque or Power Torque or Fulcrum Racing-Torq bearings

LEVEL For Campagnolo Ultra-Torque and Fulcrum Racing-Torq cranks (Fig. 11.8), you need a special puller like the Park CBP-3 to get the bearing off the spindle (Fig. 11.42B); for Campagnolo Power Torque cranks, you'll need the CBP-5 tool adapter set (Fig. 11.42A) as well as the CBP-3 puller. If you don't have such a puller, I recommend you take the cranks to a shop that does. You definitely do not want to pry with a screwdriver against a carbon crank to get a bearing off!

11.42A Park CBP-3 puller and CBP-5 tool set for removing and replacing bearings on Ultra-Torque, Power Torque, and Racing-Torq cranksets

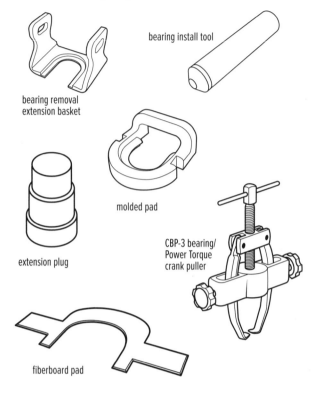

bearing install tool

bearing removal extension basket

molded pad

extension plug

CBP-3 bearing/ Power Torque crank puller

fiberboard pad

11.42B Pulling the bearing off of a Campagnolo Power Torque crank spindle

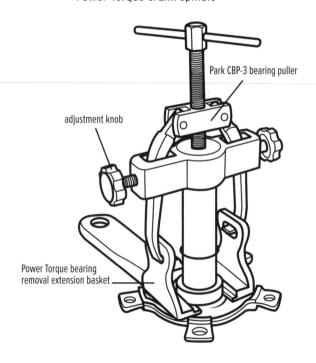

Park CBP-3 bearing puller

adjustment knob

Power Torque bearing removal extension basket

Removal

1. **Remove the crankarms as in 11-1c or d, depending on type.**

2. **Install the bearing puller.** On Ultra-Torque and Racing-Torq, hook the fingers of the CBP-3 puller under the edge of the bearing and tighten the two side knobs to remove play from the puller's fingers so they can't slip off.

   On Power Torque, first slip the CBP-5 steel extension basket under the bearing to extend the reach of the CBP-3 puller sufficiently to reach over the PT crank's spindle. Hook the CBP-3's fingers into the slots at the top edges of the basket (Fig. 11.42B) and tighten the two side knobs to remove play from the puller's fingers so they can't slip off. On a compact PT crank (with a 34- or 36-tooth inner chainring), you must first remove the chainrings in order to get the extension basket under the bearing. You must also break off the plastic tab in the Torx hole of the bolt behind the crankarm in order to insert the Torx key.

3. **Tighten the push bolt of the CBP-3 puller clockwise until the bearing pops off.**

4. **On Campagnolo Power Torque cranks, remove the left bearing.** Get it out of its cup (and install the new one) using the method in 11-14a, or replace the cup and bearing.

Installation

1. **Replace the bearing seal on the spindle.**

2. **Slide the new bearing on the spindle as far as you can by hand.**

3. **Tap the bearing into place.** Use the CBP-3 or CBP-5 bearing setter and a hammer; on Power Torque, you'll need the longer CBP-5 setter. Lacking that tool, set the old bearing on top of it, and tap it down with a tube that just fits over the spindle. Pull off the old bearing.

4. **Slide on the snapring and push it into its groove.**

5. **Reinstall the crankarms.**

OVERHAULING CARTRIDGE BOTTOM BRACKETS

Cartridge bottom brackets with a spindle integrated into them (Figs. 11.21, 11.25) are sealed units and cannot be overhauled. They must be replaced when they stop performing properly. Remove the cranks as in 11-1c. Remove the bottom bracket by unscrewing the cups with the splined cup tool (Fig. 11.34), and install a new bottom bracket as directed in 11-10.

OVERHAULING CUP-AND-CONE BOTTOM BRACKETS

Cup-and-cone bottom brackets (Fig. 11.22) can be overhauled entirely from the non-drive side, after you have removed the crankarms as described in 11-1c.

1. **Remove the lockring with the lockring spanner.** Use the tool as in Figure 11.37, except the lockring spanner and its rotation direction will be reversed.

2. **Remove the adjustable cup with the correct tool.** This is usually a pin spanner (Fig. 1.2), installed into the cup as in Figure 11.37.

3. **Check that the fixed cup is tight.** Put a fixed-cup wrench on it and try to tighten it (counterclockwise for English thread, clockwise for Italian; Fig. 11.35).

4. **Clean the cups and axle with a rag.** There should be no need for a solvent unless the parts are glazed with hardened grease.

5. **Clean the bearings with a citrus-based solvent.** Don't remove them from their retainer cages. A simple way to clean them is to drop the bearings in a plastic bottle, fill it with solvent, cap it, and shake it. A toothbrush may be required afterward, and a solvent tank is certainly handy if you have access to one. If the bearings are not shiny and in perfect shape, replace them. Balls with dull luster and/or rough spots or rust on them should be replaced.

6. **Wash the bearings in soap and water.** This will remove the solvent and any remaining grit. Towel them off thoroughly and then let them dry completely. An air compressor is handy here.

7. **Follow the installation procedure described in 11-11, starting with step 2.**

8. **Install the crankarms as in 11-2d, Figure 11.4.**

11-17

OVERHAULING OTHER TYPES OF BOTTOM BRACKETS

If any cartridge-bearing bottom bracket becomes difficult to turn, the bearing seals must be removed (Fig. 11.39) and the bearings scrubbed and flushed with solvent, dried, and regreased. If that doesn't fix the problem, the bearings must be replaced. If they are pressed into cups, you may also have to buy new cups, if you can find them.

1. **Reverse the installation procedure outlined in 11-12a to remove the bottom bracket.**

2. **Replace the bearings.**

3. **Reinstall the bottom bracket (11-12a) and crankarms (11-2).**

CYCLOCROSS CRANK SETUPS

For cyclocross, you don't need excessively large or small chainrings; sizes right in the middle are perfect. It is faster to run up short, steep, muddy dirt hills than to ride them in a superlow gear, so there is no need for a triple crank or even for a compact double with a 34-tooth inner chainring (some cyclocross bikes have compact double cranks with 34–50 chainrings, but a 36–46 compact option is preferable for racing). It is advisable to avoid a triple, since the long rear derailleur required would be a liability; it could easily get caught on shrubbery or be torn off in a crash. A compact double can save money; you can buy a complete crankset with a 34–50 chainring combination, rather than buying separate chainrings in addition to the crank. The compact's 50-tooth outer chainring is a better choice than the 53-tooth of a standard double; the downside is that the jump in size between chainrings is so big that

double shifts are common (shifting in the rear whenever you shift in the front to compensate for a big change in gear ratio with a front shift).

Cyclocrossers are split, however, between those favoring two chainrings and those favoring a single ring. A single chainring eliminates the possibility of missed front shifts in the adverse conditions that define cyclocross. It also saves some weight and eliminates another place that mud can collect, namely, on the front derailleur. And having 10 or 11 rear cogs ranging from 11 teeth to 25, 28, or 32 teeth (SRAM even has a 10 × 42-tooth cassette option, with a different freehub body) means that you still have plenty of gear choices with a single chainring.

Not all bikes have chainstay clearance for double chainring guards for a single chainring. An inner stop (a.k.a. "chain minder"; 5-45, Fig. 5.54) with an outer chainring guard is a way to run a single ring successfully if there is no room for an inner guard. Better yet, since SRAM introduced 1 × 11 systems with its X-Horizon rear derailleur coupled with its X-Sync chainring (Fig. 11.43), single-ring setups have become simpler and have no clearance issues. The X-Sync chainring has taller teeth than a chainring that is intended for shifting, and its teeth alternate fat-thin-fat to fit the wide-narrow-wide spaces in the chain (Fig. 11.43). These teeth on this stiff chainring, when combined with the tight clutch in the rear derailleur that prevents the chain from bouncing, retain the chain at least as well as double chainring guards do, without the hassle of assembling them on the crank and without chainstay clearance issues.

A downside of a single chainring is that if the chain does come off—which it almost never does if set up properly—you have to stop to put it back on, since without a front derailleur, you can't gently shift it back on. Furthermore, with chainguards, the chain can jam between the ring and a guard or even under the chain minder (chain stop); getting it back out can cost you a bunch of time and places to other riders. The poor chainline (5-46) to the big cogs from a single chainring in the big-ring position, combined with a chainring with shift-aiding ramps and tooth shapes, can lead to chain derailment.

Take your pick of downsides: complexity, weight, a cable to gum up, shifting to mess up, and the mud-catching topography of two chainrings and a front derailleur, or fewer gear choices with a single chainring and a remote possibility with a chainguard system of a firmly jammed chain.

SETTING UP A SINGLE-CHAINRING CYCLOCROSS CRANK

A standard cyclocross single chainring usually is no smaller than 38 teeth and no larger than 44 teeth.

The simplest and cleanest way to set up a "one-by" (single chainring) system is with SRAM's 1 × 11 system, which requires no chain guards (*section a*). You can also run the SRAM X-Horizon rear derailleur (which has the critical anti-chain-slap clutch) with a SRAM 10-speed shifter and a SRAM or Shimano 10-speed cassette.

Without that derailleur, though, you'll need chain guards of some sort. With the guards, you can use a standard chainring or an oval one (a "one-by" is a great application for a Rotor or other "shaped" chainring, because it doesn't need to shift between chainrings, which is the bane of oval chainrings).

You can surround the chainring with two chainring guards or with an outer chainring guard and an inner stop(s). If you have sufficient chainstay clearance for the chainring to be on the inner-ring position (best chain-line) and the inner chainring guard inboard of that, follow the steps in *section b* (Fig. 11.44). If clearance is lacking for the setup in *b*, you can still perhaps run dual chainring guards, but the inner guard would be on the inner-ring position (*section c*). If the bike still won't clear for the inner chainring guard, or if the chain derails when on the largest cog, you can put the chainring on the inner-ring position and run an outer chainring guard, and then add an inner stop(s) clamped to the seat tube.

Without chainring guards

a. With SRAM 1 × 11 system

To retain the chain without chain guards, SRAM's X-Sync chainring for 1 × 11 has taller teeth than a shifting chainring would, and every other tooth is fatter than a normal tooth to engage the wide space between outer link plates of the chain on every other link (see center illustration, Figure 11.43). For proper chain retention, it also requires the use of the SRAM X-Horizon rear derailleur, which has a clutch to prevent chain slap.

You can use this derailleur and chainring with a SRAM 10-speed shifter, a 10- or 11-speed chain, and a SRAM or Shimano 10-speed cassette.

1. **Install the X-Sync chainring.** In addition to SRAM, there are aftermarket manufacturers who offer these rings with tall, fat-thin-fat-thin teeth. Install

11.43 SRAM 1 × 11 X-Sync chainring and X-Horizon rear derailleur, and a detail view of the X-Sync tooth profile

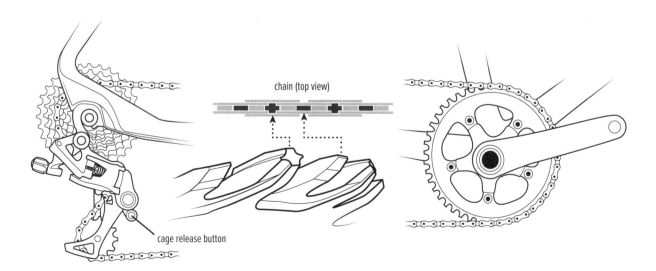

chain (top view)

cage release button

the ring in the outer-chainring position with short chainring bolts and nuts.

2. **Install and adjust the X-Horizon rear derailleur (5-2, 5-3).**

3. **Set the chain length as in 4-8.** Use method 3, but add one and a half more links of overlap for the SRAM X-Horizon rear derailleur, which is labeled Force 1 (formerly Force CX1) or Rival 1. You'll be connecting the chain with a SRAM master link, and the overlap including it will be 2.5 links.

That's it! Simple. The derailleur's clutch prevents the chain from jumping, and the optimized chainline with the tall, fat-thin-fat teeth engaging the wide-narrow-wide spaces in the chain helps keeps the chain on.

With two chainring guards

Gather the chainring, two chainring guards of the right size for the chainring, a set of extra-long chainring bolts (11 or 12mm), and five 2mm chainring spacers. The chainring is sandwiched between two toothless chainring guards just 3mm or 4mm larger in diameter than the chainring; they must be larger than the chainring or they won't keep the chain on, which is the whole idea!

Installing these parts is easier with the crankarm off and lying flat, so follow the applicable instructions in 11-1 to remove the right crankarm.

b. With two chainring guards and the chainring on the inner-chainring position

If you have enough room between the crank and the right chainstay, the chainline (5-46) is best by having the chainring at the inner-ring position (Fig. 11.44), making it less likely to derail the chain in low gear. However, many bikes can't fit the inner guard. Instead, you may have to either put the chainring on the outer chainring spider-arm shelves (*section c*) or replace the inner chainring guard with inner stops on the seat tube (see section on one chainring guard).

1. **Install the outer chain guard.** Put it in the large-chainring position on the spider arms.

2. **Insert the chainring nuts in the chainring guard.** Tape them in from the outboard side to hold them in place for assembly. Flip the crankarm over.

3. **Put the chainring on the small-chainring position on the inboard side of the spider arms.**

4. **Set a spacer on the chainring surrounding each bolt hole.**

11.44 Bolting together a single-chainring cyclocross setup: 2 chainring guards, 5 spacers, chainring, long bolts and nuts

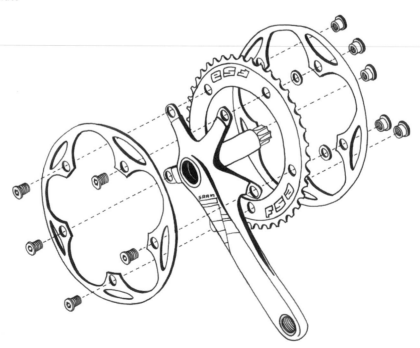

5. **Set the outer chainring guard on top of the spacers.**

6. **Screw the bolts through the outer guard, spacers, spider arm tabs, and inner guard.** Tighten them as in 11-7.

7. **Install the crankarm and make sure that the inner guard does not drag on the chainstay.** If it rubs, you have four choices:

- You can try a smaller chainring and chainring guards.
- You can remove the inner chainring guard and install a seat-tube-mounted inner stop (or two or three) as outlined for one chainring guard.
- You can use the two chainring guards, but move the chainring to the outer-ring position as in *section c*.
- If you have a three-piece (i.e., square-taper, ISIS, or Shimano Octalink) crankset, you can remove the bottom bracket, put a thin spacer between the right cup and the bottom-bracket shell, and reinstall it.

c. With two chainring guards and the chainring on the outer-chainring position

Unlike Figure 11.44, the chainring will be on the outer chainring spider-arm shelves, and the spacers will go between it and the outer chainring guard. The inner chainring guard will go in the inner-chainring position on the crank. Because 1) there is some flex to carbon-fiber chainring guards, 2) you cannot bring the inner chainring guard any closer to the chainring than the thickness of the crank spider tabs, and 3) the chainline often is too wide on the big-ring position, the chain can try to climb off of the chainring to the inside when in low gear. If the inner chainring guard is flexible enough or far enough away from the chainring to allow that, the chain can derail, so check for this before finding yourself in the middle of a race with the chain off.

1. **Flip the crank over so its backside is up.**

2. **Install the inner chain guard.** Put it in the small chainring position on the inboard side of the spider arms.

3. **Insert the chainring nuts in the chainring guard.** Tape them in from the inboard side to hold them in place for assembly. Flip the crankarm over.

4. **Put the chainring on the big chainring position on the outboard side of the spider arms.**

5. **Set a spacer on the chainring surrounding each bolt hole.**

6. **Set the outer chainring guard on top of the spacers.**

7. **Screw the bolts through the outer guard, spacers, spider arm tabs, and inner guard.** Tighten them as in 11-7.

8. **Install the crankarm and make sure that the inner guard does not drag on the chainstay.** If there is a lot of available space, you can try instead setting it up as in *section b* (Fig. 11.44). If it rubs, you have three choices:

- You can try a smaller chainring and chainring guards.
- You can remove the inner chainring guard and install a seat-tube-mounted inner stop (or two, or three) as outlined below.
- If you have a three-piece crankset, you can remove the bottom bracket, put a thin spacer between the right cup and the bottom-bracket shell, and reinstall it.

With one chainring guard

The chainring guard will prevent the chain derailing to the outside, and an inner stop (such as a Third Eye Chain Watcher, an N-Gear Jump Stop, or two or three Deda Dog Fangs, one above the other) attached to the seat tube will keep the chain from jumping off to the inside (Fig. 5.54).

The best option here is to have the single chainring at the inner chainring position inboard of the spider arms. The chainline will generally be better on average through all of the gears (5-46, 5-47, Fig. 5.55) and will be less likely to try to jump off in low gear than with the single chainring in the big ring position on the outside of the spider arms.

If the chainring is small enough to clear the chainstay, installation of a single ring and guard follows the procedure for a pair of chainrings as in 11-7, with the chainring on the inner-ring position, and the chainring guard on the outer-ring position. Otherwise, the chainring will need to go in the big ring position, and you will need spacers to install the guard as in section c.

Clamp the inner stop around the seat tube so that its top edge is a couple of millimeters higher than the chain. Rotate it on the seat tube until it is so close that it almost touches the chain when the chain is on the largest rear cog. With smaller stops, like Deda Dog Fangs, put another one above and another one below, and rotate both so their teeth project outboard farther than the first one. This will make it very difficult for the chain to derail.

11-19

SETTING UP A TWO-CHAINRING CYCLOCROSS CRANK

The standard 'cross double chainring combinations have traditionally been 39–46 or 39–44, which offer sufficient high and low gears while making the gear ratios close enough that a front shift need not mandate a rear shift. Now, with compact cranks, a 36–46 is becoming a standard setup. A 46 × 11 and even a 46 × 12 high gear is usually tall enough for even the fastest pros. A recommended cogset would be a 12–25 or a 12–28, or on faster courses, an 11–28, 11–25, or even an 11–23 might be useful.

Installing the chainrings is explained in 11-7. Set up the chain length and front-derailleur position as in Chapters 4 and 5, noting the cable-routing tips in 5-25.

TROUBLESHOOTING CRANKS AND BOTTOM-BRACKET PROBLEMS

11-20

CREAKING NOISES

Mysterious creaking noises can drive you nuts. Just when you think you have your bike tuned to perfection, a little noise comes along to ruin your ride. What's worse is that these annoying little creaks, pops, and groans can be a bear to locate.

Pedaling-induced noises can originate from almost anything connected to the crankset, including movement of the cleats on your shoes, loose crankarms on the bottom-bracket axle, loose chainrings, and poorly adjusted pedal or bottom-bracket bearings. Of course, noise could also originate from seemingly unrelated components like the seat, seatpost, frame, wheels, or

handlebar. A front derailleur with a band clamp on an unpainted titanium or stainless steel frame can creak with each pedal stroke; grease under the band usually silences it.

Before spending hours overhauling the drivetrain, spend some time trying to isolate the source of the noise. Try different pedals, shoes, and wheels. Grease the faces of the front and rear dropouts and the wheel skewer end faces and clamping mechanism. Pedal out of the saddle, and pedal without flexing the handlebar. If the source of the creak turns out to be the saddle, seatpost, pedals, wheels, or handlebar, turn to the appropriate chapter for directions to correct the problem.

If the creaking is definitely in the crank area, here are some steps to resolve it:

1. **Check to make sure that the chainring bolts are tight, and tighten them if they are not (Fig. 11.12).**
2. **Make certain that the crankarm bolts are tight (Figs. 11.3, 11.4).** If they are not, the resulting movement between the crankarm and the bottom-bracket axle is a likely source of noise. If the crank is of a different brand than the bottom-bracket, check with the manufacturers or your local shop to make sure that they are recommended for use together. Incompatible cranks and axles will never properly join and are a potential problem area.
3. **Rusting can break the glue bond between a cartridge bottom bracket and one or both of its cups (Figs. 11.21, 11.25), allowing movement between cartridge and cup.** This movement can make creaking noises when pedaling. To quiet the noises, remove the cartridge, grease the inside of the cup(s) as well as the threads, and reinstall the bottom bracket.
4. **The bottom-bracket cups can move in the frame threads, causing creaking.** Remove the cups, grease the threads, reinstall them, and tighten them to the correct torque (Appendix E).
5. **The bottom bracket can creak owing to improper alignment or adjustment, lack of grease, cracked bearings, worn parts, or loose cups.** All of these things require adjustment or overhaul procedures, outlined in 11-9 through 11-12. Many external-bearing

designs (Fig. 11.19), as well as cup-and-cone bottom brackets (Fig. 11.20), are very sensitive to being out of parallel, and creaking can occur if the bottom-bracket shell is not perfectly tapped and faced. This is a job for a good bike shop.

6. **Check to make sure the front-derailleur clamp is tight.** The noise from a loose clamp while pedaling, especially under heavy load, can seem to emanate from the crankset. Thick grease under the derailleur's band clamp can sometimes eliminate creaking.

7. **Now for the bad news.** If creaking persists, the problem could be rooted in the frame. Creaks can originate from cracks in and around the bottom-bracket shell. Or the threads in the bottom-bracket shell could be so worn that they allow the cups to move slightly. Neither of these is a good sign—unless, of course, you were hoping for an excuse to buy a new frame.

11-21

CLUNKING NOISES

1. **Crankarm play: Grab the crankarm and push on it side to side.**

   a. If there is play, tighten the crankarm bolt (Figs. 11.3, 11.4; torque spec is in Appendix E).

   b. If there is still crankarm play, and you have a cup-and-cone bottom bracket (Figs. 11.20, 11.22, 11.24) or a cartridge-bearing bottom bracket with a lockring on each side (Fig. 11.23), adjust the bottom-bracket axle-end play (11-11, steps 11–15).

   c. If bottom-bracket adjustment does not eliminate crankarm play, or you have a nonadjustable cartridge bottom bracket (Figs. 11.21, 11.25), the bottom bracket is loose in the frame threads. With a cup-and-cone bottom bracket, you can go back to 11-8 and start over, making sure that the fixed cup is very tight. Adjustable-cup lockrings also need to be tight (Fig. 11.37), once the axle-end play is adjusted properly.

   d. The lockring and the fixed-cup flange must be flush against the bottom-bracket shell all the way around (Fig. 11.20); if they are not, the bottom bracket must be removed, and the bottom-bracket shell must be tapped (threaded) and faced (cut parallel) by a shop equipped with the tools.

   e. If the crankarm play persists and the crankarm won't stay tight, the square hole is damaged due to riding it while insufficiently tight. A new crankset is in order.

   f. If the bottom-bracket fixed cup or lockring will not tighten completely, then either the bottom-bracket cups are stripped or undersized, or the frame's bottom-bracket shell threads are stripped or oversized. Either way, it's an expensive fix, especially the frame replacement option! Get a second opinion if you reach this point. If your bike has a square-taper crank and you can find a Mavic or Stronglight cartridge-bearing bottom bracket (11-12b, Fig. 11.26), you can still use a frame with stripped threads.

2. **Pedal-end play: Grab each pedal and wobble it to check for play.** If you find pedal-axle end-play, see the section on overhauling pedals in Chapter 13.

11-22

HARD-TO-TURN CRANKS

If the cranks are hard to turn, you need to overhaul the bottom bracket (see 11-13 through 11-17), unless you want to continue intensifying your workout or boost the egos of your cycling companions. The bottom bracket may be shot and need to be replaced.

11-23

INNER CHAINRING DRAGS ON CHAINSTAY

If the inner chainring drags on the chainstay, the bottom-bracket axle may be too short or the square hole in the crankarm may be so deformed that the crank slides on too far. If the inner chainring is too large, get a smaller one. A misaligned frame, with either bent chainstays or a twisted bottom-bracket shell, can cause chainring rub as well. A badly misaligned frame needs to be replaced.

With an adjustable cartridge-bearing bottom bracket (Fig. 11.23) with a lockring on each end, it is possible to fix the problem by offsetting the entire bottom bracket to the right (Fig. 11.37). If the bottom bracket axle is too short, replace it with one of the correct length. If the square hole in the crank is badly deformed, replace the crankarm. There's no other cure; it will continue to loosen up and cause problems otherwise.

NOTE: *See 5-46 through 5-47 and Figure 5.55 concerning the chainline to establish proper crank-to-frame spacing.*

STEMS, HANDLEBARS, AND HEADSETS

TOOLS

4mm, 5mm, 6mm
  hex keys
Hammer
Screwdriver
Hacksaw
Flat file
Round file
Electrical tape
Grease
Citrus solvent

OPTIONAL

32mm headset wrenches
  (two)
Star-nut installation tool
Threadless saw guide
Carbon-specific hacksaw
  blade
Carbon assembly paste
  or spray
Slip-joint pliers
Securely mounted vise
Crown-race slide punch
Crown-race remover
Headset press
Headset cup remover
Head tube reamer
Crown-race facer

On a bike, you maintain or change your direction by applying force to the handlebar. If everything works properly, variations in that pressure will result in your front wheel changing direction. Pretty basic, right? The interconnected parts between the handlebar and the wheel make that simple process possible. The parts of the steering system are illustrated in Figures 12.1 and 12.2. In this chapter, we'll cover most of that system by going over stems, handlebars, and headsets.

STEMS

The stem connects to the fork's steering tube (which is generally either 1 inch or 1⅛ inches in diameter at its top) and clamps around the handlebar, which usually has one of two standard diameters: 26.0mm or 31.8mm (Cinelli handlebars used to be 26.4mm, some 26.0mm handlebars call themselves 25.8mm, and many low-end handlebars have a 25.4mm clamp diameter). Stems come in one of two basic types: for threadless fork steering tubes (Figs. 12.2–12.5) or for threaded ones (Figs. 12.1, 12.6–12.8).

Fork steering tubes on most modern high-end road bikes are 1⅛ inch in diameter at the top, although many of them no longer maintain that diameter over the entire length of the steering tube; many forks now have a steering tube that tapers to 1⅛ inch from a larger diameter at the base (i.e., at the top of the fork crown), where the stress is highest. In the 1990s, 1-inch-diameter threadless steering tubes were the norm on road forks, and prior to that, there was a century of road bikes with 1-inch-diameter threaded fork steering tubes.

Stems for unthreaded steering tubes (Fig. 12.3) have a clamping collar to grip the tube. Because the steering tube has no threads, the top headset cup merely slides on and off when the clamping collar is loosened. In this case, the stem plays a dual role. It clamps around the steering tube to connect the handlebar to the fork, and it also keeps the headset in proper adjustment by preventing the top headset cup from sliding up the steering tube (Figs. 12.4, 12.5). If you have a 1-inch-diameter threadless steering tube (the old standard) and a stem for a 1⅛-inch threadless steering tube (the current standard), you can get a slotted aluminum reduction bushing

12.1 The components of the steering system with a threaded fork

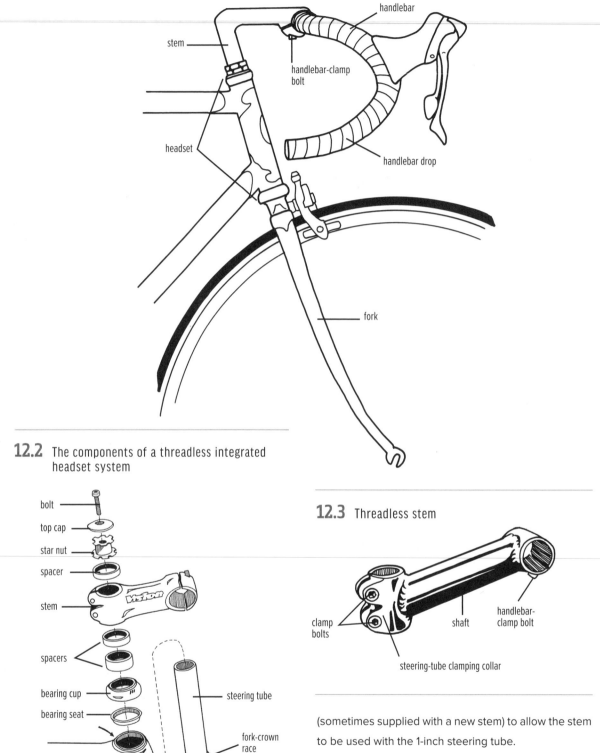

stem

handlebar

handlebar-clamp
bolt

headset

handlebar drop

fork

12.2 The components of a threadless integrated
headset system

bolt

top cap

star nut

spacer

stem

spacers

bearing cup

bearing seat

steering tube

fork-crown
race

12.3 Threadless stem

clamp
bolts

shaft

handlebar-
clamp bolt

steering-tube clamping collar

(sometimes supplied with a new stem) to allow the stem
to be used with the 1-inch steering tube.

On most bikes made before 1990, the steering
tube on the fork has external threads at the top, and the
headset screws onto it for attachment and adjustment.
Stems for threaded steering tubes (Figs. 12.6–12.8) have
a "quill" that extends into the steering tube of the fork

12.4 Threadless headset and stem cutaway

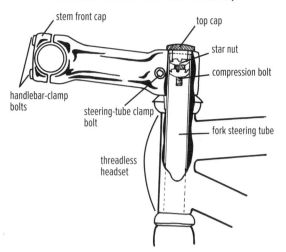

stem front cap

top cap

star nut

compression bolt

handlebar-clamp bolts

steering-tube clamp bolt

fork steering tube

threadless headset

and a shaft, or extension, that connects to the handlebar clamp. The stem binds to the inside of the steering tube by means of a conical expander plug (Fig. 12.6) or angularly truncated cylindrical wedge (Fig. 12.7) pulled up by a long stem-expander bolt that runs through the quill (Fig. 12.8).

The shaft of a traditional road bike stem extends out at an angle of about 73 degrees (negative 17 degrees) from the fork steering tube so that, when installed on the bike, the shaft is horizontal (Figs. 12.1, 12.8). Stems on track sprint bikes historically tended to be angled downward when mounted on the bike. Stems with zero- to 10-degree angles, resulting in an upward angle

12.5 Threadless headset cup held in place by stem

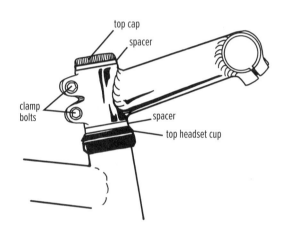

top cap

spacer

clamp bolts

spacer

top headset cup

12.6 Forged aluminum quill road stem with expander plug

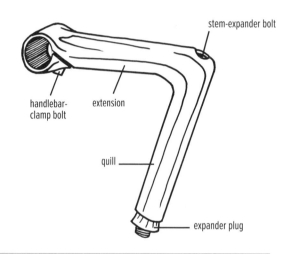

stem-expander bolt

handlebar-clamp bolt

extension

quill

expander plug

12.7 Welded quill-type stem with expander wedge

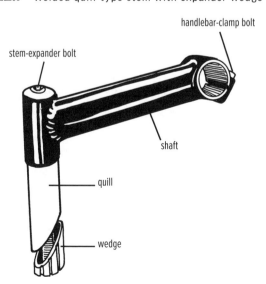

handlebar-clamp bolt

stem-expander bolt

shaft

quill

wedge

12.8 Threaded headset system cutaway: note the expander plug securing the stem inside the steering tube.

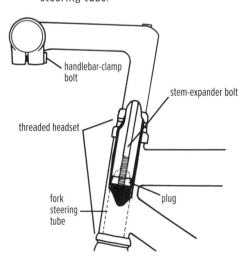

handlebar-clamp bolt

stem-expander bolt

threaded headset

fork steering tube

plug

12.9 Loosening and tightening the compression bolt on a threadless headset

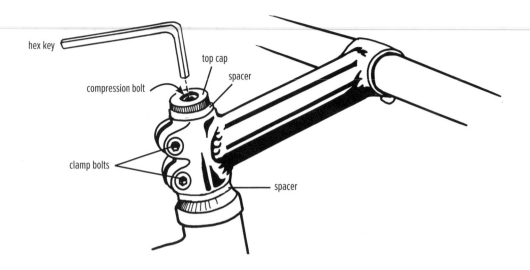

hex key

top cap

spacer

compression bolt

clamp bolts

spacer

on the assembled bicycle (Figs. 12.2–12.5, 12.7, 12.9), are becoming commonplace on road bikes and even track bikes.

12-1

REMOVING STEM FROM THREADLESS STEERING TUBE

LEVEL 1. **Loosen the horizontal clamp bolt(s) (Fig. 12.5) securing the stem around the steering tube.**

2. **Unscrew the compression bolt.** With a hex key—usually 5mm, sometimes 4mm—unscrew and remove the compression bolt (or "adjusting bolt"

because it compresses the headset into the proper bearing adjustment) in the headset top cap (Fig. 12.9). The fork can now fall out, so hold the fork as you unscrew the bolt.

NOTE: *Without a top cap, as soon as you loosen the stem, the fork can slip out.*

3. **Remove the cap and stem.** With the bike standing on the floor, or while holding the fork to keep it from falling out, pull the cap and the stem off the steering tube. Leave the bike standing until you replace the stem, or slide the fork out of the frame, keeping track of all headset parts.

4. **If the stem is stuck to the steering tube and will not budge, see 12-6a.**

PRO TIP — SPACERS WITH CARBON STEERING TUBES

IF THE FORK HAS A carbon steering tube, always place at least one spacer above the stem (Figs. 12.5, 12.9). That way, the entire stem clamp is clamped onto the steerer, and there is no chance for the upper part of the clamp to pinch the end of the steerer. This is a good idea for a steel or aluminum steering tube as well.

If you want to raise the handlebar up high, be careful about using too many spacers below the

stem; consult the owner's manual for the fork for recommendations on maximum spacer stack height. From a strength and stiffness perspective, it's preferable to use an up-angled stem, rather than a level or down-angled one with a lot of spacers below it. And, of course, make sure the support plug inside the steering tube (which prevents the stem clamp from crushing the carbon steering tube) is supporting the area under the stem clamp.

12-2

INSTALLING AND ADJUSTING HEIGHT OF STEM ON THREADLESS STEERING TUBE

 Adjusting the height of a stem on a thread-less fork is more complicated than adjusting the height of a standard stem in a threaded fork, because the stem is integral to the operation of the headset. As you can see from Figures 12.4 and 12.5, any change to the stem height will alter the headset adjustment. That's why this step is listed with a level 2 designation.

1. **Stand the bike on its wheels, so that the fork does not fall out.**

2. **Lubricate the parts.** Grease the top end of the steering tube if it is steel or aluminum, but leave it dry if it is carbon fiber (or apply carbon assembly paste or spray). Loosen the stem-clamp bolts and grease their threads. Slide the stem onto the steering tube.

3. **Set the stem height to the desired level.** If you want to place the stem in a position higher than directly on top of the headset, you must put some spacers between the bottom of the stem clamp and the top piece of the headset (Fig. 12.9). No matter what, there must be contact (either directly or through spacers) between the headset and the stem. Otherwise, the headset will be loose.

4. **Check the steering tube length.** In order to adjust the threadless headset, the top of the stem clamp (or, ideally, spacers placed above it; see the Pro Tip on spacers) should overlap the top of the steering tube by 3–5mm (⅛–³⁄₁₆ inch) (Fig. 12.10). If it does, skip ahead to step 7.

NOTE ON 1-INCH STEERERS: *Most stems now have a 1⅛-inch clamp size. Simple split shim sleeves (short pieces of tubing slotted down one side) are widely available to adapt a 1⅛-inch clamp stem to a 1-inch steering tube. Slide the sleeve over the steering tube and slide the stem over it. With this type of stem and shim on a 1-inch steering tube, you can usually use spacers under the stem sized for a 1-inch steering tube, as long as they are wide enough to contact the entire bottom edge*

12.10 Minimum steering tube length

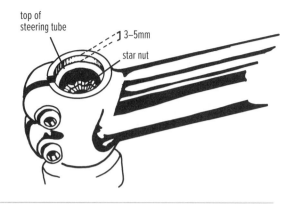

of the stem. However, above the stem, you may need to use a spacer and a headset top cap meant for a 1⅛-inch steering tube, in order to push the stem down properly and to aesthetically match the top of the stem.

5. **Check overlap.** If the top of the stem clamp overlaps the top of the steering tube by more than 5mm (³⁄₁₆ inch), the steering tube is too short to set the stem height where you have it. If you have spacers below the stem, remove some until the top edge of the stem clamp overlaps the top of the steering tube by a maximum of 5mm (Fig. 12.10). Ideally, have the steerer come all the way through the stem clamp, and put a spacer above it; see the Pro Tip. If you cannot or do not wish to lower the stem any farther, you will need a fork with a longer steering tube, a stem with a shorter clamp, or a stem that is angled farther upward to achieve the desired handlebar height. Replacing the stem is a lot cheaper and easier than replacing the fork.

6. **If the steering tube is too long, there are several steps you can take.**

   a. If the top of the steering tube is less than 3mm (⅛ inch) below the top spacer or the top edge of the stem clamp, or if the steering tube sticks up above the top of the stem clamp by a small amount, put a headset spacer or two on top of the stem clamp until the top spacer overlaps the top edge of the steering tube by 5mm.

   b. If, on the other hand, you are sure you will never want the stem any higher, you can cut off the excess tube. First, mark the steering tube along the top edge of the spacer above the stem clamp (do not cut at the top of the stem clamp itself or lower; see the Pro Tip on cutting carbon steering tubes). Remove the fork from the bike. Wrap a piece of tape around the steering tube 3mm below the mark. Place the steering tube in a padded vise or bike-stand clamp. By using the edge of the tape as a guide for cutting straight, cut the excess steering tube off with a hacksaw.

c. In steel and aluminum steering tubes that have already been installed in a bike, there is a star nut that is inserted inside the steering tube (Fig. 12.4). You screw the compression bolt through the top cap to adjust the headset bearings (but not to retain the headset; the stem-clamp bolts do that). If the star nut is already inside the steering tube and it looks like the saw is going to hit it, you must move the star nut down before cutting. See step 7 for instructions on pushing the star nut in deeper. (Carbon fiber fork steering tubes have either a glued-in support insert with a star nut inside, or an expandable steering tube support insert with an integrated anchor for the top-cap bolt [Fig. 12.27]; in either case, the insert must be removed before cutting the steering tube.)

d. Make your cut straight. Measure twice, cut once! Mark it straight by wrapping a piece of tape around the steering tube and cutting along it. If you are not sure the cut will be straight, start it a little higher and file it down flat to the tape line. If you really want to be safe, use a tool specifically designed to help you make a straight cut; Park Tool's "threadless saw guide" will do the trick (you need to put some spacers between the plates of this tool if you are using the fatter CarboCut saw blade). Remember that you can always shorten the steering tube, but you cannot make it longer! After making your cut, use a round file on the inside of the tube and a flat file on the outside to remove any metal burrs left by the hacksaw or cutter.

e. When you have completed cutting and deburring, put the fork back in, replacing all headset parts the way they were originally installed (12-14). Return to step 1 in this section.

7. **Check top cap clearance.** Check that the top of the star-shaped nut or steerer support plug does not hit the bottom of the headset top cap once the adjusting bolt is tightened. Generally, you can fix this by putting another spacer (or a thicker one) above the stem. If the nut is not in deeply enough in a steel or aluminum steerer, you can drive it deeper into the steering tube after removing the stem. In the case of a carbon steering tube with an expander plug inside, you can move it down by loosening its bolt with a hex key, tapping it in deeper, and retightening it. An expander plug or glue-in support insert is a must to prevent crushing the carbon steering tube with the stem clamp.

Metal steering tubes

a. Driving the star nut deeper into a metal steering tube is best done with a star nut installation tool (Fig. 1.3). The tool threads into the nut, and you hit it with a hammer until it stops; the star nut will now be set 15mm deep in the steering tube. If you do not have this tool, go to a bike shop and have the nut set for you. If you insist on doing it yourself, read step b. Just remember that it is easy to mangle the star nut if you do not tap it in straight.

b. Pushing the star nut in deeper without a star nut installation tool requires three steps: 1) Put the adjusting bolt through the top cap and thread it six turns into the star nut. 2) Set the star nut over the end of the steering tube and tap the top of the bolt with a mallet; use the top cap as a guide to keep it going in straight. 3) Tap the bolt in until the star nut is 15mm below the top of the steering tube.

NOTE: *If the wall thickness of the steering tube is greater than standard, the stock headset star nut will* not fit, and it will bend when you try to install it. Even pros sometimes ruin star nuts. It's not a big problem, because replacements can be purchased separately. If it goes in crooked, take a long punch or rod, set it on the star nut, and drive it to or all the way out of the bottom of the steering tube. Dispose of that star nut, and get another.

If the internal diameter (ID) of the steering tube is undersized (standard ID is 22.2mm [⅞ inch] on a 1-inch steering tube; 25.4mm [1 inch] on a 1⅛-inch steering tube; and 28.6mm [1⅛ inch] on a 1¼-inch steering tube), you cannot use the stock star nut from the headset for that size. Get a correctly sized star nut at a bike shop or from the fork manufacturer. In a pinch, you can make a big stock star nut fit by bending each pair of opposite leaves of the star nut toward each other with a pair of slip-joint pliers to reduce the nut's width. Now you can insert the nut; be aware that it may not grip as well as a properly sized one.

8. **Install the headset top cap on the top of the stem clamp (or spacers you set above it).** Grease the threads of the top-cap compression bolt and screw it into the star nut inside the steering tube with a 4mm or 5mm hex key (Fig. 12.9).

9. **Adjust the headset.** Tighten the top cap while keeping the stem lined up straight with the front wheel, and hold it in adjustment by tightening the stem-clamp bolts. The steps are outlined and possible complications are addressed in 12-16.

REMOVING QUILL-TYPE STEM FROM THREADED FORK

LEVEL 1. **Unscrew the stem-fixing bolt on the top of the stem about three turns or so.** Most stem bolts take a 6mm hex key. Some stems have a rubber plug on top that must be removed to get at the stem bolt.

2. **Tap the top of the bolt down with a mallet or hammer (Fig. 12.11).** This will disengage the plug or wedge from the bottom of the quill and will free the stem, as long as it's not rusted in place. If the head

12.11 Freeing the stem wedge

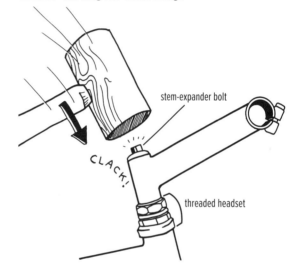

stem-expander bolt

CLACK!

threaded headset

PRO TIP — CUTTING CARBON STEERING TUBES

IF YOU ARE CUTTING a carbon fiber steering tube, cut three-quarters of the way through and then rotate the steerer 180 degrees and cut from the other side to meet your cut. This will prevent the peeling back of the last few layers of carbon at the bottom of a single cut.

You can use a standard hacksaw with a fine-tooth blade if you're careful, but better options are the Effetto Mariposa CarboCut and the Park Carbon saws. These saws have toothless, grit-edge tungsten-carbide blades for cutting carbon fiber (and Kevlar and boron fiber) without damaging the matrix surrounding the fibers. The special tungsten-carbide blades cut on both push and pull strokes and do not grab like a toothed blade. They leave a smooth cut edge that requires no sanding. Note, however, that you still need to cut through partway from each side to avoid peeling back carbon fibers as the blade exits the tube.

of the bolt is recessed down in the stem so that a hammer cannot get at it, leave the hex key in the bolt and tap the top of the hex key until the wedge is free.

3. **Pull the stem out of the steering tube.** If the stem will not budge, see 12-6b.

12-4

INSTALLING AND ADJUSTING HEIGHT OF QUILL STEM IN THREADED FORK

1. **Lubricate the parts.** Generously grease the stem quill, the expander-bolt threads, the outside of the wedge or conical plug, and the inside of the steering tube. If this is the first time you've done this, I know what you're thinking: "Why put grease on something that I want to wedge together?" Don't worry; the grease won't prevent the wedge from keeping the stem tight. Rather, it will prevent the parts from creaking, seizing, or rusting together.

2. **Assemble the parts.** Thread the expander bolt through the stem and into the wedge or plug until the bolt pulls the plug or wedge into place, but not so far as to prevent the stem from inserting into the steering tube.

3. **Install the stem.** Slip the stem quill into the steering tube (Fig. 12.8) to the depth you want. Make sure the stem is inserted beyond its height-limit line. Tighten the bolt until the stem is snug but can still be turned.

4. **Set the stem to the desired height, line it up with the front wheel, and tighten the bolt.** It needs to be tight, but don't tighten it so much that it bulges the steering tube.

12-5

STEM MAINTENANCE AND REPLACEMENT SCHEDULE

A bike cannot be controlled if the stem breaks, so make sure yours doesn't break. Aluminum has no fatigue endurance limit (the point below which a material can be stressed indefinitely), which means that any alumi-num part regularly stretched or flexed will eventually fail. Steel and titanium parts repeatedly stressed more than about one-half of their tensile strength will eventually fail as well. Carbon fiber has high fatigue resistance but is susceptible to breakage, particularly after an impact has damaged underlying fibers, even if that damage is invisible from the outside.

What this means is that stems made from any material may fail suddenly, causing a crash. Therefore, do not look at the stem on your bike as a permanent accessory. Replace it before it fails.

Clean the stem regularly. Whenever you clean it, look for corrosion, cracks, and bent or stressed areas. If you find any, replace the stem immediately. If you crash hard, especially hard enough to bend the handlebar, replace the stem and possibly the fork—and the bar, of course. Err on the side of caution.

Some stem makers recommend replacing stems every four years or less. If you rarely ride the bike, this is overkill. If you ride hard and ride often, every four years may not be frequent enough. Do what is appropriate for you, and be aware of the risks.

12-6

REMOVING A STUCK STEM

 A stem can get stuck onto (or into) the steering tube owing to poor maintenance. Regular maintenance involves periodically regreasing the stem and steering tube to enable the parts to slide freely when disassembled. The grease also forms a barrier to sweat and water.

If the stem is really stuck, be careful as you try to remove it; you can—quite easily—ruin the fork as well as the stem and the headset trying to get it out. In fact, you're better off having a shop work on it, unless you really know what you are doing and are willing to accept the risk of destroying a lot of expensive parts.

a. Removing a stuck stem from a threadless fork

1. **Remove the top cap (Fig. 12.9) and the bolts clamping the stem to the steering tube.**

2. **Thread the clamp bolts in from the other side.** Spread the stem clamp by inserting a coin into

12.12 Spreading the stem clamp to free a stuck stem

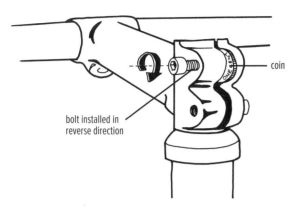

bolt installed in
reverse direction

coin

the slot between each bolt end and the opposing unthreaded half of the binder lug (Fig. 12.12). Tighten each bolt against each coin so that it spreads the clamp slot open wider. The stem should come right off the steering tube now.

NOTE: *If the stem is the type that comes with a single bolt in the side of the stem shaft ahead of the steering tube (Fig. 12.4), loosen the bolt a few turns and tap it in with a hammer to free the wedge. It might require some ammonia and perhaps some heat to expand it.*

If it still will not come free, you may have to use a vise (as long as it's not a carbon fork; see the Caution in the next section), following steps 6–7 in the next section on freeing a quill-type stem. Failing that, your last resort is to saw through the steering tube at the base of the stem clamp, and replace the stem and fork.

b. Removing a stuck stem from a threaded fork

1. **Unscrew the stem bolt on top of the stem three turns or more.** Smack the bolt (or the hex key in the bolt) with a mallet or hammer (Fig. 12.11) to disengage the wedge.

2. **Give it a twist.** Grasping the front wheel between your knees, make one last attempt to free the stem by twisting back and forth on the bar. Don't use all of your strength because you can ruin a fork and front wheel this way.

3. **If the stem didn't budge, squirt ammonia or Coca-Cola around the stem where it enters the headset.** Let the bike sit for several hours and add more ammonia every hour or so.

4. **Turn the bike over and squirt ammonia into the bottom of the fork steering tube.** You want the ammonia to run down around the stem quill. Let the bike sit for several hours and add more ammonia every hour or so.

5. **Try step 2 again.** If still stuck, repeat steps 3 and 4 but try penetrating oil as a last resort. Ammonia or Coca-Cola generally will dissolve aluminum oxide, but if one of these hasn't worked, there is a slim chance that penetrating oil might work.

6. **Shrink the parts.** If the previous steps haven't worked, remove the stem-expander bolt and discharge a tire inflator inside the stem quill to shrink it with cold. Now try step 2 again.

7. **Use a vise.** If the stem does not come free this time and you don't have a carbon fork, you'll have to go to your workbench and use that heavy-duty vise. It's solidly mounted, isn't it? Good, because it will need to be.

8. **Remove the front wheel (Chapter 2) and the front brake (Chapter 9).** Put pieces of wood on each side of the vise. Clamp the fork crown into the vise (Fig. 12.13).

12.13 Clamping the fork crown in vise

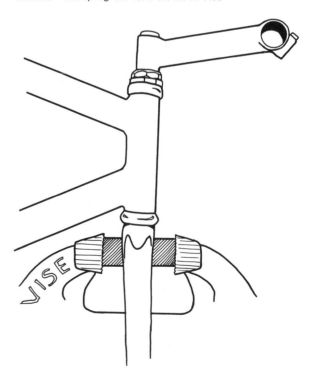

CAUTION: *Never clamp a carbon fork in a vise; you may damage it to the point that it could fail catastrophically while you are riding. Rather than risk that, get a new fork and stem if you can't remove this one (you can chop off the stem above the headset to get the fork out).*

9. **Grab both ends of the handlebar and twist back and forth.** Again, try discharging a tire inflator inside the stem quill to shrink it with cold. The stem will generally come free with a loud pop. If this doesn't work, you may have to saw off the stem just above the headset and have the bottom of the stem reamed out of the steering tube by a machine shop. In this case, unscrew the headset and remove the fork; don't take the bike to a machine shop as an assembly. I told you that you should have gone to a bike shop before getting to this point!

HANDLEBARS

LEVEL I am generally referring here to standard road bike drop bars (Fig. 12.1), but most of these comments also apply to the "cowhorn" style of handlebar common on time trial and triathlon bikes (Figs. i.2, 12.16A–B). Handlebar work is straightforward; all of the procedures that follow are level 1 jobs.

12-7

HANDLEBAR REMOVAL

a. From a stem with a removable front stem cap

Stems with a removable front stem cap are shown in Figures 12.2 and 12.4.

1. **Completely remove the bolts holding the front stem cap.** There may be two (Figs. 12.2, 12.4), three, or four bolts, depending on model.
2. **Pull off the stem cap, and the handlebar will drop off the stem.** Makes it easy, eh, having a removable front cap?

b. From a stem with a single handlebar-clamp bolt

Stems with a single handlebar-clamp bolt are shown in Figures 12.1, 12.3, and 12.5–12.9.

1. **Remove the handlebar tape (Fig. 12.14), at least from one side.**

12.14 Removing handlebar tape

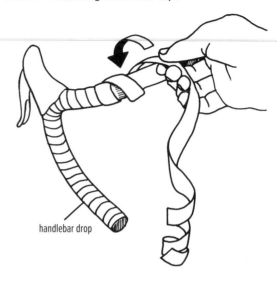

handlebar drop

2. **Remove the brake levers (see Chapter 9).**
3. **Loosen the bolt on the stem clamp surrounding the bar.** This usually takes a 5mm hex key.
4. **Pull the bar out, working the bend around through the stem.** If the bar won't budge, if it will budge but it appears that you will tear up the bar's finish working it out through the stem, or if the bend in the bar will not pass through the stem clamp, you need to open the stem clamp a bit more. On many stems, you can do this by removing the clamp bolt, threading it back in from the opposite side with a coin inserted into the clamp slot, and tightening the bolt against the coin to spread the clamp. (Opening a stem clamp in this way—but in this case the steering tube clamp—is illustrated in Fig. 12.12.) Otherwise, you can (carefully!) pry the opening wider with a large screwdriver.

12-8

HANDLEBAR INSTALLATION: DROP BAR

1. **Lubricate the parts.** Remove the handlebar stem's clamp bolt (or bolts), grease the threads, and replace it (or them). Grease the inside of the stem clamp and the clamping area in the center of the bar. Grease keeps the parts from seizing over time and prevents creaking noises from developing later. With a carbon fiber handlebar, leave it dry or

apply carbon assembly paste or spray (Fig. 14.6) on its clamping area and inside the stem clamp.

2. **Install the bar and rotate it to the position you find most comfortable.** The old-school way was to set a drop bar so that the bottom flat section (the "drop"; see Fig. 12.1) was horizontal. The style now tends to be with the drops aimed down and back toward the rear brake or the rear hub so that the top section extending forward is level, but the setting you choose is entirely a matter of personal preference.

3. **Tighten the bolt or bolts that clamp the bar.** Tighten it or them to the recommended torque (see the torque table in Appendix E). This step is particularly important with expensive, lightweight stems and bars. You can pinch and thereby weaken a lightweight handlebar by overtightening it so much that the high-strength tubing cracks right next to the stem. Light stems come with small bolts with fine threads, and overtightening can strip the threads inside the aluminum (or magnesium, etc.) stem. If you don't have a torque wrench and you have a lightweight stem with small bolts (e.g., M5

or M6 bolts, which take 4mm and 5mm hex keys, respectively), use a short hex key so that you can't get much leverage. Proper torque is even more important with carbon fiber handlebars.

Also, make sure that there is the same amount of space between the stem and the edge of the front cap on both the top and bottom of a front-opening stem. Any stem whose clamp gap or gaps get pinched nearly closed when tightened around the bar needs to be replaced, along with the handlebar.

12-9

INSTALLING A CLIP-ON AEROBAR ONTO A DROP BAR OR COWHORN BAR

Open the clamps that will attach the aerobar to the handlebar by removing the bolts with a hex key.

NOTE: *I will refer to the bike's handlebar when a clip-on is attached to it as the "base bar."*

Clip-ons (clip-on aerobars; Figs. 12.15, 12.16) generally mount on the bulge of the base bar, right next to the stem. If the clip-on you have chosen mounts on the

12.15 Installing clip-on aerobars onto a drop bar

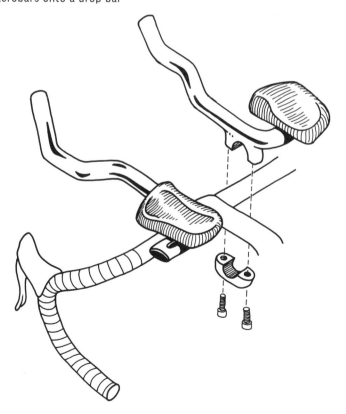

12.16A–B Installing clip-on aerobars onto cowhown bars

12.16A Aerobar above the base bar

12.16B Aerobar with fore-aft adjustability below the base bar

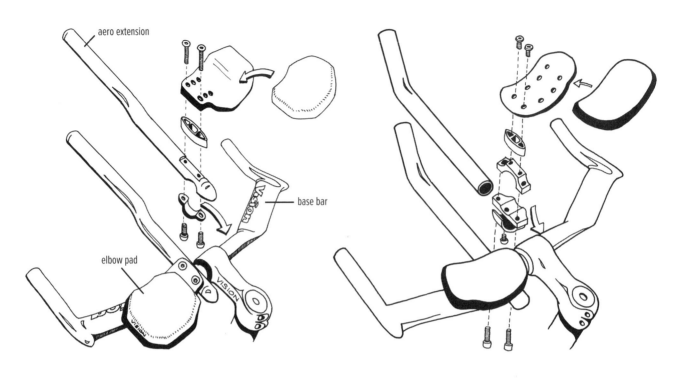

thinner-diameter section of the base bar, you will have to peel back some handlebar tape from the section adjacent to the bulge (Fig. 12.14).

CAUTION: *Do not put a clip-on bar onto a carbon handlebar unless its manufacturer specifically says that it is okay. Most carbon base bars are not rated for clip-on use; some can be downright dangerous if you attach a clip-on to them.*

For starters, set the clip-on bar level or angled upward slightly. Bolt the clip-on clamps around the base bar as shown in Figures 12.15 and 12.16. Some clip-ons mount atop the base bar (Figs. 12.15, 12.16A); others mount either above or below the base bar and allow fore-aft adjustment of the extensions (Fig. 12.16B).

Tighten the bolts enough that the clip-on bars won't slip when you hit a bump or pull on them, but be careful not to pinch or crush the base bar. See the torque table in Appendix E.

Set the elbow pads in a medium-width position. The pad is often held onto the elbow support with Velcro; pulling off the pad will reveal the adjusting bolt (Figs.

12.16A–B). Ideally, you want the elbow pad positioned under your elbow or slightly forward of it, and you want the clip-on to be of such a length that your hands grasp the ends comfortably with the elbows on the pads.

12-10

INSTALLATION OF COMPLETE AEROBAR: BASE BAR, AERO EXTENSIONS, AND ELBOW PADS

1. **Assemble the parts.** If the aerobar has an integrated stem, clamp it onto the steering tube and adjust its height and the headset as in 12-2 and 12-16. If not, first clamp the handlebar into the stem, then clamp the stem onto the steering tube, and adjust its height and the headset as in 12-2 or 12-4 and 12-16 or 12-17.

2. **Set the angle.** Adjust the angle of the base bar (if you have a separate bar clamped into a stem) to the position you like, and tighten the stem clamp.

3. **Install the aero extensions and the elbow pads.**

In many cases, the hardware holding the extensions is the same as that which holds the elbow pads (Figs. 12.16A–B). Some bars have hardware that allows you to run the extensions underneath the base bar (Fig. 12.16B) or on top of it; if you have the choice, try them both to see which you prefer. The elbow pads, of course, clamp above the base bar. You will also often have the choice of shape of the aero extensions, be they straight, single-bend, double-bend, S-bend, or the short, J-shaped bend of a Slam bar (Figs. 12.17A-E).

4. **Adjust the positions.** Set the stem height; adjust the elbow pad fore-aft, twist, tilt, and width and the aero extension width, twist, and reach to your liking and tighten all of the clamp bolts. If you will be using end plugs in the tail ends of the aero extensions, insert them before you clamp the extensions in place, or you won't be able to get them in with some bars, once they are in place.

5. **Trim to length.** If you have excess length on the extensions, pull them back out and cut them down to the right length. If they are carbon fiber, saw in first from one side and then from the other and meet in the middle to avoid fraying and peeling the carbon layers (see the Pro Tip on cutting a carbon steering tube). Clamp the extensions back in place when you're done.

6. **Install the brake levers and shifters.** Unless the bar has integrated brake levers, clamp the brake levers onto the ends of the base bar and tighten the bar-end shifters into the ends of the aero extensions.

7. **Install the cables and housings.** Route them through or along the bars and frame and to the derailleurs and brakes (see Chapters 5 and 9). If you are running the shift cables straight out of the tail ends of the extensions, make sure the housing loops entering the frame or cable stops don't hit your knees; if the loops must be long due to frame or bar configuration, you may be able to restrain them with a zip tie that holds them together.

8. **Tape the handlebars (see 12-12).**

9. **Check that all of the bolts are tight, and go ride your bike!**

12.17A-E Bend variations of aerobar extensions

12.17A Straight extensions

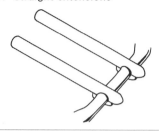

12.17B S-bend extensions

12.17C Single-bend extensions

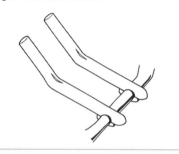

12.17D Double-bend extensions

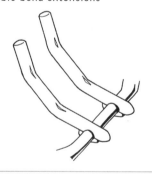

12.17E "Slam" bar extensions

12-11

HANDLEBAR MAINTENANCE AND REPLACEMENT SCHEDULE

A bike cannot be controlled without a handlebar, so you never want one to break. Do not look at the bar as a permanent accessory on your bike. All handlebars will eventually fail. The trick is not to be riding them when they do.

Keep the bar clean, and check carbon bars for a sudden increase in flexibility. Regularly inspect the bar (yes, under the tape!) for cracks, crash-induced bends, corrosion, and stressed areas. If you find any sign of new flexibility, wear, bending, or cracking, replace the bar. Never straighten a bent handlebar! Replace it! Even though the bar was bent to shape in the first place, it was done when the aluminum was in an annealed (soft) state; the bar was then heat-treated for strength, making it more brittle.

If you crash hard, consider replacing the bar even if it looks fine. If the bar has taken an extremely hard hit, it's a good idea to replace it immediately rather than gamble on its integrity.

Some manufacturers recommend replacing stems and bars every four years or less. As with a stem, if you rarely ride the bike, this is overkill. If you ride hard and often, every four years may not be frequent enough. Do what is appropriate for you, and be aware of the risks.

12-12

WRAPPING HANDLEBAR TAPE

You need both hands free and the bar rigidly held. Clamping the bike in a bike stand or holding it in a stationary trainer should do the trick, but you may need to stabilize the front wheel between your knees or with a strap around the down tube and rim or a bar-holder from the seatpost to the handlebar. Before wrapping, clean the bar and inspect it for cracks, crash-induced bends, corrosion, and stressed areas. If you find any sign of wear or cracking, replace the bar.

Tape down concealed brake and shift cables in a few places with electrical tape or strapping tape (Fig. 12.18). (Pre-1984 or so brake levers have no concealed cables; the brake cable comes out of the top of the lever.) Shimano STI levers up until 2009 always have concealed brake cables and exposed shift cables. Both the shift cables and brake cables are concealed on Campagnolo Ergopower, SRAM DoubleTap, or post-2009 Shimano Dura-Ace STI levers.

Tape the cables into place where they will be the most comfortable for your hands. Some handlebars have creases in them for the cables; tape the cables so that they stay in the creases. Some levers allow cable housings to go either in front of or behind the handlebar. My preference is to route both the brake cable and the shift cable along the front of the handlebar.

Handlebar tape sets usually come with two short pieces to cover the brake-lever-clamp bands. Peel back the edges of the rubber hood on the brake lever, wrap the little tape piece around the clamp band, and insert each end under the hood. You may want to tape the ends down with some Scotch tape. Leave the hood peeled back so that you can wrap the bar tape up onto the edge of the lever body and then cover it with the skirt of the rubber hood.

Peel back the paper backing on the tape and start wrapping at the end of the bar from the inside out. Overlap the end of the bar by more than about an inch, so that you can push the excess in with the end plug later. Lightweight bars tend to have thin walls and consequently a large inside diameter, and many end plugs will not fit tightly in them. In this case, the extra tape sticking out will fill the extra space. Alternatively, you can put the plugs in first and simply start wrapping right at the ends of the bar with no overlap; you'll need to wrap some tape around the insertion prongs of the plug until it fits tightly and won't rattle out.

To have a long-lasting tape job, you always want to wrap from the end of the bar so that each wrap holds down the inner edge of the prior one. The wrapping direction is important too. Wrapping from the center of the bar and finishing at the end plug is a mistake, because your hands will constantly peel back the edge of each tape wrap as you ride. The tape will look bad and get torn quickly. Wrap from the ends of the bar, working up to the stem.

Pull the tape tightly but don't break it. Overlap each wrap about one-quarter to one-half of its width

12.18 Taping cables to the bar before installing bar tape

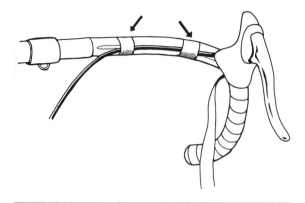

12.19 Wrapping handlebar tape

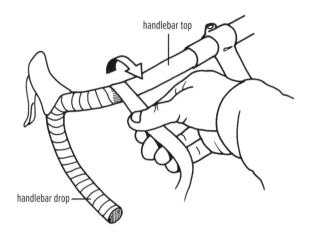

handlebar top

handlebar drop

(Fig. 12.19). Use as much overlap as you can to increase padding and decrease the chance of the tape slipping enough to reveal the handlebar. The amount of overlap will depend on the length of the tape, the width and drop depth of the bar, and the amount you stretch the tape as you wrap.

When you get to the bulged section of the bar that clamps into the stem, you should have just run out of tape. If you have more, you can rewrap part of the bar with more overlap, or you can cut off the excess. If the tape doesn't make it to the bulge, you can rewrap part of the bar with less overlap. If you want to end with only a narrow piece of sticky tape holding it down, trim the end of the bar tape to a point, and hold it down with a single width of electrical tape wrapped around a couple of times. You can follow with the decorative tape piece that came with the bar tape. Otherwise, just wrap around

the bar a number of times with electrical tape, going wide enough with it to completely cover the square-cut end of the bar tape. Cut or break the electrical tape so that it ends under the bar.

Push the plugs into the ends of the bar, using them to push in the extra tape you left sticking off the ends of the bar. Most plugs are now simply that—cylindrical plastic plugs. Old-school end plugs have an expanding device in them; you tighten a screw on the end, and it pulls a wedge into its internally tapered inner end, expanding it out against the walls of the handlebar.

12-13

SETTING STEM AND BAR POSITIONS

Setting handlebar height and reach is very personal. Much depends on your physique, your flexibility, your frame, your riding style, and a few other preferences. This subject is covered in depth in Appendix C. Here are some brief suggestions:

- The old method of setting the bar angle with a round-bend bar was to make the ends of the bars horizontal, but road handlebar shapes have changed. With modern bar shapes, which have a sharp bend followed by a longer, large-radius curve, I recommend setting the bar so that the flat section above the bend is horizontal; the ends will be aimed slightly downward toward the rear brake bridge.

- If you stand a lot when you climb, you will want the bar low enough that you can use your arms efficiently when gripping the brake levers and pulling on them, and high enough that you need not bend over to do so.

- A low, stretched-out position is aerodynamically more efficient. A low position is one with the top of the handlebar more than 6cm (2 inches) lower than the top of the saddle. With your hands on the drops, a stretched-out position places your elbow at least 2cm (¾ inch) in front of your knee at the top of the pedal stroke.

- If you are using an aerobar (clip-on) , you want to find a position that maximizes both comfort and aerodynamic efficiency. The lower and more aerodynamic you are trying to be, the more forward you will want to position the saddle to open up the angle between your torso and your thigh. When setting the reach to the bar, a good rule of thumb is to position the elbow pads so that the front of your shoulder or your ear is over the bend in your elbow (Fig. C.7). As for width, the narrower the elbow pads, the more aerodynamic you will be. Work on getting lower only after you have gotten comfortable and efficient with a narrow position.

HEADSETS

There are two main types of headsets: threadless and threaded. Nearly all modern bikes come with threadless headsets, of which there are three basic types: standard (i.e., external; Fig. 12.20), cupless internal (i.e., integrated; Fig. 12.21), and press-in internal with lipped cups (Fig. 12.22). Older bikes (and some current bikes that prefer a retro design) generally have threaded headsets (Fig. 12.23). A threaded headset requires a threaded steering tube.

Road headsets have traditionally come in the 1-inch-diameter size (originally only in threaded versions and, later, after a century, threadless), but now most road bikes take 1⅛-inch threadless headsets. However, just as with cranksets, headset bearing standards are rapidly changing to systems that have a larger lower bearing (1½-inch or 1¼-inch) paired with a 1⅛-inch upper bearing (Fig. 12.26).

12-14

HEADSET ASSEMBLY

A threadless headset (Fig. 12.20), originally dubbed "Aheadset," is a lightweight system that eliminates the stem quill, bolt, and wedge of a threaded system. The threadless system's clamped connection between the handlebar and the stem saves weight and is more rigid, too.

On a threadless headset, the top cup or cone and a conical compression ring slide onto the steering tube (Figs. 12.20–12.22, 12.25, 12.26). The stem clamps around the top of the steering tube, above the compression ring and top cover. When its conical base is pressed

12.20 External threadless headset

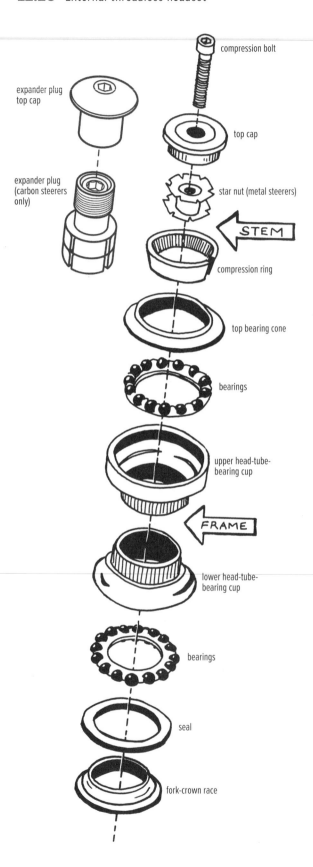

compression bolt

expander plug top cap

top cap

expander plug (carbon steerers only)

star nut (metal steerers)

STEM

compression ring

top bearing cone

bearings

upper head-tube-bearing cup

FRAME

lower head-tube-bearing cup

bearings

seal

fork-crown race

12.21 Cupless (drop-in) internal cartridge-bearing headset

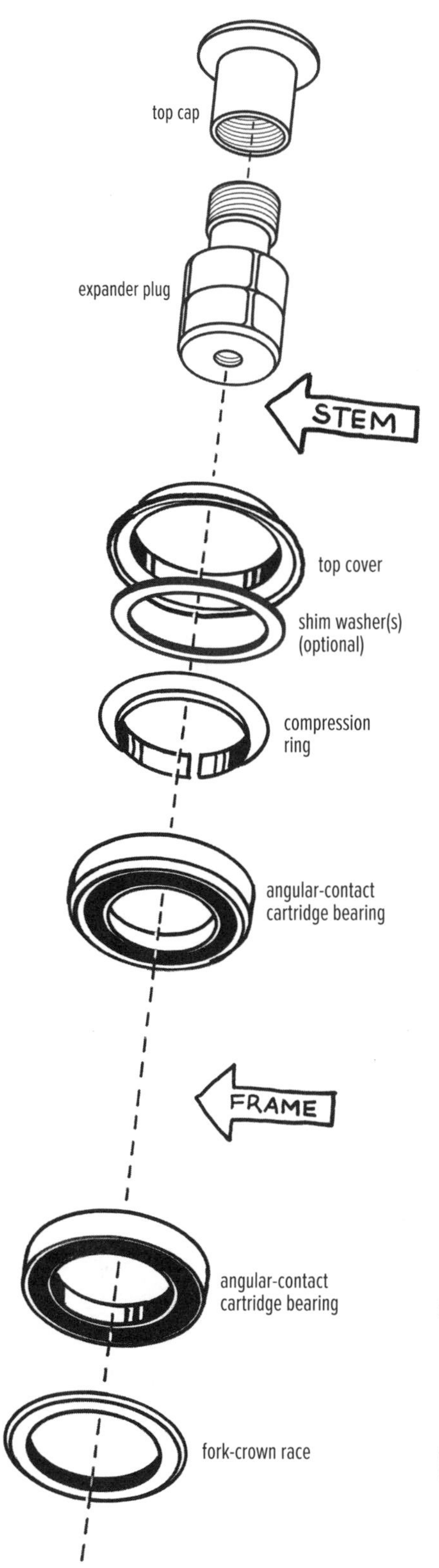

12.22 Inset (or zero stack) press-in internal headset with lipped cups

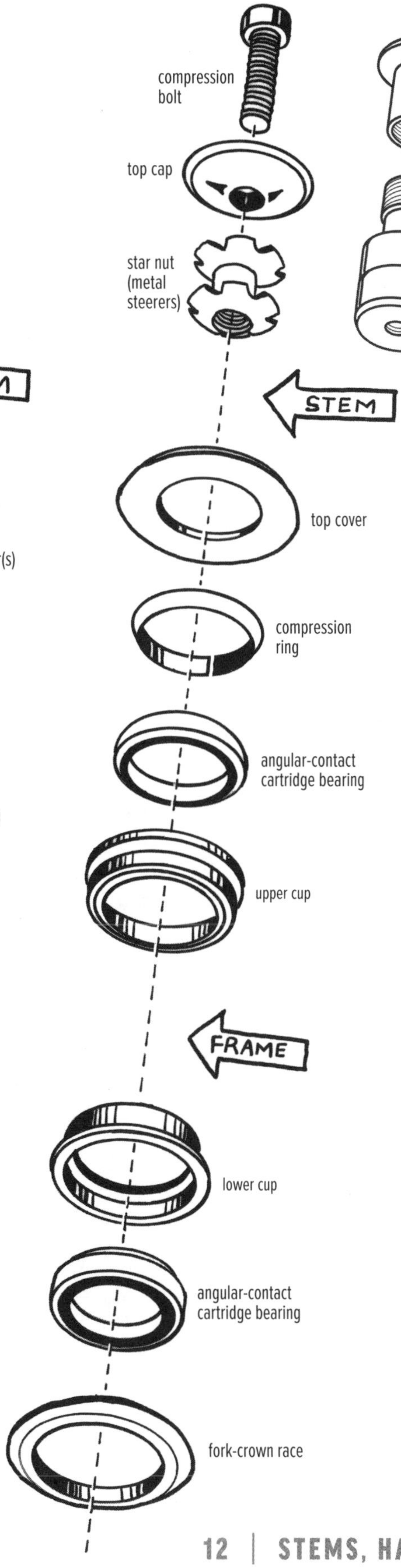

12.23 Threaded headset

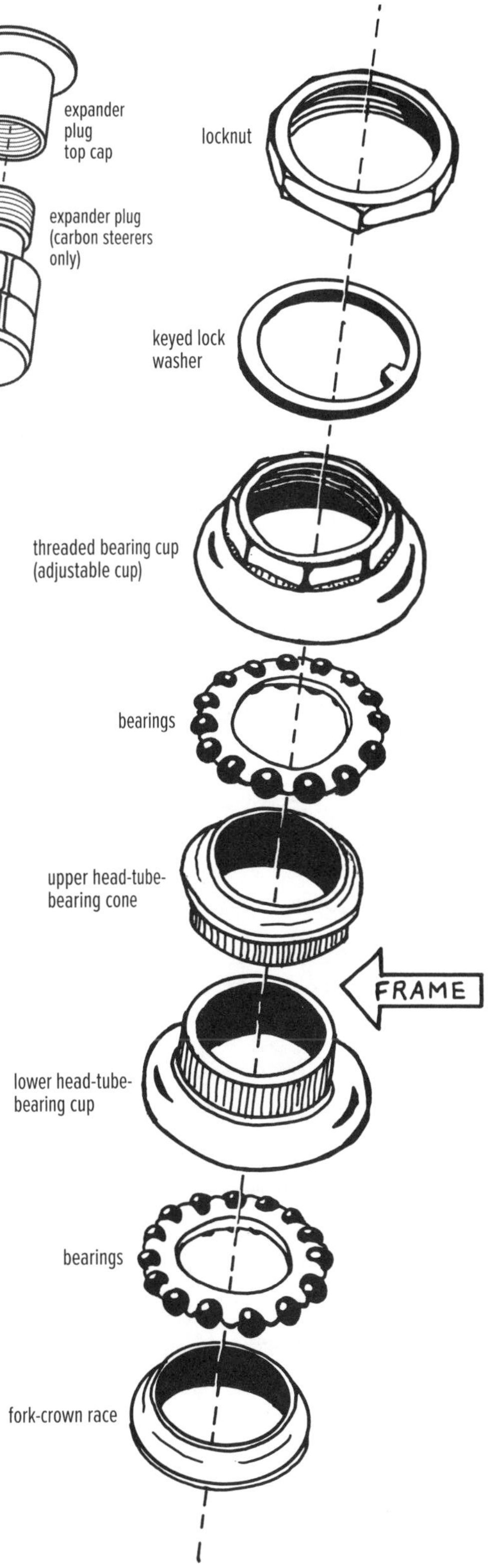

top cap

expander plug

STEM

top cover

shim washer(s) (optional)

compression ring

angular-contact cartridge bearing

FRAME

angular-contact cartridge bearing

fork-crown race

compression bolt

top cap

star nut (metal steerers)

expander plug top cap

expander plug (carbon steerers only)

STEM

top cover

compression ring

angular-contact cartridge bearing

upper cup

FRAME

lower cup

angular-contact cartridge bearing

fork-crown race

locknut

keyed lock washer

threaded bearing cup (adjustable cup)

bearings

upper head-tube-bearing cone

FRAME

lower head-tube-bearing cup

bearings

fork-crown race

12.24 Zero stack internal headset with press-in cups

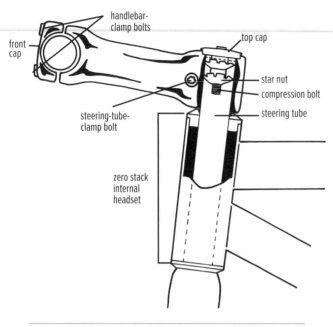

front
cap

handlebar-
clamp bolts

top cap

star nut

compression bolt

steering-tube-
clamp bolt

steering tube

zero stack
internal
headset

12.25 Campagnolo Hiddenset drop-in-style
(non-press-in) integrated headset

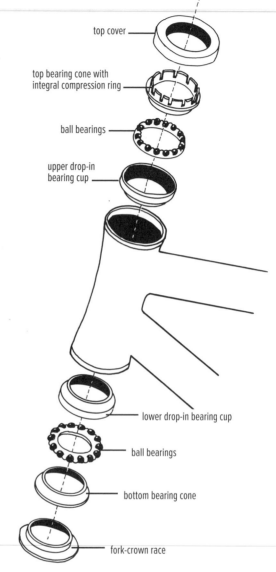

top cover

top bearing cone with
integral compression ring

ball bearings

upper drop-in
bearing cup

lower drop-in bearing cup

ball bearings

bottom bearing cone

fork-crown race

into the beveled edges of the bore of the top bearing cup or cone, the compression ring keeps the top bearing cup or cone centered on the steering tube.

For forks with metal steering tubes, whether steel, aluminum, or titanium, a star nut—a nut with two layers of sharp, spring-steel teeth sticking out from it (Fig. 12.20)—fits into the steering tube and grabs its inner walls (Fig. 12.24). For steering tubes made of carbon fiber, an expander plug (Figs. 12.20–12.22, 12.27) or a glue-in insert with a threaded hole for the top-cap bolt replaces the standard star nut and serves the dual purpose of anchoring the top cap and protecting the steering tube from being crushed by the stem clamp.

The top cap pushes the stem clamp down to adjust the headset by means of the long compression bolt threaded into the star nut (Fig. 12.24) or expander plug (Fig. 12.27). The stem clamp secured around the steering tube holds the headset in adjustment.

The latest generation of headset is the threadless internal type, or integrated headset, concealed inside the frame's head tube (Figs. 12.21, 12.22, 12.24–12.26). Whereas standard threadless and threaded headsets have bearing cups above and below the ends of the head tube (Figs. 12.20, 12.23), integrated headsets have bearings seated inside the head tube. Some types of integrated headsets have no press-in cups; either the

bearings roll on bearing cups that drop into the flared head tube and rest on machined shelves within the head tube itself (Fig. 12.25), or the headset has angular-contact cartridge bearings that drop into a flared head tube and rest without a cup on shelves machined or molded within the head tube (Figs. 12.21, 12.26). Varying the shim washer(s) between the top cover and the compression ring (Fig. 12.26) prevents the top cover from scraping the top of the head tube. On carbon forks with tapered steering tubes, the lower bearing often sits right on the fork crown without a crown race under it (Fig. 12.26).

12.26 Cupless integrated headset with tapered steering tube and differentially sized bearings

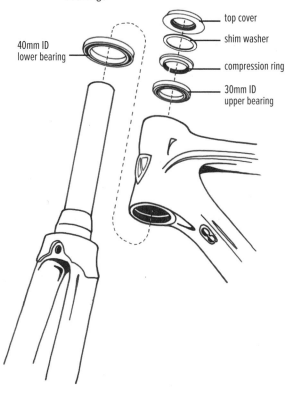

40mm ID lower bearing

top cover

shim washer

compression ring

30mm ID upper bearing

12.27 Inserting an expandable support plug into a carbon-fiber fork steering tube

12.28 Needle bearings

12.29 Lower parts of a cartridge-bearing headset

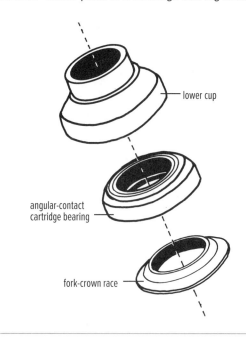

lower cup

angular-contact cartridge bearing

fork-crown race

Other types of internal headsets, called "zero stack" or "inset," have press-in cups with thin flanges that extend out to the edges of the head tube (Figs. 12.22, 12.24). Otherwise, integrated headsets are identical to, and are adjusted in the same way as, original threadless headsets.

Threaded headsets are different. The top bearing cup on a threaded headset has wrench flats, a keyed lock washer stacked on top of it, and a locknut that covers the top of the steering tube. That locknut tightens against the keyed lock washer and threaded cup (Fig. 12.23). Extra spacers may be included under the locknut.

Many headsets—threaded or threadless—use individual ball bearings held in some type of steel or plastic retainer or "cage" (Figs. 12.20, 12.23, 12.25) so that you are not chasing dozens of separate balls around when you work on the bike. A variation on this has needle bearings held in conical plastic retainers (Fig. 12.28) riding on conical steel bearing surfaces.

Cartridge-bearing headsets usually employ "angular-contact" bearings (Figs. 12.21–22, 12.26, 12.29, 12.35), since normal cylindrical cartridge bearings (Fig. 8.15) cannot take the side forces encountered by the lower bearing of a headset. Each angular-contact cartridge bearing is a separate, sealed, internally greased unit.

CHECKING HEADSET ADJUSTMENT

If the headset is too loose, it will rattle or clunk while you ride. You may even notice some play (back-and-forth movement) in the fork as you apply the front brake. If the headset is too tight, the fork will be difficult to turn or will feel rough to rotate.

1. **Check for headset looseness.** Hold the front brake and rock the bike forward and back. Try it with the front wheel pointed straight ahead and then with the wheel turned at 90 degrees to the bike. Feel for play at the lower head cup with your other hand. If there is play, you need to adjust the headset because it is too loose. If the headset is loose, skip to the appropriate adjustment section, 12-16 or 12-17.

2. **Check for headset tightness.** Turn the handlebar back and forth with the front wheel off the ground. Feel for any binding or stiffness of movement. Also, check for the chunk-chunk-chunk movement to fixed positions characterizing a pitted headset (if you feel this, you need a new headset; skip to 12-20). Lean the bike to one side and then the other; the fork should turn as the bike is leaned (be aware that cable housings can resist the turning of the front wheel). Lift the bike by the saddle so that it is tipped down at an angle with both wheels off the ground. Turn the handlebar one way and let go. See whether it returns to center quickly and smoothly on its own. If the headset does not turn easily on any of these steps, it is too tight, and you should skip to the appropriate adjustment section, 12-16 or 12-17.

3. **Check for loose parts on threaded headsets.** If the headset is a threaded model, try to turn the top nut and the threaded cup by hand. They should be so tight against each other that they can only be loosened with wrenches. If you can tighten or loosen either part by hand, even if it passed tests 1 and 2, you still need to adjust the headset; go to 12-17.

ADJUSTING A THREADLESS HEADSET

 LEVEL Adjusting a threadless headset—whether it is an internal (or "integrated") type (Figs. 12.21, 12.22, 12.24–12.26) or an external type (Figs. 12.4, 12.20)—is much easier than adjusting a threaded one. It's a level 1 procedure and usually only takes a hex key or two.

a. First steps

1. **Check the headset adjustment (12-15).** Determine whether the headset is too tight or too loose.

2. **Loosen the bolt or bolts that clamp the stem to the steering tube.**

3. **Adjust the headset by turning the top-cap compression bolt.** Be careful not to overtighten it, which will put too much pressure on the bearings and eventually pit the headset. If you're using a torque wrench, tighten this bolt to 22 in-lbs (2.5 N-m), which is a very low torque. This is a good place to start, but your headset may require a different torque for proper adjustment.

   a. If the headset is too tight, loosen the compression bolt on the top cap about ¹⁄₁₆th of a turn (Fig. 12.30). This usually takes a 5mm or 4mm hex key, but on many expander inserts for carbon-fiber steering tubes

12.30 Loosening and tightening the compression bolt on a threadless headset

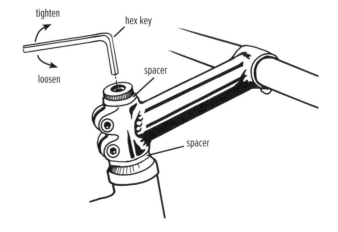

tighten

hex key

loosen

spacer

spacer

(Fig. 12.27), the top cap itself is turned with a 6mm hex key.

b. If the headset is too loose, tighten the compression bolt on the top cap about ¹⁄₁₆th of a turn (Fig. 12.30).

NOTE: *Not all threadless stems are adjusted with the top-cap system. DiaTech threadless headsets have no top cap. Instead, a clamping collar below the stem adjusts headset tension. The stem is first clamped in place. The collar is beveled on the inside from both ends, and it slides down an externally beveled ring above it as you tighten the clamp screw to put pressure on the headset. As soon as you loosen the stem, the headset comes out of adjustment.*

Adjustment problems

If the cap does not move down and push the stem down, redo step 2, making sure the stem is not stuck to the steering tube.

Another hindrance occurs if the conical compression ring (Figs. 12.20–12.22, 12.25, 12.26) is stuck to the steering tube, preventing adjustment via the top-cap bolt. Remove the top cap, stem, spacers, and headset top cover first to address this problem.

With most (i.e., non-Campagnolo) compression rings, which are simply cone-shaped pieces split on one side (Figs. 12.20–12.22, 12.26), you need only tap the steering tube down with a mallet and then push the fork back up to free the compression ring. Grease the ring and the steering tube, and reassemble.

With a Campagnolo threadless headset (either an integrated Hiddenset [Fig. 12.25] or a standard external one [Fig. 12.31]), the compression ring is plastic and is conical on both ends. Its bottom end presses into the beveled hole in the top cone, but its turreted top end is also conical and presses into the bore of the headset top cover, which is beveled toward the bottom. Pushing down on the top cover (via the compression bolt pushing down on the stem) simply pinches the compression ring tighter in place, rather than pushing it down. Instead, you must flip the top cover upside down (Fig. 12.31) so that the nonbeveled end of its through-hole is against the turreted top edge of the compression ring. Now pushing down on it will push the compression ring down to preload the headset bearings by seating the top cone into them. Then flip the top cap back over (Fig. 12.32), put it back in place, and reassemble the spacers, stem, top cap, and compression bolt.

12.31 Seating a Campagnolo threadless headset 1

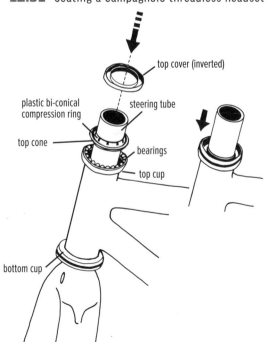

12.32 Seating a Campagnolo threadless headset 2

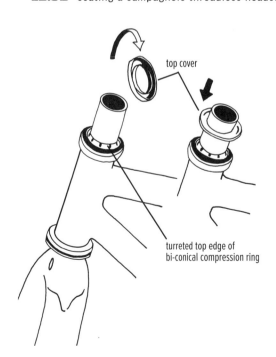

If neither the stem nor the compression ring is stuck, yet the cap still does not push the stem down, the steering tube may be so long that it is hitting the lip of the top cap and preventing the cap from pushing the stem down. The steering tube's top should be either 3–5mm below the rim of the stem clamp (Fig. 12.10) or, ideally, 3–5mm below the rim of the spacer(s) placed above the stem (see the Pro Tip on spacers in carbon steering tubes). If the steering tube is too long, add a spacer above or below the stem, or use a flat file to make the steering tube shorter. Some top caps have thicker edge lips than others and require more space down to the top of the steering tube to avoid bottoming out on it.

Another thing that can thwart adjustment is the star nut not being installed deeply enough, so that the cap bottoms out on the star nut. The highest point of the star nut should be 12–15mm below the top of the steering tube. With metal steering tubes, tap the star nut deeper with a star nut installation tool. Alternatively, put the bolt through the top cap, thread it five or six turns into the star nut, and gently tap it in with a soft hammer; include the top cap to keep the star nut going in straight. Some top caps have taller center sections than others and require deeper insertion of the star nut to avoid bottoming out on it.

With a carbon steering tube, first loosen the aluminum expander with a 5mm hex key (Fig. 12.27). Next, unscrew its top cap a turn or two with a 6mm hex key. By hand, push the assembly farther into the steering tube until the top cap stops it, and reexpand the plug with a 5mm hex key. Finally, tighten the top cap down (22 in-lbs, or 2.5 N-m, of torque is standard) against the top of the stem to adjust the headset.

Once you have fixed the cause of the adjustment problem, return to step 1.

b. Final steps

4. **Tighten the stem's steering tube clamp bolt, or bolts.** Ideally, use a torque wrench to torque spec (see Appendix E).

5. **Recheck the headset adjustment.** Repeat steps 2–4 if necessary. With some integrated headsets, you may need a 1mm shim or two under the top

bearing cover so that the edges of the cover do not drag and scrape on the top end of the head tube.

6. **Check alignment.** If the headset is adjusted properly, make sure the stem is aligned straight with the front wheel, and go ride.

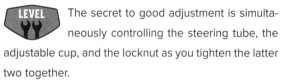

ADJUSTING A THREADED HEADSET

LEVEL The secret to good adjustment is simultaneously controlling the steering tube, the adjustable cup, and the locknut as you tighten the latter two together.

NOTE: *Perform the adjustment with the stem installed. Not only does it give you something to hold on to that keeps the fork from turning during the process, but there are slight differences in adjustment when the stem is in place as opposed to when it is not. Tightening the stem bolt inside a threaded steering tube (Fig. 12.8) can sometimes bulge the walls of the steering tube slightly, just enough for it to shorten the steering tube and tighten a previously perfect headset adjustment.*

1. **Check for proper adjustment.** Following the steps outlined in 12-15, determine whether the headset is too loose or too tight.

2. **Prepare your headset wrenches.** Put a pair of headset wrenches that fit the headset on the headset's top nut (the locknut) and top bearing cup (threaded cup or adjustable cup). Headset nuts come in a wide variety of sizes, so make sure you have purchased the proper wrench size. The standard wrench size for a road bike is 32mm. Place the wrenches so that the top one is slightly offset to the left of the bottom wrench. That way you can squeeze them together to free the nut (Fig. 12.33).

NOTE: *People with small hands or weak grip will need to grab each wrench out at the end to get enough leverage.*

3. **Loosen the locknut (upper nut).** Hold the lower wrench in place and turn the top wrench counterclockwise about one-quarter turn to loosen the locknut. Breaking it loose may take considerable force, because it is generally tight to keep the headset from loosening.

12.33 Offsetting the headset wrenches to loosen the locknut

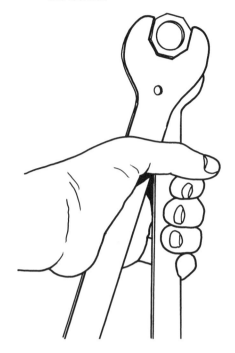

12.34 Offsetting the headset wrenches to tighten the locknut

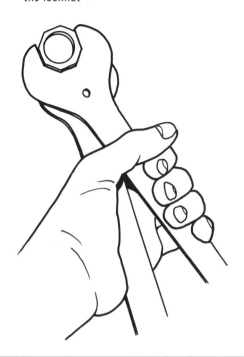

4. **Adjust the headset.** If the headset was too loose, turn the lower (threaded) cup clockwise about 1/16 of a turn while holding the stem with your other hand. Be careful not to overtighten the cup, which can ruin the headset by pressing the bearings into the bearing surfaces, making little indentations. The headset then stops at the indentations rather than turning smoothly, a condition known as a "pitted" or "brinelled" headset.

   If the headset was too tight, loosen the threaded cup counterclockwise 1/16 of a turn while holding the stem with your other hand. Loosen it until the bearings turn freely, but not to the point where play develops.

5. **Tighten the locknut.** Holding the stem, tighten the locknut clockwise with a single wrench. Make sure that the threaded cup does not turn while you tighten the locknut. If it does turn, either you are missing the keyed lock washer separating the cup and locknut (Fig. 12.23) or the washer you have is missing its key. In this case, remove the locknut (the stem has to come out first) and replace the keyed lock washer. Put the locknut on the steering tube so that the key engages the longitudinal groove in

the steering tube. Thread on the locknut, install the stem, and redo the adjustment procedure.

NOTE: *You can adjust a headset without a keyed lock washer by working both wrenches simultaneously, but it is trickier, and the headset may come loose while you are riding.*

6. **Check the headset adjustment again.** Repeat steps 4 and 5 until the headset is properly adjusted.

7. **Tighten the locknut.** Once the headset is properly adjusted, place one wrench on the locknut and the other on the threaded cup. Tighten the locknut (clockwise) firmly against the washer(s) and threaded cup to hold the headset adjustment in place (Fig. 12.34).

8. **Check the headset adjustment again.** If it is off, follow steps 2–7 again. Once it is adjusted properly, make sure the stem is aligned with the front wheel before riding.

NOTE: *If you repeatedly get what you believe to be the proper adjustment and then find it to be too loose after you tighten the locknut and threaded cup against each other, the steering tube may be too long, causing the locknut to bottom out. Remove the stem and examine the inside of the steering tube. If the top end of the* *steering tube butts up against the top lip of the locknut, the steering tube is too long. Remove the locknut and add another spacer.*

If you don't want to add another spacer, file 1–2mm from the steering tube. Be sure to deburr it inside and out after filing, and avoid leaving filings in the bearings or steering tube threads. Replace the locknut and return to step 5.

12-18

OVERHAULING THREADLESS HEADSET

LEVEL These instructions apply to both internal (i.e., integrated) threadless headsets (Figs. 12.21, 12.22, 12.24–12.26) and external (Figs. 12.4, 12.20) threadless headsets.

Like any other bike part with bearings, headsets need periodic overhauls. If you use your bike regularly, you should probably overhaul a loose-bearing headset once a year. Headsets with cartridge bearings (Figs. 12.21, 12.22, 12.26, 12.29, 12.35) need less frequent overhaul due to the bearing seals on the cartridge. Some angular-contact bearings can be disassembled and cleaned, and some cannot. With those that cannot, if a bearing fails, you either replace the bearing or, if it has press-in bearings (like Chris King; Fig. 12.35), you replace the entire cup (12-20 and 12-22).

Either place the bike upside down in the work stand or be ready to catch the fork when you remove the stem.

1. **Disconnect the front brake (Chapter 9 or 10).** If the bike has a standard caliper brake attached to the fork crown or a hydraulic disc brake, remove the front brake as well. Cable-actuated disc brake calipers and cantilever brakes can be left in place.

12.35 Chris King–style pressed-in cartridge bearing

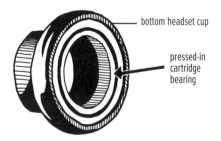

bottom headset cup

pressed-in cartridge bearing

2. **Unscrew the top-cap compression bolt (Fig. 12.30) and the stem-clamp bolt(s).** Remove the top cap and the stem.

3. **Remove the top headset cup.** Slide the top cup or cover, conical compression ring (Fig. 12.20), and any spacers off the steering tube. Freeing the compression ring may require a tap with a mallet on the end of the steering tube, followed by pushing the fork back up and the top cup or cover back down.

4. **Pull the fork from the frame.**

5. **Remove any seals that surround the edges of the cups.** Remember the position and orientation of each.

6. **Remove the bearings from the cups, if removable.** Be careful not to lose any ball bearings. Separate top and bottom sets if they are of different sizes.

7. **Check the bearings.** If the bearings are the type that will not come apart, check to see if they turn smoothly. If they do not, buy new ones and skip to step 8, or clean and regrease the bearings:

   a. With loose-ball-bearing or needle-bearing headsets, put the bearings in a jar or an old water bottle along with some citrus-based solvent. Shake. If the bearings from the top and bottom are of different sizes, keep them in separate containers to avoid confusion. Blot the bearings dry with a clean rag.

   b. Some cartridge bearings (Fig. 12.29) can be pulled apart and cleaned. Over a container to catch the balls, hold the bearing so that the beveled outer surface that fits into the cup faces down, and push up on the bearing's inner ring. The bearing should come apart—the inner ring will pop up and out with the bearings stuck to its outer surface. It may take a little rocking of the inner ring as you push up. If the bearing does not come apart, first pry off the plastic seal covering the bearings with a knife or razor blade, as in Figure 8.15, and then try again. Wipe the bearings and bearing rings and seals with a clean rag.

8. **Blot the bearings dry with a clean rag.** Plug the sink and wash the bearings in soap and water in your hands, just as if you were washing your palms

by rubbing them together. Your hands will get clean for the assembly steps as well. Rinse the bearings thoroughly and blot them dry. Let them air-dry completely; an air compressor or hair dryer may come in handy here.

9. **Wipe all of the bearing surfaces with clean rags.** Wipe the steering tube clean.

10. **Inspect all bearing surfaces for wear and pitting.** If you see pits (separate indentations made by bearings in the bearing surfaces), you need to replace the headset. If so, skip to 12-20.

11. **Apply fresh grease to all bearing surfaces.** If the headset has cartridge bearings, apply grease conservatively.

12. **Turn the bike upside down in the bike stand to begin assembly.**

   a. Place a set of bearings into the top cup and a set into the cup on the lower end of the head tube.

   b. With a Campagnolo or other integrated headset with drop-in bearing cups (Fig. 12.25), place a bearing cup into the seat in the bottom of the head tube. Place a greased set of ball bearings into the cup, with the bearing retainer oriented properly (see step 12d for tips on determining proper bearing orientation).

   c. With a cupless integrated headset (Figs. 12.21, 12.26), set a bearing into the seat in the bottom of the head tube itself. See step 12e regarding orientation.

   d. With loose-ball headsets, make sure you have the bearing retainer right side up so that only the bearings contact the bearing surfaces (note the different upper-cup styles and bearing orientations in Figs. 12.23, 12.20). If you have installed the retainer upside down, it will come in contact with one of the bearing surfaces, and the headset will not turn well. This is a bad thing, because assembling and riding it that way will turn the retainer into jagged chunks of broken metal. To be safe, double- and triple-check the retainer placement by placing each cup

pair in your hand with the bearing set in place and rolling them before proceeding. Most loose-ball headsets have the bearings set up identically top and bottom (Fig. 12.23). The top piece of each pair is a cup, and the bottom piece is a cone; the bearing retainer rides the same way in both sets. Some headsets, however, place both cups (and hence the bearing retainers) facing outward from the head tube (Fig. 12.20).

NOTE: *If the ball bearings are loose with no bearing retainer, stick the balls into the grease in the cups one at a time, making sure that you replace the same number you started with in each cup.*

   e. With angular-contact cartridge bearings, the beveled end faces into the cup (Fig. 12.29) or into the seat machined or molded inside of the head tube (Figs. 12.21, 12.26).

13. **Reinstall any seals that you removed from the headset parts.**

14. **Drop the fork into the head tube so that the lower headset bearing set seats properly (Fig. 12.36).**

12.36 Setting the fork in the head tube to seat bearings

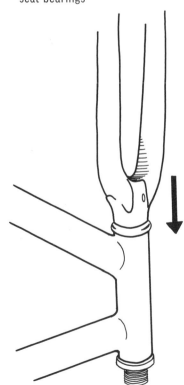

15. **For integrated headsets only:** With a Campagnolo or other integrated headset with drop-in cups (Fig. 12.25), place a bearing cup into the seat in the top of the head tube. Otherwise, skip this step.

16. **Install the upper bearing.** Slide first the bearing and then the top cup or cone (not required with a cartridge bearing) onto the steering tube. Keep the bike upside down at this point to keep the fork in place and prevent grit from falling into the bearings as you slide on the parts above it.

17. **Install the compression ring.** Grease the compression ring and slide it onto the (greased) steering tube, so that the narrower end slides into the conical space in the top of the top cup or cone or angular-contact cartridge bearing (Figs. 12.20–12.22, 12.25, 12.26).

NOTE: *On a Campagnolo threadless headset (integrated or external), there is a plastic biconical compression ring inserted into the top bearing cone (Fig. 12.25), and it acts like a normal split compression ring to center the top cone over the bearings. The upper edge of the plastic compression ring is notched like a turreted castle tower. Above this part comes the top bearing cover, whose inner edge is beveled for the turreted top conical edge of the plastic compression ring. To preload the bearings, you must first install the top bearing cover upside down (Fig. 12.31) and then push down on it. This preloads the bearings by pushing the top cone down. If you install the top bearing cover in its standard orientation before the top cone and plastic compression ring are slid down far enough to preload the bearings, the beveled inner edge of the top bearing cover will pinch the turreted upper conical edge of the plastic compression ring in place and not allow it to slide down farther. Once you have pushed the bearing cone down fully in this manner, flip the top bearing cover right side up and put it in place over the cone and compression ring (Fig. 12.32).*

18. **Install the top cover.** If there was a washer between the compression ring and the top cover (Fig. 12.26), install that first to prevent the top cap from scraping the brim of the head tube.

    On recent Chris King headsets, the top cover, compression ring, and a rubber O-ring must be assembled together before installation. Hold them tightly together when sliding the assembly onto the steering tube; if they become separated, you won't be able to slide the top cover down into place.

19. **Slide on any spacers you had under the stem.**

20. **Slide on the stem.** Tighten one stem-clamp bolt to hold it in place.

21. **Turn the bike over and inspect your work.** Check that the stem clamp or, ideally, the top spacer above the stem extends 3–5mm above the top of the steering tube (Fig. 12.10) and that the star nut is 12–15mm down in the steering tube.

22. **Install the top cap.** Screw the compression bolt into the star nut (Fig. 12.30).

23. **If the steering tube is too long, add a spacer.** Remove the stem and add a spacer that extends 3–5mm above the top of the steering tube.

24. **If the steering tube is too short, remove spacers from below the stem.** If there are no spacers to remove, try a new stem with a shorter clamp.

25. **Adjust the headset (12-16).**

26. **Reconnect and adjust the front brake (Chapter 9 or 10).**

12-19

OVERHAULING THREADED HEADSET

LEVEL Like any other bike part with bearings, headsets need periodic overhauls. If you use your bike regularly, you should probably overhaul a loose-bearing headset once a year. Headsets with sealed cartridge bearings usually never need to be overhauled; if a bearing fails, you either replace the bearing (such as the standard type shown in Fig. 12.29) or, if the headset has pressed-in bearings (like Chris King; Fig. 12.35), you replace the entire cup. If you have a standard cartridge-bearing headset, continue with these instructions. If you are replacing a Chris King or other headset cup with a pressed-in bearing, move on to the instructions for headset removal and installation (12-20 and 12-22).

A bike stand is highly recommended when overhauling a headset.

1. **Disconnect the front-brake cable (Chapter 9 or 10).**

2. **Remove the stem.** Loosen the stem bolt three turns, tapping the bolt down with a hammer to free the wedge (Fig. 12.11), and pulling it out.

3. **Either turn your bike upside down or be prepared to catch the fork as you remove the upper part of the headset.** To remove the top headset cup, unscrew the locknut and threaded cup with headset wrenches: Place one wrench on the locknut and one on the threaded cup. Loosen the locknut by turning it counterclockwise. It's easiest if the top wrench is angled just to the left of the lower wrench, and you squeeze them together (Fig. 12.33). Unscrew the locknut and the cup from the steering tube. The headset washer or washers will slide off the steering tube as you unscrew the threaded cup.

4. **Follow steps 4–13 in 12-18.**

5. **Drop the fork into the head tube so that the lower headset bearing set seats properly (Fig. 12.36).**

6. **Screw the top cup, with the bearings in it, onto the steering tube.** Keeping the bike upside down at this point keeps the fork in place and prevents grit from falling into the bearings as you thread the cup on.

7. **Turn the bike upright.** Slide on the keyed lock washer (Fig. 12.23). Align the key in the groove of the steering tube threads. Screw on the locknut with your hand.

8. **Grease the stem quill and insert it into the steering tube (Fig. 12.8).** Make certain that it is in at least as deep as the imprinted limit line, and preferably a bit deeper. Align the stem with the front wheel and tighten the stem bolt.

9. **Reconnect the front brake (Chapter 9 or 10).**

10. **Adjust the headset as outlined in 12-17.**

12-20

REMOVING THE HEADSET

1. **Open the headset and remove the fork.** Follow steps 1–3 in either 12-18 or 12-19, depending on headset type.

2. **Remove the fork.**

3. **Remove the bearings.**

   a. If you have a cupless integrated headset (Figs. 12.21, 12.26) with bearings seated on steps machined or molded into the head tube itself, just pull the bearings out and skip to step 5. If the drop-in headset is a non-press-in integrated headset with bearing cups (Fig. 12.25), pull out both bearing cups as well as the bearings from either end of the head tube and skip to step 5.

   b. For a headset with cups pressed into the head tube (Figs. 12.20, 12.22, 12.23), just pull out the bearings, unless they are pressed into the cups as in Figure 12.35.

4. **On nonintegrated headsets, remove the cups.** For a headset with cups pressed into the head tube (Figs. 12.20, 12.22–12.24), slide the solid end of the headset-cup remover (sometimes called a headset "rocket," a wonderfully evocative name, as you'll see) through one end of the head tube (Fig. 12.37). As you pull the headset-cup remover through the head tube, the splayed-out tangs on the opposite end of the tool will pull through the cup and spread out.

12.37 Inserting a headset-cup removal tool or "rocket"

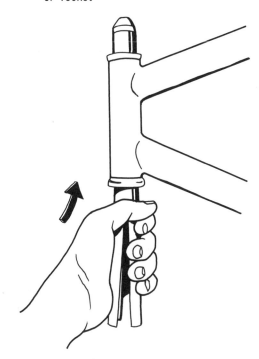

a. Remove the first cup. Strike the solid end of the cup remover with a hammer, and drive the cup out (Fig. 12.38). Be careful as you do this, as the remover, or rocket, is liable to launch the cup and itself across the room if hit with sufficient force.

b. Remove the other cup. Place the cup-remover rocket into the opposite end of the head tube and repeat step 4a on the opposite end of the end tube.

5. **Remove the fork-crown race.** Some of the new integrated-headset forks have no crown race; the bearing just sits right atop the fork crown (Fig. 12.26); with one of these, just lift the bearing off. Your headset removal is done. Easy peasy.

Old steel road bike fork crowns with 1-inch steering tubes in them are narrower than the diameter of the headset fork-crown race, so that you can elegantly remove the crown race with a U-shaped crown-race remover (step 5a, Fig. 12.39) or an appropriately sized bench vise (step 5b, Fig. 12.40). For forks whose crown race is not larger in diameter than the crown, skip to steps 5c and 5d.

a. To use an old-school U-shaped crown-race remover, stand the fork upside down on the steering tube. Place the U-shaped crown-race remover so that it straddles the underside of the fork crown and its ledges engage the front and back edges of the crown race. Smack the top of the crown-race remover with a hammer to knock the race off (Fig. 12.39).

b. To use a bench vise, slide the fork into the vise, straddling its center shaft. Tighten the vise so that its faces ever so lightly contact the front and back of the fork crown with the lower side of the crown race sitting on top of them. Put a block of wood on the top of the steering tube to pad it. Strike the block with a hammer to drive the fork down and knock the crown race off (Fig. 12.40).

c. If the fork crown is larger in diameter than the fork-crown race and you don't have the slick tool mentioned in step 5d, you will need to

12.38 Removing a headset cup

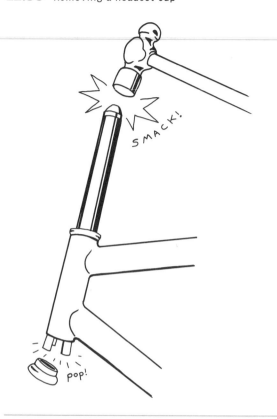

12.39 Removing the fork-crown race with a crown-race remover

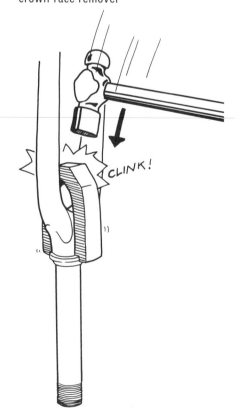

12.40 Removing the fork-crown race with a vise

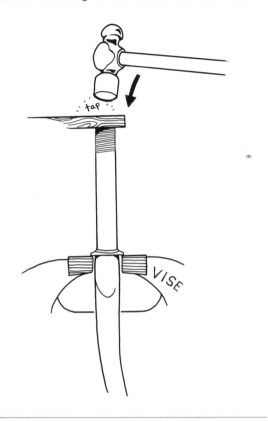

12.41 Removing the fork-crown race with a screwdriver

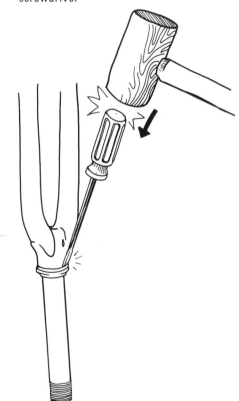

knock the race off with a screwdriver—preferably an old, cheap screwdriver that you no longer use to drive screws. Turn the fork upside down so that the top of the steering tube is sitting on the workbench, or clamp the steering tube horizontally in a bike stand or in a vise between a pair of V-blocks. If there are notches at the front and back of the fork crown under the bearing race, place the blade of a large screwdriver into the notch on one side of the crown so that it butts against the bottom of the headset fork-crown race. If there is no notch, work the screwdriver blade under the race however you can; you may need to first drive a thin knife or razor blade between the race and fork crown to open a gap. Tap the handle of the screwdriver with a hammer to drive the crown race up the steering tube a bit (Fig. 12.41). Move the screwdriver to the other side of the crown and tap it again to move that side of the crown race up a bit. Continue in this way, alternately tapping either side of the crown race up the steering tube, bit by bit, until it gets past the enlarged section of the steering tube and slides off.

d. If you are fortunate enough to have a Park Universal Crown-Race Remover (Fig. 1.4), use it! First back off the screw on the top so that the tool will slide down over the steering tube until the blades are below the fork-crown race. Using the screws at its base, finger-tighten the blades under the fork-crown race until they stop. Then tighten the handle of the long screw on top to pull the crown race off its seat.

12-21

FRAME AND FORK PREPARATION PRIOR TO INSTALLATION OF HEADSET

LEVEL The frame and fork need to be properly prepared for the headset prior to installation. If new frames and forks are not properly prepared prior to sale, the headset will bind up at some steering angles and be too loose at others. Proper frame and fork preparation requires tools that only some shops possess.

NOTE: *You can do nothing to prep the head tube on a carbon frame, unless it has internal aluminum sleeves.*

1. **If this is a new frame (or one that has "eaten" headsets in the past):** Ream and face the head tube, if it is a frame for external-cup headsets (Figs. 12.1, 12.20, 12.23) or for an internal headset with cups (Fig. 12.24). If you do not have the tools for this, have a bike shop equipped with the proper tools do it for you. Reaming makes the head-tube ends perfectly round inside and of the correct diameter for the headset cups to press in. Facing makes the ends of the head tube parallel so that the bearings can turn smoothly and uniformly. The tool that simultaneously reams and faces the end of the head tube is pictured in Figure 1.4.

2. **If this is a metal bike frame for a cupless (non-press-in) integrated headset (Fig. 12.21), like a Campagnolo Hiddenset (Fig. 12.25):** These frames will eat bearings in a hurry if the bearing seats are so badly machined inside the head tube that the bearings are not parallel. Fortunately, many shops have a tool to recut the bearing seats so that they are parallel.

3. **If the fork is a metal one:** The base of the steering tube also needs to be turned down to the correct diameter for the crown race. And the crown-race seat on the fork crown must be faced in a way that places the crown race parallel to the head-tube cups and perpendicular to the steering tube.

NOTE: *With a carbon-fiber fork, do not run a cutter over the crown-race seat; doing so could cut the carbon fibers. You can be fairly confident that the crown race is perpendicular to the steering tube, whose base should also be the correct diameter. As carbon forks are molded rather than welded or brazed together, it is easier to control these dimensions during manufacturing than machining them later, as on metal forks. The same goes for any type of suspension fork, as their parts are generally machined before assembly. If the crown-race seat is oversized or untrue, many shops have a tool that simultaneously machines the outer dimension of the base of the steering tube and cuts the crown-race seat flat and perpendicular to the steering tube axis.*

4. **The fork steering tube (threaded or threadless) must also be cut to the proper length.** Remember, you can always go back and cut off more. You can't go back and add any, so be careful! You can wait until the headset (and stem and spacers, in the case of a threadless headset) is installed. Or you can figure out the length first.

Threadless headsets

a. The safest way to make sure you don't cut the steering tube too short is to install the headset, stem, and spacers onto the frame and fork first (12-22). Then cut the steerer to 3–5mm below the top of the spacer placed atop the stem (see the Pro Tip on cutting carbon steerers). With a steel or aluminum steerer, you can cut it as short as 3–5mm below the top edge of the stem (Fig. 12.10).

b. If you don't want to or can't install the headset yet, add the headset stack height (measure it or find it from the manufacturer or from Barnett's or Sutherland's manuals) to the length of the head tube, the spacers under the stem, and the stem clamp, and add 2mm to the total so you can put a 5mm spacer above the stem. This is the length the steering tube should be from fork crown to top. I recommend not cutting until the headset is assembled and the stem is installed so that you can see if you want more spacers under the stem to raise the bar higher.

c. Cut off the steerer, and smooth and straighten its end as instructed in 12-2, steps 6b–e.

Threaded headsets

a. The safest way to make sure you don't cut the steering tube too short is to install the headset first (12-22). Once a threaded headset is assembled, you can measure the amount of excess length as in Figure 12.42, remove the top nut, and add keyed lock washers (Fig. 12.23), adding up to 2mm taller than that gap in total. Or you can trim

12.42 Measuring amount of steering tube to cut

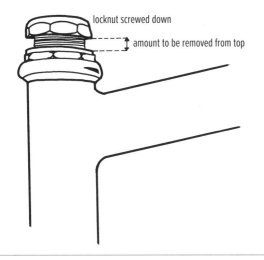

locknut screwed down

amount to be removed from top

that much length from the top of the steering tube, deburring it inside and out afterward. Determining the steering tube length for an already installed threadless headset is detailed in 12-2.

b. If you choose to cut the steering tube before installing the headset and are using a threaded headset, you need to know the headset's stack height. It is often listed in the headset owner's manual or online, or a bike shop can look it up in Barnett's manual or Sutherland's manual. Armed with this number, measure the length of the frame's head tube and add the headset stack height to this length. If you are adding extra spacers or brake-cable hangers between the headset nuts, add their thickness in as well. The resulting value represents the length that the steering tube must be. If the steering tube is already more than 5mm shorter than this sum, you need to find another headset with a shorter stack height (or, if you have included spacers, remove as many as needed).

c. If the steering tube is longer than this sum, you can cut the tube down to size. Measure twice and mark the cut line well, so you only have to cut once.

To get your cut straight, wrap tape around a threadless steering tube to line up the saw with; on a threaded steering tube,

thread the headset adjustable cup onto the steering tube well below the cut point, and then cut the steering tube to the correct length with a hacksaw, following a thread for straightness. When you unscrew the adjustable cup, it will dress the threads as it comes off the steering tube.

Use a flat file to square off the cut, and a round file to remove the burrs the hacksaw left on the inside and outside edges of the steering tube end.

12-22

INSTALLING THE HEADSET

LEVEL With the exception of the super-simple assembly of a modern carbon frame with molded-in bearing seats and a tapered fork steering tube and head tube (Fig. 12.26), there is really no good way to install a headset without at least a fork-crown-race punch (Fig. 12.43) and, for pressed-in cup frames,

12.43 Setting the fork-crown race

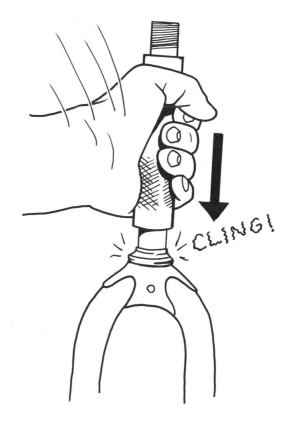

CLING!

12.44 Pressing in headset cups with a headset press

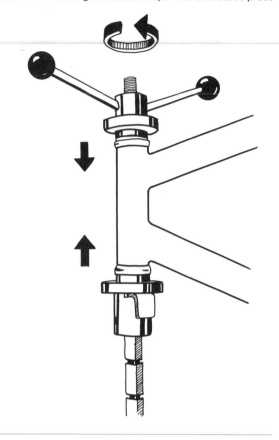

a headset cup press (Fig. 12.44). For most integrated headsets (Figs. 12.21, 12.25), you need only the former tool. If you do not have the necessary tools, it is better to take the parts to a bike shop for installation.

a. Headset installation in a frame with drop-in bearings and no races to press in the frame or on the fork (Fig. 12.26)

1. **Clean the parts.** Clean inside the ends of the head tube and around the base of the steering tube atop the fork crown; apply a thin layer of grease to those surfaces.

2. **Install the lower bearing.** Slide the bigger bearing down the (tapered) steering tube onto the top of the fork crown (Fig. 12.26). Check bearing orientation; if the bearing has a chamfered bore, place that side down onto the fork.

3. **Install the fork.** Insert the steering tube up into the head tube from the bottom.

4. **Install the upper bearing.** Slide the smaller bearing down the steering tube into the bearing surface in

the top end of the head tube. If the bearing has a chamfered bore, have it facing up.

5. **Install the compression ring.** Slide the compression ring down the steering tube, narrow end first, and push it into the bore of the bearing.

6. **Install shims, if present.** Some frames will require 1mm shim washers between the compression ring and the bearing cover to prevent the edge of the bearing cover from rubbing on the edge of the head tube as the fork turns. If yours has shims, install them.

7. **Install the bearing cover.** Slide the bearing cover down the steering tube against the compression ring or shim.

8. **Check clearances.** If the edges of the bearing cover rub on the edges of the head tube, pull the cover off and put in a 1mm shim. If the edge of the bearing cover is more than 1mm, remove a shim. The gap between the top edge of the head tube and the bottom edge of the bearing cover should be 0.5mm.

9. **Install the preload adjuster.** Install a steering tube support plug (Fig. 12.27) if the fork has a carbon steering tube. Install a star nut (Figs. 12.20, 12.24) if the fork has a metal steering tube.

10. **Install the headset spacers, stem, top cap, and compression bolt.**

11. **Adjust the headset (12-16).**

12. **Align the stem with the front wheel.**

13. **Tighten the stem.**

b. Installation in a frame with a press-on fork-crown race, with or without pressed-in headset cups (external or internal)

1. **Lubricate the parts.** Put a thin layer of grease on the ends of the headset cups that will be pressed into the head tube, in the fork-crown-race bore, inside the ends of the head tube itself, and on the base of the steering tube.

2. **Install the crown race.** Slide the fork-crown race down on the fork steering tube until it hits the enlarged section at the bottom. Slide the crown-race punch up and down the steering tube, pounding the crown race down until it sits flat on top of its seat on the fork crown (Fig. 12.43). Some crown-race

punches are longer and closed on the top and are meant to be hit with a hammer rather than slid up and down by hand. Hold the fork up against the light to see if there are any gaps between the crown race and the crown.

NOTE: *Thin crown races can be bent or broken by the crown-race punch. Chris King, Park Tool, and Shimano all offer support tools that sit over the race and distribute the impact from the punch. The Park punch, for instance, has three interchangeable ends (for each steering tube diameter). Hold the crown race against each of the three to see which one will best support the race while you strike the punch with a hammer.*

NOTE ON INTEGRATED HEADSETS: *For integrated headset frames with a press-on fork-crown race, there are no bearing cups requiring installation in the frame, so there is no sense in continuing with these instructions. Instead, continue with the assembly following the instructions in 12-22a, except note that the bearings will probably require a specific side to be up, since they won't be beveled on both ends. Make sure you install them right side up; the beveled edge of the bearing must sit in the beveled seat in the frame. You can tell by looking at them and also by the way the headset turns.*

3. **Install the headset cups.** By hand, place the headset cups into the ends of the head tube. Slide the headset press shaft through the head tube. Press the button on the detachable end of the tool and slide it onto the shaft until it clicks into the groove nearest the cups (Fig. 12.44). Or, with the press shown in Figures 11.28 and 11.29, screw the arms on as far as they need to go. This same method, and often the same press, can be used for internal (Fig. 12.22) and external (Figs. 12.20, 12.23) headsets with cups. You must make sure with internal cups that the press makes contact only with the outer cup flange and not the bearing seat. Some headset presses use a system of spacers and cones on both ends of the cups. Follow the instructions to set yours up properly. Whatever you do, be certain that the parts that make contact with the cups are not touching the precision surfaces the bearings roll in.

NOTE: *Chris King headsets have bearings that are pressed into the cups and cannot be removed (Fig. 12.35). If you use a headset press that pushes on the center of the cups, you will ruin the bearings. You need a press that pushes the outer part of the cup and does not touch the bearings. Chris King makes tool inserts that fit most headset presses, and Park has a headset press with large, flat ends for the purpose. On the other hand, some thin aluminum headset cups can be mashed by pushing on the outside of the cup with the flat surface of a headset press; stop pressing as soon as they reach the ends of the head tube. Otherwise, these cups need press inserts pushing on the edges of the bearings to support them under high loads.*

4. **Press the headset cups fully into place.** Hold the lower end of the cup press shaft with a wrench. That will keep the tool from turning as you press in the cups. Tighten the press by turning the top handle clockwise (Fig. 12.44). Keep tightening the tool until the cups are fully pressed into the ends of the head tube. Examine them carefully to make sure there are no gaps between the cups and the ends of the head tube.

NOTE: *You can easily crush thin headset cups with a flat-surface headset press, so be careful and stop when the cups reach the head tube.*

5. **Liberally apply grease to all bearing surfaces.** If you are using sealed cartridge bearings, a thin film will do.

6. **Assemble and adjust the headset.** Follow the directions in 12-18 and 12-16 for a threadless headset and those in 12-19 and 12-17 for a threaded headset.

12-23

HEADSETS, STEMS, AND BARS FOR CYCLOCROSS

Given the mud bath a 'cross bike gets, the headset must be well sealed. The front tire is constantly flinging mud straight up at the bottom headset bearing, so the seal there has to be particularly good. The headset is no place to scrimp on a 'cross bike.

If the bike has cantilever brakes, you'll need a front-brake cable stop somewhere in the steering system.

Although there are other ways to stop the front-brake cable housing before the bare cable runs to the straddle cable yoke (Figs. 9.32–9.41), the most common way is with a headset spacer that has a cable hanger on it (Fig. 9.9).

Regarding stem choice, your position on your 'cross bike will be similar to that on your road bike, so select a stem length and angle that give you a similar fit, at least to start with. You can adjust from there.

Regarding handlebar choice, the bar should be similar to the bar on your road bike. If your 'cross bar differs at all, it should be wider and have a shallower drop, so the height of your hands in the drops is closer to their height when riding on the lever hoods. The bar should have a round cross-section on the tops in case you decide to mount bar-top levers (Fig. 9.44).

12-24

TROUBLESHOOTING STEM, HANDLEBAR, AND HEADSET PROBLEMS

a. Bar slips

Tighten the pinch bolt on the stem that holds the bar, but not beyond the maximum allowable torque (see Appendix E). With a front-opening stem, make sure that there is the same amount of space between the stem and the front cap on both edges of the front cap. With any stem, if the clamp closes on itself without holding the bar securely, check to see if the bar is deformed or of a smaller diameter than the stem was made to fit and if the stem clamp is cracked or stretched. Replace any questionable parts.

Especially with a carbon handlebar, smear or spray carbon assembly compound on the bar and inside the stem clamp (this often works to increase clamping force with an aluminum bar as well). Whether it's a paste or a spray, it expands against the clamp and holds the bar tighter.

Failing that, you can slide a shim made from a beer or soda can between the stem and bar to hold it better, but never do this on a carbon or superlight aluminum handlebar; even with a heavy bar, replacing parts is a safer option. There is always a reason why parts that are meant to fit together no longer do! With superlight stems and bars, you cannot just keep tightening the small clamp bolts as you can the larger bolts on heavy stems because you will strip threads and/or cause bar and stem failures.

b. Bar makes creaking noise while you are riding

Loosen the stem clamp, grease the area of the bar that is clamped in the stem, slide the bar back in place, and tighten the stem bolt. Or, better yet, use carbon assembly paste or spray as recommended in 12-24a. Creaks are caused by two parts moving relative to each other, so filling those spaces with grease or assembly compound will quiet them down.

Lightly sanding the hard anodized surface inside the stem clamp and on the clamping area of the bar can sometimes eliminate creaking. Use a very fine grit sandpaper or emery cloth.

If the bar has a sleeved center section rather than a bulged section, the bar could be creaking inside the sleeve. There's no cure for this; replace the bar.

Less commonly, creaking can also emanate from between the steering tube and the rear stem clamp. Apply grease (or, with carbon parts, carbon assembly paste or spray) inside the stem clamp and on the steerer.

c. Clip-on bar slips

Tighten the clip-on's clamp bolts.

d. Stem not pointed straight ahead

Loosen the bolt (or bolts) securing the stem to the fork steering tube, align the stem with the front wheel, and tighten the stem bolt (or bolts) again. With a threaded headset, the bolt you are interested in is a single vertical bolt on top of the stem; loosen it about two turns, and tap the top of the bolt with a hammer to disengage the wedge on the other end from the bottom of the stem (Fig. 12.11). With a threadless headset, there are one (Fig. 12.4), two (Fig. 12.5), or, rarely, three horizontal bolts pinching the stem around the steering tube that need to be loosened to turn the stem on the steering tube. Do not loosen the bolt on the top of the stem cap (Fig. 12.30); you'll have to readjust the headset if you do.

Line up the stem by eyeballing it with the front wheel, and tighten the stem bolt(s).

e. Fork and headset rattle or clunk when you are riding

The headset is too loose. Adjust the headset (see 12-16 or 12-17).

f. Stem-bar-fork assembly does not turn smoothly but instead stops in certain fixed positions

The headset is pitted and needs to be replaced (see 12-20–12-22).

g. Stem-bar-fork assembly does not turn freely

The headset is too tight. The front wheel should swing easily from side to side when you lean the bike or lift the front end. Adjust the headset (see 12-16 or 12-17, depending on type).

h. Stem is stuck in or on fork steering tube

See 12-6.

i. Turning the fork, you hit a tight spot; it turns freely through some of the range and is too tight at other points

The bearings are not running parallel to each other and perpendicular to the axis of the steering tube and head tube. It is possible that the fork-crown race and/or headset cup(s) are not pressed in fully against their seats. If you see gaps under any of these, try pressing them in again (see 12-22). More likely, however, the surfaces on the frame and/or fork onto which the bearing cups and/or bearings sit are not cut parallel and concentric with the fork steerer and head tube. The fork and/or frame needs to be faced (see 12-21). This only works with metal frames and forks, however. If you have these symptoms with a carbon frame and fork, there is nothing you can do to correct it. In that case, take the frame and fork to a bike shop or send them back to the manufacturer.

PEDALS

TOOLS

15mm pedal wrench
2.5mm, 3mm, 4mm, 5mm, 6mm, 8mm hex keys
Phillips-head screwdriver
Small flat-blade screwdriver
Snapring pliers
8mm, 9mm, 10mm, 12mm, 19mm, 20mm, 22mm combination wrenches
13mm cone wrench
Shimano, Look splined pedal axle tools
Grease
Fine-tipped grease gun
Chain lubricant

OPTIONAL

Splined Campagnolo pedal cap tool
Torx drivers
8mm socket wrench
Speedplay grease fitting
Crank Brothers grease fitting
Threadlock compound

To best serve its purpose, a bicycle pedal only needs to be attached to the crankarm and provide a stable platform for the shoe. A simple enough task, but you'd be amazed at the different approaches that have been taken to achieve this goal.

There are two basic types of road pedals: 1) The standard cage-type pedal, with or without a toeclip and strap, is the simplest and cheapest (Fig 13.1). A "quill" pedal is a variation of the standard pedal in which the cage is asymmetrical on the two sides (Fig. 13.2). The cage is thinner on the bottom side and curves up on the outboard end to improve cornering clearance. 2) "Clip-in"–type pedals (Fig. 13.3) retain the foot with spring-loaded clips (like a ski binding) and are almost universal on mid- to high-end road bikes. Clip-in pedals are commonly called "clipless," because they have no toeclip.

Cage-type pedals are fairly common on lower-end bikes. They are relatively unintimidating for the novice rider, and the frame (or "cage") that surrounds the pedal provides a large, stable platform. A symmetrical, mountain bike–style, cage-type pedal has an identical top and bottom, and it can be used with just about any type of shoe. If you install toeclips on these without straps, your feet won't slide forward and will release easily in almost any direction.

One-sided road bike quill pedals (Fig. 13.2) are designed to be used exclusively with toeclips, because they cannot be pedaled upside down very well. A toe strap keeps your foot on the pedal and also allows you to pull on the upward part of the pedal stroke, giving you more power, a fluid pedal stroke, and balanced muscle development. Of course, as you add clips and straps, the pedal becomes harder to enter and to exit, and running shoes with aggressive tread become difficult to use. A tab on the cage plate opposite the toeclip is there so that you can flip the pedal up with your toe in order to slide your foot into the toeclip.

In addition to toeclips and straps, quill pedals are designed to be used with slotted cleats, which are attached to the soles of cycling shoes; the slot in the cleat engages with the rear of the cage, and helps hold the foot tight when climbing and sprinting. But almost no one uses this system these days; instead, clip-in pedals are almost universal.

Clip-in pedals (Figs. 13.3–13.5) offer the tight connection of the old cleat/toeclip/strap system, while also allowing easier entry and exit from the pedal. Clip-in pedals are more expensive and require special shoes and accurate mounting of the cleats. Your choice of shoes is limited to stiff-sole models that

13.1 Standard cage-type pedal

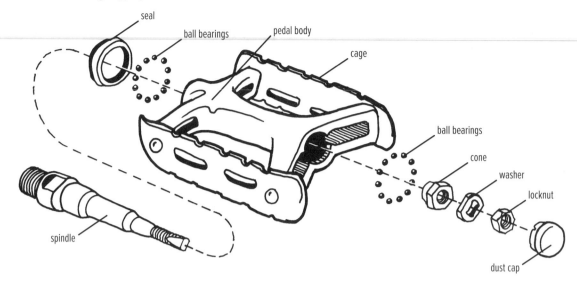

13.2 Standard cage-type "quill" pedal with toeclip and strap

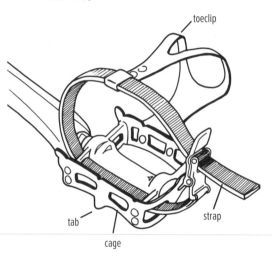

13.3 Clip-in pedal

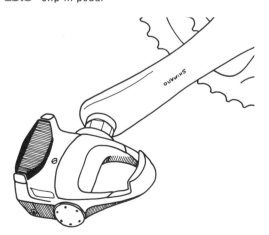

accept cleats for your particular pedal. Once properly mounted and adjusted, clip-in pedals waste less energy through flex and slippage and allow more direct power transfer to the pedals.

Clip-in models for cyclocross and mountain bikes have an open design to help clear mud. The cleat is small, and it mounts into a recess in the knobby outsole of a treaded shoe that is easier to run and walk in than a road shoe with an attached cleat, which protrudes from the bottom of the sole.

This chapter explains how to remove and replace pedals, how to mount the cleats and adjust the release tension with clip-in pedals, how to troubleshoot pedal problems, and how to overhaul and replace spindles on almost all road pedals. Incidentally, I use the terms "axle" and "spindle" interchangeably, as you are likely to hear either one when visiting bike shops for spare parts.

13-1

PEDAL REMOVAL AND INSTALLATION

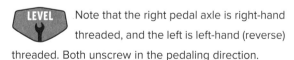

LEVEL Note that the right pedal axle is right-hand threaded, and the left is left-hand (reverse) threaded. Both unscrew in the pedaling direction.

There's an interesting bit of history behind the threading of pedal axles this way. In the early days of cycling, fixed-gear bikes were the norm, and it was decided that if the pedal bearings were to seize up, the

13.4 Removing or installing a pedal with a 15mm wrench

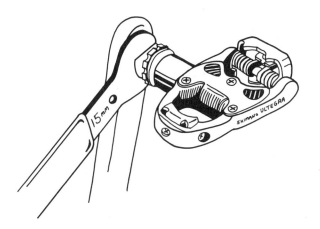

13.5 Removing or installing a pedal with a hex key

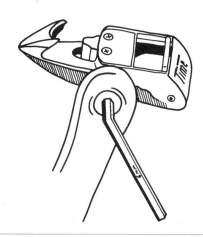

pedal should unscrew from the crank rather than tear up the rider's strapped-in feet.

a. Removal

1. **Slide a 15mm pedal wrench onto the wrench flats of the pedal axle (Fig. 13.4).** Or, if the pedal axle is designed to accept it, you can use a 6mm or 8mm hex key from the backside of the crankarm (Fig. 13.5). The latter is particularly handy on the road, because you probably won't be carrying a 15mm wrench. But if you are at home and the pedal is really tight, it will be easier to use a pedal wrench, assuming the pedal axle has wrench flats.

2. **Unscrew the pedal in the appropriate direction.** The right, or drive-side, pedal unscrews counterclockwise when viewed from that side. The left-side pedal is reverse threaded, so it unscrews in a clockwise direction when viewed from the left side of the bike. Once loosened, either pedal can be unscrewed quickly by turning the crank forward with a 15mm pedal wrench engaged on the pedal spindle and the rear wheel off the ground. You can remember that they unscrew in the pedaling direction from the earlier story about the concern with seized pedal bearings in the early days of cycling!

b. Installation

1. **Clean the threads.** Use a rag to wipe the threads clean on the pedal axle and inside the crankarm.

2. **Apply a light coat of grease to the pedal threads.**

3. **Install washer(s), if supplied.** If your cranks came with pedal washers, install one on each spindle, between the pedal and the crankarm. The crank manufacturer supplied those washers for a reason, so don't throw them out!

    You can also install up to 3mm of spacers between the pedal and the crankarm if you wish to obtain a wider pedaling stance (see 13-9 or Appendix C).

4. **Start screwing the pedal in with your fingers.** Pedals go in clockwise for the right pedal, counterclockwise for the left.

5. **Tighten the pedal.** Use a 15mm pedal wrench (Fig. 13.4) or a 6mm or 8mm hex key (Fig. 13.5). This can be done quickly by turning the cranks backward with a 15mm pedal wrench engaged on the pedal spindle.

SETTING UP CLIP-IN PEDALS

Setting up clip-in pedals involves installing and adjusting the cleats on the shoes and adjusting the pedal-release tension.

There are a number of different mounting platforms for road pedals, and your shoe sole must be compatible with your pedal cleats. The original clip-in road bike pedal system was the Look, which has three M5-threaded holes arranged in a triangular pattern (Fig. 13.6) to accept a three-hole cleat (Fig. 13.11). The original Time pedal system required a flat surface with four smaller threaded holes (Fig. 13.7), and all Speedplay

13.6 Three-hole (Look) cleat drill pattern

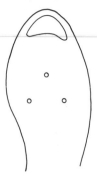

13.7 Original Time cleat drill pattern

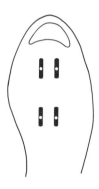

13.8 SPD cleat drill pattern

13.9 SPD-R cleat drill pattern

pedal cleats can be mounted on these as well. Shimano Pedaling Dynamics, or SPD, began as a system for mountain bikes with tiny cleats (Fig. 13.10) that were easy to walk in and less likely to clog with mud. SPD cleats mount with two side-by-side M5-thread screws, spaced 14mm apart. They screw into a movable threaded cleat-mounting plate behind two longitudinal grooves in the sole (Fig. 13.8). Crank Brothers cleats mount on this system, as do all mountain bike cleats. Some riders prefer an SPD-compatible system for their road bike that uses either a single-sided road bike pedal or a double-sided mountain bike pedal so that they can use a mountain bike shoe with a recessed cleat, which is far easier to walk in than a road bike cycling shoe. And for cyclocross, mountain bike pedals and shoes are a must.

Shimano's SPD-R pedal (since abandoned) required a shoe having a single lengthwise slot in the sole with an M5-threaded hole at either end moving on a threaded backing plate behind the slot (Fig. 13.9). (Diadora pedals have yet another mounting pattern, but they are long gone from the market.)

After a long shakeout, Look's original three-hole mounting system (Fig. 13.6) has emerged victorious on road bikes. The two-adjacent-hole system pioneered by Shimano is the only game in town when it comes to cyclocross and mountain bikes, and, as already noted, many riders and at least one pedal system use it on the road as well. Now Shimano SPD-SL, Time, Speedplay, and all other road pedals save for Crank Brothers Quattro (Fig. 13.25), Ritchey V4 road, and a few others mount on the standard three-hole system.

13-2

INSTALLING AND ADJUSTING PEDAL CLEATS ON THE SHOES

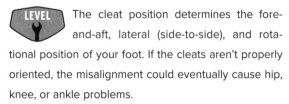 The cleat position determines the fore-and-aft, lateral (side-to-side), and rotational position of your foot. If the cleats aren't properly oriented, the misalignment could eventually cause hip, knee, or ankle problems.

1. **Put your shoe on and then mark the position of the ball of your foot (the big bump behind your big toe) on the outside of the shoe.** This mark will

13.10 Cleat centered 1cm behind the ball-of-foot line

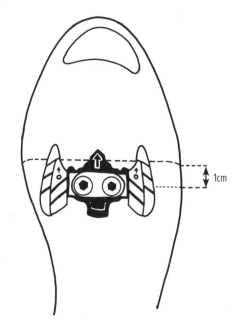

help you position the cleat so that the ball of your foot will be straight above or in close proximity to the pedal spindle. Take the shoe off, and continue drawing the line straight across the bottom of the shoe.

NOTE: *On SPD-R cleats, the pontoon mounts on the rear bolt, pointing back.*

2. **Grease the cleat screw threads, and screw the cleat that came with the pedals onto the shoe.** This usually requires a 4mm hex key or a Phillips-head or standard screwdriver. If the cleats come with adhesive-backed sandpaper cut to the shape of the cleat, adhere it to the bottom of the cleat so that it faces the sole.

   Make sure you orient the cleat in the appropriate direction. Some cleats have an arrow indicating forward (Fig. 13.10); if yours do not, the instructions accompanying the pedals will specify which direction the cleat should point, and in some cases, on which shoe an asymmetrical cleat should be mounted. If mounted on a road shoe, SPD and SPD-R cleats require rubber "pontoons" on a plate mounted under the cleat (Fig. 13.10). The pontoons guide the small cleat into the pedal. The pontoons are not necessary on a mountain bike shoe, as the recessed area in its tread will guide the cleat.

3. **Position the cleat.** Temporarily place it in the middle of its lateral- and rotational-adjustment range. Setting the fore-and-aft position requires knowing where the pedal spindle is positioned relative to the cleat. Many cleats have a mark on the side indicating the spindle position (Fig. 13.11). If your cleat has such a mark, line it up 0–1cm behind the line you drew in step 1 across the shoe sole. With an SPD pedal, line up the mounting screws 0–1cm behind the mark you made in step 1 (Fig. 13.10). With a Speedplay cleat, place the center of the hole in the middle of the cleat 0–1cm behind the mark you made in step 1. If you're not sure, tighten the screws and set the shoe in the pedal. When the shoe is level, the standard is for the ball of the foot to be between 0cm and 1cm forward of the pedal spindle. Putting the ball farther forward is usually helpful to develop power, while high-cadence spinning is usually enhanced with the ball of the foot farther back. If you know which type of rider you are, you can set the shoe as appropriate; a gear-masher will like the cleat farther back than will a spinner. Very small feet sometimes do better with the cleat farther forward on the shoe, placing the ball of the foot behind the spindle. Riders with large feet often prefer the cleat all the way back, so the foot goes as far over the pedal as possible. The long lever that is the rider's foot and which must be controlled from ankle to foot attachment point on the pedal will be reduced. And pedaling force from

13.11 Look cleat with a mark for the pedal center

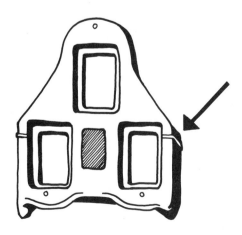

a large rider, concentrated on the same-size cleat as for a small rider, is better distributed over the shoe if the cleat is located behind the ball of the foot, resulting in less pain under the metatarsals. Speedplay offers a cleat extender base plate kit to offset the cleats either 14mm farther rearward or 2mm farther forward than the standard black, plastic base plates.

NOTE: *If you have an old-style Time pedal (Fig. 13.5), make sure you don't put an old-style Time rear cam on the wrong shoe, or you will not be able to release by twisting outward.*

4. **Snug the screws down.** Tighten them enough to prevent the cleat from moving when clipped in or out of the pedals, but don't tighten them fully. Follow the same steps with the other shoe. Check that the curvature of the cleat matches that of the shoe. If it does not, the cleat may bow when tightened, which will make it hard to clip in or release. Some pedal manufacturers offer cleat shims to fill spaces and keep the cleat flat; see your bicycle dealer for assistance.

5. **To set the lateral cleat position, put the shoes on, sit on the bike, and clip into the pedals.** Ride around a bit. Notice the position of your feet. Generally, the closer your feet are to the plane of the bike, the more efficient your pedaling will be, but you don't want them in so far that your ankles bump the cranks. Take the shoes off and adjust the cleats laterally, if necessary, to move the feet side to side. Get back on the bike and clip in again.

   Speedplay Zero cleats offer independently adjustable fore-and-aft, side-to-side, and rotational foot positions; each can be set or changed without affecting the position of the other two adjustments. Note that early Time pedals have no lateral cleat adjustment; recent Time models offer it by means of interchanging the left and right cleats.

6. **To set the rotational cleat position, ride around and notice if your feet feel twisted and uncomfortable.** You may feel pressure on one side of your heel from the shoe. If so, remove your shoes and rotate the cleat slightly in the direction that relieves that pressure.

NOTE: *Most pedals now offer free-float, allowing the foot to rotate freely for a few degrees before releasing. Precise rotational cleat adjustment is less important if the pedal is free-floating.*

I recommend starting with the greatest amount of free-float angle the system allows. You can reduce the float later if you desire.

Some pedals have a dial on the back of the clip to set the amount of free-float rotation, and some cleats (Speedplay Zero) can be adjusted to set the amount of float. Many companies also offer a number of cleat styles having increased or reduced (or eliminated) free-float range.

SPD-R cleats (for the discontinued pedals in Figs. 13.4, 13.14, and 13.18) come in three styles: one with a wide tip for fixed operation, and two narrower-tip models for different amounts of free-float. Vertical cleat play can be eliminated by raising rubber bumpers on the pedal body. Dura-Ace SPD-R pedals have a 3mm nut on the bottom of the pedal to push the bumper up, and Ultegra SPD-R pedals require removing three screws on the face of the pedal to interchange the two pads with thicker ones.

7. **Once your cleat position feels right, trace the cleats with a pen or a scribe.** That way, you can tell if the cleat stays put.

8. **While holding the cleat in place, tighten the bolts down firmly.** Hold the hex key close to the bend to minimize your leverage so that you do not apply too much torque and strip the bolts. There is little danger of overtightening with a screwdriver, but do take care that the blade (or Phillips tip) fits well in the screw slot (or Phillips cross). Push down firmly while tightening to avoid stripping the head of the screw.

NOTE: *If you have a small torque wrench, tighten the cleat screws to 35–43 in-lbs (4–5 N-m); see Appendix E.*

9. **Check the screw length.** Remove the insole and feel around inside above the cleat to ensure that the screws are not too long and pushing up on the cardboard lasting sole. Get shorter screws or shorten the longer ones if that is happening. You don't want bumps sticking up into the balls of your feet while riding!

10. **When riding with new shoes or pedals, bring cleat-tightening tools along.** You may want to fine-tune the cleat adjustment over the course of a few rides.

11. **Retighten the cleat bolts after every ride for the first few rides.** After that, the cleat will have pushed itself into the shoe sole as far as it can go. This is particularly important with a mountain bike shoe, where the cleat is harder and smaller and the surface of the outsole is generally softer than a road shoe outsole. This is the key to keeping the cleat bolts from falling out as well as preventing the cleats from slipping. Threadlock compound on the bolts also can help. Once the bolts stop turning at the same torque setting, you can stop doing this daily, but do check them from time to time.

13-3

ADJUSTING RELEASE TENSION OF CLIP-IN PEDALS

LEVEL If you find the factory-set release adjustment to be too loose or too restrictive, you can change it on many clip-in pedals; notable exceptions without a spring-tension adjustment are Crank Brothers (Figs. 13.25, 13.30) and some Speedplay (Fig. 13.23), Look Quartz mountain pedals, and some Time models (Figs. 13.5, 13.29). The adjusting screws are usually located on or near the spring-loaded rear clip (Figs. 13.12, 13.13). The screws are usually operated with a small (usually 3mm) hex key or a small screwdriver.

1. **Locate the tension-adjustment screws.** Older Looks have a screw (either slotted or 2.5mm or 3mm hex head) on top of the platform (Fig. 13.12); Look Anatomics and Campagnolo ProFits (Fig. 13.15) have a 3mm hex screw on the side. Look Keos and Shimano SPD-SLs have a 3mm hex screw on the top of the rear clip. Ritchey, Shimano SPD road (Fig. 13.13), SPD-R (Fig. 13.14), and SPD MTB/cyclo-cross (Fig. 13.32) pedals have a 3mm hex screw on the back of the clip, as do clones of these pedals by other brands.

NOTE: *There are many SPD- and Look-style pedal clones under various brand names on the market. The*

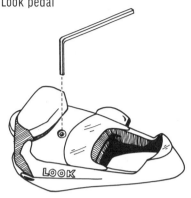

13.12 Release-tension adjustment screw on a Look pedal

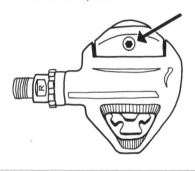

13.13 Release-tension adjustment screw on a Shimano SPD pedal

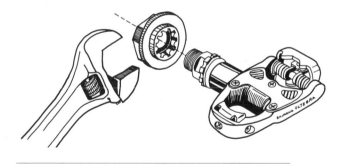

13.14 Tool for removing a Shimano pedal-axle assembly

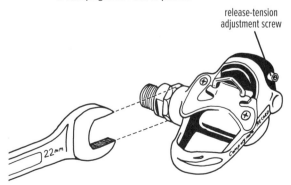

13.15 A 22mm wrench fits the axle assembly of a Campagnolo ProFit pedal.

release-tension adjustment screw

22mm

cleat-mounting and tension-adjustment instructions for SPD or Look pedals generally apply to these pedals as well.

2. **To loosen the tension adjustment, turn the screw counterclockwise; to tighten it, turn it clockwise (Figs. 13.12, 13.13).** It's the classic lefty loosey, righty tighty approach. There usually are click stops in the rotation of the screw. Tighten or loosen one click at a time (one-quarter to one-half turn), then ride the bike to test the adjustment. Many types include an indicator that moves with the screw to show relative adjustment. Make certain that you do not back the screw out so far that it comes out of the spring plate or so far that it can vibrate loose; feel for at least the first "click" to hold it in place.

NOTE: *With some pedals, you will decrease the amount of free-float in the pedal as you increase the release tension.*

OVERHAULING PEDALS

 Like a hub or bottom bracket, pedal bearings and bushings need to be cleaned and regreased periodically.

There is a wide variation in road bike pedal designs. This book is not big enough to go into great detail about the inner workings of every model. Speaking in general terms, pedal guts fall into two broad categories: those that have loose ball bearings (Figs. 13.18, 13.19, 13.26–13.28, 13.32), and those that have cartridge bearings (Figs. 13.16, 13.17, 13.20–13.25, 13.29, 13.31, 13.33). Furthermore, there are two other broad categories that overlap the above two categories, namely, pedals that are closed on the outboard end and have a nut or a snapring surrounding the axle on the inboard end holding the assembly together (Figs. 13.14–13.22, 13.29, 13.32), and pedals that have a dust cap on the outboard end with a nut on that end that holds the assembly together (Figs. 13.23, 13.24, 13.27, 13.28, 13.30, 13.31). I've organized the following instructions based on these divisions rather than on whether the pedals have loose balls or cartridge bearings. There is also a small category of pedals that come apart like a clamshell (Fig. 13.33); I have lumped these in with pedals that have a dust cap on the outboard end.

NOTE: *Whether the pedals have ball bearings or cartridge bearings, you can return a pedal to like-new performance by replacing rusted or otherwise compromised bearings with new ones. You can also replace loose steel balls with ceramic balls or steel cartridge bearings with ceramic cartridge bearings. Ceramic ball bearings are harder, stiffer, generally rounder, and more uniform in size than steel balls; can't rust; and spin with less resistance. They are, however, more expensive than steel bearings.*

Getting started on pedal overhaul

1. **Remove the pedal from the bike (Figs. 13.4, 13.5).**
2. **Examine the pedal style.** Before you start, figure out how the pedal is put together so that you will know how to take it apart; the following paragraphs and the illustrations on subsequent pages should help. In a few cases, the workings of the pedal guts may not be clear until you have completed step 1 in the overhaul process.

   Most Shimano pedals have two sets of loose bearings and a bushing (Figs. 13.19, 13.32). The bearings and bushing will come out with the complete axle assembly (Fig. 13.26); you can see the tiny ball bearings at the small end of the axle behind the wrenches in Figure 13.26. Shimano's Dura-Ace model SPD-R (Fig. 13.18) and SPD-SL pedals have a set of ball bearings on each end of the spindle and a set of 6mm-inside-diameter (ID) needle bearings (not shown) in between them.

   Speedplay X/3 pedals have an inboard 10mm ID Teflon bushing, and an outboard 6mm ID cartridge bearing. Speedplay X/5, Light Action Chrome-Moly, and Frog (Fig. 13.33) pedals have a clamshell body with an inboard needle bearing and an outboard cartridge bearing. Speedplay X/1, X/2, Zero, and high-end Light Action pedals have an inboard pressed-in needle bearing (not shown) and an outboard pair of cartridge bearings (Fig. 13.23).

   The Campagnolo Record ProFit (Fig. 13.15) pedal has one inboard and two outboard 17mm-outside-diameter (OD) cartridge bearings.

   Older Look, Diadora, and older Time pedals have an inboard cartridge bearing (19mm, 24mm, and

13.16–13.22 Exploded views of clip-in pedals closed on the outboard end

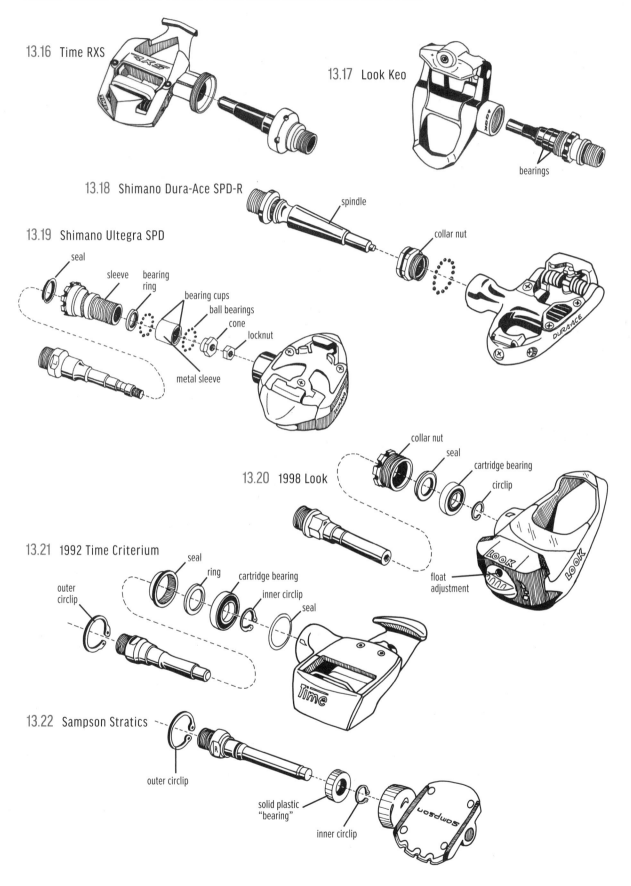

13.16 Time RXS

13.17 Look Keo

bearings

13.18 Shimano Dura-Ace SPD-R

spindle

collar nut

13.19 Shimano Ultegra SPD

seal

sleeve

bearing ring

bearing cups

ball bearings

cone

locknut

metal sleeve

collar nut

seal

cartridge bearing

circlip

13.20 1998 Look

float adjustment

13.21 1992 Time Criterium

seal

ring

cartridge bearing

inner circlip

seal

outer circlip

13.22 Sampson Stratics

outer circlip

solid plastic "bearing"

inner circlip

13.23–13.25 Exploded views of clip-in pedals openable from the outboard end

13.23 Speedplay Zero, Light Action Ti or SS, X/1 or X/2 pedal

13.24 Ritchey SPD-style

13.25 Crank Brothers Quattro

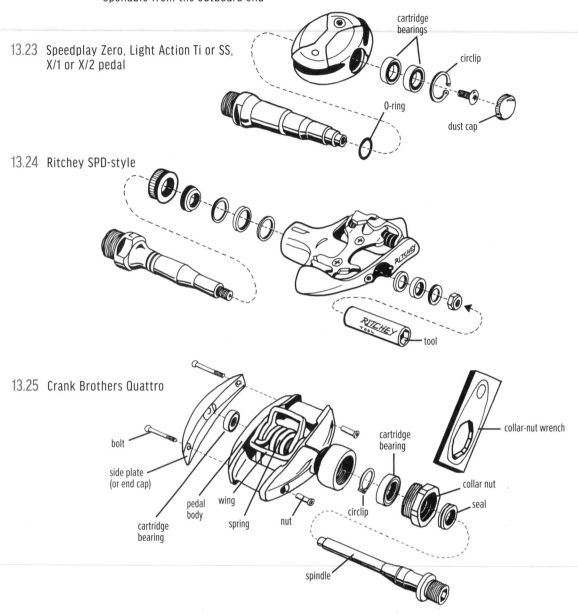

24mm OD, respectively) and one or two pressed-in outboard needle-bearing sets (not shown) (Figs. 13.20, 13.21).

Of the newer, carbon-composite-body road pedals, Look Keos (Fig. 13.17) have a pair of 15mm OD inboard cartridge bearings and an 8mm ID outboard needle bearing, whereas Time RXS pedals (Fig. 13.16) have a 21mm OD inboard cartridge bearing and an 8mm ID outboard needle bearing.

Older Time mountain pedals were closed on the outboard end and accessed by means of a snapring (Fig. 13.29) or a threaded collar. They had a needle bearing deep inside and a cartridge bearing at the inboard opening like Time road pedals (Figs. 13.16, 13.21). Later Time mountain pedals have an outboard dust cap and cartridge bearings or a bushing and a cartridge bearing (Fig. 13.31).

Crank Brothers Quattros (Fig. 13.25) constituted a brief foray into road pedals for the company, but they have been discontinued. Quattros have two cartridge bearings: a large one on the inboard side and a smaller one on the outboard side.

Depending on model, some Crank Brothers mountain bike pedals, like the Eggbeater (Fig. 13.30), Candy, Mallet, and Acid, either share this spindle arrangement or have one cartridge ball bearing (outboard) and one bushing (inboard). Higher-end, numbered (i.e., Eggbeater 3 or 11 or Candy 3 or 11) Crank Brothers pedals have an inboard needle bearing instead of a bushing. While these spin with less friction when properly cleaned and greased, the needle bearing, unlike the brass bushing in the lower-end and older Crank Brothers pedals, will rust and seize if used in wet and muddy conditions (cyclocross!) with insufficiently frequent overhaul. When the pedal seizes up, it will unscrew while pedaling and will come right off, attached to your shoe!

Sampson Stratics (Fig. 13.22) pedals have a 24mm OD, solid-plastic "bearing" on the inboard side, and a plastic bushing inside the pedal body.

Ritchey SPD-style road pedals (Fig. 13.24) have two sets of pressed-in needle bearings, one with an ID of 10mm and the other with an ID of 7mm.

13-4

OVERHAULING PEDALS CLOSED ON THE OUTBOARD END

 1. Make sure the pedal does not have a dust cap or screw cover on the outboard end. If it does, skip to 13-5. The exception is the Crank Brothers Quattro (Fig. 13.25), which has a removable end cap but still is overhauled by unscrewing the inboard collar nut.

2. **Remove the pedal body.** Unless you have an old Time (Fig. 13.21), Diadora, or Sampson (Fig. 13.22) pedal, remove the axle assembly by unscrewing the nut surrounding the axle (collar nut) where it enters the inboard side of the pedal (Figs. 13.14, 13.15). You can usually hold the pedal in your hand and unscrew the collar nut, but you may want to hold the pedal body in a padded vise while unscrewing the nut. The collar nut is often made of plastic and can crack if you turn it the wrong way, so be careful. Hold the pedal body with your hand or in a padded vise while you unscrew the assembly. The fine threads take many turns to unscrew.

NOTE: *The threads on the pedal body are reversed compared with the crankarm threads on the axle. That means the right-axle assembly unscrews clockwise, and the left-axle assembly unscrews counterclockwise.*

a. Most Shimano pedals disassemble with a special plastic, splined tool (Fig. 13.14); some Looks (Figs. 13.17, 13.20) and Crank Brothers Quattros (Fig. 13.25) also have their own special tools. Use a large adjustable wrench or a vise to hold the tool (Fig. 13.14). Most other pedals take a 19mm, 20mm, or 22mm open-end wrench (Fig. 13.15). The collar nut on a Time RXS (Fig. 13.16) or Xpresso requires a special tool, but in its absence, the large-diameter nut is easy to unscrew with a pair of pliers wrapped in cloth to avoid marring the nut's surface. In a pinch, you can do this with any of these pedals.

b. Campagnolo ProFit (Fig. 13.15) and many Look pedal-axle assemblies unscrew with a 22mm open-end or box wrench. Removal of the Dura-Ace SPD-R (Fig. 13.18) and SPD-SL axle assemblies requires a 20mm wrench, and some Look Keos (Fig. 13.17) take a 19mm wrench (Keos often require a standard wrench for the left pedal and a Look splined tool for the right pedal). Some older Look axles are accessed with a special Look splined tool similar to the one that unscrews most Shimano pedals (Fig. 13.14). Note that original Shimano clip-in road pedals are actually Looks with Shimano axle assemblies, and Campagnolo clip-in pedals prior to 1997 (other than an unfortunate attempt by Campagnolo itself in the late 1980s) are also Looks with Campagnolo axle assemblies.

c. Older Time, Diadora, and Sampson pedal axles are retained by a snapring on the crank side (Figs. 13.21, 13.22, 13.29). Popping the snapring out usually requires inward-squeezing snapring pliers (Fig. 11.32), but the snapring on a Time Impact requires only a thin screwdriver to remove, once you move the end of the snapring under the little notch in the inboard pedal-body face so that you can pry it up with the screwdriver. Skip to step 3 after removing the snapring.

3. **Examine the axle assembly and the bore of the pedal body.** Is there grease inside? Is it relatively clean? Are all internals free of rust? If there is a sleeve surrounding the spindle (Fig. 13.19), does it spin smoothly on the ball bearings without back-and-forth play? If the answer is no to any of these questions, thorough overhaul is recommended; skip to step 5 for that. Otherwise, you can get away with a quick wipe-down and blob of grease inside of the pedal bore; continue to step 4.

4. **Perform quick clean and lube:**

   a. **With a rag, wipe the outside of the spindle assembly.** Do not remove any nuts from the spindle or disassemble anything attached to it.

   b. **Clean inside of the pedal bore.** Shove the thin end of a rag in there and twist it around.

   c. **Grease inside of the pedal bore.** Put a glob of grease inside of the pedal body about equal to a third or half of the volume of the pedal bore.

   d. **Push the spindle assembly back in.** If it won't go all of the way in, you may have too much grease or an air bubble behind the grease stopping it, so remove some grease and try again. Pushing the axle assembly into the hole provides enough pressure on the grease to squeeze through bearings like those in Figures 13.18 and 13.19, as well as through some cartridge bearings.

   e. **Tighten the collar nut.** Avoid cross-threading, something easy to do with a plastic collar nut. You're done! Skip to step 10.

5. **Disassemble the axle assembly.** You will notice zero, one, or two nuts on the thin end of the axle that serve to hold the bearings and/or bushings in place. Remove the nut or nuts as follows:

   a. No nut? Skip to step 6.

   b. If the axle has two nuts on the end (Figs. 13.19, 13.32), they are tightened against each other. To remove them, hold the inner nut with one wrench while you unscrew the outer nut with another (Fig. 13.26). On Shimano pedals, the inner nut does double duty as the bearing cone; be careful not to lose the tiny ball bearings as you unscrew the cone!

   c. If the axle has a single nut on the end, simply hold the axle's large end with the 15mm pedal wrench and unscrew the little nut with a 9mm or 12mm wrench (or whatever fits it). The nut will be tight, because it has no locknut.

6. **Clean all of the parts as follows:**

   a. If it is a loose-bearing pedal, use a rag to clean the ball bearings, the cone, the inner ring that the bearings ride on at the end of the plastic sleeve (it looks like a washer), the bearing surfaces on either end of the little steel cylinder, the axle, and the inside of the plastic or metal axle sleeve (Fig. 13.19). To get the bearings really clean, wash them in the sink in soap and water with the sink drain plugged; the motion is the same as washing your hands, and results in both the bearings and your hands being clean for a sterile reassembly. Blot dry.

   b. On a pedal with a cartridge bearing (Figs. 13.16, 13.17, 13.20–13.22, 13.29), if the bearing is dirty or worn out and has steel covers that cannot be pried off without damaging them, then replace it. Pry off plastic cartridge-bearing covers that can be pried off (Fig. 11.37), and clean and grease the ball bearings inside.

   c. Needle bearings (not visible in the figures because they are pressed inside) on Dura-Ace SPD-R/PD-7700 and SPD-SL/PD-7800, Look, Time, and Diadora (Figs. 13.18, 13.20, 13.21, 13.29) can be cleaned with solvent and a bottle brush or a thin toothbrush slipped inside the pedal-body bore. Unless you waited too long to open the pedals after heavy-duty rain riding or power-washing, the needle bearings usually need little cleaning, because they are well isolated inside.

d. On a Sampson (Fig. 13.22), just wipe down the axle, the plastic bearing, and the pedal-body bore. Do the same for an inexpensive bushing-only pedal.

7. **Grease and reassemble.** This is a simple process with all pedals closed on the outboard end except in the case of pedals with loose bearings that are too rusty, dirty, or maladjusted to perform step 4 on!

a. With a loose-bearing pedal, you have some exacting work to place the bearings on their races and screw the cone on while they stay in place. For most Shimano guts (Fig. 13.19), grease the bushing inside the axle sleeve, and slide the axle into the sleeve. Slide the steel bearing ring, on which the inner set of bearings rides, down onto the axle and against the end of the sleeve. Make sure that the concave bearing surface faces away from the sleeve. Coat the ring with grease and stick half of the bearings (usually 12) onto the outer surface of the ring. Slip the steel cylinder onto the axle so that one end rides on the bearings. Make sure that all of the bearings are seated properly and that none are stuck inside of the sleeve.

b. To prevent the bearings from piling up on each other and ending up inside the sleeve instead of on the races, grease the cone and start it on the axle a few threads. Place the remaining half of the bearings on the flanks of the cone. Being careful not to dislodge the bearings, screw the cone in until the bearings come close to the end of the cylinder without touching it. While holding the axle sleeve, push the axle inward until the bearings seat against the end of the cylinder. Make sure that the first set of bearings is still in place. Screw the cone in without dislodging the

inboard bearings by avoiding turning the axle or the cylinder. Tighten the cone with your fingers only, and loosely screw on the locknut.

c. Pre-1997 Look-style Campagnolo pedal guts are similar to Shimano's, except that the bearing race is machined into the axle (rather than being a separate ring), and there are two sleeves, not one. Orient the sleeves so that their bearing races face outward, and then follow the previous steps.

d. With Dura-Ace SPD-R/PD-7700 (Fig. 13.18) or Dura-Ace SPD-SL/PD-7800 pedals, you needed to push back on the bearing cup (on the end of the 20mm nut that holds the axle into the pedal body) to remove the ball bearings in the first place. Grease the cup and push back on it again to allow enough space between the cup and the cylinder to set each of the 17 balls onto the edge of the cup with a small screwdriver.

8. **Adjust the axle assembly.** (For cartridge-bearing pedals with no end nut, skip this step.)

a. On pedals with two nuts on the end of the axle, hold the cone or inner nut with a wrench and tighten the outer locknut down against it (Fig. 13.26). Check the adjustment for freedom of rotation, and be sure there is no lateral play. Readjust as necessary by

13.26 Most Shimano axles have a cone and a locknut, used to adjust bearing play.

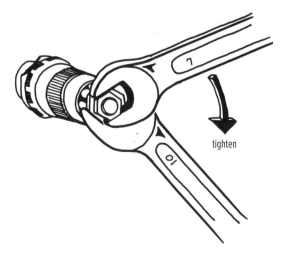

tighten

tightening or loosening the cone or inner nut and retightening the locknut.

b. Pedals with a small cartridge bearing and a single nut on the end of the axle, such as Campagnolo Record ProFit (Fig. 13.15), require that you tighten the nut against the cartridge bearing while holding the other end of the axle with the 15mm pedal wrench. Tighten it enough to remove play but not enough to bind the axle.

9. **Replace the axle assembly in the pedal body.** Smear grease on the inside of the pedal hole; this will ease insertion and act as a barrier to dirt and water. Screw the sleeve in with the same wrench you used to remove it (Figs. 13.14, 13.15).

NOTE: *Pay attention to proper thread direction (see the Note in step 2)! Tighten carefully; it is easy to cross-thread or overtighten, which can strip or crack a plastic nut.*

10. **Put the pedals back on your bike.** Go ride.

13-5

OVERHAULING PEDALS WITH A DUST CAP ON THE OUTBOARD END

NOTE: *Assess the value of the pedals and your time before continuing. Well-made, older, classic quill-racing pedals like Campagnolo (Figs. 13.2, 13.27) deserve careful attention, but many non-clip-in pedals may not be worth the effort of overhaul.*

1. **Remove the dust cover from the outboard end of the pedal with the appropriate tool.** This could be a pair of pliers, a flat or Phillips screwdriver, a coin, a hex key, or a splined tool made especially for the pedals; it's pretty easy to figure out which one is needed to remove the cap. Some Time ATAC pedals require a tool with two thin pins; a pair of snapring pliers may do the trick. Dig the dust cap out from SPD-style Ritcheys (Fig. 13.24) and Speedplay X/1, X/2, Zero, and Light Action Ti and SS (Fig. 13.23) with a sharp pick. (Ritchey pedals first require removal of a 2.5mm hex screw holding down the corner of the dust cap.)

2. **If you can just squirt grease inside, do it.** Speedplay bearings can be regreased without removing the axle and on newer models without removing the dust cap. This is far preferable to pulling the pedal apart, and it generally works really well. It is so easy that there is no reason to not do it frequently and spare yourself the need to ever pull these pedals apart. On current Zero, X/1, X/2, X/5, and Light Action, and recent X/3, after removing the screw from the outboard end, pump grease in with a fine-tip grease gun while slowly turning the spindle until grease squirts out the opposite end. On an older X/1 or X/2, remove the dust cap as just described in step 1, insert Speedplay's Speedy Luber grease-injection fitting, and squirt grease in with a fine-tip bicycle grease gun until it squirts out the other end.

Older Crank Brothers Eggbeater and Candy pedals (not the Quattro) have a similar feature, and the screw-in grease adapter (which screws in where the dust cap was and accepts the grease gun tip) is included with every pair of pedals. Improved internal seals and sealed cartridge bearings that were phased into Crank Brothers pedals in 2006 made the system too tight to flush grease through with a grease gun. You need to remove the end cap as well as the nut on the end of the spindle, and slide the pedal body off the spindle to regrease it.

You're done; skip to step 13.

3. **Unscrew the locknut.** Hold the end of the axle with a 8mm or 6mm hex key or 15mm pedal wrench, and unscrew the locknut with the appropriate-size socket wrench.

a. The locknut may require a deep, thin-wall 8mm socket; Ritchey makes a double-ended, thin 8mm socket (Fig. 13.24) for the purpose that you can turn with an 8mm hex key in the other end.

b. On a Speedplay Zero, X/1, or X/2 pedal, remove the Torx screw on the outboard end under the dust cap (Fig. 13.23) with the appropriate Torx driver. Heat the bolt first with a soldering iron to soften the threadlock compound to avoid stripping the shallow bolt head.

c. On a Speedplay X/3, X/5, Frog (Fig. 13.33), or Light Action Chrome-Moly, carefully pry the halves of the pedal apart with a knife or razor blade after removing the 2.5mm hex pedal-body screws from either side.

4. **Remove the axle.**

a. With a loose-bearing pedal (Fig. 13.27), hold the pedal over a rag to catch the bearings and then unscrew the cone. Keep the bearings from the two ends separate in case they differ in size or in number. Count them so that you can put the right numbers back in when you reassemble the pedal. The guts should look like Figure 13.27. If the pedal does not have loose bearings (Figs. 13.23, 13.24), the procedure is different, as detailed next.

b. With a sealed-bearing pedal (Figs. 13.24, 13.30, 13.31), once you have removed the locknut, you can pull the axle out.

c. With a Speedplay Zero, X/1, X/2, or high-end Light Action pedal (Fig. 13.23), pull out the axle. Remove the little snapring from the outboard end of the pedal bore with inward-squeezing snapring pliers (Fig. 11.32). With the axle or a hex key, carefully push the cartridge bearings out. The cartridge bearings are easily replaceable, but if the needle

bearings are in bad shape, you will have to buy a new pedal body from Speedplay with the needle bearings already pressed in.

d. On a Speedplay X/3, X/5, Frog (Fig. 13.33), or Light Action Chrome-Moly, lift out the axle assembly and remove the 9mm locknut from the end of the spindle. Pull the bearings and bushing (all located in an alloy sleeve) and O-ring off the spindle (Fig. 13.33).

5. **Clean the bearings, bushings, and bearing races.** If there is a dust cover on the inboard end of the pedal body, you can clean it in place, or pop it out with a screwdriver and clean it separately.

a. Use a rag to clean the inside of the pedal body by pushing the rag through with a screwdriver; if there are bronze bushings inside, get those really clean, and if there are needle bearings inside, skip to step d.

b. If you want to get loose bearings really clean, wash them in a plugged sink with soap and water. The motion is the same as washing your hands, and it results in both the bearings and your hands being clean for a sterile reassembly. Blot dry.

c. On a pedal with a cartridge bearing on the end (Figs. 13.23–25, 13.30, 13.31, 13.33), if the bearing is dirty or worn out and has

13.27 Loose-bearing "quill" pedal

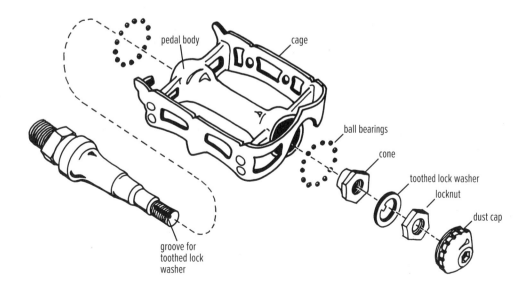

pedal body

cage

ball bearings

cone

toothed lock washer

locknut

dust cap

groove for toothed lock washer

steel bearing covers that cannot be pried off without damaging them, then replace it. Pry off plastic cartridge-bearing covers that can be pried off (Fig. 11.37) to clean and grease the bearing.

d. Ritchey, Speedplay, and some high-end Crank Brothers pedals have pressed-in needle bearings inside. Scrub them with solvent and a rag or thin bottle brush, if they are dirty. Unless you waited too long to open the pedals after heavy-duty rain riding or power-washing, the needle bearings usually need little cleaning, because they are well isolated inside. Removal of bad needle bearings requires a special tool to pull them out; you're probably looking at a new pedal or pedal body if yours have fallen apart.

6. **Press the inboard dust cover back into the pedal body (if you removed it earlier).**

7. **Grease the bearings, bushings, and bearing races.**

   a. With loose bearings, smear a thin layer of grease in the inboard bearing cup and replace the bearings. Once all of the bearings are in place, there will be a gap equal to about half the diameter of one bearing.

   b. Grease bushings and needle bearings in the bore of the pedal body.

   c. Push (greased) cartridge bearings back into the pedal body with your finger.

8. **Drop the axle in and turn the pedal over so that the outboard end is up.**

   a. With loose bearings, smear grease in that end and replace the bearings (Fig. 13.28).

   b. With cartridge bearings, skip to step 10.

9. **Screw the cone in until it almost contacts the loose ball bearings, then push the axle straight in to bring the cone and bearings together.** This prevents the bearings from piling up and getting spit out as the cone turns down against them. Without turning the axle (which would knock the inboard bearings about), screw the cone in until it is finger-tight.

10. **Slide on the washer, if included, and screw on the locknut.**

13.28 Dropping in bearings

a. With loose bearings, while holding the cone with a cone wrench, tighten the locknut (similar to Fig. 13.26, but you will be holding the cone with a 13mm or similar cone wrench, not the pictured 10mm standard open-end wrench).

b. With cartridge bearings (Figs. 13.24, 13.30, 13.31, 13.33), tighten the locknut with a socket wrench. Do not overtighten the Ritchey locknut (Fig.13.24); remove bearing play, but don't bind the axle.

c. On a Speedplay Zero, X/1, X/2, or high-end Light Action pedal (Fig. 13.23), put threadlock compound on the Torx end bolt and tighten it. Install the snapring with inward-squeezing snapring pliers (Fig. 11.32).

d. On a Speedplay X/3, X/5, Frog (Fig. 13.33), or Light Action Chrome-Moly pedal, tighten the locknut snugly against the bearing (35–40 in-lbs/4–5 N-m). When you reassemble the pedal, seal it from water by caulking the inside edges of the pedal-body halves and putting on a new O-ring (Fig. 13.33).

11. **Check that the pedal spins smoothly without play.** Readjust as necessary by tightening or loosening the cone and retightening the locknut. On Crank Brothers (Fig. 13.30), check bearing adjustment only after replacing the dust cap (step 12); the dust cap is what prevents the axle assembly from moving laterally inside of the pedal body.

12. **Replace the dust cap.** Again, check Crank Brothers (Fig. 13.30) for lateral play.

13. **Install the pedals.**

14. **Go for a ride.**

CYCLOCROSS PEDALS

Mountain bike pedals with high mud-clearing ability are the ticket for cyclocross. The first pedal to offer exceptional mud performance was the Time ATAC (Fig. 13.29). Low-priced Time models (Fig. 13.31) are as good at clearing mud as the original ATAC.

When Crank Brothers came along with the Eggbeater (Fig. 13.30), into which you can clip on any of the four sides, it became the 'cross pedal of choice.

More and more pedals offer great mud clearing, though it's hard to imagine any exceeding the Eggbeater. Other pedals using the loop spring design include the Crank Brothers Candy—an Eggbeater with a small platform around it—the newer, more open and angular Time ATAC models, and the Look Quartz.

Shimano pedals continue to use steel plate clips (Fig. 13.32), but they and similar pedals like the Ritchey V4 have become more open to allow mud to pass through more easily, with the latest Shimano XTR and XT having a minimalist design excelling in that capacity. Stepping into a Speedplay Frog pushes accumulated mud away, making this a great 'cross performer as well.

With carbon-sole shoes, install "shoe shields," which are thin steel plates that go between the cleat and the shoe sole. They prevent carbon soles from becom-

13.29–13.33 Exploded cyclocross pedals

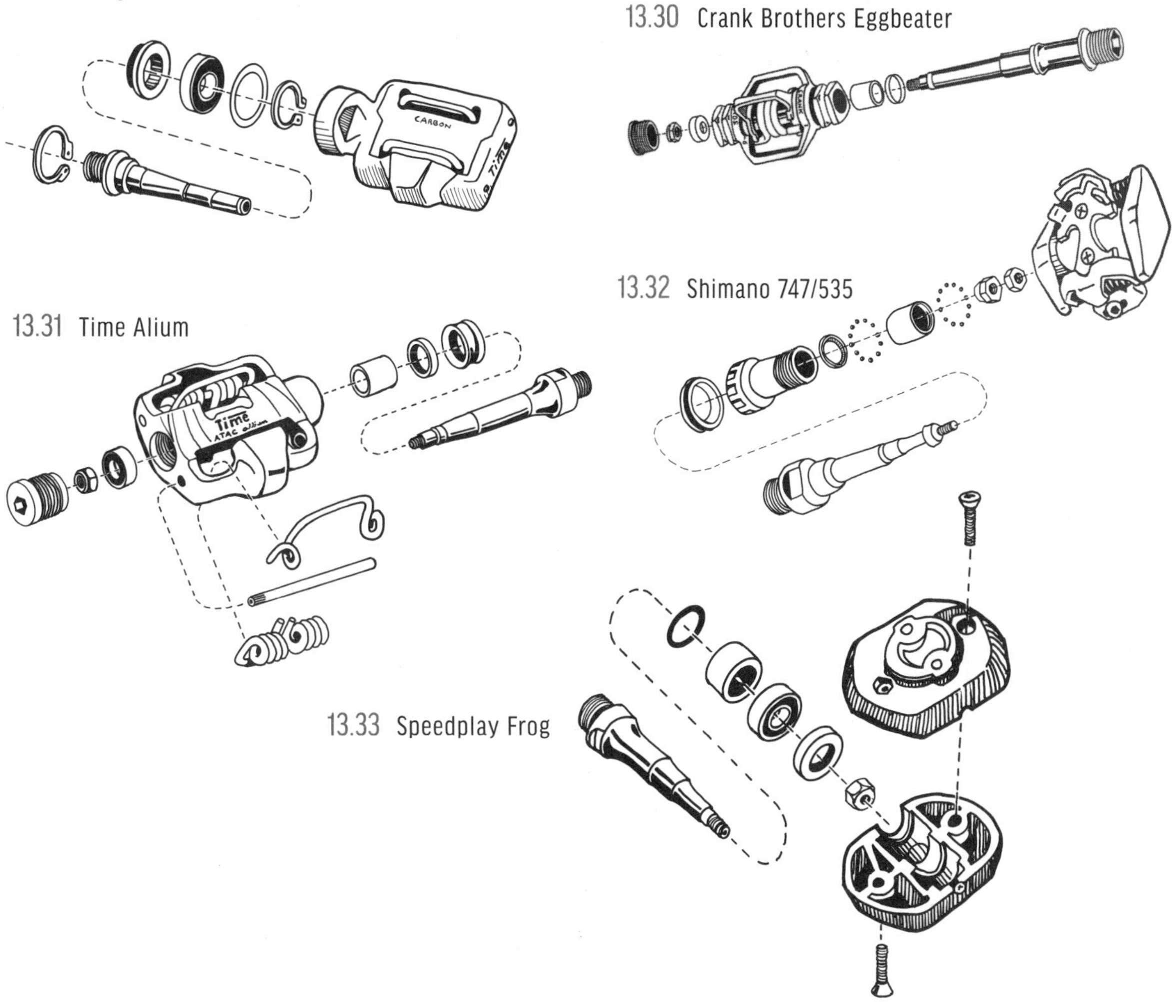

13.29 Original Time ATAC

13.30 Crank Brothers Eggbeater

13.31 Time Alium

13.32 Shimano 747/535

13.33 Speedplay Frog

ing deeply indented and eventually cracking due to pressure from the wire loops on Crank Brothers, Time, and Look mountain/cyclocross pedals.

If the shoe rocks side to side, it is due to insufficient tread contact with the ends of the pedal body. Another symptom of poor tread contact can be the foot popping down farther under high pedaling efforts on pedals with wire-loop cleat engagement systems; the spring can open farther under high pressure from the shoe, since the soft tread is not preventing the shoe from dropping down farther toward the pedal, allowing the sole to pry the loops apart until the cleat hits the center of the pedal and stops. On the upstroke, the loops spring back, lifting the shoe higher off of the pedal again, and the cycle will recommence on the next hard downstroke.

There are two things you can do to fix this. The first is to replace the shoe with one with taller tread (or if you have a shoe with replaceable tread, then replace the worn tread sections). The second is to shim the pedal ends to bring them into contact with the shoe tread when the cleat is engaged. On Eggbeaters, riders often did this in the past by wrapping tape or heat-shrink tubing around the ends. Today, Crank Brothers offers tread contact sleeves for both Eggbeater and Candy models; you have to pull the axles out (13-5) to install them.

Washing off the pedals and cleats after they get muddy is critical to good clip-in/clip-out performance, and so is lubricating the springs (with chain lube). Apply a dry lubricant like Pedro's Extra Dry or Ice Wax to the cleat tips on shoes. Doing so will help them clip in and out, won't pick up dirt, and won't leave oily stains on carpets.

TROUBLESHOOTING PEDAL PROBLEMS

13-7

CREAKING NOISE WHILE PEDALING

1. **Grease pedal threads in crank and tighten fully.** A pedal washer between the crank and pedal can help with this, too. So can plumber's tape around the pedal threads.
2. **Cleats need attention.** The shoe cleats need grease on the tips, or they are loose and need

to be tightened, or they are worn and need to be replaced (13-2).

3. **Pedal bearings and pedal-body threads need cleaning and lubrication.** See the "Overhauling Pedals" section in this chapter and especially 13-4 and 13-5.
4. **The noise is originating from somewhere other than the pedals.** See "Troubleshooting Cranks and Bottom-Bracket Problems" in Chapter 11 or Appendix A.

13-8

RELEASE OR ENTRY WITH CLIP-IN PEDALS IS TOO EASY OR TOO HARD

1. **Release tension needs to be adjusted.** See 13-3.
2. **The pedal-release mechanism needs to be cleaned and lubricated.** Clean off mud and dirt, and drip chain lubricant on the springs (Fig. 13.34) and a dry lubricant (like Pedro's Extra Dry or Ice Wax) on the cleat-contact surfaces of the clips.
3. **The cleats themselves need to be cleaned and lubricated.** Clean off dirt and mud, oil the springs, and put a dry chain lubricant like Pedro's Extra Dry or Ice Wax on the contact ends of the cleats.
4. **The cleats are worn out.** Replace them (13-2).
5. **The clips on the pedal are bent or the guide plates on top of the pedal are worn, bent, broken, or missing.** Straighten bent clips or replace them. If you can't repair the clips, you may have to replace the entire pedal. On Speedplays (Fig. 13.23), the top and bottom metal plates may need to be replaced.
6. **The pedal body is so worn that a new cleat is loose in it.** This can be a particular problem with all-carbon pedals. Replace the pedal.
7. **The cleat guide needs attention.** If it is hard to clip in, check the metal cleat guide plate at the center of an SPD-type pedal. It is held with Phillips screws (Figs. 13.19, 13.24, 13.32). They may be loose or missing, or the guide plate may be bent or broken. Tighten loose screws and replace missing or damaged guide plates.
8. **The cleat is not flat.** If you have small feet and have difficulty getting in and out of a pedal that has

13.34 Lubricating the springs and cleat contact areas

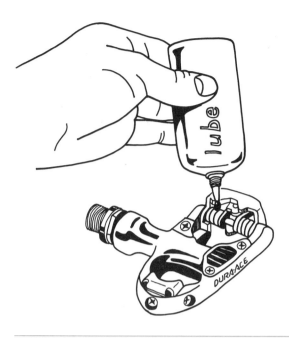

a large cleat or a large adapter plate, the curvature of your shoe sole may be so extreme that the center of the cleat hits the center of the pedal before the ends have clipped in. Try removing the little rubber plug from under the Look cleat (Fig. 13.11). You may also need to shim the front and rear of the cleat away from the shoe or file down the center of the cleat to make the cleat flat.

13-9

EXPERIENCING KNEE AND JOINT PAIN WHILE PEDALING

1. **The cleats need to be realigned.** Rotational misalignment often causes pain on the sides of the knees. Loosen and realign the cleats the way your feet want to be oriented when pedaling (13-2).
2. **You need more rotational float.** Consider a pedal that offers more float (or replace fixed cleats with floating ones, or adjust pedals with a float adjustment; Fig. 13.20). Alternatively, knees can also be sensitive to too much float.
3. **You need orthotics.** If your foot naturally needs to tip inward for proper pedaling mechanics, yet your shoe and cleat tip your feet farther out (this correction is built into some shoes), then there is likely to be an increase in the tension on the iliotibial (IT) band, which is the tendon connecting the hip and calf. This tension will eventually cause pain on the outside of the knee. You need to see a specialist about custom orthotics for your shoes to correct the problem.
4. **Your foot is tipped incorrectly on the pedal.** You may need wedges under the cleat or inside the shoe under the insole to tip your foot one way or the other so that the knee lines up over it; see bikefit.com for more on this.
5. **Your pedaling stance is too narrow or too wide.** The former is much easier to correct; if your hips want one or both of your feet to be out farther from the bike and it or they are not, your knee(s) will flare out. Putting washers (maximum of 3mm) between the pedal and crank, getting longer pedal spindles, or installing 20mm pedal spacers from bikefit.com or "Knee Savers" can get the pedal(s) out under your knee(s).

   It is harder to bring your feet in closer to the bike. You can move the cleats as far outboard on the shoes as possible without the shoes dragging on the cranks. After that, you can search for narrow-stance crankarms; consult your bike shop.
6. **You need to adjust saddle height.** Fatigue and improper seat height can contribute to joint pain. Pain in the front of the knee right behind the kneecap can indicate that your saddle is too low. Pain in the back of the leg behind the knee suggests that your saddle is too high.

CAUTION: *If any of these problems result in chronic pain, consult a specialist.*

14

I do most of my work sitting down.
That's where I shine.

—ROBERT BENCHLEY

SADDLES AND SEATPOSTS

After a few hours on your bike, you will be most keenly aware of one of its components: the saddle. It is the part of your bike with which you are most . . . uh . . . intimately connected. Nothing can ruin a good ride faster than a poorly positioned or uncomfortable saddle.

And the seatpost is critical, as it connects the saddle to the frame. It must hold the saddle firmly in the proper position without letting it tilt or slide down or back, and it may also be designed to provide some compliance to smooth out the ride.

14-1

SADDLES

Most bike saddles are made of a flexible plastic shell suspended like a hammock between attachment points with the rails at the tip and tail, along with some padding and a cover (Fig. 14.1). Not much to it, which perhaps explains why there are countless variations on this theme. To make the shell suspended on the rails cheaper to produce, stiffer, lighter, or conformable to the rider over time, it can be made of nylon, carbon-filled nylon, carbon fiber, or thick leather. To reduce pressure on sensitive areas, the shell can have depressions, holes, or splits in it. To reduce weight, the rails can be made of solid titanium rod, hollow titanium tubing, hollow steel tubing, or braided carbon fiber. Or, to reduce manufacturing cost, add or remove suspension, or fit a new saddle-clamping standard, the rail and shell can be molded out of the same piece of nylon and may even require a unique seatpost by virtue of being shaped like a single I-beam rather than like two pieces of thick wire. To create more comfort, the padding can be extra thick or have high-tech gel cushions within it. The cover over the padding can be leather, synthetic leather, Kevlar, or any of an infinite array of materials.

You can expect to spend anywhere from $20 to $300 and more for a decent saddle, yet comfort is the best indicator of what makes a saddle really good. My best advice is to ignore weight, fashion, and looks, and choose a saddle that is comfortable. High-zoot gel padding, scientifically designed shells that support some parts, don't contact others, and flex just right, as well as all sorts of factors that engineers consider when designing a saddle, mean nothing if a saddle turns out to be a giant pain in the rear. People are different and saddles are different.

14.1 Modern lightweight saddle

14.2 Saddle designed not to contact the perineum

14.3 Brooks leather saddle

Try as many as necessary until you find one you like.

The marketing war raging over saddles that are designed to prevent male impotency (Fig. 14.2) can ruin a consumer's ability to select appropriately. If you buy a saddle out of fear and it is uncomfortable, you have done yourself a disservice. Don't take it on faith or accept without question that such a saddle must be protecting you, even if you don't particularly like it; if it hurts or you get numb, it isn't working for you. What works for one person won't necessarily work for another.

There are lots of saddle-fitting systems worth trying from various saddle brands. Determine which saddle shape and design are the most comfortable for your body. Then—and only then—start looking at things like titanium rails, fancy covers, and all of the other things that improve a saddle and add to its cost. Some people can only find comfort on 400-gram saddles with tons of thick padding. Others can ride for hours on a skinny little sub–100-gram saddle. It's a matter of preference. Any decent bike shop worth its weight in carbon fiber should let you try a saddle for a while before locking you into a sale. And keep in mind that the position of the saddle can be as important as the shape.

Brooks and similar saddles have no plastic shell, foam padding, or cover. They are constructed from a single piece of thick leather attached to a steel frame with large brass rivets (Fig. 14.3). This was the main type of saddle up until the 1980s. Brooks still makes them this way. This type of saddle requires a long break-in period and frequent applications of a leather-softening compound. As is the case with a lot of old or retro bike parts, either you love 'em or you hate 'em.

A saddle with a plastic shell and foam padding (Figs. 14.1, 14.2) requires little maintenance, except to keep it clean. Check the rails periodically for bends or cracks (clear signs that you need to replace the saddle).

14-2

SADDLE POSITIONS

Even the perfect saddle will feel like a medieval torture device if it isn't properly positioned. Saddle placement is the most important part of finding a comfortable riding position. Not only does saddle position affect how you feel on the bike, but it affects your control and efficiency as well. With the saddle in the right place, you become a better rider. See Appendix C, C-3, for a detailed explanation of setting saddle and handlebar positions. The following are some brief guidelines.

There are three basic elements to saddle position: tilt, fore-and-aft position, and saddle height (Fig. 14.4). Proper saddle height is key to effectively transferring power to the pedals. The ideal height on a road bike places your leg in around 90 percent extension (knee bend of 25–35 degrees) when your foot is at bottom dead center in riding position (Fig. 14.5). If your frame is the correct size, you should have no trouble achieving this without pulling the seatpost beyond its height-limit line. Appendix C has more detail on this.

A common cause of numb crotch and butt fatigue (and even sore arms and shoulders) is an improperly

14.4 Saddle adjustments

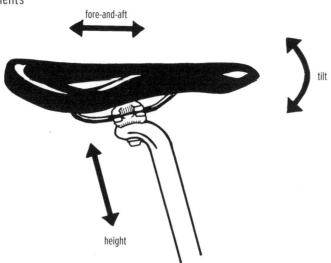

fore-and-aft

tilt

height

tilted saddle. The rule of thumb is that the saddle should be level when you are sitting on it. If you have a lot of seatpost sticking out of the frame or the saddle is pushed way back on its rails, you may need to tilt it slightly down (maybe 2 degrees or so) at the nose so that it will come to level when you sit on it, flexing the seatpost and saddle rails. After a while, some people find that they prefer a slight upward or downward tilt to their saddles. I strongly recommend against making that tilt much more than ¼ inch. Too much upward tilt places too much of your body weight on the nose of the saddle. Too much downward tilt will cause you to slide down the saddle as you ride. That puts unnecessary pressure on your arms, back, and shoulders as they fight to oppose the forward slide.

Fore-and-aft position (Fig. 14.4) determines where your butt sits on the saddle, the position of your knees relative to the pedals, and how much of your weight is transferred to your hands. Except when riding aerobars, saddles are generally designed to have your butt centered over the widest part. If this is not where you sit, reposition the saddle, or get a different one. You want to have a comfortable bend in your arms, without feeling cramped or stretched out. If you find that your neck and shoulders feel tighter than usual and your hands are going numb, redistributing your weight by moving the saddle back and leveling it could make a difference.

Fore-and-aft saddle position also affects how your legs are positioned relative to the pedals. Ideally, your knee should push straight down on the forward pedal

14.5 Knee bend at bottom dead center should be 25–35 degrees from straight

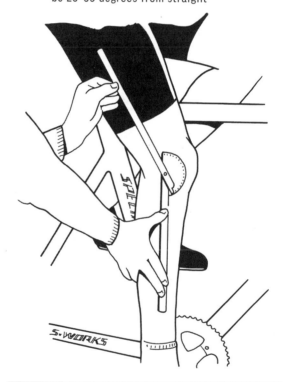

when the crankarms are in a perfectly horizontal position. Appendix C explains how to determine this precisely.

Butt pain is intimately connected to handlebar position, as are other aches and pains. The shorter the upper-body reach and higher the handlebar, the more weight will go on the butt. On the other hand, the longer the reach and lower the bar, the more the pelvis rotates forward and the pressure point moves from the sit bones to the soft tissue of the perineum and genital area. As a general rule, a novice rider will want a shorter reach and higher bar, and perhaps a correspondingly wider saddle, than an experienced rider. Once again, consult Appendix C.

14-3

SEATPOST MAINTENANCE

A standard seatpost requires little maintenance other than removing it from the frame every few months. Wipe down the seatpost, regrease it, dry out and grease the inside of the frame's seat tube (turn the bike upside down to pour out any trapped water), and then reinstall the seatpost. This procedure keeps it clean and able to move freely if you want to adjust it. It also should prevent the seatpost from getting stuck in the frame (a very nasty and potentially serious problem), and it will prevent a steel seat tube from rusting out from the inside.

The procedures for installing a new seatpost and for removing a stuck seatpost are outlined later in this chapter.

If you have a carbon seatpost, make sure that you read the Pro Tip on the subject.

Regularly check the seatpost for cracks or bends so that you can replace it before it breaks with you on it.

Because suspension and dropper seatposts are rare on road bikes, maintenance of them is not covered here, but it is covered in *Zinn & the Art of Mountain Bike Maintenance*.

PRO TIP — CARBON SEATPOSTS

CARBON-FIBER SEATPOSTS CAN BREAK if the pinch-bolt assembly digs in and cuts some fibers or makes a notch in the post. Point-loading on a thin carbon part is not a good idea. Some seat-binder clamps are designed to distribute the clamping force. If the post comes with such a clamp, make sure you use it. In the absence of one, at a minimum you should reverse the binder clamp so that its slot is not lined up with the seat-tube slot (Fig. 14.6); that way, it will pull more evenly around the circumference and not push the slot into the seatpost. And if you have a slotted seat-tube shim that some bikes have to fit the seatpost, you want to offset that slot from the other two slots as well. If you were to crank down on the bolt with the slots of the binder clamp and seat tube and perhaps a seat-tube shim all aligned, you can imagine how the top corners of the slot could dig into the back of the seatpost (where the load on it is also highest). This is where carbon seatposts break.

Carbon seatposts often slip, which of course also leads to further tightening of the bolt, increasing the probability of snapping it off. If the post slips, apply some carbon assembly paste or spray (Fig. 14.6) to the post shaft. This stuff has small, gritty plastic spheres in the solution that press back against the clamp to increase the pressure uniformly and avoid point-loading. In the absence of carbon assembly paste, you can try not greasing the post. Unlike an aluminum or steel seatpost, a carbon post itself cannot corrode, so grease to prevent seizure is less of an issue. However, corrosion of a steel or aluminum seat tube or seat-tube sleeve can nonetheless seize it.

14.6 Spraying CarboGrip carbon assembly compound on a carbon post

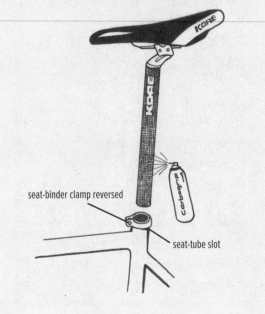

seat-binder clamp reversed

seat-tube slot

14-4

INSTALLING A SADDLE

LEVEL Remember the heavy steel posts that tapered to a skinnier section on top that you had on your bike as a kid (or on a cheap adult bike)? Those seatposts had a single horizontal bolt that pulled together a number of knurled washers with ears to hold the saddle rails. They are cheap to make but do not hold up well to adult use. If your bike has that style of post and it fails, upgrade the post with a stronger style, but note that you may have to upgrade the saddle as well if it has flat rails rather than round ones.

Much more secure (and generally lighter) is a post with one or two vertical bolts holding an aluminum clamshell together. Most posts have either one or two bolts for clamping the saddle. A single-bolt post can have a vertical bolt pulling the clamshell clamp halves together (Figs. 14.4, 14.7), or it can have a horizontal bolt pulling two ears of the clamp toward each other to clamp the rails; the ears can even slide on angled ramps so that tightening the bolt pulls the rails down onto the saddle cradle as well as clamps them from the sides. Two-bolt posts usually rely on one of three systems. In one, the two bolts work together by pulling the saddle into the clamp (Fig. 14.9). On others, a smaller second bolt holds the tilt angle (Fig. 14.8). Yet another type has two side-by-side bolts holding the saddle clamp together. It is reasonably easy to figure out how to remove, install, and adjust the saddle, no matter what kind of post you have.

14-5

SADDLE INSTALLATION ON SEATPOST WITH A SINGLE VERTICAL BOLT

Posts with a single vertical bolt (Figs. 14.4, 14.7) usually have a two-piece clamp that fastens onto the saddle rails. On most models, saddle tilt is controlled by moving the clamp and saddle along a curved platform. Before you tighten the clamp bolt, make sure there is not a second, smaller bolt (setscrew) that adjusts seat tilt. If one is present, skip to the next section (14-6).

1. **Loosen the bolt until there are only a couple of threads holding the upper clamp.**

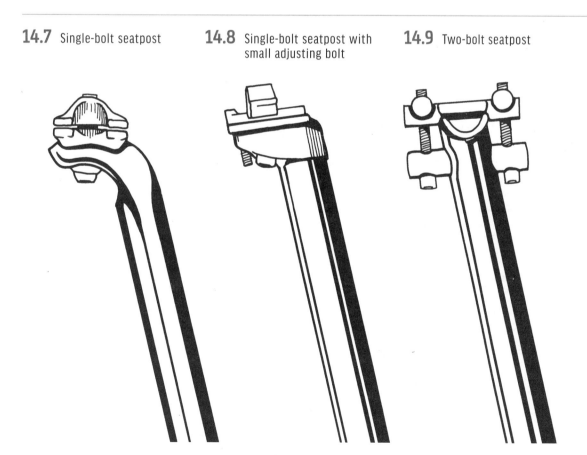

14.7 Single-bolt seatpost

14.8 Single-bolt seatpost with small adjusting bolt

14.9 Two-bolt seatpost

14.10 Saddle installation on single-bolt seatpost

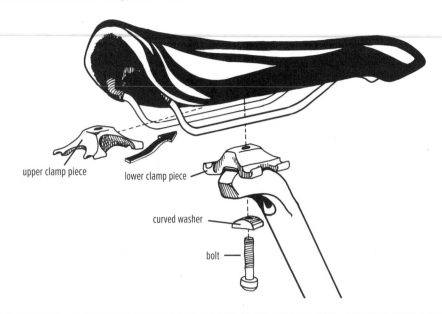

upper clamp piece

lower clamp piece

curved washer

bolt

2. **Turn the top half of the clamp 90 degrees.** Put a dab of grease on each clamp contact point to prevent squeaking. Slide the saddle rails into position. Do it from the back where the space between the rails is wider. You may need to remove the top clamp piece completely from the bolt if it is too large to fit between the rails. If you disassemble the clamp, pay attention to the orientation of the parts so that you can put it back together the same way.

3. **Set the seat rails into the grooves in the lower part of the clamp.** Then set the top clamp piece on top of the rails (Fig. 14.10). Slide the saddle to the desired fore-and-aft position.

4. **Tighten the bolt and check the seat tilt.** Readjust if necessary. Wipe off excess grease.

14-6

SADDLE INSTALLATION ON SEATPOST WITH LARGE CLAMP BOLT AND SMALL SETSCREWS

This post type is illustrated in Figure 14.8.

1. **Loosen the large bolt.** Unscrew it until the top part of the clamp can be moved out of the way or removed so that you can slide the saddle rails into place.

2. **Put a dab of grease on the clamp contact areas to prevent squeaking.** Set the saddle rails between the top and bottom sets of grooves in the seat clamp. Slide the saddle to the desired fore-and-aft position. Tighten the large bolt. Wipe off any excess grease.

3. **Adjust the tilt.** To change saddle tilt, loosen the large clamp bolt, adjust the saddle angle as needed by turning the setscrew, and retighten the clamp bolt. Repeat until the desired adjustment is reached.

CAUTION: *Do not use the setscrew to make the clamp tight! Do not adjust the setscrew unless the clamp bolt is loose!*

NOTE: *On this type of seatpost, the setscrew may be vertical or horizontal. On posts with a vertical setscrew, the screw is usually adjacent to the clamp bolt, as in Figure 14.8.*

A horizontal setscrew would be at the top front of the seatpost, pushing back on the clamp. When setting saddle tilt, push down on the back of the saddle with the clamp bolt loose to make sure the clamp and horizontal setscrew are in contact.

Another type of post has a horizontal tilt-adjusting bolt that passes crosswise through a curved slot in the seatpost clamp. On this type, the saddle can be fully tightened down, yet the tilt-adjust screw can be loosened, the saddle tilt adjusted, and the screw retightened without adjusting the clamp bolt.

14-7

SADDLE INSTALLATION ON SEATPOST WITH TWO EQUAL-SIZED CLAMP BOLTS

This post type is illustrated in Figure 14.9.

1. **Loosen or remove one or both of the bolts.** Unscrew them sufficiently to open the clamp enough to slide the saddle rails into the grooves of the clamp. By hooking your fingers over one rail and prying with the heels of your hands along the centerline of the saddle, sometimes you can flex the rail wide enough to get it to snap down over the top clamp piece and into its slot without needing to open the clamp completely.

2. **Slide the saddle to the desired fore-and-aft position.** Tighten down one or both of the clamp bolts completely.

3. **Adjust tilt.** Loosen one clamp bolt and tighten the other to change the tilt of the saddle (Fig. 14.11). Repeat as necessary.

4. **Finish by tightening both bolts.**

   Seatposts that instead have two identical side-by-side bolts (rather than fore-aft bolts) work in the same basic way as a seatpost with a single vertical bolt (14-5). The saddle clamp slides over a curved platform to vary the saddle tilt. Follow the directions in 14-5, except you may need to remove one or both of the bolts completely to get the saddle rails into the clamp.

14.11 Saddle installation on two-bolt seatpost

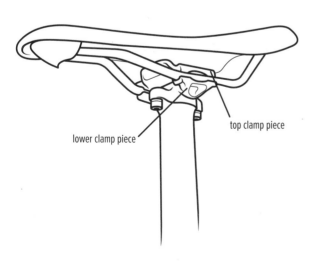

lower clamp piece

top clamp piece

14-8

SEATPOST INSTALLATION INTO THE FRAME

LEVEL 1. **Inspect the seat tube.** Check for irregularities, burrs, and other problems inside the seat tube visually and with your finger; if there are some, you may need to sand them by hand or clean up inside the seat tube with a flex hone (see 14-11, step 7). A bike shop may be able to ream the tube if the correct-size post won't go in.

2. **Grease the seatpost and the inside of the seat tube.** If you have a carbon-fiber seatpost, use carbon assembly paste on the post, especially if it slips with grease on it (see the Pro Tip on carbon seatposts earlier in this chapter). Grease the seatpost binder bolt. If you are using a sleeve or shim to adapt an undersized seatpost to fit the frame, grease it inside and out, and insert it.

3. **Insert the seatpost (Fig. 14.12) and tighten the binder bolt.** Some binder bolts are tightened with a single hex key, and some require two wrenches (usually two 5mm hex keys or two open-end

14.12 Seatpost installation into the frame

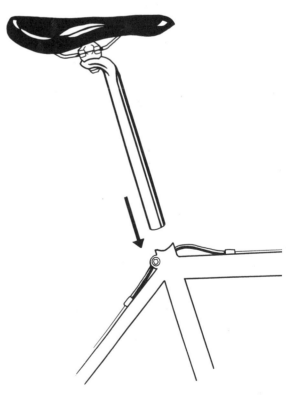

wrenches). If you have a carbon-fiber seatpost, make sure the binder clamp surrounding the seat tube either is an angled-slot type designed for the seatpost or that you rotate the binder clamp 180 degrees so that its slot and the seat-tube slot are not lined up over each other (see the Pro Tip).

4. **After the saddle is attached, adjust the seat height to your desired position.** If you mark this height on the post with an indelible marker or a piece of tape, you can just slide it right back into the proper place if you remove it.

IMPORTANT: *Periodically remove the seatpost, invert the bike to drain water out of the seat tube, and let it dry out. The frequency depends on riding conditions. With a steel frame, spray oil (or Frame Saver, a product made for the purpose that is available in bike shops) into the seat tube to arrest the rusting process. Regrease the post and the inside of the seat tube, and then reinstall it (use carbon assembly compound if it's a carbon post; see the Pro Tip).*

14-9

INTEGRATED SEAT MASTS

Some lightweight carbon frames do not have a seatpost that inserts into the frame at the seat-tube/top-tube junction. Rather, the seat tube continues past the intersection with the top tube and forms a mast onto which the saddle attaches (Fig. 14.13). Reasons for incorporating a seat mast into the frame include reducing weight, improving aerodynamics, increasing strength or stiffness or otherwise changing the physical characteristics of that area of the frame, and producing a unique look to the bike. Downsides include difficulty fitting the bike into a travel case, likelihood of damage if clamped in a standard bicycle work stand, and, for bikes with masts requiring cutting to length, reduced resale value once cut.

When you work on an integrated-seat-mast frame, finding a way to hold it in a conventional workstand can be difficult. You cannot use the stand clamp on the seat mast or on any of the tubes without risking cracking the frame. Probably the best solution is the one already mandated by aero bikes without a round seat-

14.13 Integrated seat mast on a Trek Madone

post: Use a Euro-style race team mechanic's bike stand that supports the bottom bracket and has a long arm with a quick-release clamp to hold the fork ends or the rear dropouts (Fig. 1.4). Or you may be able to obtain a special adapter clamp from the frame manufacturer intended to allow you to clamp the seat-mast cap in the arm of a standard bike stand.

There are a number of variations on the integrated seat-mast theme, and the method for clamping the saddle to the mast varies. Some have a short, hollow seatpost split at the bottom with a binder bolt or two to clamp it around the mast; I'll call this a seat-mast cap. Others have a short seatpost with a standard binder clamp around the seat mast. Yet others have a seat clamp that inserts inside the seat mast and is held in place either with a bolt going in radially through the seat mast or with an expander wedge expanding the bottom of the seat clamp against the interior walls of the seat mast. Many systems require the end user to cut the seat mast to the correct height. The obvious concern here is cutting it too short or not cutting it straight. The clamp usually has a small amount of height adjustment, sometimes with a fixing bolt and sometimes by means of shims, but it may not be enough to make up for a big cutting error. And of course, resale options for the bike are limited to people with legs no longer than allowed by the cut length of the seat mast.

Some integrated seat-mast frames do not require cutting; skip to 14-9c.

a. Cutting a seat mast

Measure twice; cut once!

1. Record your current seat height. On your current bike, measure and record your saddle height from the bottom bracket (or from the bottom pedal if the new bike will have a different crank length).

2. Set up your new bike. Install the saddle on the clamp atop the uncut seat mast and measure and record the saddle height from the bottom bracket (or from the bottom pedal if this bike has a different crank length than your existing bike).

3. Calculate how much to cut. Subtract the step 1 measurement from the step 2 measurement. This is the amount you will cut from the top of the seat mast.

4. Remove the seat clamp from atop the seat mast.

5. Measure from the top of the uncut seat mast the length you found in step 3.

6. Prepare the cut. If you have a cutting guide for the frame, put that on the seat mast up against your mark. If not, wrap tape around the mast at your mark. Cut the seat mast along the cutting guide or tape, but don't cut all of the way through or the fibers will peel back when you come through the opposite side. Instead, cut halfway from one direction and halfway from the other, meeting in the middle. You can use a standard hacksaw with a fine-tooth blade if you're careful, but a better option is a saw blade like the Effetto Mariposa CarboCut, a toothless, grit-edge tungsten-carbide blade for cutting hard materials, including ceramics, titanium, and steel, as well as carbon fiber. The blade cuts on both push and pull strokes and does not grab like a toothed blade, leaving a smooth cut edge that requires no sanding.

7. Check the cut surface. If your cut is not straight and/or smooth, carefully file or sand it smooth and flat.

8. Create a height adjustment hole. Some systems require you to also drill a hole in the seat mast for a bolt to allow some height adjustment and to hold the seat clamp at that height. If you have a cutting guide for your seat mast, it will have a guide hole for drilling through it in the right spot. If not, you will have to measure carefully using the guidelines in the owner's manual. If that is not available, measure from the seat clamp itself. Carefully drill through it, using the drill bit size specified by the manufacturer.

9. Install the seat clamp, bottoming it out on the top of the seat mast.

   a. Some systems have a cap with a pinch bolt that slides down over the seat mast. Make sure it is pushed on beyond the minimum insertion point, and tighten the bolt to torque spec.

   b. Some systems have a normal seat binder clamp surrounding the top of the seat mast, which is round in cross-section. The seat clamp is simply a short, little seatpost; insert it beyond the minimum insertion point and tighten the binder to torque spec.

   c. Some systems (Look, for example) have an elastomer that goes between the clamp and the seat mast, so insert that first. The Look seat clamp has an expander wedge inside that expands against the seat mast when you tighten a bolt atop the clamp.

10. Install the saddle, and check that your seat position is correct.

11. Adjust saddle height. If the seat is too high and you have no shims under the seat clamp, you will need to cut the seat mast a bit shorter. Again, measure twice and cut once!

b. If the seat mast is cut off too short

With some frames, this is not a problem. Certain Time integrated-seat-mast models allow you to cut the mast anywhere you want, slot it, and install a binder clamp and a 27.2mm seatpost. Similarly, Wilier Triestina's Cento 1 integrated seat mast has a Ritchey single-bolt seat-mast cap atop it, but you can cut the mast completely off where you wish, slot it, slap on a binder clamp, and stick in a 31.6mm seatpost.

With some systems, you can get a longer seatpost or seat-mast cap. If not, you may be able to insert more shims or elastomers, as long as you don't exceed the minimum insertion mark (for the seat mast itself, not for the shims) on the post or cap.

c. Seat masts that don't need to be cut

The BH Global Concept has a binder clamp atop the seat mast and a short seatpost; install it like a standard seatpost.

Installing the seat-mast cap

The post-2007 Trek Madone (Fig. 14.13) mast is not to be cut, and a seat-mast cap slips over it and clamps onto it. Different cap lengths are available, each in three different setback options (5mm or 20mm backward or 10mm forward). The cap has a ball-and-socket clamp: A horizontal clamp bolt pulls two ears against the saddle rails, which sit in grooves atop two hemispherical supports (adjusting balls), into mating concave socket ends on the large, crosswise through-hole in the cap (Fig. 14.13).

1. **Lubricate the parts.** Grease the bolt and the concave and convex mating surfaces in the cap and on the adjusting balls.

2. **Assemble the saddle onto the clamp parts as in Figure 14.13.** Tighten the bolt only enough to hold everything together.

3. **Wipe excess grease from the clamp.**

4. **Loosen the clamp bolts on the seat-mast cap and slip it over the seat mast.** Trek recommends not greasing the cap interior or the mast exterior.

5. **Slide the cap up or down as necessary.** Do not fully tighten until step 7.

6. **Set the saddle's fore-aft adjustment and tilt, and tighten the saddle-clamp bolt.** If the clamp is already tight and the saddle won't move even with the bolt loose, you will have to knock the adjusting balls free. Loosen the bolt until the saddle-clamping ear will move aside enough that you can insert a 4mm hex key through the hole in the adjusting ball. Push on the hex key against the opposite adjusting ball to free it. Repeat from the other direction.

7. **Set the seat height.** Make sure the seat-mast cap covers the minimum insertion line.

8. **Tighten the clamp bolt.** Tighten to the torque specification embossed on the clamp (7 N-m).

9. **Fine-tune the adjustment.** If you cannot achieve the height or fore-aft adjustment you desire, get a different Trek seat-mast cap that will hold your saddle where you want it.

14-10

SADDLES AND SEATPOSTS FOR CYCLOCROSS

In cyclocross, you are constantly jumping onto the saddle. Seatpost strength is imperative; imagine the consequences if the post were to break when you were jumping on the saddle with all of your weight.

Bending or breaking a seat rail or two while jumping on the bike is neither safe nor comfortable. If you keep riding on a saddle with a bent or broken seat rail, you obviously risk the saddle breaking off completely. But before that happens, you will screw up your pedaling mechanics by riding on a tilted saddle and potentially injure your back in the process.

Thus the first consideration in saddles and posts for cyclocross is strength. Unless you are a light person, opt for an aluminum seatpost of good quality, but not a superlight one, and get a saddle with chromoly steel rails. Of course, in cyclocross, you do want the bike to be as light as possible, because you are constantly hefting it up and running up hills and jumping over barriers with it. So don't overdo it. Get strong seatposts and saddles for your 'cross bikes that are also reasonably light.

Avoid saddles that are pointed at the tail. When you swing your leg over your saddle to jump up on it at speed, you want no long saddle tail (or spare-tire bag) to catch and deflect your leg.

Since you are leaping onto the saddle so often, it's not a bad idea to choose a saddle with some padding. You should land on the saddle with the inside of your thigh rather than on your crotch, but still, a superhard saddle may not be ideal. Although a more padded saddle does not necessarily equate to a more comfortable saddle, seek out a saddle you find comfortable. After all, you will be riding as fast as you can on bumpy surfaces without any suspension apart from the tires.

14-11

REMOVING A STUCK SEATPOST

LEVEL You are having this difficulty because you did not follow the important Note in 14-8. This is a level 3 job because of the risk involved. It may be best to entrust this job to a shop because if you make a mistake, you run the risk of destroying your frame. If you're not 100 percent confident in your abilities, go to someone who is—or at least to someone who will be responsible if it gets screwed up.

1. **Remove the seat-lug binder bolt.**

2. **Squirt penetrating oil around the seatpost and let it sit overnight.** To get the most penetration, remove the bottom bracket (Chapter 11), turn the bike upside down, squirt more penetrating oil in from the bottom of the seat tube, and let it sit overnight.

3. **The next day, stand over the bike and twist the saddle.** No luck? For an aluminum post, repeat step 2 but use ammonia or Coca-Cola instead to dissolve the aluminum oxide.

4. **If step 3 does not free the seatpost, warm the seat-lug area with a hair dryer to expand it.** Discharge the entire contents of a CO_2 tire inflator at the joint of the seatpost and the seat collar to freeze the post and shrink it. (Alternatively, ice the exposed seatpost with a plastic bag filled with crushed ice.) Now try twisting as in step 3.

5. **If step 4 does not free the seatpost, you will need to move into the difficult and risky part of this procedure.**

   a. You will now sacrifice the seatpost. Remove the saddle and all of the clamps from the top of the seatpost. With the bike upside down, clamp the top of the seatpost into a large bench vise that is bolted to a very secure workbench.

   b. Congratulations, you have just ruined the seatpost. Don't ever ride it again.

   c. Perform the heat/ice or CO_2 cartridge trick in step 4. Grab the frame at both ends and apply twisting pressure. You can easily apply enough force to bend or crack the frame,

so be careful. If the seatpost releases, it can make such a large "pop" that you will think that you have broken many things!

6. **If step 5 does not work, you will need to cut the seatpost.** Cut off the seatpost a few inches above the seat lug and clamp the top of it in a vise. Warm up the seat-lug area with a hair dryer to expand it. Discharge the entire CO_2 cartridge of a tire inflator down inside the seatpost to freeze it and shrink it. Now try twisting as in step 4.

7. **If step 5 or 6 does not work, you need to go to a machine shop and get the post reamed out of the seat tube.**

If, at this point, you still insist on getting the post out yourself, you should really sit down and think about it for a while. Will the guy at the machine shop really charge you so much money that it is worth the risk of completely destroying your frame yourself?

Still insist on going it alone? Okay, but don't say I didn't warn you.

Take a hacksaw and cut off the post a little more than an inch above the frame. Remove the blade from the saw and wrap a piece of tape around one end. Hold the taped end and slip the other end into the center of the post. Carefully—very carefully—make two outward cuts about 60 degrees apart. Your goal is to remove a pie-shaped wedge from the hunk of seatpost stuck in the frame. Be careful; this is where many people cut too far and go right through the seatpost into the frame. Of course, you wouldn't do that, would you?

Once you've made the cut, pry or pull this piece out with a large screwdriver or a pair of pliers. Be careful here too. A lot of overenthusiastic home mechanics have damaged their frames by prying too hard.

Once the wedge is out, work the remaining section of the post out by curling in the edges with the pliers to free more and more of it from the seatpost walls. It should eventually work its way out.

With the post out of the frame, clean the inside of the seat tube thoroughly. A flex hone, sold in auto parts stores (or loaned at rental stores) for reconditioning brake cylinders, is an excellent tool for the purpose. Put the hone in an electric drill, and be sure to use plenty of honing fluid or cutting oil as you work it up and down inside the seat tube as it spins. An alternative is to use sandpaper wrapped around your fingers, although you will not be able to reach very far into the frame.

Inasmuch as removing a stuck post is so miserable that no one wants to do it twice, I'm certain that I do not need to remind you to grease any metal seatpost thoroughly before inserting it in the frame, and check it regularly thereafter as previously outlined. Carbon seatposts have a soft, clear-coated exterior and can mechanically lock into the frame or be held in by corrosion of the seat tube or seat-tube sleeve, so I recommend carbon assembly spray or paste to reduce these dangers (see the Pro Tip earlier in this chapter).

TROUBLESHOOTING SADDLE AND SEATPOST PROBLEMS

14-12

LOOSE SADDLE, CREAKS AND NOISES, STUCK SEATPOST

a. Loose saddle

Check the clamp bolts. They are probably loose. Tighten the bolts and set the desired saddle tilt, after setting fore-and-aft saddle position (14-2). Check for any damage to the clamping mechanism and replace the post if necessary. If parts of the saddle clamp are bent, they can bottom out on each other before clamping the saddle rail adequately.

b. Stuck seatpost

This can be a serious problem. Follow the instructions in 14-11 to avoid damaging the frame.

c. Saddle squeaks with each pedal stroke

This problem can come from the seat moving against metal parts or from grit in the points where the saddle rails enter the saddle base.

1. Contact between the leather under the saddle shell and the seatpost clamp or rails is a likely culprit. Greasing the contact area will eliminate the noise.

Also, roughing up the leather where it contacts metal will quiet it down. Or sprinkle the squeaky leather with talcum powder.

2. Also try squirting chain lube into the three points where the rails are inserted into the plastic shell of the saddle, in case some grit working at the rails is making the noise.

d. Creaking noises from the seatpost

A seatpost can creak from movement of the clamp holding the saddle. The shaft moving back and forth against the sides of the seat tube while you ride can also creak. Pull the post from the frame and regrease the post and seat tube, unless you have a carbon seatpost; carbon assembly paste is the ticket there (see the Pro Tip earlier in this chapter).

1. Some frames use a sleeve to adapt the seat tube to a certain seatpost diameter. Remember that the internal diameter of the seat tube is larger below the collar. I have seen bikes that creaked because the bottom of the seatpost rubbed against the sides of the seat tube below the extension of the collar. You can solve that problem by shortening the seatpost with a hacksaw to just reach the bottom of the sleeve. If you do saw off the post, make sure that you still have at least 3 inches of seatpost inserted in the frame for security (and a sleeve at least that long).

2. Similarly, movement between the frame, sizing shims, and the post can cause creaking. Grease all of these parts well.

3. If the creaking originates from the post head where the saddle is clamped, check the clamp bolts. Lubricate the bolt threads, and you will be able to tighten them a bit more.

4. The saddle rails can creak within the clamp areas. Loosen the clamp, oil or grease the rails and the clamp valleys, and retighten the parts. Wipe off any excess lubricant.

5. Shock-absorbing seatposts can squeak as they move up and down. Try greasing the sides of the inner shaft. Grease the elastomers inside, if it has them.

e. Seatpost slips down

Tighten the seat-lug binder bolt.

If the seat lug is pinched closed and the post still slips, you may be using a seatpost with an incorrect diameter, or the seat tube may be oversized or has stretched. Double-check the seat-tube diameter with calipers, or ask your local shop to do so. Try putting a larger seatpost in the frame, and replace it if you find one that fits better.

Try carbon assembly spray or paste on the post, even if it is not carbon (see the Pro Tip earlier in this chapter). The stuff expands against the inside of the seat tube and can increase clamping pressure at the same bolt torque.

If the next seatpost size up is too big and the assembly paste did not do the trick, you may need to shim the existing post. Cut a 1-by-3-inch piece out of an aluminum pop can. Pull the seatpost out, grease it and the pop-can shim (or put assembly paste on them), and insert both back into the frame. Bend the top lip of the shim over to prevent it from disappearing inside the frame. You may need to experiment with various shim dimensions until you find a piece that will go in with the seatpost and will also prevent slippage. Fortunately, they're cheap.

If you must tighten the clamp further around a carbon-fiber seatpost, make sure it is not forcing the corners of the seat-tube slot into the seatpost, creating damage and the possibility of breakage. Use an offset-clamp binder, or turn the binder around so its slot does not line up with the seat tube's slot (Fig. 14.6).

On a titanium frame or steel frame with an integral seat binder (one that is welded to the seat tube, as opposed to an external clamp), the seat tube can stretch if the binder is chronically overtightened. If the post is slipping because the binder slot has closed up, the shim method may work. If not, contact the frame manufacturer for assistance. You may be able to remedy the situation by filing the binder slot wide enough to keep it from pinching closed. Plug the seat tube with a greasy rag to keep metal filings from falling into the bottom bracket. This may not be a perfect solution, however, because while filing the slot wider will prevent the ears from hitting, the bolt will have to bend as the ears of the binder pull in. It may eventually break. By the way, don't file the frame until you speak to the frame builder, who may have a different solution or reasons not to do it.

15

> *If you think you can or think you can't, you're right.*
>
> —HENRY FORD

WHEELBUILDING

Congratulations! You have arrived at the task most often used to gauge the talent of a bike mechanic. Next to building a frame or fork, building a good set of wheels is a mechanic's most critical and creative task. Despite the air of mystery surrounding the art of wheelbuilding, the construction of a good set of bicycle wheels is actually straightforward. I have not labeled any of the procedures in this chapter as level 1, 2, or 3 tasks, because apart from a few special tools, the art of wheelbuilding isn't terribly difficult. It only requires concentration, patience, and a love of mechanical objects.

Wheels are the central component of a bike. For any bike to perform well, its wheels must be well made and properly tensioned. Turning a pile of small parts into a set of strong, light wheels on which you can corner and descend with confidence is quite rewarding, once you learn how. You will be amazed at what they can withstand, and you will no longer go through life thinking that building wheels is something only the "experts" do. With practice and patience, you can build wheels at your house that are as good as any custom-made set, and far superior to those built by machine.

This is not meant to be an exhaustive description of how to build all types of wheel spoking patterns. Entire books are devoted to the subject—justifiably so, because the bicycle wheel is an artful engineering miracle that deserves thoughtful exegesis. (If you are interested in a more comprehensive treatment of the subject of wheelbuilding, I recommend *Barnett's Manual* by John Barnett, *The Art of Wheelbuilding* by Gerd Schraner, or *The Bicycle Wheel* by Jobst Brandt.)

You can, however, build great wheels following the methods presented here. The first section describes how to build a wheel laced in the classic "three-cross" spoking pattern, in which each spoke crosses three other spokes (Fig. 15.1). A later section (15-7) details how to build a radially spoked front wheel or a rear wheel spoked radially on one side and three-cross on the other side. Other sections detail the construction of disc-brake wheels (15-9) and cyclocross wheels (15-11). So let's get started.

15-1

PARTS AND TOOLS

Gather the parts you need: a rim, a hub (make sure that the hub has the same number of holes as the rim), and properly sized spokes and nipples to

15.1 The complete wheel with a three-cross spoke pattern

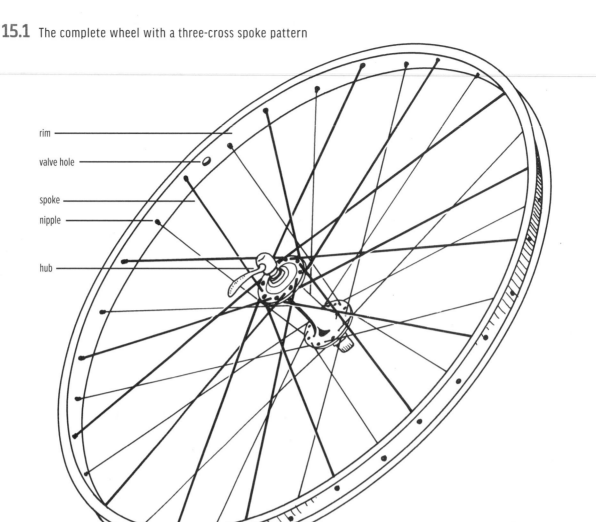

rim

valve hole

spoke

nipple

hub

15.2 Spoke and nipple

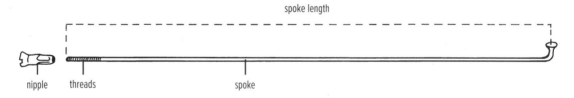

spoke length

nipple   threads   spoke

match. Make sure you have the right-size spoke wrench for those nipples as well.

a. Spokes and nipples

I suggest getting the spokes from your local bike shop. That way, a mechanic can help make sure you get the right spoke lengths (Fig. 15.2) and can counsel you on which gauge (thickness) of spoke to select, as well as which rim makes sense for your weight, budget, and the kind of riding you do. If you will not be getting the spokes from a local shop, you can find the spoke lengths you need by using a number of online spoke calculators. The one I always use is at dtswiss.com. Remember when you calculate spoke length to specify that you will be using a three-cross spoking pattern (unless you are building a radial wheel [15-7] or other spoke pattern).

NOTE ON USING OLD SPOKES: *If you are replacing a rim on an old wheel, do not use the old spokes unless the wheel had limited mileage before you trashed the rim, and never reuse the nipples. Saving money by reusing the old parts is a false economy. Once the nipples get rounded out, your spoke wrench can no longer turn them as you're building the wheel, and the weakened spokes will start to fail when you're riding.*

b. Tensioning tools

You only need a spoke wrench. If you are building lots of wheels, though, a bent-handle nipple screwdriver (Fig. 15.3A) speeds up tensioning.

c. Tools for aero (bladed) spokes

If you will be using bladed spokes, you will need a slotted tool to grab the flat of the spoke to keep it from twisting. DT Swiss makes a nice nesting pair of tools for this: a spoke wrench with a cone-shaped longitudinal groove that mates with the cone-shaped end of a slotted, L-shaped tool (Fig. 15.3B). With these tools, you can hold the spoke way down at the end of its flat section, close to the nipple, as you turn the spoke; with other slotted tools you will be grabbing the spoke above the spoke wrench, far from the nipple, and, with a thin spoke, you can easily leave a permanent kink—a twist in the middle of the flat section of the spoke.

d. Tools for deep rims

If you will be building onto a deep-section rim, you will need another special tool. Which tool that is will depend on whether the rim has standard-size spoke holes so the nipples extend out of the rim in the normal way, or the rim has tiny spoke holes so the nipples will be internal to the rim. In the former case, you need a way to hold the nipple as you extend it through the large hole in the rim bed to the hole in the inner wall without losing it in the open cavity of the rim. Special tools for this exist, but for building a few deep-section wheels, an extra spoke with another nipple threaded on it will be a good enough tool. Spin the nipple on to leave about 6mm of thread exposed, and then fix it in place by crimping it onto the spoke. You'll thread this "tool" into the top of a nipple until it stops against

15.3A Bent-handle nipple screwdriver

15.3B DT Swiss spoke wrench and antitwist tool for aero spokes

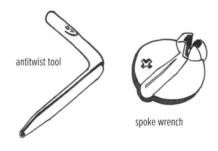

antitwist tool

spoke wrench

15.3C Three-way spoke wrench for internal nipples

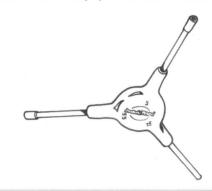

the crimped nipple; reach through the rim with the tool and thread the nipple a few turns onto a spoke coming from the hub. This tool will also speed tensioning later, because the crimped-on nipple will only allow each nipple to thread onto its spoke a set number of turns before it stops. Having each nipple threaded on the same amount will eliminate tension disparities as you begin the tensioning process (you'll understand the importance of this when we reach that section).

If the deep rim has only small holes that will not let a nipple through, you will need a deep-wall socket with which to turn the internal nipples. Park makes a three-way square/5mm/5.5mm specialty spoke wrench for this purpose (Fig. 15.3C); the square drive is for using standard nipples upside down in the rim, and the 5mm and 5.5mm deep-wall sockets are for turning long hex nuts, including 3⁄16-inch ones (equaling 4.7mm, which is close enough to 5mm to work fine).

e. Spoke prep

For the sake of brevity and clarity, I do not mention using spoke prep compound with every instruction to thread a nipple onto a spoke. Although the use of thread compound is not mandatory, I think that the wheel is improved: It encourages the nipples to thread on smoothly, it takes up some of the slop between the spoke and nipple threads, and its thread-locking ability discourages the nipples from vibrating loose. Instead of spoke prep compound, you can use DT Pro Lock nipples, which contain a two-component adhesive in the nipple thread to prevent the spoke-nipple connection from loosening under the effect of operating loads (loading and unloading of the wheel during riding), thus ensuring constant spoke tension.

If you use spoke prep, apply it to the spoke threads before putting on the nipples. You do not want too much, as it will be hard to adjust the nipples months and years down the road; you just want the spoke prep in the valleys of the threads. You can get the right amount if you dip the threads of a pair of spokes into the prep compound, and then take two more dry spokes and roll the threads of all four spokes together with your fingers.

With DT Pro Lock nipples, you don't have to do any of this; just thread them onto the spoke. However, you will want to complete the wheel in one sitting. As you thread the nipples, you will be bursting little beads of the two glue components inside the nipple, like epoxy glue. If you finish the wheel while the glue is viscous, the nipples will hold better than if you let them harden and then turn them again in ensuing days. The nipples still offer benefits upon later wheel truing, since not all of the spheres will burst during initial wheel-building, but the efficacy of the thread-locking will be reduced each time.

In the absence of spoke prep, at least dip the threads of each spoke in grease. Grease accomplishes everything spoke prep does, save for locking the threads.

A little linseed oil around the head of a nipple where it contacts the rim makes turning it easier as tension increases.

15-2

LACING THE WHEEL

1. **Divide the spokes into two separate groups, one set for each side of the hub flange, and rubber-band each set together.** If you are building a rear wheel, you should be working with two different spoke lengths, because spokes on the right-hand side, or drive side, are almost always shorter. For a radial front wheel, skip to 15-7.

2. **Hold the rim on your lap with the valve hole away from you.** Note that the holes may alternate being offset upward or downward from the rim centerline.

NOTE: *If you are building a rear wheel onto a rim with spoke holes drilled asymmetrically off center, make sure that you orient the rim so that the spoke holes are offset to the left (non-drive) side (Fig. 15.4). The asym-*

15.4 Asymmetrical, off-center-drilled rear rim laced correctly

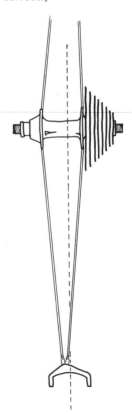

15.5 First half of right-side spokes placed in a hub

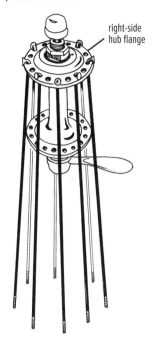

right-side
hub flange

15.6 First spoke, right-side up

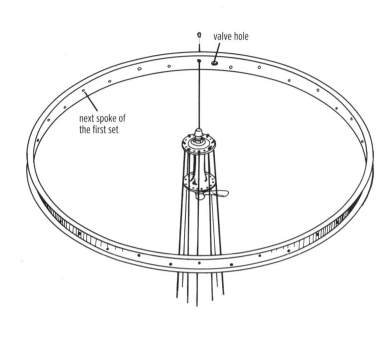

valve hole

next spoke of
the first set

metrical rim is meant to reduce wheel dish, so that offsetting the nipples to the left reduces the otherwise very steep angle at which drive-side spokes normally hit the rim. The balanced left-to-right spoke tension should increase the lifetime of the wheel, and the lower spoke angle moves the drive-side spokes away from the rear derailleur. So with an asymmetrical rim (Fig. 15.4), have the spoke holes offset downward, toward your lap.

3. **Hold the hub in the center of the rim, with the right side of the hub pointing up.** On a rear hub, the right side is the drive side. Front hubs are symmetrical; pick a side to be the right side. In the illustrations, the right side has the nut end of the quick-release.

a. First set of spokes

4. **Drop a spoke down into every other hole in the top (right-side) hub flange, so that the spoke heads are facing up (Fig. 15.5).** Make sure if it's a rear wheel that you put the shorter spokes on the right (drive) side. On some (older) hubs, half of the holes you are looking at will be countersunk deeper into the hub flange to provide a radius less stressful on the spoke elbow, so don't use those

holes—use their neighbors. That said, most hubs these days have the same countersinking on all holes to prepare for the eventuality of building a completely symmetrical, radially spoked wheel (15-7).

5. **Put a spoke into the first hole counterclockwise from the valve hole, and screw the nipple on three turns (Fig. 15.6).** Notice that this hole is offset upward. On an asymmetrically-drilled rim (Fig. 15.4), this means that the hole is offset upward from the centerline of the spoke holes, not the centerline of the rim. If the first hole counterclockwise from the valve hole isn't offset upward, you have a misdrilled rim, and you must offset all instructions one hole.

6. **Working counterclockwise, put the next spoke on the hub into the hole in the rim four holes away from the first spoke, and screw a nipple on three turns.** There should be three open rim holes between these spokes, and the hole you put the second spoke into should also be offset upward.

7. **Continue counterclockwise around the wheel in the same manner.** You should now have used half of the rim holes that are offset upward, and there

15.7 First set of spokes laced

15.8 Spoke-hole offset

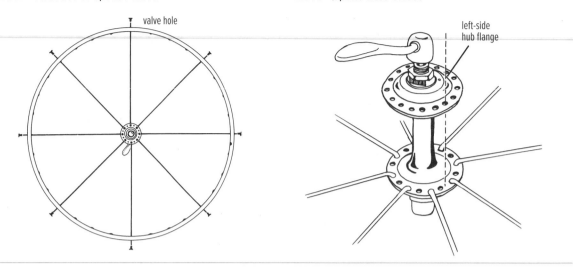

15.9 Lacing second set

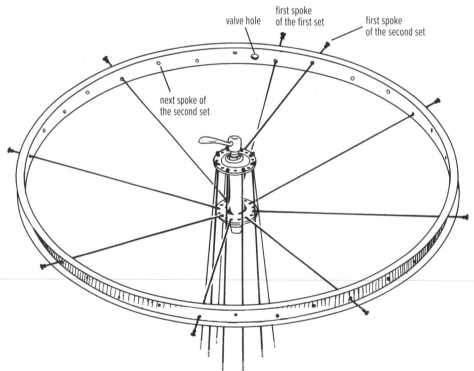

should be three open holes between the spokes (Fig. 15.7).

8. **Flip the wheel over.**

b. Second set of spokes

9. **Sight across the hub from one hub flange to the other.** Notice that the holes in one flange do not line up with the holes in the other; each hole lines up between two holes on the opposite flange (Fig. 15.8).

10. **Drop a spoke down through the hole in the top flange that is immediately clockwise from the first spoke you installed (the spoke that is adjacent to the valve hole).**

11. **Put this new spoke into the second hole clockwise from the valve hole, next to the first spoke you installed (Figs. 15.9, 15.10).** This hole will be offset upward from the rim centerline.

12. **Thread the nipple on three turns.**

15.10 "Diverging parallel" spokes

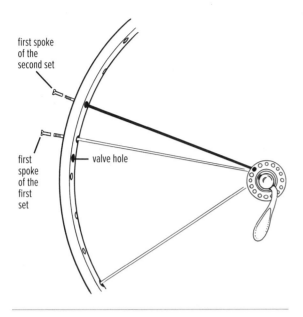

first spoke of the second set

first spoke of the first set

valve hole

15.11 Second set of spokes laced

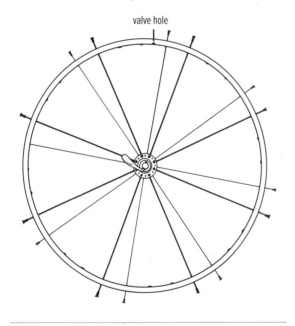

valve hole

13. **Double-check your work.** Make sure that the spoke you just installed starts at a hole in the hub's top (left-side) flange that is a half-hole space clockwise from the hole in the lower flange where the first spoke you installed started. When you look at the wheel from the side, these two spokes (your first spoke and the one you just installed) should not cross each other and should look like they are trending slightly away from each other (Fig. 15.10).

In wheel-building parlance, these two spokes are called "diverging parallel" spokes.

14. **Drop a spoke down through the hole in the top (left-side) hub flange two holes away in either direction.** Continue around until every other hole has a spoke hanging down through it (Fig. 15.9).

15. **Working counterclockwise, take the next spoke from the hub and put it in the rim hole that is three holes counterclockwise from the valve hole.** This hole should be offset upward and four holes to the left of the spoke you just installed. Thread the nipple on three turns.

16. **Follow this pattern counterclockwise around the wheel (Fig. 15.11).** You should have now used half of the rim holes that are offset upward, as well as half of the total rim holes. The second set of spokes should all be in upwardly offset holes, one hole clockwise from each spoke of the first set.

c. Third set of spokes

17. **Drop spokes through the remaining holes on the right side of the hub, from the inside out (Fig. 15.12).** Remember, if it's a rear wheel, these spokes should be shorter than the spokes used on the left side.

18. **Flip the wheel over, grabbing the spokes you've just dropped through to keep them from falling out.**

19. **Fan the spokes out.** Now they cannot fall back down through the hub holes.

20. **Grab the hub shell and rotate it counterclockwise as far as you can (Fig. 15.13).**

21. **Pick any spoke on the top (right-hand) hub flange that is already laced to the rim. Now find the spoke five hub holes away in a clockwise direction.**

22. **Take this new spoke, cross it under the spoke you counted from (the one five holes away), and stick it into the rim hole two holes counterclockwise from that spoke (Fig. 15.14).** Thread a nipple on three turns.

23. **Continue around the wheel, doing the same thing (Fig. 15.15).** You may find a spoke or two that don't reach quite far enough. If that's the case, at a point about an inch from the spoke elbow, push down on the spoke to help it reach.

24. **Make sure that every spoke coming out of the upper side of the top flange (the spokes that come out toward you with their spoke heads hidden from view) crosses over two spokes and under a third.** All three of these crossing spokes come from the underside of the same flange and have their spoke heads facing toward you. These crossing spokes begin one, three, and five hub holes counterclockwise from the spoke that you just inserted into the rim (Fig. 15.15). This is called a "three-cross" pattern because every spoke crosses three others on its way to the rim (over, over, under). Every upwardly offset hole should now be occupied on the rim.

15.12 Placing the third set of spokes in hub

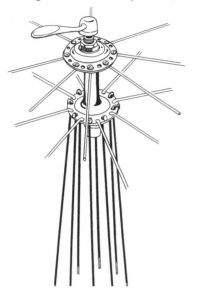

15.13 Rotating the hub counterclockwise

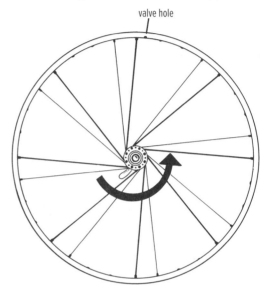

15.14 Lacing the third set of spokes

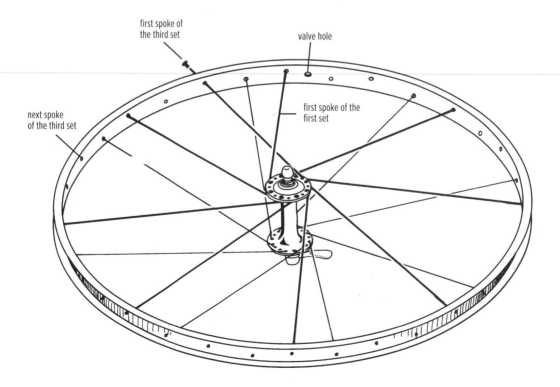

15.15 Third set of spokes laced

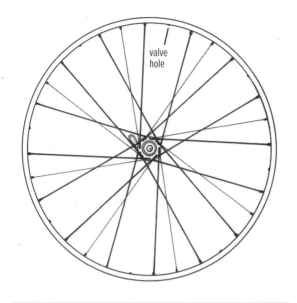

valve
hole

d. Fourth set of spokes

25. **Drop spokes down through the remaining hub holes in the bottom flange from the inside out.** (They should look like Figure 15.12, but with the other side of the hub up.)

26. **Flip the wheel over.** Grab the spokes to keep them from falling back down through the holes.

27. **Fan the spokes out.**

28. **Pick any spoke on the top (left-hand) hub flange that is already laced to the rim.** Now find the spoke five hub holes away in a counterclockwise direction.

29. **Take that spoke, and cross it over two spokes and under the spoke you counted from.** Stick the spoke into the rim hole two holes clockwise from the spoke it crosses under (Fig. 15.16). Thread a nipple on three turns.

30. **Continue around the wheel, doing the same thing until the wheel is laced (Fig. 15.1).** You may find that some spokes don't reach far enough. In that case, at a point about an inch from the spoke elbow, push down on the spoke to help it reach.

31. **Make sure that every spoke coming out from the upper side of the top flange (the spokes that come out toward you with their spoke heads hidden from view) crosses over two spokes and under a third (Fig. 15.1).** All three of these crossing spokes come from the underside of the same hub flange and have their spoke heads facing you. The crossing spokes begin one, three, and five hub holes clockwise from each spoke emerging from the top

15.16 Lacing the fourth set of spokes

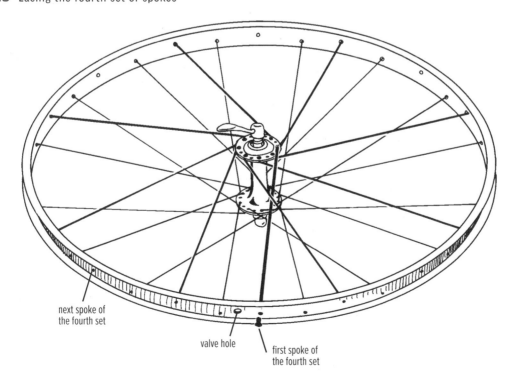

next spoke of
the fourth set

valve hole

first spoke of
the fourth set

15.17 "Converging parallel" spokes

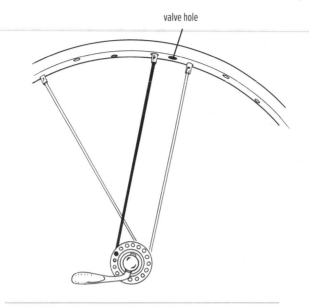

valve hole

of the upper (left) hub flange (Fig. 15.1). Every hole should now be occupied on the rim. When you look at the wheel from the side, the valve hole should be between two spokes (your first spoke and the first spoke of the fourth set) that do not cross each other but whose trajectories look like they are trending slightly toward each other. In other words, if these spokes were to continue infinitely outward, their trajectories would eventually cross far beyond the rim (Fig. 15.17). In wheel-builder speak, these two spokes are called "converging parallel" spokes. Lacing the spokes this way around the valve hole will provide the maximum possible space between the spokes for the pump head when inflating the tire.

NOTE: *If this is a rear wheel, the spokes coming out of the outside of the hub flange on both sides oppose the clockwise twist the chain applies on the cogs. See 15-6 for more on this subject.*

15-3

TENSIONING THE WHEEL

1. **Put the wheel in the truing stand.**
2. **Tighten each nipple first with a screwdriver and then with a spoke wrench until only three threads**

are visible beyond the bottom of the nipple. See Figure 15.18 for tighten rotation direction. The bent-shaft nipple screwdriver shown in Figure 15.3A speeds this process up immeasurably. The shaft spins in the handle, which you just turn like a crank; it's much faster than twisting a screwdriver.

From now on, every time you tighten or loosen a spoke nipple, turn it back the opposite direction one-eighth turn afterward. This unwinds the twist in the spoke that your tightening or loosening just caused. If you are using aero spokes, keep them from twisting with the tool shown in Figure 15.3B.

3. **Using your thumb, press the spokes coming outward from the outer side of the hub flanges down at the elbow to straighten their line to the rim.** Spokes coming out of the inner side of the flange do not need this.
4. **Go around the wheel, tightening each nipple a half turn.** Do this uniformly, and only a half turn, so that the wheel is not thrown out of true.
5. **Check to see if the spokes are tight enough to give a tone when plucked.** Squeeze pairs of spokes together and compare with a good wheel with spokes of the same gauge; your new wheel should have considerably less tension at this point.
6. **Repeat steps 4 and 5 until the spokes all make a tone but are under less tension than an existing, good wheel.** Final tensioning will come with the remainder of the truing process.

NOTE ON BLADED SPOKES: *If you are using flat spokes, you will need to hold the spoke flats (15-1, Fig. 15.3B) to keep them from twisting.*

15-4

TRUING THE WHEEL

a. Lateral true

Side-to-side trueness is the most obvious wheel parameter when you spin a wheel.

1. **Make sure the hub axle has no end play.** If play is present, adjust the hub (see hub adjustment in 8-6d).
2. **Optionally, put a drop of linseed oil around the top of each nipple where it seats in the rim.** The

oil will lubricate the contact area between it and the inside of the rim hole.

3. **Set the truing stand feelers so that one of them scrapes the side of the rim at the worst lateral wobble (Figs. 15.18, 15.19).**

4. **Ending a few spokes on either side of where the rim scrapes, tighten the spokes coming from the opposite flange of the hub, and loosen the spokes coming from the same-side flange of the hub (Figs. 15.18, 15.19).** Start with a one-quarter turn on nipples at the center of the scraping area, and decrease the amount you turn each nipple as you move away in either direction. This step pulls the rim away from the feeler. If it does the opposite, you are turning the nipples in the wrong direction. Remember, you normally turn something to the right to tighten and to the left to loosen, but tightening and loosening spoke nipples at the bottom of the wheel are the opposite of what you would normally do (Figs. 15.18–15.21). This is because the nipple head is underneath your spoke wrench. Try opening a jar that is upside down, and you will immediately understand the principle involved.

5. **Work around the wheel in this way, bringing in the feelers as the wheel gets truer.**

b. Radial true

While not as obvious visually as side-to-side trueness, out-of-roundness is more noticeable when riding, and it is more important to the longevity of the wheel, because a wheel with uniform tension lasts longer. Radial truing, however, can be somewhat slow and frustrating work. If you find yourself running out of patience for this job, step away for a while and then start again when you feel fresh and ready.

6. **Set the truing-stand feelers so that they now contact the circumference of the rim, rather than the sides.**

7. **Bring the feelers in until they scrape against the highest spot on the rim (Fig. 15.20).**

8. **Tighten the spokes a one-quarter turn where the rim scrapes.** This will pull the rim inward. Decrease the amount of each turn (to a one-eighth turn and less) as you move away from the center of the scraping area.

9. **Work around the wheel this way, bringing the feelers in as the wheel becomes rounder.**

10. **Wherever there is a dip in the rim, loosen the spokes (Fig. 15.21).** If the spokes are too tight at this point, they will be hard to turn and will creak and groan as you do. When the spokes become

---

15.18 Adjusting spokes to pull the rim to the right

15.19 Adjusting spokes to pull the rim to the left

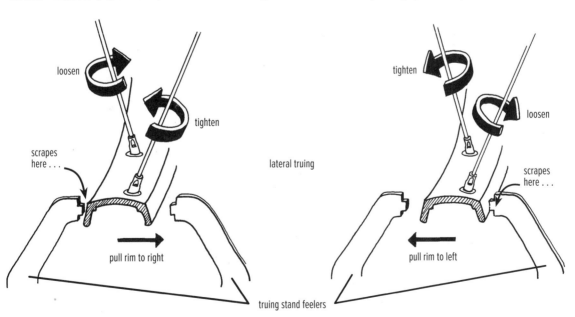

loosen

tighten

scrapes here . . .

pull rim to right

lateral truing

tighten

loosen

scrapes here . . .

pull rim to left

truing stand feelers

15.20 Radial truing: Pull the rim in.

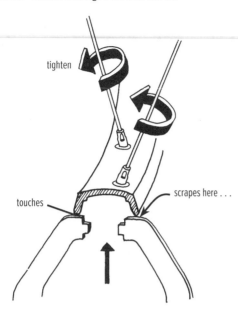

tighten

touches

scrapes here . . .

15.21 Radial truing: Let the rim out.

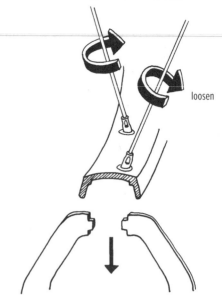

loosen

hard to turn (i.e., the nipples feel on the verge of rounding off), loosen all of the spokes in the wheel a one-quarter turn before continuing. Compare tension with a good wheel with the spokes of the same gauge; tension at this point should still be lower in the wheel you are building.

15-5

DISHING (CENTERING) THE WHEEL

1. **Place the dishing tool across the right side of the wheel, bisecting the center (Fig. 15.22).**
2. **Tighten or loosen the dishing gauge screw until the gauge contacts the outer face of the axle end nut (Fig. 15.22).**
3. **Flip the wheel over.**
4. **Place the dishing tool across the other side of the wheel.**
5. **Check the gap of the dishing gauge with this axle end nut face (Fig. 15.23).** Any gap between the dishing gauge and the axle end nut face indicates the amount the rim is offset from the centerline of the wheel. If there is no gap but an overlap instead, reset the dishing gauge on this side (the previously overlapped side). Flip it over and check the other side (i.e., repeat steps 3–5 on the opposite side).
6. **Put the wheel back in the truing stand.**

7. **Pull the rim toward the center (reducing the gap between the dishing tool and the axle end face) by tightening the spokes on the opposite side of the wheel from the axle end that had the gap between it and the dishing gauge.** Tighten a half turn each—no more. If the spokes are getting really tight (they will creak a lot when tightening, the nipples will start rounding off, and the spokes will feel much tighter than the spokes in a comparable wheel), then loosen the spokes uniformly on the opposite side of the wheel.

NOTE ON BLADED SPOKES: *You will need to hold the flats of bladed spokes with a slotted tool (Fig. 15.3B), as explained in 15-1, to keep them from twisting. Turning back one-eighth after each nipple tightening or loosening (15-3, step 2) may reduce the twist somewhat, but a bladed spoke has so much less torsional stiffness than a round spoke (twist a round vs. flattened paper towel tube to see what I mean) that it may just twist back and forth without actually turning in the nipple if you don't prevent the spoke from twisting.*

8. **Recheck the wheel with the dishing gauge by repeating steps 1–5.**
9. **If the dish is still off (there is still a gap between the dishing gauge and the end nut when you flip it over), repeat steps 6–8.** Continue until the dish is correct (the gap is zero).

10. **Stress the spokes by squeezing each pair together with your hands (Fig. 15.24).** They will make a "ping" noise as they unwind. If you followed my recommendation in 15-3, step 2, to turn the nipple back the opposite direction one-eighth turn after each time you tighten or loosen it, the spokes should not be wound up much, and prestressing the wheel will be unnecessary (you'll know because the spokes won't ping when you stress them).

   a. Leaning on the wheel is a quick way to prestress it, but you can wreck the wheel if you are not careful. To proceed, set the axle end on the workbench and carefully press down on the rim with your hands at the nine o'clock and three o'clock positions. This pressure will affect an area of about three spokes on each side, so rotate the wheel three spokes, press down again, rotate three more spokes in the same direction, press down again, and so on. After you finish one side, flip the wheel over and do the other side. Do not press down with all your might; too much pressure can destroy your work.

   b. If prestressing throws the wheel way out of true, the spokes are probably too tight. Loosen them all one-eighth turn. Note, though, that some loss of wheel "trueness" is normal. If the loss is minor, you can overlook it and continue with step 11.

11. **Repeat the process.** Repeat truing the wheel (15-4), followed by dishing the wheel (15-5), prestressing the spokes (and turning the nipples back a one-eighth turn from the rotation direction on each adjustment of one) frequently as you go. Keep improving the accuracy of the build this way.

12. **Bring up the tension to that of a comparable wheel by making small tightening adjustments to every nipple.** Adjust dish and true after each time around, until the wheel is as you want it.

13. **If the rim is oily, wipe it down with a citrus-based biodegradable solvent.**

14. **Congratulate yourself on building your wheel, and show it off to your friends.**

15.22 Using the dishing tool to check the centering of the rim relative to the axle ends

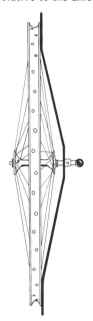

15.23 Checking the wheel dish on the other side of the hub

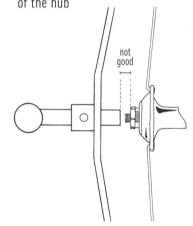

15.24 Relieving tension

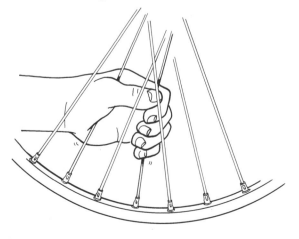

15-6
COMMENTS ON WHEELBUILDING

Your wheel has some features that you won't find on machine-built wheels. Most significantly, on your rear wheel, the "pulling spokes" are to the outside. This means that you have a spoking pattern that best resists the twisting force on the hub produced by pedaling forces on the chain.

In the wheel you've built, half of the spokes are called "pulling" or "dynamic" spokes, and the other half are called "static" spokes (this is true of any spoking pattern except radial). The pulling spokes are the ones directed in such a way that a clockwise twist on the hub increases the tension in them. If you look at the wheel from the drive side, you will see what I am talking about.

You will also see that the static spokes do not oppose a clockwise twist on the hub. In fact, their tension decreases when you stomp on the pedals.

By placing all of the pulling spokes so that they come out to the outside of the hub flanges (i.e., the spoke heads are on the inward side of the flanges), we have attached the spokes doing the most work the farthest outward on the hub, reducing the fatigue on them and increasing their ability to oppose forces acting on the rim. The reasoning is that the spokes whose tension changes the most during the working of the wheel should be the ones that are lying across the hub flange with the heads on the inside of the flange. Tension changes lead to spoke breakage due to fatigue, and the weakest part of a spoke is the elbow. If there is more contact between the spoke elbow and the flange, there is less stress on the spoke elbow. Also, the spokes under the most stress should have the widest "stance" (if you want to resist being knocked over, you plant your feet farther apart), because they come from the outside of the flanges and are thus farthest apart at their elbows.

If you choose the appropriate parts for your weight and riding style, and have the proper spoke tension, you should have a strong wheel that will last a long time. Congratulations!

15-7
BUILDING RADIALLY SPOKED WHEELS

With the advent of stronger rim materials and stiffer, deep-section rim cross-sections, radially spoked wheels (Fig. 15.25) have become popular. They are simple to build, and radial spoking offers a number of advantages.

A radially spoked wheel is vertically stiffer than a crossed one, because radial spokes allow little opportunity for spokes to absorb energy in the spoking pattern. The radial wheel can be stiffer laterally too, because the spoke length is shorter, increasing their pulling angle to the rim.

A radial wheel is lighter because the spokes are shorter. Further weight can be removed with fewer spokes (made possible by deep, stiff rims), and radial spoking allows any even spoke count to be used (with nonradial patterns, the spoke count must be a multiple of four). And radial spoking allows the use of direct-pull hubs and nail-head spokes (straight spokes without elbows), eliminating a potential weak spot in the spokes.

Radially laced spokes line up behind each other and thus improve the aerodynamics of the wheel. Aero-shaped spokes can improve the aerodynamics even further, but using wide aero-shaped spokes in a standard hub often requires slotting the hub holes with a jeweler's file to get the spoke through. If you do this your-

15.25 Radially spoked front wheel

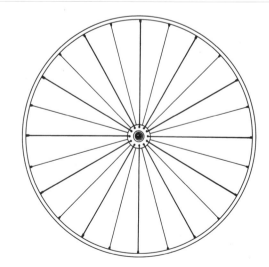

15.26 Radial/three-cross rear wheel

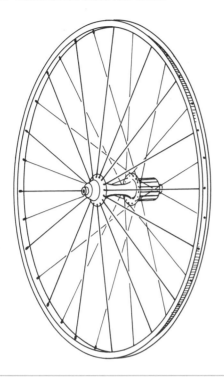

self, make sure you only file downward from the hole, toward the meat of the flange. Slotting upward toward the edge greatly weakens the hub and invites the spoke to rip through.

Speaking of torn hub flanges, the warranty of some hubs is voided when they are spoked radially; Shimano, for one, has this stipulation. The stress is greater on hub holes with radial spoking because the spoke tension in a radial wheel is often higher and because there is less material resisting the hub's tearing out when the spoke is pulling straight outward than when it is pulling at an angle along the hub flange.

A completely radial wheel can only be used as a front wheel for rim brakes. On the rear, or on a front disc-brake wheel, at least one side of the wheel must still have a crossing pattern to oppose the twist on the hub caused by the chain (Fig. 15.26) or disc brake.

a. How to lace a radial front wheel

Drop all of the spokes from the inside of each flange outward, or from the outside of the flange inward, and lace them straight to the rim.

b. How to lace a rear wheel with a radial left side and a three-cross drive side

First lace the drive side following the instructions in 15-2a, steps 4–7, and 15-2c, steps 17–24. Now lace the left-side spokes outward through the hub flange (inward through the hub flange is okay too, but slightly less laterally stiff) and straight to the rim.

The tensioning and truing steps are the same as for standard three-cross wheels, but radial-spoke tension must be high to help prevent the spokes from vibrating loose.

c. How to lace a rear wheel with a radial drive side and a three-cross left side

Advantages

If the hub is stiff enough not to twist when the only thing opposing the twist created by power applied on the drive side via the chain and cogs will be crossed spokes on the opposite end of the hub, then this method has advantages over the pattern in step b, as well as over three-cross and two-cross patterns on both sides.

As in the pattern in step b, the radial spokes will improve the aerodynamics and reduce the weight of the wheel. And, as in the step b pattern, the chain winding up the hub always tightens all of the spokes on the radial side, rather than loosening half of them while tightening the other half as happens with a crossing pattern.

But there is another benefit in terms of spoke tension. The Achilles' heel of most rear wheels is the difference in tension on the two sides, due to the greater width from hub flange to axle end on the drive side to make space for the cogs. If you have radial spokes on the drive side that come out to the outside of the hub flange (i.e., the spoke heads are on the inboard side of the hub flange), you have made the side-side angle from the spoke elbow to the rim the widest possible. Even more important, since these spokes are so much shorter than the crossed spokes on the non-drive side, the relative angle to the rim of the spokes on the two opposing sides is lower than in a symmetrically crossed wheel, and lower yet than in the wheel in step b.

That is because the spokes on the left side in this arrangement travel a longer distance to move the same distance laterally inward to their attachment point on the rim, making their angle to the rim steeper, and vice versa on the drive side. So, if the angle to the rim of the spokes on the left is steeper than normal, and that of the spokes on the right is shallower than normal, then the angles are closer to being the same on both sides, making the spoke tension more even. This makes for a stiffer, longer-lasting wheel.

Method

First lace the non-drive side following the instructions in 15-2b, steps 9–16, and 15-2d, steps 25–30. Now lace the drive-side spokes outward through the hub flange and straight to the rim.

15-8

BUILDING TWO-CROSS WHEELS

Two-cross wheels are stiffer vertically and slightly lighter than three-cross wheels. And two-cross (or radial) may be the only good way to go with deep-section rims with internal spoke nipples, because three-cross will probably not be possible to build. That is because the spoke-nipple access holes in the rim bed will not be wide enough to allow you to get at the nipple with the long nipple-driver socket (see tool in Fig. 15.3C), since the spoke will be coming in at such an angle from the hub flange. The straighter fore-aft spoke angle from flange to rim hole on a two-cross wheel will allow you to get the nipple on the end of the spoke and turn it.

To build a two-cross pattern, follow the lacing instructions in 15-2 exactly, with the following changes, starting at step 21:

21. **Pick any spoke on the top (right-hand) hub flange that is already laced to the rim.** Now find the spoke that is three hub holes away in a clockwise direction.

22. **Take this new spoke, cross it under the spoke you counted from (the one three holes away), and stick it into the rim hole two holes counterclockwise from that spoke.** Thread a nipple on three turns.

23. **Continue around the wheel, repeating the pattern.** You may find a spoke or two that doesn't reach quite far enough. If that's the case, at a point about an inch from the spoke elbow, push down on the spoke to help it reach.

24. **Make sure that every spoke coming out of the upper side of the top flange (the spokes that come out toward you with their spoke heads hidden from view) crosses over one spoke and under a second.** Both of these crossing spokes come from the underside of the same flange and have their spoke heads facing toward you. These crossing spokes begin one and three hub holes counterclockwise from the spoke that you just inserted into the rim. This is called a "two-cross" pattern because every spoke crosses two others on its way to the rim (over, under). Every upwardly offset hole should now be occupied on the rim. Follow steps 25–27 as in 15-2.

28. **Pick any spoke on the top (left-hand) hub flange that is already laced to the rim. Now find the spoke three hub holes away in a counterclockwise direction.**

29. **Take that spoke, and cross it over one spoke and under the spoke you counted from.** Stick the spoke into the rim hole two holes clockwise from the spoke it crosses under. Thread a nipple on three turns.

30. **Continue around the wheel, doing the same thing until the wheel is laced as in Figure 15.27.** You may find that some spokes don't reach far enough. If that's the case, at a point about an inch from the spoke elbow, push down on the spoke to help it reach.

31. **Make sure that every spoke coming out from the upper side of the top flange (the spokes that come out toward you with their spoke heads hidden from view) crosses two spokes (Fig. 15.27).** Both of these crossing spokes come from the underside of the same hub flange and have their spoke heads facing you. The crossing spokes begin one and three hub holes clockwise from each spoke emerging from the top of the upper (left) hub flange. Every hole in the rim should now

15.27 Two-cross wheel

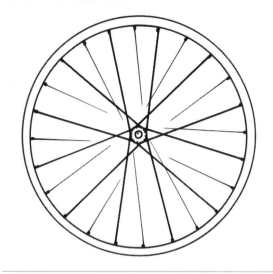

be occupied. When you look at the wheel from the side, the valve hole should be between two spokes (your first spoke and the first spoke of the fourth set) that do not cross each other but whose trajectories look like they are trending slightly toward each other. In other words, if these spokes were to continue infinitely outward, their trajectories would eventually cross far beyond the rim (Fig. 15.17). In wheel-builder speak, these two spokes are called "converging parallel" spokes. Lacing the spokes this way around the valve hole will provide the maximum possible space between the spokes for the pump head when inflating the tire.

NOTE: *If this is a rear wheel, the spokes coming out of the outside of the hub flange on both sides oppose the clockwise twist the chain applies on the cogs. See 15-6 for more on this subject.*

15-9

LACING REAR THREE-CROSS DISC-BRAKE WHEELS

Disc-brake wheels need to have crossed (non-radial) spokes to oppose the twist applied to the hub by the brake pads grabbing the rotor. Three-cross spoking makes sense with shallow-section rims, but with deep-section rims, particularly when coupled with tall hub flanges, three-cross spokes come in at too low of an angle for the rim holes and can also interfere

with spoke heads at the hub flange; instead, use a two-cross pattern (15-8).

As explained in 15-6 regarding the chain force on a rear wheel, the spokes coming out of the outside of the hub flange should ideally be oriented to oppose the twist the rotor being grabbed by the brake pads applies on the hub. This pattern makes for a stronger wheel, by having the wider-angle spokes doing more of the work. On the front, you would set the spokes coming from the inside of the hub flange out on both sides to head counterclockwise toward the rim when viewed from the brake (non-drive) side (Fig. 15.28) to oppose a clockwise twist on the hub by the rotor.

However, on the rear wheel, while the brake applies a clockwise twist on the hub viewed from the brake side, the chain still applies a clockwise twist on the hub viewed from the drive side. So, if you are building a rear disc-brake wheel (Fig. 15.30), you want the drive-side outer spokes opposing the chain force on the cogs, but you still want the left-side outer spokes opposing the braking force on the rotor. And with all disc-brake wheels, I recommend 14/15-gauge double-butted spokes and brass nipples (see 15-10a and 15-10b for more on this).

15.28 Completed front disc-brake wheel

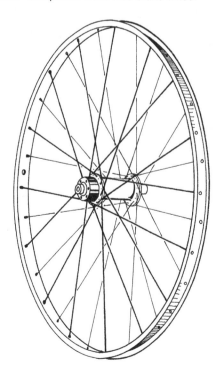

The drive side will be laced in just the same way as described in the lacing instructions in 15-2, but the non-drive side will be laced in the opposite way that the left side turns would be laced in 15-2.

a. First set of spokes

1. **Follow steps 1–9 from 15-2.**

b. Second set of spokes

2. **Push a spoke up through the hole in the top flange that is immediately clockwise from the first spoke you installed (the spoke that is just clockwise from the valve hole).**

3. **Follow steps 11, 12, and 13 from 15-2 and then step 4 below.**

4. **Drop one spoke down through each of the adjacent hub holes on either side of the newly laced spoke.** Skip a hole and continue around the hub flange, dropping a spoke down into every other hole.

5. **Rotate the hub shell clockwise as far as you can.**

6. **Find the spoke that is five hub holes counterclockwise from the single spoke coming up out of the flange that you installed in steps 2 and 3.**

7. **Take this new spoke, cross it over the spoke you counted from (the one five holes away), and stick it into the rim hole two holes clockwise from that spoke.** Thread the nipple onto the spoke three turns.

8. **Find the next spoke counterclockwise on the hub flange.** Put it in the rim hole four holes counterclockwise from the spoke you just installed in step 7. Thread the nipple onto the spoke three turns.

9. **Continue counterclockwise around the top (brake-side) hub flange until all spokes whose heads stick out of the top flange are one rim hole counterclockwise from the first set of spokes installed in the rim.**

c. Third set of spokes

10. **Follow steps 17–27 in 15-2.** After you complete step 22, the wheel should look like the wheel shown in Figure 15.29 (except the other fanned-out, unlaced spokes coming out of the top flange are not shown).

15.29 Rear disc-brake wheel: first two sets of spokes completed, first spoke of third set installed

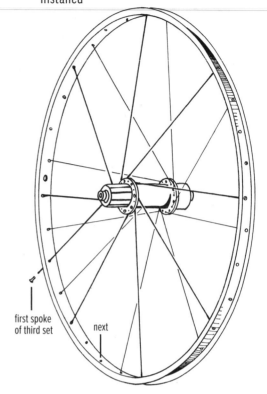

first spoke
of third set   next

15.30 Completed rear disc-brake wheel

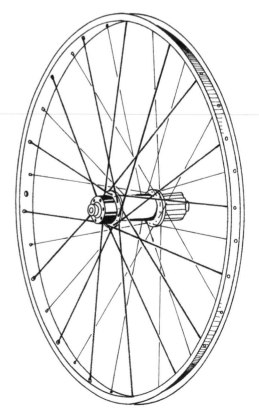

d. Fourth (and final) set of spokes

11. **Pick any spoke on the top (rotor-side) flange whose head is facing up and is already laced to the rim. Now find the spoke five clockwise hub holes away.**

12. **Follow steps 22–24 in 15-2.**

13. **Your wheel is now laced.** Note that the drive-side outer spokes oppose the chain pull, and the rotor-side outer spokes oppose the braking force on the rotor (Fig. 15.30). Give yourself a big pat on the back and then begin tensioning and truing your wheel, starting with 15-3.

15-10

BUILDING WHEELS FOR BIG RIDERS

Building wheels for heavy and tall riders requires greater lateral and vertical stiffness. The weight of the rider can bend and flex the rim, but it creates another problem as well. The heavier rider reduces the tension of the spokes at the bottom of the wheel more by making the rim more D-shaped at the bottom as it rolls. If the tension drops to the point that the nipple flanges periodically lose contact with the bases of the rim holes, the nipples can unscrew, or the rim can fatigue and crack at each spoke hole. To achieve the higher strength required, you can add the following characteristics.

a. Spoke count and thickness

The spoke count needs to be high: Thirty-six or more spokes is highly preferable for riders over 190 pounds. The spokes need to be heavier, as thicker spokes are less prone to breakage. Although 14/15-gauge (2.0mm, or 14-gauge, on each end, and 1.8 mm, or 15-gauge, throughout the center section) double-butted spokes are thinner than straight 14-gauge spokes, DT Swiss testing has shown that the wheel will probably last longer with them. Because most breakage occurs at the nipple or the elbow, and butted spokes are the same thickness there, spoke breakage will not increase. But butted spokes will stretch more, allowing the spoke nipples to stay in contact with the rim better as the rim changes shape while rolling.

b. Nipple type

Brass nipples are preferable to aluminum ones, due to the extra stress a big rider puts on the wheel. And ones with threadlock compound inside them, like DT Pro Lock nipples, will be far less likely to loosen up over time.

c. Rim section and drilling

The deeper the rim, the higher its hoop strength (vertical stiffness and strength). Very deep V-section rims work with low spoke counts because of this high hoop strength. The strongest wheel would be from a deep-section rim drilled for more spokes. Unfortunately for heavy riders, many deep V-section rims are also thinner to reduce weight and hence lose some strength.

d. Spoking pattern

With 8-, 9-, 10-, and 11-speed rear wheels, dish is high (one side of the wheel is flatter than the other), meaning that there is a great tension difference between spokes on the two sides. The loose spokes on the left can unscrew, especially under high pedaling forces, and the tight spokes on the right can break. As the chain twists the cogs clockwise, the spokes opposing the twist (the "pulling spokes") get tighter, while the "static spokes" are under reduced tension and can unscrew (especially without locked nipples).

An off-center rim can help by reducing the wheel dish. The rim holes are offset to the left side (Fig. 15.4), so that the drive-side spokes come to the rim at a lower angle and can work with lower tension and more even tension between the two sides. The left-side spokes come to the rim at a higher angle and can be under higher tension without forcing the use of dangerously high tensions on the drive side. Before lacing an off-center rim, make sure you read the Note in step 2 of 15-2.

Using radial spokes on the left side (see 15-7b) can counteract the problem of grossly uneven tension. With a radial left side, the chain twisting the hub forward always tightens all of the left-side spokes, rather than loosening half of them as it would with a crossing pattern.

There is also an argument that you want the radial spokes on the drive side and the crossing spokes on the left side (Mavic wheels use this philosophy). The idea is that the shorter drive-side spokes, by taking a shorter distance to the rim, increase the effective spoke bracing angle. Similarly, the longer crossed spokes on the left, which would normally have a much bigger bracing angle than the drive-side spokes, will have the bracing angle reduced due to their increased length. Thus, spoke tension on each side will be more balanced. However, the hub shell must be stiff enough to carry the drive-force twist from the cog side to the non-drive side.

15-11

BUILDING WHEELS FOR CYCLOCROSS

a. Firm courses

Wheels for cyclocross races on hard ground need to be light (since you're carrying the bike so much as well as repeatedly accelerating it), strong (due to the abuse they receive), and vertically compliant (due to the harsh ride on the firm, bumpy surface without any suspension apart from the tires). A strong, light rim with a shallow cross-section built up three-cross with double-butted spokes (thin in the middle, thick on the ends) will make for a great hard-course wheel. Choose one that has some curvature to the nipple side (Fig. 15.31); a domed or teardrop shape will help shed mud. If you are using disc brakes, be sure to build the wheel with brass nipples. Disc brakes place more stress on the spokes and nipples because the braking force must be transmitted through them to reach the rim and tire.

b. Soft courses

Wheels for muddy and/or sandy courses need to be light, but, most important, they must steer well through the mud or sand. They can be stiff vertically because the soft ground absorbs much of the impact.

A deep-section rim (Fig. 15.32) will steer straighter through mud and deep sand than will a shallow-section one. It will also shed mud better, so you won't be lifting or accelerating as much weight or pedaling as much glop into the brakes and into the junctions of the fork, chainstays, and seatstays.

Deep aluminum rims are heavy; deep carbon tubular rims are a better choice (Fig. 7.24), and be sure to use carbon-specific brake pads as well.

Get a rim with full-size holes for the nipples so that you can use external nipples rather than nipples hidden inside the rim. You can end up truing 'cross wheels frequently, and removing the tire to do so is a pain, particularly with a tubular tire—which is what you will choose if you want to minimize rim and tire weight.

A two-cross (15-8) or three-cross (15-2) pattern will give the wheel a bit more vertical compliance, but on soft ground when using cantilever brakes, you may prefer radial spokes (15-7a); if so, go with three-cross on the non-drive side of the rear wheel (15-7c) and radial front (15-7a).

c. Tire choice determines rim type

If you're using tubular tires, you will be able to choose among lighter rims than for clinchers or tubeless clinch-

15.31 Shallow-section clincher rim cross-section (with a bit of a domed shape)

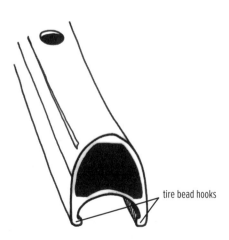

tire bead hooks

15.32 Deep-section clincher rim cross-section

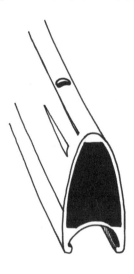

ers, because tubular rims don't have the extra rim wall needed for the clincher bead hook (Fig. 15.31).

d. Build lots of wheels

To be competitive in cyclocross on muddy courses, you need multiple sets of wheels for racing. The reason, of course, is that on muddy courses, you need a second bike that a buddy cleans for you and exchanges with you at the service pit every lap or so. That bike needs to have similar wheels to the bike you start on, for the reasons listed earlier, and you will want to bring extra wheels in case you get flat tires. It makes sense, therefore, to have a number of lightweight wheelsets for different race conditions and bikes. You will also want to have at least one set of sturdy clincher wheels for training.

Unless you are running disc brakes, be sure to select rims that have the same width for all of your 'cross wheelsets. Doing so will eliminate the hassle of adjusting brakes when changing wheels.

16

Someday we'll look back on this
moment and plow into a parked car.

—EVAN DAVIS

FORKS

TOOLS

2mm, 2.5mm, 5mm, 8mm
   hex keys
8mm open-end wrench
Small screwdriver

OPTIONAL

Ruler
True front wheel
Dropout alignment tools
V-blocks
Vise
Threadlock compound

The fork serves a number of purposes. It connects the front wheel to the handlebar, allows the bike to be steered, and supports the front brake. The fork also offsets the front hub some distance forward of the steering axis (Fig. 16.1). This offset distance (fork rake, R in Fig. 16.1), combined with the steering axis (the head angle, Ø in Fig. 16.1) and the wheel size (the radius, r in Fig. 16.1), determines the fork trail (T in Fig. 16.1), which largely dictates how your bike is going to handle and steer.

All forks—even rigid road forks (Fig. 16.2)—provide at least a minimum amount of suspension by allowing the front wheel to move up and down. The steering axis angles the fork forward from vertical, while the front hub is offset farther forward yet, and these things allow for a fork to flex along its length and absorb vertical shocks.

Virtually every road bike fork is made of the components illustrated in Figure 16.2: the steering tube, the fork crown, the fork legs (sometimes called "blades"), and the fork ends (also called "dropouts" or "fork tips"). Road bike forks are manufactured from carbon fiber, steel, aluminum, or titanium, with carbon being standard on today's mid- to high-end road bikes.

Prior to the early 1990s, all forks had threaded steering tubes (Fig. 16.2); today they are all unthreaded (Figs. 16.3–16.6). Until the changeover, threaded steering tubes for road bikes were one inch in diameter, as were early threadless steerers. Up until 2010 or so, all threadless steering tubes had a constant diameter, with a diameter of 1⅛ inch having replaced the previous one inch diameter. However, high-end bikes now often have a larger lower headset bearing than the upper one (Chapter 12), so that the steering tube tapers from a larger diameter at its base up to the standard 1⅛-inch diameter at the top (Figs. 16.3, 16.5, 16.6).

Since standard road brakes that sit above the tire (Figs. 9.1–9.3) limit the tire's diameter, forks designed for larger tires—as on cyclocross bikes, gravel-road bikes, and some touring and general-purpose road bikes—have mounts for cantilever (Fig. 16.4) or disc brakes (Figs. 16.5, 16.6). The latest disc-brake forks take yet another cue from mountain bikes; they forgo slotted dropouts in favor of "through axles" (Fig. 16.6). The left dropout has a large hole to allow the axle to slide right through, and the right dropout has a threaded hole to screw the axle into.

16.1 Front-end geometry of a bicycle

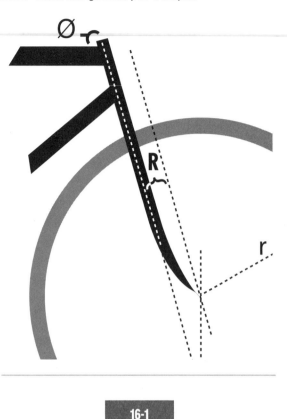

16.2 Threaded steel road fork

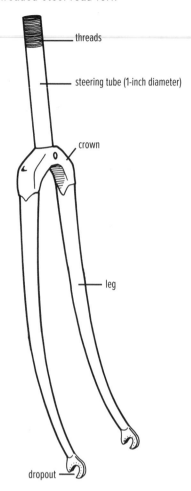

16-1

FORK INSPECTION

Forks are generally pretty durable, but they do break occasionally. A fork failure can ruin your day, since you wouldn't be able to control the bike, which would likely lead to the rapid acceleration of your body downward onto the road.

Ever since I first opened my frame-building shop, people have regularly brought in broken forks of all types to show me, sometimes in the hope that I can repair them. Some forks break with catastrophic consequences. Some have steering tubes broken at the fork crown or in the threads. Others have fork crowns that broke or separated (releasing a fork leg or two), fork legs that folded, cantilever posts that snapped, and front dropouts that bent over, pulled out, or broke off. You can go a long way toward preventing problems like these by regularly inspecting your bike's fork.

With that in mind, get into the habit of checking the fork regularly for signs of impending failure. If you find something amiss, read the next section (16-2) to see whether there is a remedy.

Obvious things to look for include bends, cracks, and stressed paint. On carbon forks, look for cut or torn fibers on the legs and crown, or loose glue bonds on the dropouts, fork crown, or steering tube. Feel for a change in stiffness and listen for strange noises. Try tapping on the fork legs with a coin over their length, listening for a change in sound from point to point and comparing it with the other side; this is a way to discover delamination or cracking in underlying carbon layers that is not visible from the outside.

If you have crashed your bike, give the fork an especially thorough inspection. If you find any indication that the fork has been damaged, replace it. A new fork is cheaper than emergency room charges, brain surgery, or an electric wheelchair.

When you inspect a fork, remove the front wheel, wipe any dirt off the legs and crown, and look under the crown and between the fork legs. Carefully examine all

16.3 Trek Madone carbon road fork with tapered steering tube

16.4 Threadless rim-brake cyclocross fork

16.5 Carbon fiber road or cyclocross fork with disc-brake mounts and tapered steering tube

16.6 Disc-brake road or cyclocross fork with through axle

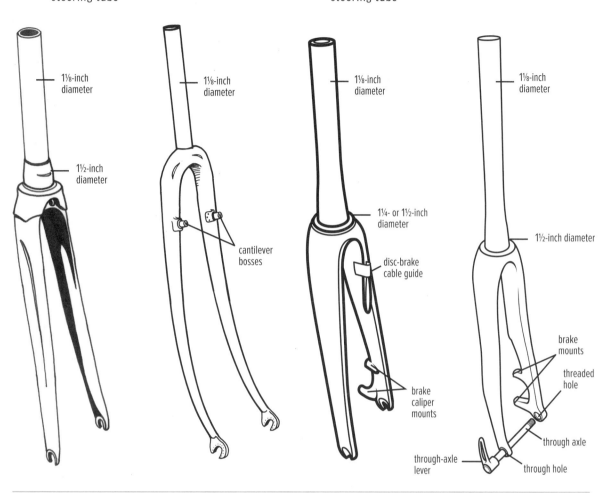

of the outside areas. Look for any areas where the paint or finish looks cracked or stretched. Look for bent parts, from little ripples in fork legs to bent dropouts (Fig. 16.7). Skewed or broken cantilever posts are something to look for on cyclocross and touring forks (Fig. 16.4).

Put the wheel back in and watch to see if the fork legs twist when you tighten the hub into the dropouts. Check to make sure that a true wheel centers under the fork crown. If it doesn't, turn the wheel around and put it back in the fork to determine whether the misalignment is in the fork or the supposedly true wheel. If the wheel lines up off to one side when it is in one way, and off the same amount to the other side when it is in the other way, the wheel is off and the fork is straight. If the wheel is skewed off to the same side in the fork

no matter which way you install the wheel, the fork is misaligned. A misaligned fork can cause problems like front-end shimmy, especially at high speed; uneven tire wear; and inconsistent steering.

I recommend overhauling the headset annually (12-18 and 12-19); when you do, carefully examine the steering tube for any signs of stress or damage. Check for bent, cracked, or stretched areas, stripped threads (Fig. 16.7), a bulging threaded steering tube where the stem expands inside (on an older-style threaded steerer), or a crimped threadless steering tube where the stem clamps around its top. On a carbon steering tube, look for cracks where the stem is clamped on, and make sure that there is an expandable or glued-in support inside the steering tube under the stem clamp. Check also to

16.7 Types of fork damage

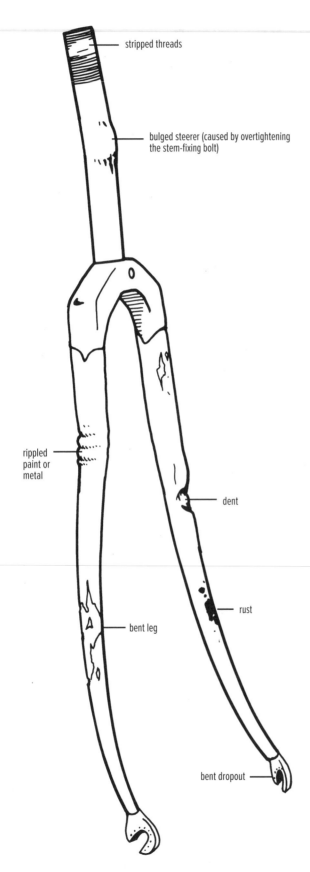

stripped threads

bulged steerer (caused by overtightening the stem-fixing bolt)

rippled paint or metal

dent

rust

bent leg

bent dropout

see that the steering tube shows no signs of pulling up out of the fork crown.

With a threaded fork, hold the stem up next to the steering tube to make sure that, when the stem is inserted to the depth you have been using it, the bottom of the stem is always more than an inch below the bottom of the steering tube threads. If the stem expands in the threaded region, you are asking for trouble; the threads cut the steering tube wall thickness down by about 50 percent, and each thread offers a sharp breakage plane along which the tube can cleave.

16-2

FORK DAMAGE

If your inspection uncovers damage that does not automatically require fork replacement, here are some guidelines to help you.

a. Dents

Not all fork dents threaten the integrity of the fork. A small dent in a steel fork usually poses little risk; a large dent (Fig. 16.7) demands attention (replace the fork). Carbon forks don't tend to dent. If yours has a dent, that is a cause for immediate replacement, especially if you see cracking in the clear coat and/or separated fibers in the same area. In a carbon fork, a dent that holds its shape can indicate delaminated (separated) carbon layers underneath. Tap on the fork leg with a coin at that spot and compare it with the sound in surrounding areas and the other side; you'll be able to hear how the delamination or cracking in underlying carbon layers deadens the nice "clack" noise you should hear.

b. Fork misalignment

Within limits, a steel fork can be realigned if it is slightly off-center (16-3b). Aluminum, carbon fiber, and titanium forks cannot be realigned. Don't try it!

c. Stripped steering tube threads

If the threads on the steering tube are damaged (Fig. 16.7) so that the headset slips when you try to tighten it, you need to replace the fork. Same story if the steering tube is bulged; replace the fork, because it can split.

You can have a frame builder replace the steering tube on a steel fork, but getting a new fork makes more economic and safety sense.

d. Obvious bend, ripple, or crease in fork legs

Replace the fork if ripples and bends are obvious (Fig. 16.7). The poor handling and potential breakage threaten your safety. Save a few bucks on something else.

e. Bent or stripped cantilever bosses

The pivot studs on some cantilever bosses (Fig. 16.4) thread into the boss and can be unscrewed with an 8mm open-end wrench and replaced. Use a thread-locking compound on the threads of the new stud.

On many touring and cyclocross forks, the entire cantilever boss is welded on (Fig. 16.4). Bent or stripped cantilever bosses on such forks usually mean that you have to buy a new fork. If you have a frame builder in your area, he or she may be able to weld or braze a new one on a steel, titanium, or aluminum fork. If it's steel, you will also need to repaint the fork; all that work may cost more than a new fork, by the way.

| **16-3** |

MAINTAINING ROAD AND CYCLOCROSS FORKS

 LEVEL Beyond touching up the paint on steel forks and performing regular inspections, the only maintenance procedure for a road or cyclocross fork is to check the alignment (16-3a) if your bike is handling badly or has a shimmy at high speeds or when riding with no hands. You can perform minor realignment on a steel fork if you find that it is off-center, but note that it is risky enough to qualify as a level 3 job. Do not try to realign carbon fiber or aluminum forks. If the alignment is correct and the headset is correctly adjusted but shimmy is a problem, a fork with more rake can sometimes reduce shimmy problems.

a. Check fork alignment

You will need a ruler, a true front wheel, and dropout-alignment tools (Fig. 1.4). If you have an aluminum, titanium, or carbon fiber fork, this procedure is diagnostic only, because you should not try to realign any of these forks. Checking the alignment may help explain bike-handling problems and may indicate that a different fork could reduce or solve them.

If you find the alignment to be off more than a couple of millimeters in any direction with any fork other than a steel unsuspended one, you will need a new fork. If the fork is new, misalignment should be covered by the warranty.

If a steel fork is more than 8mm off in any direction, you ought to get a new fork. If the dropouts of a steel fork are slightly bent, you can realign them. You can also take a moderately bent (less than 8mm off) steel fork to a frame builder for realignment. Make sure that whoever you take it to is properly equipped with a fork jig or alignment table and is well versed in the art of "cold setting" (a fancy term for bending) steel forks.

1. **Remove the fork from the bike (12-18 and 12-19).**
2. **With the front wheel out, measure the spacing between the faces of the dropouts (Fig. 16.8).** Adult bikes should have a spacing of 100mm between the inner surfaces of the dropouts. (Some low-end kids' bikes have narrower spacing—about 90mm or so.) Remember that you are measuring the distance between the flat surfaces that meet the hub-axle faces (and not between wheel-retaining nubs that protrude inward from the dropouts on some forks). Dropout spacing as wide as 102mm and as narrow

16.8 Measuring dropout spacing

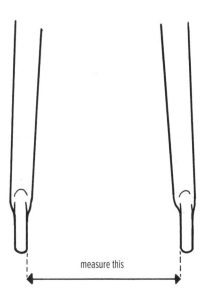

measure this

16.9 Installing a dropout-alignment tool and bending the dropout with it

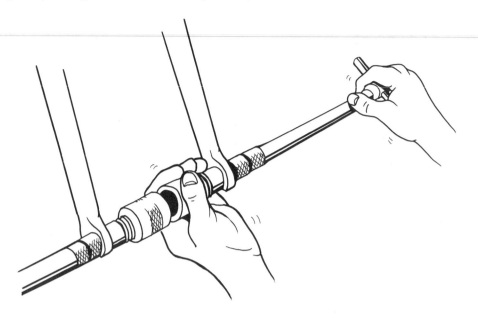

as 99mm is acceptable. Beyond that in either direction means a new fork. If you have a steel fork, you can take it to a frame builder for alignment.

3. **Clamp the steering tube of the fork.** Use a bike stand or two V-blocks in a vise.

4. **Install the dropout-alignment tools (Fig. 16.9).** They can be used on either the fork or the rear triangle of the bike, and thus they have two axle diameters and spacers for use in the wider rear dropouts. For use on the fork, move all of the spacers to the outside of the dropouts so that only the cups of the tools are placed inboard of the dropouts. Install the tools so that the shafts are seated up against the tops of the dropout slots. Tighten the handles.

5. **Check the positions of the alignment tool cups.** Ideally, the ends of the cups on the dropout-alignment tools should be parallel and lined up with each other (Fig. 16.10). The cups of Campagnolo dropout-alignment tools are nonadjustable and are nominally 50mm in length; the ideal space between their ends is 0.1–0.5mm. The cups on Park dropout-alignment tools (illustrated in Figs. 16.9–16.11) are adjustable in length, so that you can bring the faces up close to each other no matter what the dropout spacing. If they are lined up with each other and the

16.10 Correct dropout alignment

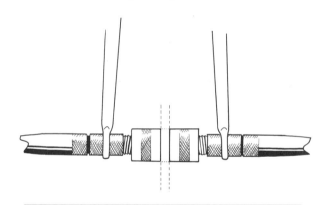

16.11 Incorrect dropout alignment (dropout is twisted or right fork leg is bent back)

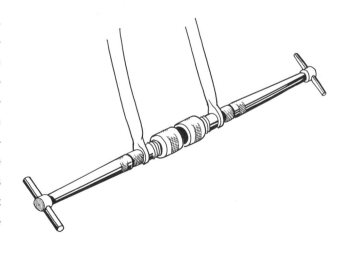

dropouts are spaced between 99mm and 102mm apart, continue to step 6. If a steel rigid fork's dropouts are not lined up straight across with each other (Fig. 16.11) and the dropouts are within the 99–102mm spacing range, skip to 16-3b to align them before returning to this point for the next steps.

NOTE: *The dropout faces must be parallel before you continue with step 6, or the rest of the alignment procedures will be a waste of time. Clamping the hub into misaligned dropouts will force the fork legs to twist. If the dropouts are misaligned, any measurement of the side-to-side and fore-and-aft alignment of the fork legs will not be accurate.*

6. **Remove the tire from the front wheel.** Make sure the wheel is true and properly dished (15-4 and 15-5).

7. **Install the wheel in the fork.** Make sure the axle is seated against the top of the dropout slot on either side, and make sure the quick-release skewer is tight. Lightly push the rim from side to side to make certain that there is no play in the front hub. If there is play, you first must adjust the hub (Chapter 8).

8. **Look down the steering tube and through the valve hole to the bottom side of the rim (Fig. 16.12).** The steering tube should be lined up with this line of sight through the wheel (Fig. 16.13).

NOTE ON CARBON FORKS: *Carbon forks generally are closed under the fork crown, so you cannot sight down through them. In that case, the best you can do is compare the rim's position (flip the wheel around and install it the other way as well) with the brake hole and with the fork legs on either side of the rim to determine if it is centered in the fork. This will tell you if the fork legs are symmetrical but not if the steering tube is in alignment with them.*

    a. When you are sighting through the steering tube and the valve hole, you should see the same amount of space between either side of the rim and the sides of the steering tube. You should also see the center of the bottom side of the rim through the valve hole.

    b. Turn the wheel around and install it again so that what was the right end of the axle is now the left and vice versa. Sight through the steering tube and the wheel valve hole again.

16.12 Correct alignment of the wheel valve hole in a straight fork

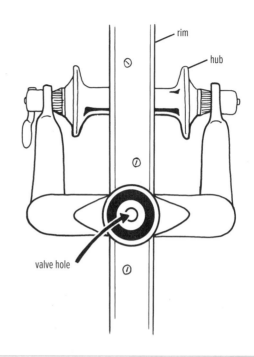

16.13 Sighting through the steering tube to check fork alignment

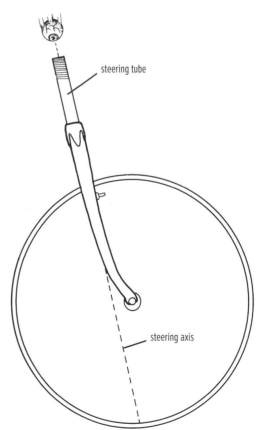

c. Placing the wheel in the fork both ways corrects for deformation in the axle or any wobble in the wheel. If the wheel is true and dished properly, and the axle is in good shape, the wheel should line up exactly as it did before. If it does not line up but is off by the same amount to one side as it is to the opposite side, when the wheel is turned around, the wheel is off and the fork is fine side to side.

d. If this test indicates that the fork is as much as 2–3mm off to the side, that is close enough; continue on. If it is off by more than 3mm, get a new fork or have it aligned by a frame builder (if it is steel, that is; do not try to realign suspension, carbon fiber, or aluminum forks).

NOTE: *If you are sighting through the wheel in this way and you cannot see the bottom side of the rim through the valve hole because the hub is in the way, the fork has big problems. In order for the bike to handle properly, the fork must have some forward offset of the front hub from the steering axis. This offset, or "rake," is usually around 4–5cm. If you sight through the steering tube and see the front hub, the fork is bent backward so much that it has little or no offset! If this is the case, you need a new fork!*

9. **With the wheel still in the fork, place a ruler on edge across the fork blades with its flat side resting on the rim (Fig. 16.14) and its length perpendicular to the steering tube.**

10. **Holding the ruler in place, lift the fork toward a light source so that you are sighting across the ruler and the front hub toward the light.** The ruler's edge should line up parallel with the fronts of the dropouts (or with the axle ends sticking out of either end of the hub; Fig. 16.14). This test will tell you whether one fork leg is bent back relative to the other one. If the two line up parallel or very close to that, the fork alignment has checked out completely, and you can put the fork back in the bike. If one fork leg is considerably behind the other, you need to get a new fork or have this one aligned (only possible if it is a steel fork). If the dropout-alignment tools also indicated that one dropout was ahead of the other (Fig. 16.11), then the fork legs alone could be bent, and not the dropouts.

b. Align dropouts on steel fork

Dropouts are easy to tweak out of alignment; simply pulling the bike off a roof rack and failing to lift it high enough to clear the rack skewer will do it. Forks can also come with misaligned dropouts to start with. Dropouts can be aligned, but only on a steel, nonsuspension fork.

If the dropout is bent more than 7 degrees or so, or if the paint is cracked at the dropout where it is bent, bending it back may be dangerous. Replace the fork.

1. **Install dropout-alignment tools and check alignment as described in steps 4 and 5 of 16-3a. If**

16.14 Checking fore-and-aft alignment of forks; x should equal y.

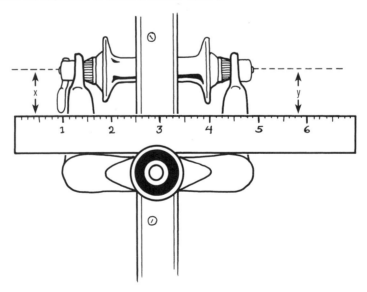

one alignment tool is ahead of the other (Fig. 16.11), it could indicate 1) that the dropouts are bent, 2) that one fork leg is ahead of the other (which you checked for in 16-3a, steps 9 and 10), or 3) a combination of both problems.

2. **If the dropouts are not aligned and the fork spacing is between 99mm and 102mm, and step 10 of 16-3a indicated that both fork legs are parallel, then you can align the dropouts.** (You'll have to go through all of the steps in 16-3a again to check the fork alignment again after you align the dropouts, because that will change how the wheel sits in the fork.) If the fork spacing is wider than 102mm or less than 99mm, there is no point in aligning the dropout faces, because you must bend the fork legs as well to correct the spacing. Without an alignment table or fork jig, you cannot do this accurately. You should get a new fork or have a qualified mechanic or frame builder align the steel fork. If the fork spacing is between 99mm and 102mm, clamp the crown or unicrown of the fork very tightly between two wood blocks in a well-anchored vise.

3. **Grab the end of the dropout-alignment tool handle with one hand and the cup of the tool with the other (Fig. 16.9).** Bend each dropout until the open faces of the dropout-alignment tools are parallel and the edges align with each other (Fig. 16.10).

4. **Remove the tools and continue with 16-3a, step 6.**

16-4

FORK UPGRADES

You may be able to improve the ride of your bike by replacing the fork. There are a number of reasons to do this. To lighten the bike and add gee-whiz value, you could get a carbon fiber fork. To stiffen the ride, you could get a steel fork or a beefier carbon fork. To lighten the bike and get a more rigid fork-to-bar connection on your old bike with a gooseneck stem and threaded fork (Fig. 12.1), you could switch to a threadless system (see Chapter 12 on headsets for the difference between threaded and threadless systems). To switch to a disc brake, you could get a disc-brake-specific fork (Fig. 16.5). And to reduce aerodynamic drag, you could get an aero fork.

Make sure you get a fork with a steering tube of the same diameter and length as your old fork, unless you are also switching from a threaded to a threadless system, in which case you will just get a long, unthreaded steering tube that must be considerably longer than the threaded steerer was. Chapter 12 covers the installation of the headset.

The crowns on many carbon forks are so deep that they require an extra-long brake nut to reach the brake bolt. The longer nut should be supplied with the fork.

For cyclocross, weight is as much or more of an issue than it is for a road bike, because you throw the bike onto your shoulder, accelerate it often, and carry it while running. So a carbon cyclocross fork with a carbon steering tube will be a benefit. It of course needs to have cantilever brake bosses or disc-brake mounts on it (Figs. 16.4, 16.5) so it can handle tires at least up to 700 × 33mm. It also must have extra room under the crown and between the fork legs for good mud clearance around the tire.

With road bikes and particularly with cyclocross bikes, given the regular bashing they take, replacing the fork after a few years is prudent for safety reasons, and you may as well upgrade it and get more performance out of your bike while you're at it.

Duct tape is like The Force.
It has a light side, a dark side,
and it holds the universe together.

—CARL ZWANZIG

17

FRAMES

TOOLS

2.5mm, 3mm, 4mm, 5mm hex keys
A true rear wheel
Oil
Grease

OPTIONAL

Derailleur-hanger-alignment tool
String
Ruler
Dropout-alignment tools
Metric taps
Bottom bracket tap set
Electric drill
Drill-bit set
16mm wrench
8mm open-end wrench

Pay close attention to the frame because it holds your entire bike together. It is nearly impossible to fix on the road, and if it fails, the consequences can be serious.

17-1

FRAME DESIGN

a. Road

The traditional "double-diamond" design of a road bike frame relies on a "front triangle" and a "rear triangle" (Fig. 17.1); never mind that the front triangle is not actually a triangle—or much of a diamond, for that matter.

Referring to Figure 17.2, the angle of the seat tube relative to the horizontal (the "seat angle") determines the fore-and-aft position of the rider relative to the pedals. It also plays a role in determining the weight distribution on the wheels. And seat angle partially dictates the length of the chainstays, because the more tipped back the seat tube is, the farther back the rear wheel will have to be to avoid hitting it.

For a given top-tube length and front-end geometry, the seat angle also dictates whether or not your feet hit the front wheel when pedaling around a tight, low-speed turn (the interference is called, quaintly these days, "toeclip overlap"). And unless the frame tubing is altered to compensate, the vertical and lateral compliance of the rear of the bike will increase with decreasing seat angles and correspondingly longer chainstays.

The angle of the head tube relative to the horizontal (the head angle)—in combination with the fork offset, or "rake" (explained at the beginning of Chapter 16), and wheel diameter—determines much of the steering and handling characteristics of the bike. The head angle and fork rake also dictate in large part how much shock is absorbed by the fork; the shallower the head angle and/or greater the fork offset, the more suspension the bike will offer. Most road frames will have seat angles and head angles in the 72- to 75-degree range.

The height of the bottom bracket above the ground determines how much clearance you will have for the pedals when rounding a turn. Low bottom brackets lower the center of gravity of the rider and impart a feeling of stability. Frame builders often make compromises between cornering clearance and stability, especially in pro-level bikes where the rider is assumed to have sufficient experience to always corner with the inner crank up.

17.1 The frame and its parts

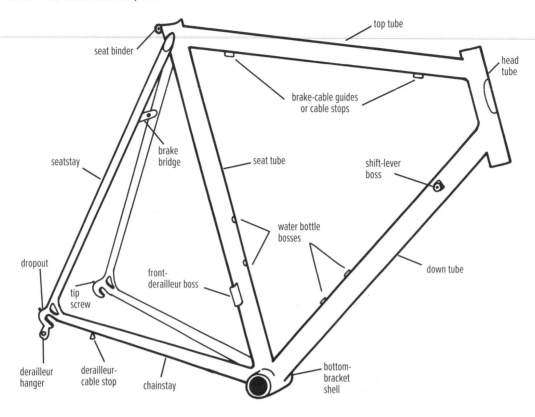

seat binder

top tube

head tube

brake-cable guides
or cable stops

seatstay

brake
bridge

seat tube

shift-lever
boss

dropout

water bottle
bosses

tip
screw

front-
derailleur boss

down tube

derailleur
hanger

derailleur-
cable stop

chainstay

bottom-
bracket
shell

17.2 Frame dimensions

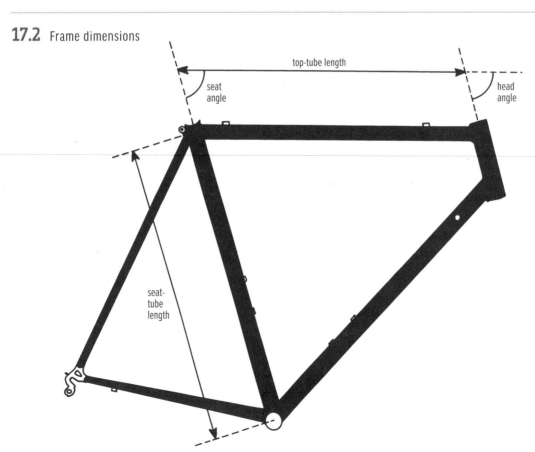

top-tube length

seat
angle

head
angle

seat-
tube
length

Along with the seat-tube length and angle, the bottom-bracket height also helps determine the standover clearance your crotch has over the top tube. A typical bottom-bracket height for a road bike is around 265mm.

The top-tube length—along with the stem length, seat angle, and seat fore-aft position on the seatpost—determines your reach to the handlebar.

The seat-tube length (or frame size) determines the amount of seatpost extension you will require to attain a given seat height, as well as the minimum seat height possible on the bike. It also is one of the variables determining standover height.

The wheelbase is the distance between the wheel axles. It determines the minimum possible turning radius.

On modern road bike frames, the shift-lever boss on the down tube shown in Figure 17.1 is generally replaced by a threaded shift-cable stop to accept a barrel adjuster, and this is usually located either on the head tube or on the down tube near the intersection of these two tubes.

b. Cyclocross

Cyclocross frames have either disc-brake mounts (Fig. 17.3) or cantilever-brake posts and a slotted cable hanger attached to the seatstays (Fig. 17.4); the cable hanger may even have a threaded barrel adjuster on it for adjusting the tension on the rear brake cable. Many 'cross frames have slotted cable guides on the top tube not only for the rear brake cable, but also for both shift cables to keep them up out of the muck they would encounter with standard routing under the bottom bracket. Cyclocross frames will generally also have a threaded hole on the back of the seat tube about an inch up from the bottom bracket. This is to screw on a cable roller (Fig. 5.46) if the frame has over-the-top cable routing and the rider will be using a front derailleur. 'Cross frames may or may not have bottle mounts.

To achieve greater stability and slower steering, thanks to increased fork trail (Fig. 16.1), cyclocross frames have shallower head angles than road frames, generally in the 71- to 72.5-degree range. The seat angle on a 'cross frame will usually be a degree or so shallower as well; 72–74 degrees is pretty common. The bottom-bracket height will often be 10–20mm higher than a road frame. A 280mm+ bottom-bracket height will provide better pedaling clearance than a 265mm one, allowing the rider to keep pedaling hard over hummocks and other localized topography variations. The rear end will be longer, and the chainstays and seatstays will generally be bent in the area near the tire; all of this will be done to improve mud-clearing ability.

17.3 Carbon-fiber road or cyclocross frame and fork with disc-brake mounts and tapered head tube and steering tube

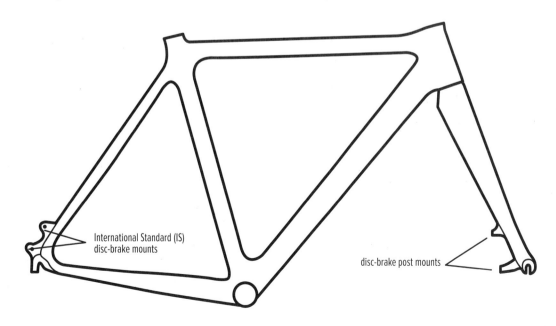

International Standard (IS) disc-brake mounts

disc-brake post mounts

17.4 Carbon-fiber cyclocross frame and fork with cantilever brake mounts

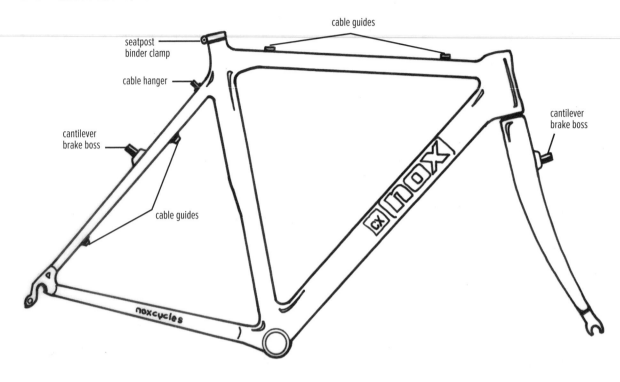

Frame materials for cyclocross are the same as for road bikes, but some of the reasons for them and ways that they are employed are different. Aluminum or superthin steel frames were de rigueur until the turn of the millennium because they offered low weight, sufficient strength, and durability. A problem with both of those designs is that when the rider crashed, the handlebar could easily dent the top tube. Carbon, titanium, and magnesium 'cross frames offer the low weight, strength, and toughness required for the sport, and they can resist denting without being any heavier than a superlight steel or aluminum frame. Carbon and titanium frames will not rust or otherwise oxidize in the horrendous environmental conditions to which they can be exposed.

c. Time trial/triathlon

Some frames designed for time trials and triathlons are designed for improved aerodynamic performance and have wing-shaped tubes and a low-profile design to reduce air drag (Fig. i.2). They generally have much steeper seat angles than standard road frames; 78 degrees is common.

d. Touring

The single most distinguishing feature of touring frames is the presence of mounts for front and rear racks. Touring frames are also sturdier—meaning heavier—in order to carry high loads. Compared with road bikes, they have longer rear ends for more vertical compliance and to make room for fenders and fatter tires; the clearance under the brake bridge and fork crown is also higher to accommodate fenders. They may be built with disc-brake mounts (Fig. 17.3), cantilever brake mounts (also called "bosses," Fig. 17.4), or standard brake holes, but if the latter, the holes will be higher to fit fenders and larger tires, so longer-reach brakes will be required. The head and seat angles are shallower, as described for a cyclocross frame, for many of the same reasons.

17-2

FRAME MATERIALS

Bicycle materials have evolved continually over time. Wood was the material of choice for the first bikes but was soon replaced by steel, aluminum, and even bamboo. Carbon fiber composites, aluminum, and steel are the materials most commonly used to build frames today, but titanium and magnesium account for a share at the more expensive end.

Carbon fiber, boron fiber, and similar composite frame materials consist of fibers embedded in a resin (plastic) matrix. These materials can be very light, very strong, and very stiff. Bikes can be built by gluing carbon fiber tubes into lugs (usually made of carbon fiber or aluminum); by gluing several large, molded subassemblies together; or by molding the frame in a single piece (monocoque construction).

The big advantage of carbon composites is that extra composite fabric can be added into sections of the mold to add thickness precisely where extra strength is needed. The tricky part is holding the composite parts together in a frame that won't come apart.

Metals used in road bike frames come in a variety of grades with varying costs and physical properties, but in the following discussion I am talking about the highest grades used in bicycles. For example, the aluminum used in pop cans and window frames is much weaker than the 6061 and 7000 series aluminum used in high-end bicycle frames.

Steel has the highest modulus of elasticity (a principal determinant of stiffness) as well as the highest density and tensile strength of any of the metals commonly used in frames. Aluminum has a much lower modulus, density, and tensile strength than steel; titanium has a modulus, density, and tensile strength between the two. With good frame design and construction combined with intelligent selection of tube properties, diameters, shapes, and wall thicknesses, long-lasting frames with comparable stiffness-to-weight and/or strength-to-weight ratios can be built from any of these metals.

Butting of metal tubing reduces weight by putting thicker material at the tube ends and thinning the center sections. "Double-butted" means that both ends are thicker than the center section, whereas "triple-butting" and "quad-butting" refer to gradation steps in the thickness at the ends.

The tensile strength of most metals used for bikes is boosted by the addition of alloys into the pure base metal, by heat treatments, or both. Low-carbon steel (like gas pipe) is soft and easy to bend and break. High-carbon steels alloyed with chromium, molybdenum, and other materials are far stronger; heat treating makes them stronger yet. The same goes for aluminum. One improvement in aluminum for bicycles is alloying it with the element scandium, which raises aluminum's strength considerably. Most aluminum frames require a postweld heat-treatment step or they will be soft and breakable.

Titanium alloyed with 3 percent aluminum and 2.5 percent vanadium (3Al/2.5V) is far stronger than commercially pure (CP) titanium, which is 98 percent titanium. Titanium alloyed with 6 percent aluminum and 4 percent vanadium (6Al/4V) is stronger yet but is rarely drawn into tubing, so 6/4 bike tubes are generally made from rolled and welded sheet, which can reduce ultimate strength somewhat. Titanium, like steel, requires no postweld heat treatment, but it must be welded in an inert-gas atmosphere or it will oxidize and become extremely brittle.

Advertising claims touting one frame-tubing material over another can be misleading, because you may not know whether a manufacturer is comparing its material with the high- or low-end forms of competitors' materials.

17-3

FRAME INSPECTION

You can avoid potentially dangerous frame failures by inspecting your frame frequently. If you find damage and you are not sure if the bike may be dangerous to ride, take it to a bike shop for advice.

1. **Clean the frame every few rides or when it gets dirty.** It's easier to spot problems on a clean frame.
2. **Inspect all tubes for cracks, bends, buckles, dents, and paint stretching or cracking, especially near the joints where stress is highest.** With a carbon frame, use the "coin test" to check for underlying damage. Tap on the tube with a coin in questionable areas and compare it with the sound on other tubes, in surrounding areas, and on the opposite side. If delamination or cracking exists in underlying carbon layers, especially in central areas away from the joints, you'll be able to hear the difference; the damaged fibers deaden the nice "clack"

sound you hear when tapping on an undamaged tube. If in doubt, take it to an expert for advice.

Tubes in metal frames can be replaced. Some types of damage in carbon frames can be repaired by specialists such as Calfee Design (www.calfeedesign.com). Otherwise, carbon frames must be replaced if they have large dents, buckles, cracks, bends (sometimes indicated by stretched or fractured paint), or delamination.

3. **Inspect the rear dropouts.** Check for cracks around the welds or glue joints and around the brake bridge and chainstay bridge (the little cross-tube between the chainstays just behind the bottom bracket on some frames; see Fig. 17.1 for names and locations of frame parts). Check to be sure the dropouts (and brake posts and cable hangers on cyclocross and touring frames) are not bent. Some dropouts and brake posts bolt on and are replaceable, and some cable hangers are glued in and replaceable. Otherwise, badly bent or broken dropouts, brake posts, and cable hangers may be repairable by a frame builder.

4. **Remove the seatpost every few months and after riding in the rain.** Invert the bike to remove any water that has collected in the seat tube and let it dry out. On steel frames, look for deeply rusted areas. Look and feel for rust inside or for rust falling out. I recommend squirting oil or a rust-preventive spray for bicycle frames (like Frame Saver) inside the tubes periodically. With thin-walled steel tubing, the time from when the rust starts until the frame rusts through can be short. Often you will see bubbles in the paint, for example, around the bottom-bracket joints or on the back of the seat tube. Although bubbles can indicate paint problems, they often indicate that the seat tube has pinholes rusted through it under the paint and needs to be replaced.

Also on steel frames, remember to grease both the seatpost and the inside of the seat tube when you reinsert the seatpost. After sanding off the rust, touch up any external areas where the paint has come off with touch-up paint or nail polish (hey, it's available in lots of cool colors).

5. **Check that a true and properly dished rear wheel sits straight in the frame.** It should be centered between the chainstays and seatstays and lined up in the same plane as the front triangle. Check that tightening the hub skewer does not result in bowing or twisting of either the chainstays or the seatstays.

17-4

CHECK AND STRAIGHTEN REAR-DERAILLEUR HANGER

 LEVEL 1. **If you have a derailleur-hanger-alignment tool (Fig. 1.4), thread it into the derailleur hanger on the right dropout (Fig. 17.5).**

2. **Install a true rear wheel in the rear dropouts.**

3. **Swing the tool around, measuring the spacing between its arm and the rim all the way around.** The arm of the tool should be the same distance from the rim at all points. Some tools, like the one shown in Figure 17.5, have an indicator rod extending from the arm that you can adjust to check the spacing; others require you to measure the gap with a ruler or caliper.

4. **If the tool has play in it, keep it pushed inward lightly as you perform all of the measurements, or you will get inconsistent data.**

5. **If the spacing between the tool arm and the rim is not consistent (within a millimeter or two all the way around), carefully bend the hanger by pulling outward lightly on the arm of the tool where it is closest to the rim.** You can do this on replaceable aluminum derailleur hangers as well as on steel, aluminum, or titanium ones that are a single piece with the dropout. A titanium hanger may require considerable force to align; if you are not confident about rebending it correctly, take the frame to a shop or frame-builder. Do not heat the hanger with a torch to soften it for alignment; heating will weaken the metal.

6. **If the derailleur hanger is severely bent, you may not be able to align it without breaking it.** You may even have trouble threading the tool in,

17.5 Checking derailleur-hanger alignment

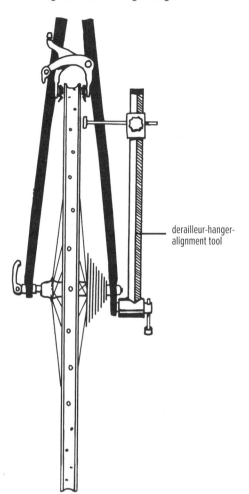

derailleur-hanger-
alignment tool

17.6 Checking frame alignment with a string

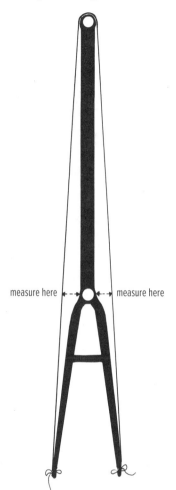

measure here ←→ measure here

because the threaded hole will be ovalized. If the dropout bolts to the frame, remove it and get an exact replacement from Wheels Manufacturing or a dealer of your bike brand.

7. **If the threads or the hanger itself is really screwed up and you do not have a replaceable dropout, see 17-7b for other derailleur-hanger options.**

17-5

CHECK FRAME ALIGNMENT AND ADJUST DROPOUT ALIGNMENT

LEVEL Exacting alignment checks require a precision surface plate, an uncommon tool in the home workshop. Thus the following methods for determining frame alignment are inexact, but sufficient for determining gross alignment woes.

If you find problems more severe than moderately bent dropouts or a misaligned derailleur hanger, do not attempt to correct them. Adjusting frame alignment, if it can be done at all, should only be performed with an accurate frame-alignment table by someone who is practiced in its use.

1. **With the frame clamped in a bike stand, tie the end of a string to one rear dropout.** Stretch it tightly around the head tube, and tie it symmetrically to the other dropout (Fig. 17.6).

2. **Measure from the string to the seat tube on each side (Fig. 17.6).** The measurement should be the same to within 1mm.

3. **Put a true and properly dished rear wheel in the frame and check that it lines up in the same plane as the front triangle.** Make certain that the wheel is centered between the seatstays and chainstays. If

you have an old steel frame with dropouts with tip screws that thread in from the back of the dropout, turn one or the other of them so that the true wheel lines up straight behind the seat tube (centered between the chainstays).

The hub should slide easily into the dropouts without requiring you to pull outward or push inward on the dropouts. Tightening the hub quick-release should not result in bowing or twisting of frame members.

4. **Remove the wheel and measure the spacing between the dropouts (Fig. 17.7).** For 8-, 9-, 10-, or 11-speed rear hubs, this spacing should be 130mm; it should be 126mm for 5-, 6-, or 7-speed rear hubs; and it should be 135mm for a disc-brake hub with a

quick-release skewer. If the spacing on the frame is 1mm less or 1.5mm more than nominal, it is acceptable. For instance, if you have a frame whose rear spacing should be 130mm, acceptable spacing is 129–131.5mm.

5. **If you have dropout-alignment tools, put them in the dropouts so that their shafts are fully seated into the dropouts (Fig. 17.8).**

If you have an old steel frame with dropouts with tip screws that thread in from the back of the dropout, you can remove them and seat the dropout-alignment tools all the way to the rear of the dropout slots. Alternatively, you can leave the tip screws installed, and before installing the dropout-alignment tools, turn one or the other of them until a true wheel lines up straight behind the seat tube.

Arrange the tool spacers (and the cups, if they are adjustable) so that the faces of the cups are within a millimeter of each other. Tighten the handles on the tools. The tool cups should line up straight across from each other, with their faces exactly parallel. If the tools do not align, one or both dropouts are bent. If the frame has replaceable dropouts, go ahead and replace one or both of them. If the frame has a composite (i.e., carbon) or bonded rear triangle of any kind, there is nothing you can do about the problem.

If the bike has a steel, aluminum, or titanium rear triangle, you can align the dropouts by bending them carefully with the dropout-alignment tools. Hold the cup of the tool with one hand and

17.7 Measuring dropout width

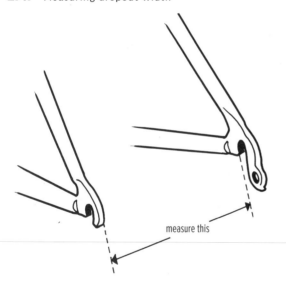

measure this

17.8 Using dropout-alignment tools on rear dropouts

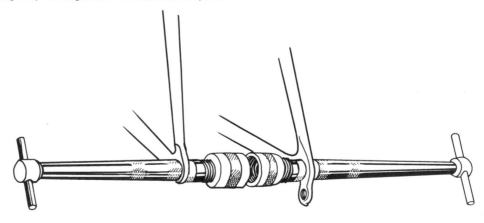

push or pull on the handle with the other. Steel is strong yet still bendable, while titanium is hard to bend because it keeps springing back. Aluminum poses the biggest challenge; you run a great risk of breaking aluminum by trying to bend it.

CAUTION: *Never heat the dropouts (or any part of the frame) for alignment purposes. Doing so could irreparably change the strength, temper, or hardness of the part and lead to failure.*

17-6

CORRECTING FRAME DAMAGE

Apart from the alignment items already covered in this chapter, the only frame problems you can correct are damaged threads, chipped paint, and small dents. Broken braze-ons and bent, broken, or deeply dented tubes call for a new frame or require a frame builder to perform the repair.

17-7

FIXING DAMAGED THREADS

a. Retapping and using a new bolt

LEVEL

A road bike frame has threaded water bottle bosses and a threaded rear-derailleur hanger, and it may have threads in the bottom-bracket shell and in a small hole in the bottom of the bottom-bracket shell to which a derailleur-cable guide is bolted. In addition, some bikes have a threaded seat binder rather than a replaceable seat-post binder clamp. Cyclocross frames (and some road touring frames) have threaded cantilever brake posts (Fig. 17.4), and many 'cross frames also have a threaded hole for a front-derailleur cable roller on the back of the seat tube, near its base (Fig. 5.46). Cyclocross and road frames with disc-brake mounts generally have threaded post mounts on the front (Fig. 16.5) and unthreaded transversely drilled International Standard (IS) mounts on the rear (Fig. 17.3), but they can also have threaded post mounts on the rear.

1. **If any threads on the frame are stripped or cross-threaded, try chasing the threads with the appropriate thread tap.** Then replace the bolt or bottom-

bracket cup with a new one. Whenever you retap any threads, use oil on the tap (use canola vegetable oil on titanium threads). The following tap sizes are commonly found on most road bikes:

Water bottle bosses and the hole for a plastic shift-cable guide:	**M5 (5mm × 0.8)**
Hole under bottom-bracket shell for derailleur-cable guide:	**M5 (5mm × 0.8)**
Cyclocross front-derailleur roller:	**M5 (5mm × 0.8)**
Seat binders, disc-brake mounts, and cantilever brake posts:	**M6 (6mm × 1)**
Derailleur hanger:	**M10 (10mm × 1)**
Bottom-bracket shells, English thread:	**1.37 inches × 24 tpi (threads per inch)**
Bottom-bracket shells, Italian thread:	**36mm × 24 tpi**

NOTE: *1) The drive-side (right-side) English bottom-bracket threads are left-hand threaded; the other side is right-hand threaded; 2) Italian bottom-bracket shells are right-hand threaded on both sides.*

2. **Turn the tap forward (clockwise) a bit, then turn it back, then forward (two steps forward and one back), and so on, to prevent the tap from binding and possibly breaking.** Taps are made of hard, brittle steel. If you put any side or twisting force on small taps, they can easily break. If the tap breaks, you'll have a real mess; the broken tap in the hole is harder than the frame, and it's impossible to drill the broken tap out. If you break off a tap in the frame, do not try to get it out yourself. Take it to a bike shop, a machine shop, or a frame builder before you break off what little is left sticking out. Unless you put the tap in crooked, breaking one should not be a problem when retapping damaged frame threads because these threads will be so worn; getting the tap to find any metal to bite into will probably instead be your biggest problem.

IMPORTANT NOTE: *Tapping a bottom-bracket shell requires expertise. If you have never done it before and want to do it yourself, get some expert supervision. In addition to making sure that you place the correct tap in the correct end of the shell, you must be certain that* *the taps go in straight. Most bottom-bracket taps have a shaft between the two taps to keep them parallel to each other (Fig. 1.4). They must both be started at the same time from both ends. If you mess up the threads, you can ruin the frame. So if in doubt, ask an expert.*

b. If retapping the threads fails:

1. Damaged water bottle bosses or threaded cyclo-cross seat-tube front-derailleur roller bosses (Fig. 5.46): Some bike shops have a tool that rivets bottle bosses into the frame. Check for this possibility first, since you can avoid a new paint job that way. But note that these riveted bosses tend to loosen up over time, especially if the bottle-cage bolts are overtightened. Otherwise, take the bike to a frame builder to get a new boss welded, riveted, bonded, or brazed in.

2. Damaged threads in rear-derailleur hanger: The threads can be so stripped or cross-threaded that a tap (17-7a) will not rethread it properly, or the hanger can be so bent or twisted that the threads will not work, even if the hanger is bent back straight. Some bikes have replaceable rear dropouts or derailleur hangers that bolt onto the frame. If your bike doesn't, one option is to use a Dropout Saver derailleur-hanger backing nut (Fig. 17.9), made by Wheels Manufacturing and available at bike shops. The Dropout Saver is simply a sleeve threaded the same as the derailleur hanger, with 16mm wrench flats. You drill out the hole in the damaged derailleur hanger with a ¹⁴⁄₃₂-inch drill bit, push the Dropout Saver in from the backside, and screw in the derailleur. Dropout Savers come in two lengths, depending on the thickness of the hanger. After installing it, realign it as in 17-4, Figure 17.5.

   Another option is to saw off the derailleur hanger with a hacksaw and use a separate derailleur hanger from a cheap bike that fits flat against the outside of the dropout and is held in by the hub-axle bolts or quick-release. Or, you could have a frame builder weld or braze in a new dropout, or have a carbon repairer replace a carbon dropout.

17.9 Inserting Dropout Saver in damaged derailleur hanger

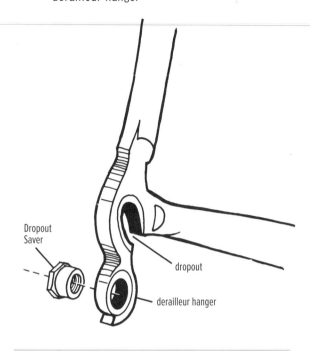

3. Damaged seat binders: Drill out the threads and install a bolt and nut or saw off the binder and use a wraparound seat binder clamp (Fig. 17.4).

4. Damaged bottom-bracket shell threads: If you are using old-school square-taper cranks, you can use an old-style Mavic or Stronglight bottom bracket (11-12b, Fig. 11.26), if you can still find one, because it does not depend on the threads in the shell to anchor it. You must have a shop bevel the ends of the bottom-bracket shell with a special cutting tool.

5. Damaged bottom-bracket cable-guide threads: A new hole in the bottom of the bottom bracket can be drilled and tapped, or the stripped hole can be tapped out with larger threads for a larger screw. Make sure the screw you use is short enough that it does not protrude into the inside of the bottom-bracket shell.

6. Damaged cantilever brake post on cyclocross (or touring) frame (Fig. 17.4): Some brake posts are replaceable; they have wrench flats (usually 8mm) at the base, and they thread into a boss welded onto or molded into the frame. If the brake posts are not of this type, you will have to ask a frame

builder to install a new boss, if it's a metal frame. With a carbon frame, if Craig Calfee can't fix it, it's toast.

17-8

REPAIRING CHIPPED PAINT AND SMALL DENTS

Fixing paint chips is simply a matter of cleaning the area and applying a bit of touch-up paint. Sand any chipped paint or rust completely away before repainting. Use a touch-up paint made for your bike, model paint of a similar color, or fingernail polish (durability is low with the latter).

Small dents can be filled with automotive body putty, but there is little point to filling them if you are only doing a paint touch-up, because the repaired area probably won't look that great anyway.

There are plenty of frame painters you can find online who can fill dents, repaint frames, and even match original decals.

17-9

HIGH-SPEED SHIMMY

More typical with tall frames and heavy riders, the bike developing a shake that builds rapidly in amplitude at high speed or when riding with the hands off the handlebars is an alarming and dangerous occurrence. If it happens to you, immediately clamp your knees against the top tube to damp it and slow down. Another method that usually stops the shimmy but that takes a larger leap of faith is to push your butt far off the back of the saddle (as when going off a steep drop on a mountain bike) to concentrate as much weight as possible on the rear wheel and as little as possible on

the front wheel and thus reduce the vibrational feedback between them; it's like letting the string go slack between tin-can telephones. When you get home, do something about the bike; you don't want that to happen again! You can imagine the horror stories of when it becomes uncontrollable.

As I discussed in Chapter 16, replacing a misaligned fork with an aligned one can help a lot. Sometimes a fork with more rake can help, too. So can a stiffer frame and stiffer wheels. A loose headset or loose wheel bearings can also cause shimmy.

17-10

COUPLED TRAVEL FRAMES

Due to the rising cost of traveling by air with a bike, travel road bikes with standard 700C wheels that can be broken down to fit into a suitcase (thus avoiding oversize baggage charges) are becoming commonplace.

These frames generally have one of two types of connection systems. One consists of screw-together couplers made by S and S Machine on the top tube and down tube; the couplers have a toothed Hirth joint inside and tighten with a special wrench. Alternatively, Ritchey coupled bikes have a seatpost binder in the seat tube as well as on the top tube, so that the seatpost holds the upper part of the frame together, and a small ringlike coupler with a pinch bolt connects the down tube to the bottom bracket.

If you are using a bike with one of these systems, check the couplers frequently to ensure that they are tight and that there is no cracking in the tubes surrounding them. Coming apart for travel is good; coming apart while you're riding is not.

APPENDIX A
TROUBLESHOOTING INDEX

This index is intended to assist you in finding and fixing problems. If you already know wherein the problem lies, consult the table of contents for the chapter covering that part of the bike. If you are not sure which part of the bike is affected, this index can be of assistance. It is organized alphabetically but, because people's descriptions of the same problem vary, you may need to look through the entire list to find your symptom.

This index can assist you with a diagnosis and can recommend a course of action. Additional troubleshooting tips and diagnoses can be found at the end of most chapters.

TABLE A.1 — TROUBLESHOOTING BIKE PROBLEMS

SYMPTOM	LIKELY CAUSES	ACTION	CHAPTER
bent wheel	1. maladjusted spokes	true wheel	8
	2. broken spoke	replace spoke	8
	3. bent rim	replace rim	15
bike pulls to one side	1. wheels not true	true wheels	8
	2. tight headset	adjust headset	12
	3. pitted headset	replace headset	12
	4. bent frame	replace or straighten	17
	5. bent fork	replace or straighten	16
	6. loose hub bearings	adjust hubs	8
	7. low tire pressure	inflate tires	2, 7
bike shimmies at high speed or when hands off of handlebars	1. frame cracked	replace frame	17
	2. frame bent	replace or straighten	17
	3. wheels way out of true	true wheels	8
	4. loose hub bearings	adjust hubs	8
	5. wheels too flexible	build stiffer wheels	15
	6. headset too loose	tighten headset	12
	7. misaligned fork	replace fork	16
	8. soft frame/heavy rider	replace frame	17
	9. poor frame design	replace frame	17
bike vibrates when braking	see "chattering and vibration when braking" under "STRANGE NOISES"		
brake doesn't stop bike	1. maladjusted brake	adjust brake	9
	2. worn brake pads	replace pads	9
	3. wet rims	keep braking	9
	4. greasy rims	clean rims	9

Continues >>

TABLE A.1 — TROUBLESHOOTING BIKE PROBLEMS, CONTINUED

SYMPTOM	LIKELY CAUSES	ACTION	CHAPTER
brake doesn't stop bike (cont.)	5. sticky brake cable	lube or replace cable	9
	6. steel or carbon rims in wet weather	use aluminum rims	15
	7. brake damaged	replace brake	9
	8. sticky or bent brake lever	lube or replace lever	9
	9. wrong pads for rim	get correct pads	9
	10. grease on disc rotor and pads	clean rotor, replace pads	10
brake pad rubs	1. brake misaligned	adjust brake	9
	2. untrue wheel	true wheel	8, 15
	3. warped or bent disc rotor	true rotor	10
	4. cable pull insufficient for V-brake	get V-brake lever	9
chain falls off in front	1. maladjusted front derailleur	adjust front derailleur	5
	2. chainline off	adjust chainline	5
	3. chainring bent or loose	replace or tighten	11
chain jams in front between chainring and chainstay (called chain suck)	1. dirty chain	clean chain	4
	2. bent chainring teeth	replace chainring	11
	3. chain too narrow	replace chain	4
	4. chainline off	adjust chainline	5
	5. stiff links in chain	free links, lube chain	4
chain jams in rear	1. maladjusted rear derailleur	adjust derailleur	5
	2. chain too wide	replace chain	4
	3. small cog not on spline	reseat cogs	8
	4. poor frame clearance	return to dealer	17
chain skips	1. tight chain link	loosen tight link	4
	2. elongated (worn) chain	replace chain	4
	3. maladjusted derailleur	adjust derailleur	5
	4. worn rear cogs	replace cogs and chain	4, 8
	5. dirty or rusted chain	clean or replace chain	4
	6. bent rear derailleur	replace derailleur	5
	7. bent derailleur hanger	straighten hanger	17
	8. loose derailleur jockey wheel	tighten jockey wheel	5
	9. bent chain link	replace chain	4
	10. sticky rear shift cable	replace shift cable	5
	11. upside-down master link	reset link	4
chain slaps chainstay	1. chain too long	shorten chain	4
	2. weak rear-derailleur spring	replace spring or derailleur	5
	3. road very bumpy	use large chainring	—
derailleur hits spokes	1. maladjusted rear derailleur	adjust derailleur	5
	2. broken spoke	replace spoke	8
	3. bent rear derailleur	replace derailleur	5
	4. bent derailleur hanger	straighten or replace hanger	17

Continues >>

APPENDIX A: TROUBLESHOOTING INDEX

TABLE A.1 — TROUBLESHOOTING BIKE PROBLEMS, CONTINUED

SYMPTOM	LIKELY CAUSES	ACTION	CHAPTER
knee pain	1. poor shoe cleat position	reposition cleat	13
	2. saddle too low or high	adjust saddle	14
	3. clip-in pedal has no float	get floating pedal	13
	4. foot rolled in or out	replace shoes or get orthotics	—
pain or fatigue when riding, particularly in the back, neck, and arms	1. incorrect seat position	adjust seat position	App. C
	2. stem too low	raise stem	App. C
	3. too much riding	build up miles gradually	—
	4. incorrect stem length	replace stem	App. C
	5. poor frame fit	replace frame	App. C
pedal entry difficult (with clip-in pedals)	1. spring tension set high	reduce spring tension	13
	2. cleat guide loose or gone	tighten or replace	13
pedal release difficult (with clip-in pedals)	1. spring tension set high	reduce spring tension	13
	2. loose cleat on shoe	tighten cleat	13
	3. dry pedal spring pivot	oil spring pivots	13
	4. dirty pedals	clean and lube pedals	13
	5. bent pedal clips	replace pedals or clips	13
	6. dirty cleats	clean and lube cleats	13
pedal release too easy (with clip-in pedals)	1. release tension too low	increase release tension	13
	2. cleats worn-out	replace cleats	13
pedal(s) move laterally or clunk, click, or twist while pedaling	1. loose crankarm	tighten crank bolt	11
	2. pedal loose in crank	tighten pedal to crank	13
	3. bent pedal axle	replace pedal or axle	13
	4. loose bottom bracket	adjust bottom bracket	11
	5. bent bottom bracket axle	replace bottom bracket or crankset	11
	6. bent crankarm	replace crankarm	11
	7. loose pedal bearings	adjust pedal bearings	13
rear shifting working poorly	1. maladjusted derailleur	adjust derailleur	5
	2. sticky or damaged cable	replace cable	5
	3. loose rear cogs	reseat and tighten cogs	8
	4. worn rear cogs	replace cogs and chain	4, 8
	5. worn/damaged chain	replace chain	4
	6. see also "chain jams in rear" and "chain skips."		
resistance while coasting or pedaling	1. tire rubs frame or fork	adjust axle; true wheel	2, 8
	2. brake pad drags on rim or rotor	adjust brake	9, 10
	3. tire pressure too low	inflate tire	2, 7
	4. hub bearings too tight	adjust hubs	8
	5. hub bearings dirty/worn	overhaul hubs	8
	6. mud packed around tires	clean bike	2

Continues >>

TABLE A.1 — TROUBLESHOOTING BIKE PROBLEMS, CONTINUED

SYMPTOM	LIKELY CAUSES	ACTION	CHAPTER
resistance while pedaling only	1. bottom bracket too tight	adjust bottom bracket	11
	2. bottom bracket dirty/worn	overhaul bottom bracket	11
	3. chain dry/dirty/rusted	clean/lube or replace	4
	4. pedal bearings too tight	adjust pedal bearings	13
	5. pedal bearings dirty/worn	overhaul pedals	13
	6. bent chainring rubs frame	straighten or replace	11
	7. true chainring rubs frame	move crank out; adjust chainline	5, 11
stiff steering	tight headset	adjust headset	12
tire bulged	1. broken casing threads	replace tire	7
	2. slipped tubular tire	reglue tire and line up valve stem	7
tire pinch flats	1. insufficient pressure	pump tire higher	7
	2. tire diameter too small	replace with larger tire	7
tire valve stem angled sharply	1. tube slipped in tire	deflate and slide tire around rim	7
	2. slipped tubular tire	reglue tire and line up valve stem	7

STRANGE NOISES

Weird noises can be hard to locate; use this to assist in locating them.

SYMPTOM	LIKELY CAUSES	ACTION	CHAPTER
chattering and vibration when braking; fork shudder when braking	1. bent or dented rim	replace rim	15
	2. loose headset	adjust headset	12
	3. brake pads toed out	adjust brake pads	9
	4. wheel way out of round	true wheel	8
	5. standard pads on carbon rim	get correct brake pads	9
	6. greasy sections of rim	clean rim	8
	7. loose brake pivot bolts	tighten brake bolts	9
	8. rim worn through and ready to collapse	replace rim ASAP!	15
	9. steering tube flex with cantilever brake	fork-crown cable hanger, stiffer fork, mini V-brake instead of cantilever	9
clicking noise	1. cracked shoe cleats	replace cleats	13
	2. cracked shoe sole	replace shoes	13
	3. loose bottom bracket	tighten bottom bracket	11
	4. loose crankarm	tighten crankarm	11
	5. loose pedal	tighten pedal	13
clunking from fork	headset loose	adjust headset	12

Continues >>

TABLE A.1 —| TROUBLESHOOTING BIKE PROBLEMS, CONTINUED

SYMPTOM	LIKELY CAUSES	ACTION	CHAPTER
creaking noise	1. dry handlebar/stem joint	grease inside stem clamp	12
	2. loose seatpost	tighten seatpost	14
	3. loose shoe cleats	tighten cleats	13
	4. loose crankarm	tighten crankarm bolt	11
	5. cracked frame	replace frame	17
	6. dry, rusty seatpost	grease seatpost	14
	7. headset cups moving in frame	replace headset or frame	12, 17
	8. bottom bracket moving in frame	grease threads or wrap with plumbing tape	11
	9. wheel axle or quick release moving against dropouts	grease axle faces	8
	10. see "squeaking noise."		
rubbing or scraping noise when pedaling	1. crossed chain	avoid extreme gears	5
	2. front derailleur rubbing	adjust front derailleur	5
	3. chainring rubs frame	install longer bottom bracket or move over	11
rubbing, squealing, or scraping noise when coasting or pedaling	1. tire dragging on frame	straighten wheel	2, 8
	2. tire dragging on fork	straighten wheel	2, 8
	3. brake dragging on rim or rotor	adjust brake	9, 10
	4. dry hub dust seals	clean and lube dust seals	8
squeaking noise	1. dry hub or crank bearings	overhaul hubs or bottom bracket	8, 11
	2. dry pedal bushings	overhaul pedals	13
	3. squeaky saddle	grease edge of leather	14
	4. rusted or dry chain	lube or replace chain	4
	5. squeaky seatpost clamps	tighten seatpost clamps	14
	6. seatpost squeaking inside seat tube	shim or shorten seatpost	14
squealing noise when braking	1. brake pads toed out	adjust brake pads	9
	2. greasy rims	clean rims and pads	9
	3. loose brake arms	tighten brake pivot bolt(s)	9
	4. improper pads on carbon rim	get correct pads	9
ticking noise when braking	1. glue on rim sidewall	clean rim with solvent	7
	2. gouge in rim sidewall	sand rough spot or replace rim	7, 15
	3. high rim seam junction	ignore, or sand seam	7
	4. dented rim from impact	replace rim	15
ticking noise when coasting	1. wheel magnet hits sensor	move computer sensor	—
	2. badly glued tubular	reglue tire	7
	3. valve stem moving in deep rim	tape valve to rim	7

APPENDIX B
GEAR CHART

The following gear table is based on a 700C × 23mm tire (671mm diameter). Your gear-development numbers may be slightly different if the diameter of the fully inflated rear tire, with your weight on it, is not 671mm. Unless your bike has 650C, 24-inch, or some other non-standard-size wheels, these numbers will be very close.

To obtain accurate gear-development numbers for the tire you happen to have on at the time, at a certain inflation pressure, measure the tire diameter very precisely using the following procedure. You can come up with your own gear chart by plugging your tire diameter into the gear-development formula below the chart on the following pages or by multiplying each number in this chart by the ratio of the tire diameter divided by 671mm (the tire diameter I used). Even easier, go to Tom Compton's interactive gear chart at www.analyticcycling.com/GearChart_Page.html.

MEASURING TIRE DIAMETER

1. Sit on the bike with the tire pumped to your desired pressure.
2. Mark the spot on the rear rim that is at the bottom, and mark the floor adjacent to that spot.
3. Roll forward one wheel revolution, and mark the floor again where the mark on the rim is again at the bottom.
4. Measure the distance between the marks on the floor; this is the tire circumference at pressure with your weight on it.
5. Divide this number by π (π = 3.14159) to get the diameter.

NOTE: *This rollout procedure is also the method for measuring the wheel size with which to calibrate your bike computer, except that the procedure will be done on the front wheel for most computers.*

TABLE B.1 | GEAR CHART

		NUMBER OF TEETH ON FRONT CHAINRING													
		27	28	29	30	31	32	33	34	35	36	37	38	39	40
	10	71	74	76	79	82	84	87	90	92	95	98	100	103	106
	11	65	67	70	72	74	77	79	82	84	86	89	91	94	96
	12	59	62	64	66	68	70	73	75	77	79	81	84	86	88
	13	55	57	59	61	63	65	67	69	71	73	75	77	79	81
	14	51	53	55	57	58	60	62	64	66	68	70	72	73	75
	15	47	49	51	53	55	56	58	60	62	63	65	67	69	70
	16	45	46	48	49	51	53	54	56	58	59	61	63	64	66
	17	42	43	45	47	48	50	51	53	54	56	57	59	61	62
NUMBER OF TEETH ON REAR COG	18	40	41	42	44	45	47	48	50	51	53	54	56	57	59
	19	37	39	40	42	43	44	46	47	49	50	51	53	54	56
	20	36	37	38	40	41	42	44	45	46	47	49	50	51	53
	21	34	35	36	38	39	40	41	43	44	45	46	48	49	50
	22	32	34	35	36	37	38	40	41	42	43	44	46	47	48
	23	31	32	33	34	36	37	38	39	40	41	42	44	45	46
	24	30	31	32	33	34	35	36	37	38	40	41	42	43	44
	25	28	30	31	32	33	34	35	36	37	38	39	40	41	42
	26	27	28	29	30	31	32	33	34	36	37	38	39	40	41
	27	26	27	28	29	30	31	32	33	34	35	36	37	38	39
	28	25	26	27	28	29	30	31	32	33	34	35	36	37	38
	29	25	25	26	27	28	29	30	31	32	33	34	35	35	36
	30	24	25	25	26	27	28	29	30	31	32	33	33	34	35
	31	23	24	25	26	26	27	28	29	30	31	31	32	33	34
	32	22	23	24	25	26	26	27	28	29	30	30	31	32	33
	36	20	21	21	22	23	23	24	25	26	26	27	28	29	29
	42	17	18	18	19	19	20	21	21	22	23	23	24	24	25

GEAR FORMULA

Gear = (number of chainring teeth) × (tire diameter) ÷ (number of cog teeth)

If you want the gear in inches, put in the tire diameter in inches.

To find out how far you get with each pedal stroke (gear rollout), multiply the gear by π (3.14159).

	41	42	43	44	45	46	47	48	49	50	51	52	53	54	55	56
						NUMBER OF TEETH ON FRONT CHAINRING										
	108	111	113	116	119	121	124	127	129	132	135	137	140	142	145	148
	98	101	103	106	108	110	113	115	118	120	122	125	127	129	132	134
	90	92	95	97	99	101	103	106	108	110	112	114	117	119	121	123
	83	85	87	89	91	93	95	97	99	101	103	106	108	110	112	114
	77	79	81	83	85	87	89	90	92	94	96	98	100	102	23	106
	72	74	76	77	79	81	83	84	86	88	90	91	93	95	97	98
	68	69	71	73	74	76	77	79	81	82	84	86	87	89	91	92
	64	65	67	68	70	71	73	74	76	78	79	81	82	84	85	87
	60	62	63	64	66	67	69	70	72	73	75	76	78	79	81	82
	57	58	60	61	62	64	65	67	68	69	71	32	74	75	76	78
	54	55	57	58	59	61	62	63	65	66	67	69	70	71	73	74
	51	53	54	55	57	58	59	60	62	63	64	65	67	68	69	70
	49	50	52	53	54	55	56	58	59	60	61	62	64	65	66	67
	47	48	49	50	52	53	54	55	56	57	58	60	61	62	63	64
	45	46	47	48	49	51	52	53	54	55	56	57	58	59	60	62
	43	44	45	46	47	49	50	51	52	53	54	55	56	57	58	59
	42	43	44	45	46	47	48	49	50	51	52	53	41	55	56	57
	40	41	42	43	44	45	46	47	48	49	50	51	43	53	54	55
	39	40	41	41	42	43	44	45	46	47	48	49	50	51	52	53
	37	38	39	40	41	42	43	0	45	45	46	47	48	49	50	51
	36	37	38	39	40	40	41	42	43	44	45	46	47	47	48	49
	35	36	37	37	38	39	40	41	42	43	43	44	45	46	47	48
	34	35	35	36	37	38	39	40	40	41	42	43	44	45	45	46
	30	31	32	32	33	34	34	35	36	37	37	38	39	40	40	41
	26	26	27	28	28	29	30	30	31	31	32	33	33	34	35	35

APPENDIX C
ROAD BIKE FITTING

If you are getting a new bike, get one that fits you properly. Fit should be the primary consideration when selecting a bike; you can adapt to heavier bikes and bikes not painted your favorite color, but your body will soon protest on one that doesn't fit. The simple need to protect your most sensitive parts should keep you away from a bike without sufficient standover clearance (Fig. C.1), but there are a lot of other factors to consider as well, including top tube length, handlebar width and drop, stem length, crank length, pedaling stance width, and toe overlap with the front wheel. An improperly sized bike will cause you to ride with less efficiency and more discomfort. Finding a bike with the right fit isn't difficult; just follow the guidelines in this appendix.

I've outlined two methods for finding your frame size. The first is a simple method of checking your fit on fully assembled bikes at a bike shop. The second method is a bit more elaborate, since it involves taking body measurements. This more detailed approach will allow you to calculate the proper frame dimensions whether the bike is assembled or not.

One other thing: If you are racing triathlons or time trials, aerodynamics and efficient positioning on aero handlebars will be important, as will compliance with technical rules in UCI-sanctioned time trials. See the following sections for more information on those topics.

C-1

SELECTING THE SIZE OF AN ASSEMBLED BIKE

1. Standover height

Stand over the bike's top tube and lift the bike straight up until the top tube hits your crotch. The wheels should be at least 1 inch off the ground to ensure that you can jump off of the bike safely without hitting your crotch. On a bike with sloping top tube, there is no maximum clearance. On a bike with a level top tube, unless the frame has been built with a head tube with extra extension above the top tube to lift the stem higher, you probably don't want any more than 3 or 4 inches of standover.

NOTE: *If you have 2 inches of standover clearance on a bike, do not assume that another bike with the same listed frame size will offer the same standover clearance. Manufacturers use different methods to measure frame size. They also slope their top tubes differently and use different bottom-bracket heights (Fig. C.1), all of which affect the final standover height.*

Most manufacturers measure the frame size up the seat tube from the center of the bottom bracket, but the top of the measurement varies. Some manufacturers measure to the center of the top tube ("center-to-center" measurement), some measure to the top of the top tube ("center-to-top"), and others measure to the top of the seat tube (also called "center-to-top"), even though there is wide variation in the length of the seatpost collar above the top tube. And some sloping-top-tube bikes (often called "compact geometry"), which may come sized in S, M, L, etc., may list an "effective" frame size or seat tube length, which would correspond to the size of the frame if the top tube were horizontal rather than sloped. Obviously, each of these methods will have a different "frame size" for the same frame.

No matter how the frame size is measured, the standover height of a bike depends on the slope of the top tube. Nowadays, most road bikes have sloping top tubes that slant up to the front, and standover clearance above a sloping tube is obviously a function of where you are standing. With an up-angled top tube, stand over it with the nose of the saddle an inch or two behind you, and then lift the bike up into your crotch to measure standover clearance.

C.1 Standover clearance and bottom-bracket height

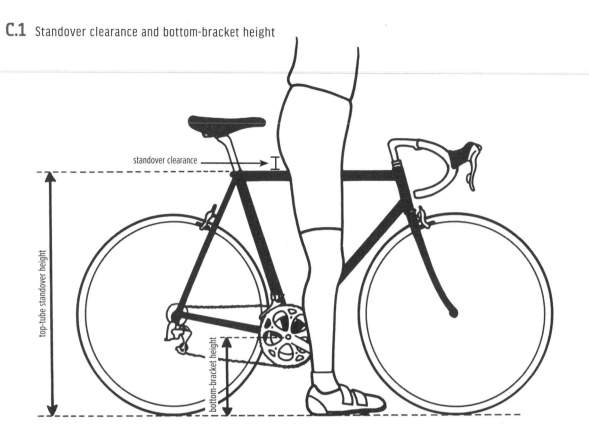

standover clearance

top-tube standover height

bottom-bracket height

C.2 Knee-to-handlebar clearance

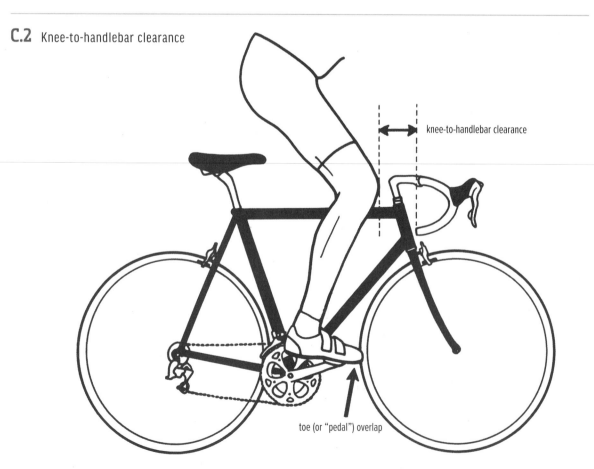

knee-to-handlebar clearance

toe (or "pedal") overlap

APPENDIX C: ROAD BIKE FITTING

Standover height is also a function of bottom-bracket height above the ground, but there is normally not substantial variation in this height between sizes and brands of standard road bikes.

NOTE: *Unless the manufacturer lists the standover height in its brochure and you know your inseam length, you need to stand over the actual bike.*

ANOTHER NOTE: *If you are short and cannot find a frame size small enough to get at least one inch of standover clearance, consider a bike with 650C (26-inch) wheels rather than 700C.*

2. Knee-to-handlebar clearance

Make sure your knee cannot hit the handlebar (Fig. C.2). Do this standing out of the saddle as well as seated and with the front wheel turned slightly, to make sure that the knee will not hit when you are in the most awkward pedaling position you might use. If your knee hits, you need a longer stem or a frame with a longer top tube.

3. Handlebar reach and drop

Ride the bike. See if the reach feels comfortable to you when holding the bars on the flat section adjacent to the stem clamp, on top of the brake hoods, and in the drops. Make sure it is easy to grab the brake levers. Make sure your knees do not hit your elbows as you pedal (Fig. C.6). Make sure that the stem can be raised or lowered enough to achieve a comfortable handlebar height.

NOTE: *Threadless headsets (the standard on all bikes today) allow very limited adjustment of stem height (12-2). Large changes in height require a change in stems.*

4. Toe overlap

In bike shops, this measurement is often called "pedal overlap;" that's a misnomer, since you are actually interested in whether your toes, not the pedals, can hit the front tire when turning sharply at low speeds. Sitting on the bike with the crankarms horizontal and your foot on the pedal, turn the handlebars and check that your toe does not hit the front tire (Fig. C.2). Toe overlap is to be avoided for any kind of slow-speed riding, since making a slow, tight turn in a parking lot can put you on your nose. Toe overlap is not an issue for most other road riding, since the speeds are high enough that turning the bike does not require turning the front wheel at enough of an angle to hit the foot.

C-2

CHOOSING A FRAME SIZE FROM BODY MEASUREMENTS

By taking three easy measurements (Fig. C.3), most people can get a very good frame fit (you will need someone to assist you in taking the measurements). When designing a custom frame, I go through a more complex procedure than this, involving more measurements. For picking an off-the-shelf bike, though, the following method works well. To avoid the trouble of making these calculations yourself, you can use the free Bicycle Fit Calculator at www.zinncycles.com and it will automatically calculate your frame size from these measurements.

C.3 Body measurements

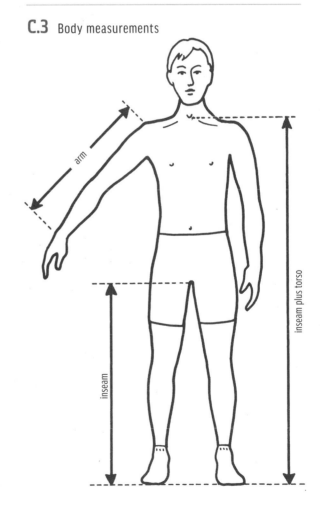

1. Measure your inseam

Spread your stocking feet about 2 inches apart, and measure from the floor up to a broomstick or dowel held level and lifted firmly up into your crotch (Fig. C.4). You can also use a large book and slide it up a wall to keep the top edge horizontal—as you pull it up as hard as you can—into your crotch. You can mark the top of the book on the wall and measure up from the floor to the mark.

2. Measure your inseam-plus-torso length

Hold a pencil horizontally in your jugular notch, the U-shaped bone depression just below your Adam's apple. Standing up straight in front of a wall, mark the wall with the horizontal pencil. Measure up from the floor to the mark.

3. Measure your arm length

Hold your arm out from your side at a 45-degree angle with your elbow straight. Measure from the sharp bone point directly behind and above your shoulder joint (the lateral tip of the acromion) to the wrist bone on your little finger side.

4. Find your frame size

Subtract 27.5 to 32cm (10.8 to 12.6 inches) from your inseam length. This length is your frame size (also known as seat-tube length) measured along the seat tube from the center of the bottom bracket to the top of the top tube (Fig. C.5). If the frame you are interested in has a sloping top tube, you need a bike with a shorter seat tube. In the case of a sloping top tube, project a horizontal line back to the seat tube (or seatpost) from the top of the top tube at the center of its length (Fig. C.5). Mark the seat tube or seat post at this line. Measure from the center of the bottom bracket to this mark; this length should be 27.5–32cm less than your inseam measurement.

Also, if the bike has a bottom bracket higher than 27cm (10.6 inches), subtract the additional bottom-bracket height from the seat-tube length as well.

Generally, smaller riders will want to subtract close to 27.5cm from their inseam measurement, while taller riders will subtract closer to 32cm. Since an average

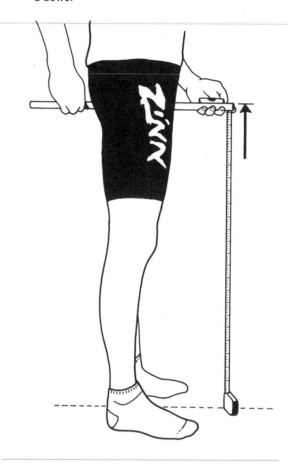

C.4 Measuring inseam using a bubble level on a dowel

bottom-bracket height on a road bike is 26.5cm, subtracting any less than 27.5cm could result in less than one inch (2.5cm) of standover clearance on a bike with a level top tube. But there is considerable range here. The top-tube length is more important than the frame size, and, if you have short torso and arms, you can use a small frame to get the right top-tube length, as long as you can raise your bars as high as you need them.

If you are short and cannot find a bike small enough for you to get at least an inch of standover clearance, consider one with 650C (26 inches) or even 24-inch wheels rather than 700C.

NOTE: *A step-through frame (i.e., "women's" frame, "mixte frame," or "girl's bike") having a steeply up-angled top tube meeting near the bottom-bracket shell makes seat-tube length irrelevant for determining standover clearance. With a step-through bike, the only considerations will be horizontal and vertical reach to the bars.*

5. Find your top-tube length

To find your torso length, subtract your inseam measurement (found in step 1) from your inseam-plus-torso measurement (found in step 2). Add this torso length to your arm length measurement (found in step 3). To find the top-tube length, multiply this arm-plus-torso measurement by a factor in the range between 0.47 and 0.485. If you are a casual rider, use 0.47; if you are a very aggressive rider, use 0.485; and, if you are in between, use a factor in between. This top-tube length is measured horizontally from the center of the seat tube to the center of the head tube (Fig. C.5).

If this is a bike you plan to set up exclusively with aero handlebars and race in time trials or drafting-prohibited triathlons, you will generally want a longer top tube. Use 0.495 as the multiplier to find the top tube in this case. If the seat angle on the bike you will be using for this purpose is not very steep—less than 75 degrees—you will likely need to add more length yet to the top tube. This would occur if you will be pushing your saddle all of the way forward or using a forward-offset seatpost to position yourself in a forward position. The forward-set saddle will consume much of your reach to the bars, so you will need more top-tube length to stretch out properly. The longer top tube will also ensure a more even weight distribution over the wheels and prevent you from using such a long stem and bar that you will be hanging way out over the front of the bike with too much weight on the front wheel.

Use 0.49–0.5 as the multiplier if you're using a straight (mountain bike–style) handlebar.

NOTE: *On a sloping-top-tube bike, the actual horizontal top-tube length is less than the length found by measuring along the top tube.*

6. Find your stem length

Multiply the arm-plus-torso length you found in step 5 by 0.09 to 0.11 to find the stem length. Again, a casual rider will multiply by 0.09 or so, while an aggressive rider will multiply by closer to 0.11. This is a starting stem length and is dependent on the top tube being the length you figured above; the stem length needs to increase or decrease accordingly if the top tube is shorter or longer than your calculation. Finalize the stem length (Fig. C.5) once you are sitting on the bike and see what feels best.

C.5 Bike dimensions on sloping-top-tube road bike

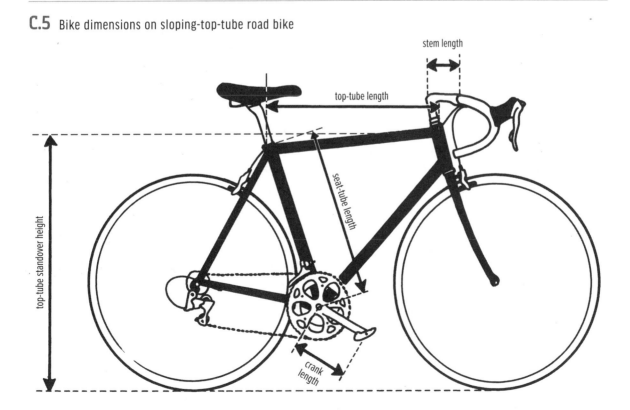

7. Determine crankarm length

Generally, road crankarms come in 2.5mm length increments (Fig. C.5) from 165mm to 180mm (although 167.5mm is often hard to find). Longer than 175mm can usually only be found on high-end cranks, if even then. It is possible to find 185–220mm as well as 130–160mm from Zinn Cycles and a few other custom crank manufacturers.

There is no general consensus on ideal crank length. Here is a simple selection method that works reasonably well. If the frame size you determined in step 4 is less than 45cm, use 165mm cranks; if your frame is between 46 and 49cm, use 167.5mm; 50–53, use 170mm; 54–57, use 172.5mm; 58–61, use 175mm; 61–64, use 177.5mm, and if your frame size is 65cm or bigger, use 180mm or longer. If your riding is focused on time trialing, triathlon, or hill climbing, try 2.5mm longer than the recommendations above.

Another way to look at this is to use a factor multiplied by your leg or thigh length, like leg length times 0.21 to 0.216. I think this and similar multiplication methods make more sense and result in more efficient pedaling than the almost one-size-fits-all approach by most bike and crank manufacturers. That said, I must warn you that for short-legged and long-legged people, the formulas will result in crank lengths considerably shorter or longer than are generally available. But tall riders will often be better off with custom cranks longer than 175mm or 180mm and short riders with custom cranks shorter than 165mm or 170mm. See www.zinncycles.com for formulas and custom crank availability.

8. Choose handlebar width and drop

Road drop handlebars should be the same width or slightly wider than the distance from the center of the top of one upper-arm bone—humerus—to the other. You can hold the front of the bar up to your shoulders and see if each side meets in the center of the top of each humerus or slightly overlaps the outside of your arms. That way, your arms will support your shoulders straight in line, and your chest will be able to open for efficient breathing.

If you are a small person, you will want a handlebar with a shallow drop, while a big person may want a deeper drop. Modern handlebar designs allow a shallower drop due to a shorter reach to the bar bend.

If you have small hands, look for a bar with a bend specifically made to reduce the reach to an STI or Ergopower brake lever.

9. Adjust pedaling stance width

If your knees swing out at the top and fall in on the downstroke, you might benefit from longer pedal spindles to move your feet farther out underneath your knees. BikeFit.com, Kneesavers, Look, Shimano, and Speedplay offer options for pedal spindles.

C-3

POSITIONING OF SADDLE AND HANDLEBARS

The frame fit is only part of the equation. Except for the standover clearance, a good frame fit is relatively meaningless if the seat setback, seat height, handlebar height, and handlebar reach are not set correctly for you.

1. Saddle height

When your foot is at the bottom of the stroke and clipped into the pedal, lock your knee without rocking your hips. Do this sitting on your bike on a trainer with someone else observing. Your foot should be level, or your heel should be slightly higher than the ball of your foot. If you have a goniometer to measure knee angle (Fig. 14.5), set it at the bottom of the stroke in the 25- to 35-degree angle range (see section 14-2). Another way to determine seat height is to take your inseam measurement (found in step 1 under "Choosing a frame size from body measurements") and multiply it by 1.09; this is the length from the center of the pedal spindle (when the pedal is down) to the top of the saddle where your butt bones (ischial tuberosities) contact it. Adjust the seat height (Chapter 14) until you get the proper distance.

NOTE: *These methods yield similar results, although the measurement-multiplying method is dependent on the thickness of your shoe sole and the pedal. They all yield a biomechanically efficient pedaling position.*

C.6 Saddle and stem positioning

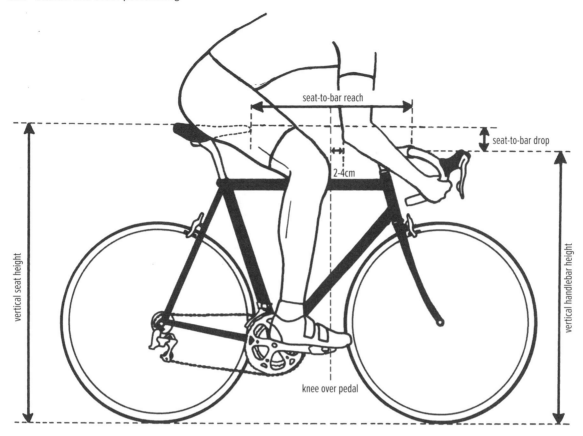

2. Saddle setback

Sit on your bike on a stationary trainer with cranks horizontal and forward foot at the angle it normally assumes at that point when pedaling. An easy way to check that the center of rotation of the knee is over the center of rotation of the pedal is to drop a plumb line from the front of the kneecap (Fig. C.6). It should drop right over the front end of the crankarm.

You also will want to make sure that your cleat position (13-2) is set properly. Generally, you will want your foot deep enough into the pedal that the ball of the foot is right over the pedal spindle or up to 2cm ahead of it; riders with big feet will want their cleats further back, and vice versa.

Slide the saddle back and forth on the seatpost (Chapter 14) until you achieve the desired fore-aft saddle position. Set the saddle level or very slightly tipped down. Re-check the seat height in step 1, since fore-aft saddle movements affect seat-to-pedal distance as well.

For time trial and draft-prohibited-triathlon purposes, most riders using aerobars will want to position their saddle considerably farther forward. Being forward will allow the shoulders to drop low and out of the wind without constraining the hips and having the knees hit the chest—or forcing the knees to swing outward to avoid contact. If it is a frame built for triathlon or time trials, it will generally have a steep seat angle (76 to 78 degrees), making the forward seat position easy to accomplish. If not, you may need to get a forward-position seatpost, being careful that such a post does not make your reach to the bars too short or require such a long stem that you will be hanging dangerously far out over the front wheel. Set the fore-aft seat position so that you are not constricted at the hips when your shoulder joint is the same level as your hip joint. You may need to tip the saddle very slightly downward or get an ISM-type horeshoe-shaped saddle to find some more crotch comfort. Speed can be painful.

If you are competing in UCI-sanctioned time trials, your bike will be subject to the UCI's technical rules and will most likely be checked by a race official for compliance. The UCI rules limit how far forward the saddle and handlebars can be positioned, among other things. See section C-4 for how to ensure that your bike is UCI-compliant.

3. Handlebar height

Measure the handlebar height relative to the saddle height by measuring the vertical distance of the saddle and bar up from the floor (Fig. C.6). How much higher the saddle is than your bar, or vice versa, depends on flexibility, riding style, overall size, and type of riding you prefer.

Aggressive and/or tall riders will prefer to have their saddle 10cm or more higher than the bars. Shorter riders will want proportionately less drop, as will less aggressive riders. Generally, people beginning at road riding will like their bars high and can lower them as they gain flexibility and become more comfortable with the bike.

If in doubt, start with 4cm of drop and vary it from there. The higher the bar, the more weight is carried on your butt, and the more wind resistance you can expect. Change the bar height by raising or lowering the stem (Chapter 12), or by switching stems and/or bars.

Again, threadless headsets allow only limited stem-height adjustment without substitution of a differently angled stem.

For optimal aerodynamic positioning for time trials and draft-prohibited triathlons on flat or rolling terrain, you will want your aerobar elbow pads low enough to get your back close to parallel with the ground (Fig. C.7).

C.7 Aerobar position

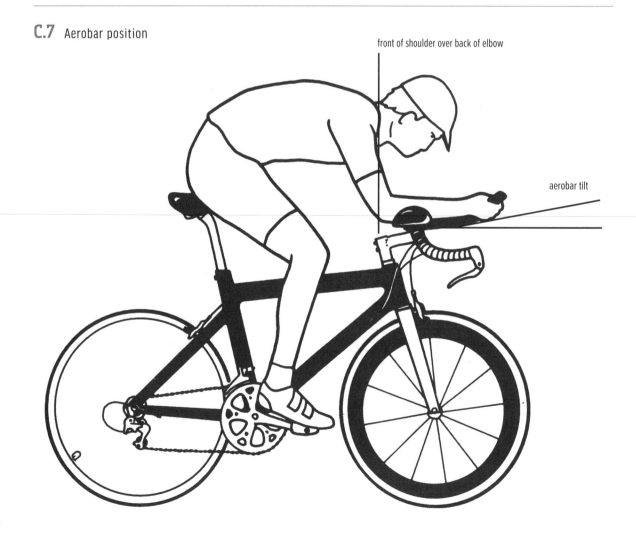

front of shoulder over back of elbow

aerobar tilt

This will make significant aerodynamic difference, but it may take a while to get used to it, so you should work the handlebar height down gradually. Also, you may find such a low position impossible to maintain for an Ironman distance or other long event.

4. Setting handlebar reach

The reach from the saddle to the handlebar is also dependent on personal preference. Aggressive riders will want a more stretched position than will casual riders. This length is subjective, and I would need to look at the rider on the bike to get a feel for how to make him or her comfortable and efficient.

A useful starting place is to drop a plumb line from the back of your elbow with your arms bent in a comfortable riding position. This plane determined by your elbows and the plumb line should be 2 to 4cm horizontally ahead of each knee at the point in the pedal stroke when the crankarm is horizontal and forward (Fig. C.6). The idea is to select a position you find comfortable and efficient; listen to what your body wants.

Vary the saddle-to-bar distance by changing stem length (Chapter 12), not by changing the seat fore-aft position, which is based on pedaling efficiency (C-3, step 2).

On aerobars, set the reach so that a plumb bob from your ear comes out over the crook of your elbow (Fig. C.7), or with the front of your shoulder over the back of your elbow. This will position your upper arms to be angled slightly forward. Again, consult section C-4 if you will be racing this bike in UCI-sanctioned time trials.

NOTE: *There is no single formula for determining handlebar reach and height. Using the all-too-common method of placing your elbow against the saddle and seeing if your fingertips reach the handlebar is close to useless. Similarly, the oft-suggested method of seeing whether the handlebar obscures your vision of the front hub is not worth the brief time it takes to look, since it is so dependent on elbow bend and front end geometry. Another method involving dropping a plumb bob from the rider's nose is dependent on the handlebar height and elbow bend and thus does not lend itself to a proscribed relationship for all riders.*

5. Other settings for aero handlebars

The elbow pads should be positioned for comfort, except in the case of racing in time trials and draft-prohibited triathlons on flat or rolling terrain, in which case they should be placed close to each other—no more than knee-width apart.

The farther forward the pads are from your elbows, the more leverage you will have smashing the pads against your forearms. You will be more comfortable with the pads farther back.

Narrow elbows can make a big difference in aerodynamic efficiency. Wind tunnels often show that having the elbows as narrow as possible is most efficient, but not always; sometimes knee-width apart is fastest. But if you find it hard to breathe well, to pull hard, or to handle the bike well with narrowly spaced elbow pads, move them out until you can.

The tilt of the aerobar is a matter of personal preference. Wind tunnel tests have shown that many different angles appear to be equally efficient aerodynamically. Start with a moderate up-angle to the bar, perhaps 5 or 10 degrees (Fig. C.7).

| C-4 |

POSITIONING A TIME TRIAL BIKE FOR UCI RULE COMPLIANCE

Figure C.8 is a visual representation of the UCI technical rules that must be followed when setting up a bike for a UCI-sanctioned time trial. UCI commissaires at a UCI race inspect every rider's bike before each time trial for saddle and handlebar position. After weighing the bike to ensure compliance with the UCI-mandated minimum bike weight of 6.8 kilograms (15 pounds), a commissaire sets the bike against a jig that has two vertical, 5cm-wide standards whose adjacent edges are 75cm apart, as in Figure C.8. The commissaire lines up the center of the bottom bracket with the forward leading edge of the rear vertical standard.

UCI rule 1.3.013 states that "the peak of the saddle shall be a minimum of 5cm to the rear of a vertical plane passing through the bottom-bracket spindle." So, if the nose of the saddle overlaps the trailing edge of the rear vertical member shown in Figure C.8, the bike

C.8 UCI technical rules diagram

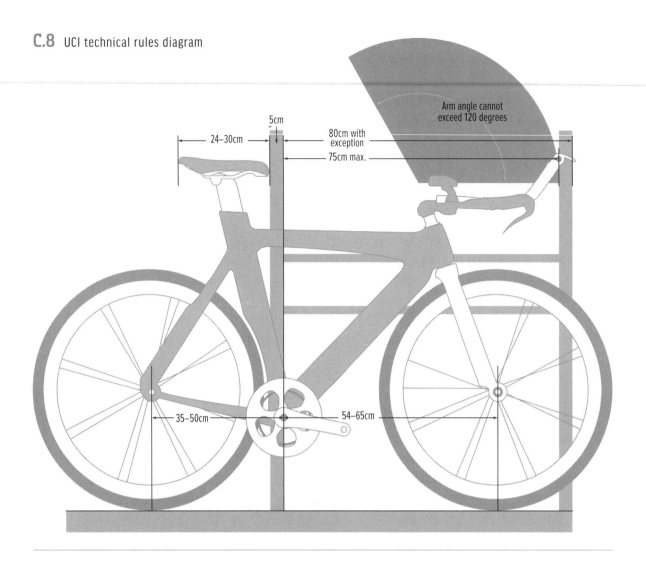

violates UCI rule 1.3.013 (see rules sidebar). If the tip of the handlebar (often defined as the center of rotation of the bar-end shift lever) overlaps the trailing edge of the forward vertical member shown in Figure C.8, indicating that it is over 75cm forward of the center of the bottom bracket, it violates UCI rule 1.3.023 (see rules sidebar). And if the distance from the bottom bracket to the plane of the front hub (a.k.a. the "front center dimension"; it is shown in Fig. C.8) exceeds 65cm (this would generally only be an issue with a very large bike), the bike is out of compliance with rule 1.3.016 (see rules sidebar).

If the tip of your saddle is ahead of the bottom bracket, you cannot race the bike unless you move it back or get a shorter saddle (at least 24mm long). However, if the tip of your saddle is between zero and 5cm behind the center of the bottom bracket (it comes out over the rear vertical member of the jig shown in Fig. C.8), you may be able to

qualify for the morphological exception permitting you to ride with this setup. To do so, you must demonstrate while sitting on the bike in front of the commissaires who just rejected your bike for non-compliance that a plump bob dropped from the front of your knee does not come out ahead of the pedal spindle.

It is worth establishing this ahead of time if you know your saddle falls within this 5cm range. Notice that the plumb bob moves back if you drop your heel. Also check whether one knee is farther back than the other and see to it that the commissaire checks with the plumb bob on that knee.

If the tip of your aerobars (or the center of the pivot of your shift lever; the interpretation of "the extremity of the handlebar" varies from country to country and commissaire to commissaire, and the measurement is generally to the end of the elec-

BELOW IS AN EXCERPT of the UCI technical rules. I have listed here the most problematic rules. The full rules can be found at www.uci.ch.

1.3.007 Bicycles and their accessories shall be of a type that is or could be sold for use by anyone practicing cycling as a sport. The use of equipment designed especially for the attainment of a particular performance (record or other) shall be not authorized.

1.3.013 The peak of the saddle shall be a minimum of 5cm to the rear of a vertical plane passing through the bottom-bracket spindle (1).

(1) The distances mentioned in footnote (1) to articles 1.3.013 and 1.3.016 may be reduced where that is necessary for morphological reasons. By morphological reasons should be understood everything to do with the size and limb length of the rider.

Any rider who, for these reasons, considers that he needs to use a bicycle of lesser dimensions than those given shall inform the commissaires' panel to that effect when presenting his license. In that case, the panel may conduct the following test. Using a plumb-line, they shall check to see whether, when pedaling, the point of the rider's knee when at its foremost position passes beyond a vertical line passing through the pedal spindle [see Fig. C.6; the line touching the front of the knee must pass through the center of the pedal spindle or behind it].

1.3.014 The saddle support shall be horizontal. The length of the saddle shall be 24cm minimum and 30cm maximum.

1.3.016 The distance between the vertical passing through the bottom-bracket spindle and the front wheel spindle shall be between 54cm minimum and 65cm maximum [see Fig. C.8].

1.3.018 In order to be granted approval wheels must have passed a rupture test as prescribed by the UCI in a laboratory approved by the UCI.

1.3.019 b) Weight The weight of the bicycle cannot be less than 6.8 kilograms.

1.3.020 c) Configuration For road competitions other than time trials and for cyclocross competitions, the frame of the bicycle shall be of a traditional pattern, i.e. built around a main triangle. It shall be constructed of straight or tapered tubular elements.

1.3.021 For road time trials and for track competitions, the elements of the bicycle frame, including the bottom-bracket shell, shall fit within a template of the "triangular form" defined in article 1.3.020.

1.3.023 For road time trial competitions and for the following track competitions: individual and team pursuit, kilometer and 500m, an extension may be added to the steering system. The distance between the vertical line passing through the bottom-bracket axle and the extremity of the handlebar may not exceed 75cm [see Fig. C.8], with the other limits set in article 1.3.022 (b, c, d) remaining unchanged. A support for the elbows or forearms is permitted (see diagram "Structure (1B)").

For road time trial competitions, controls or levers fixed to the handlebar extension may extend beyond the 75cm limit as long as they do not constitute a change of use, particularly that of providing an alternative hand position beyond the 75cm mark.

For the track and road competitions covered by the first paragraph, the distance of 75cm may be increased to 80cm to the extent that this is required for morphological reasons; "morphological reasons" should be taken as meaning anything regarding the size or length of the rider's body parts. A rider who, for this reason, considers that he needs to make use of a distance between 75 and 80cm must inform the commissaires' panel at the moment that he presents his license. In such cases the commissaires' panel may carry out the following test: ensuring that the angle between the forearm and upper arm does not exceed 120° when the rider is in a racing position.

[Ed. note: The UCI's accompanying "Structure 1B" diagram also indicates that the entire handlebar must be below the top of the saddle.]

Continues >>

1.3.024 Any device, added or blended into the structure, that is destined to decrease, or which has the effect of decreasing, resistance to air penetration or artificially to accelerate propulsion, such as a protective screen, fuselage form of fairing or the like, shall be prohibited.

A fuselage form shall be defined as an extension or streamlining of a section. This shall be tolerated as long as the ratio between the length L and the diameter D does not exceed 3.

A fairing shall be defined as the use or adaptation of a component of the bicycle in such a fashion that it encloses a moving part of the bicycle such as the wheels or the chainset. Therefore it should be possible to pass a rigid card (like a credit card) between the fixed structure and the moving part.

tronic shifter tips) is more than 80cm forward horizontally from the center of your bottom bracket, you cannot race the bike without shortening the bar. However, if that distance is between 75 and 80cm (it comes out over the forward vertical member of the jig shown in Fig. C.8), you may be able to race the bike as is if you qualify for the morphological exception. To qualify, you must demonstrate in front of the commissaires that when you are sitting on it in your riding position that the bend in your elbows is not more open than 120 degrees. This should also be checked ahead of time, if you know that your shifter pivot comes out between 75 and 80cm forward of the bottom bracket. Notice that you can reduce your elbow angle if you slide forward on the saddle and/or rotate your pelvis forward and flatten your back.

If the commissaires check your bike's front center dimension (which they seldom do) and you know it is a bit over 65cm, practice turning the front wheel slightly while holding the bike so that it measures 65cm without the commissaires noticing the slight twist you're giving the front wheel.

The commissaires generally are interested in you being able to race in the event and are not wishing to create trauma and hassles. They want you to pass the inspection, so make their job easier by understanding the rules and what you need to do for you and your bike to comply with the measurement standards. Remember that the rules can dictate the structure of the bike, but they cannot dictate how you have to sit on the bike, so if you know how to sit on it to pass, then do that if asked to.

adjustable cup the non-drive-side cup in the bottom bracket. This cup is removed for maintenance of the bottom-bracket spindle and bearings, and it adjusts the bearings. The term is sometimes applied to the top headset cup as well.

Aheadset a style of headset that allows the use of a fork with a threadless steering tube. Also called a "threadless headset." The name is a trademark of Dia-Compe and Cane Creek.

Allen key (or Allen wrench or hex key) a hexagonal wrench that fits inside a hexagonal hole in the head of a bolt.

anchor bolt (cable anchor bolt, cable-fixing bolt) a bolt securing a cable to a component.

axle the shaft about which a part turns, usually on bearings or bushings.

axle overlock dimension the length of a hub axle from dropout to dropout, referring to the distance from locknut face to locknut face.

ball bearings a set of balls, generally made out of steel or ceramic, rolling in a track to allow a shaft to spin inside a cylindrical part. May also refer to the individual balls.

barrel adjuster a threaded cable stop that allows for fine adjustment of cable tension. Barrel adjusters are commonly found on rear derailleurs, shifters, and brake levers.

BB (see "bottom bracket").

bearing (see "ball bearing").

bearing cone a conical part with a bearing race around its circumference. The cone presses the ball bearings against the bearing race inside the bearing cup.

bearing cup a polished, dish-shaped surface inside of which ball bearings roll. The bearings roll on the outside of a bearing cone that presses them into their track inside the bearing cup.

bearing race the track or surface the bearings roll on. It can be inside a cup, on the outside of a cone, or inside a cartridge bearing.

binder bolt a bolt clamping a seatpost in a frame, a bar end to a handlebar, a handlebar inside a stem, or a threadless steering tube inside a stem clamp.

bonk 1) v. to run out of fuel for the (human) body so that the ability to continue further strenuous activity is impaired. 2) n. the state of having such low blood sugar from insufficient intake of calories that the ability to perform vigorous activity is impaired.

bottom bracket (or BB) the assembly that allows the crank to rotate. Generally the traditional bottom-bracket assembly includes bearings, an axle (or spindle), a fixed cup, an adjustable cup, and a lockring.

bottom-bracket drop the vertical distance between the center of the bottom bracket and a horizontal line passing through the wheel-hub centers. Drop is equal to the wheel radius minus the bottom-bracket height.

bottom-bracket shell the cylindrical housing at the bottom of a bicycle frame through which the bottom-bracket axle passes.

brake the mechanical device that decelerates or stops the motion of the wheel (and hence of the bicycle and rider) through friction.

brake boss (or brake post or pivot; or cantilever boss, post, or pivot) a fork- or frame-mounted pivot for a brake arm.

brake bridge the cross-tube between the seatstays to which a rear road brake is bolted.

brake caliper brake part fixed to the frame or fork containing moving parts attached to brake pads that stop or decelerate a wheel.

brake pad (or brake block) a block of rubber or similar material used to slow the bike by creating friction on the rim, hub-mounted disc, or other braking surface.

brake post (see "brake boss").

brake shoe the metal pad holder that holds the brake pad to the brake arm.

braze-on boss a generic term for most metal frame attachments, even those welded or glued on.

brazing a method commonly used to construct steel bicycle frames. Brazing involves the use of brass or silver solder to connect frame tubes and attach various "braze-on" items, including brake bosses, cable guides, bottle bosses, and rack mounts, to the frame.

bushing a metal or plastic sleeve that acts as a simple bearing on pedals, suspension forks, suspension swing arms, and jockey wheels.

butted tubing a common type of frame tubing with varying wall thicknesses. Butted tubing is designed to accommodate high-stress points at the ends of the tube by being thicker there.

cable (or inner wire) wound or braided wire strands used to operate brakes and derailleurs.

cable anchor (see "anchor bolt").

cable anchor bolt (see "anchor bolt").

cable end a cap on the end of a cable to keep it from fraying.

cable-fixing bolt an anchor bolt that attaches cables to brakes or derailleurs.

cable hanger cable stop on a stem, headset, fork, seat binder, or seatstay used to stop the brake-cable housing for a cantilever brake.

cable housing a metal-reinforced exterior sheath through which a cable passes.

cable stop (or cable-housing stop) a fitting on the frame, fork, or stem at which a cable-housing segment terminates.

cage two guiding plates through which the chain travels. Both the front and rear derailleurs have cages. The cage on the rear also holds the jockey pulleys. Also, a water bottle holder.

caliper (see "brake caliper" and "measuring caliper").

Campagnolo Italian bicycle-component company.

Cane Creek American bicycle-component company and originator of the threadless headset. Originally known as Dia-Compe USA.

cantilever boss (see "brake boss").

cantilever brake a cable-operated rim brake consisting of two opposing arms, pivoting on frame- or fork-mounted posts. Pads mounted to each brake arm are pressed against the braking surface of the rim via cable tension from the brake lever.

cantilever pivot (see "brake boss").

cantilever post (see "brake boss").

carbon pad brake pad intended for use on carbon-fiber wheel rims.

cartridge bearings ball bearings encased in a cartridge consisting of steel inner and outer rings, ball retainers, and, sometimes, bearing covers.

cassette the group of cogs that mounts on a freehub.

cassette hub (or freehub) (see "freehub").

chain a series of metal links held together by pins and used to transmit energy from the crank to the rear wheel.

chainline the imaginary line connecting the center of the middle chainring with the middle of the cogset. This line should, in theory, be straight and parallel with the vertical plane passing through the center of the bicycle. The chainline is measured as the distance from the center of the seat tube to the center of the middle chainring of a triple crank or, in the case of a double crank, to the center plane midway between the two chainrings.

chain link a single unit of bicycle chain consisting of four plates with a roller on each end and in the center.

chainring a multiple-tooth sprocket attached to the right crankarm.

chainring-nut tool (or chainring-nut spanner) a tool used to secure the chainring nuts while tightening the chainring bolts.

chainstays frame tube on a bicycle connecting the bottom-bracket shell to the rear dropout (and hence to the rear-hub axle).

chain suck the dragging of the chain by the chainring past the release point at the bottom of the chainring.

The chain can be dragged upward until it is jammed between the chainring and the chainstay.

chain whip (or chain wrench) a flat piece of steel usually attached to two lengths of chain. This tool is used to remove the rear cogs on a freehub or freewheel.

chase, wild goose (see "goose chase, wild").

Chris King American bicycle component manufacturer.

circlip (or snapring or Jesus clip) a C-shaped snap-ring that fits in a groove to hold parts together.

clincher rim a rim with a high sidewall and a "hook" facing inward to constrain the bead of a clincher tire.

clincher tire a tire with a "bead" to hook into the rim sides. A separate inner tube is inserted inside the tire.

clip-in pedal (or clipless pedal) a pedal that relies on spring-loaded clips to grip a cleat attached to the bottom of the rider's shoe, without the use of toeclips and straps.

clipless pedal (see "clip-in pedal").

cog a sprocket located on the drive side of the rear hub.

cogset (see "cassette").

cone a threaded conical nut that serves to hold a set of bearings in place and also provides a smooth surface upon which those bearings can roll. Can refer to the conical (or male) member of any cup-and-cone ball-bearing system (see also "bearing cone").

crankarm the lever attached at the bottom-bracket spindle and to the pedal used to transmit a rider's energy to the chain.

crankarm anchor bolt (or crank bolt or crankarm-fixing bolt) the bolt attaching the crank to the bottom-bracket spindle on a cotterless drive train.

crank bolt (see "crankarm anchor bolt").

crank length the distance measured along the crank between the centerline of the bottom-bracket spindle and the centerline of the pedal axle.

crankset the assembly that includes a bottom bracket, two crankarms, chainring set, and accompanying nuts and bolts.

cross three (see "three-cross").

cup a cup-shaped bearing surface that surrounds the bearings in a bottom bracket, headset, or hub (see "bearing cup").

derailleur a gear-changing device that allows a rider to move the chain from one cog or chainring to another while the chain is in motion.

derailleur hanger a metal extension of the right rear dropout through which the rear derailleur is mounted to the frame.

Di2 Shimano's electronic shifting system.

diamond frame the traditional bicycle frame.

disc brake a brake that stops the bike by squeezing brake pads attached to a caliper mounted to the frame or fork against a circular disc attached to the wheel.

dish a difference in spoke tension on the two sides of the rear wheel adjusted such that the rim is centered in the frame or fork.

dishing centering the rim by adjusting spoke tension in a wheel.

dishing tool a tool to check the centering of a rim on a wheel.

double a two-chainring drivetrain setup (as opposed to a three-chainring, or "triple," one).

DoubleTap an integrated road brake/shift lever manufactured by SRAM.

down tube the frame tube that connects the head tube and bottom-bracket shell together.

drivetrain the crankarms, chainrings, bottom bracket, front derailleur, chain, rear derailleur, and freewheel (or cassette).

drop 1) the vertical distance between the center of the bottom bracket and a horizontal line passing through the wheel-hub centers (see also "bottom-bracket drop"). 2) the difference in height between two parts. 3) a terrain discontinuity you may or may not want to ride off of. 4) something not to do with your tools.

dropouts the slots in the fork and rear triangle where the wheel axles attach.

DT (a.k.a. DT Swiss) manufacturer of spokes, other bicycle components, and tools.

dual-pivot sidepull brake a sidepull brake whose arms pivot at two points rather than one.

dust cap a protective cap keeping dirt out of a part.

elastomer a urethane spring sometimes used in suspension forks and rear shocks. Also called an "MCU."

EPS Campagnolo's electronic shifting system.

Ergopower an integrated road brake/shift lever manufactured by Campagnolo.

eTap SRAM's electronic shifting system.

expander bolt a bolt that, when tightened, pulls a wedge up inside or alongside the part into which the bolt is anchored to provide outward pressure and secure said part inside a hollow surface. Expander bolts are found inside quill stems and some handlebar-end plugs and handlebar-end shifters.

expander wedge a part threaded onto an expander bolt and usually used to secure a quill stem inside the fork steering tube or handlebar-end plugs or handlebar-end shifter inside a handlebar. An expander wedge is threaded down its center axis to accept the expander bolt and is either cylindrical in shape and truncated along an inclined plane or conical in shape and truncated parallel to its base.

ferrule a cap for the end of cable housing.

fixed cup the nonadjustable cup of the bottom bracket located on the drive side of the bottom bracket.

flange the largest diameter of the hub where the spoke heads are anchored.

fork the part that attaches the front wheel to the frame.

fork crown the cross-piece connecting the fork legs to the steering tube.

fork ends (see "dropouts").

fork rake (or rake) the perpendicular offset distance of the front axle from an imaginary extension of the steering-tube centerline (see "steering axis"). Also called "wheel offset" or simply "offset."

fork tips (see "dropouts").

frame the central structure of a bicycle to which all of the parts are attached.

freehub a rear hub that has a built-in freewheel mechanism to which the rear cogs are attached.

freewheel the mechanism through which the rear cogs are attached to the rear wheel on a derailleur bicycle. The freewheel is locked to the hub when turned in the forward direction, but it is free to spin backward independently of the hub's movement, thus allowing a rider to stop pedaling and coast as the bicycle is moving forward.

friction shifter a traditional (nonindexed) shifter attached to the frame or handlebar. Cable tension is maintained by a combination of friction washers and bolts.

front triangle (or main triangle) the head tube, top tube, down tube, and seat tube of a bike frame.

FSA component manufacturer, Full Speed Ahead.

girl's bike (see "step-through frame").

goose chase, wild (see "wild goose chase").

granny ring the lowest gear on the bike in which the chain is on the inner (of three) front chainring and the largest rear cog.

Grip Shift a trademarked twist shifter from the SRAM Corporation that is integrated with the handlebar grip of a bike. The rider shifts gears by twisting the grip (see also "twist shifter").

handlebar the curved tube, connected to the fork through the stem, that the rider grips in order to turn the fork and thus steer the bicycle. The brake levers and shift levers are attached to it.

head angle the acute angle formed by the centerline of the head tube and the horizontal.

headset the bearing system consisting of a number of separate cylindrical parts installed into the head tube and onto the fork steering tube that secure the fork and allow it to spin and swivel in the frame.

headset cup (see "bearing cup").

headset top cap (see "top cap").

head tube the front tube of the frame through which the steering tube of the fork passes; attached to the top tube and down tube and contains the headset.

hex key (see "Allen key").

hub the central part of a wheel to which the spokes are anchored and through which the wheel axle passes.

hub brake a disc, drum, or coaster brake that stops the wheel with friction applied to a braking surface attached to the hub.

hydraulic brake a type of brake that uses oil pressure to move the brake pads against the braking surface.

index shifter a shifter that clicks into fixed positions as it moves the derailleur from gear to gear.

inner wire (see "cable").

integrated headset a headset in which the bearing seats are integrated into the head tube (rather than requiring separate headset cups) and the bearings are completely concealed inside of the head tube.

Jesus clip (see "circlip").

jockey wheel (or jockey pulley) a circular, cog-shaped pulley attached to the rear derailleur that is used to guide, apply tension to, and laterally move the chain from rear cog to rear cog.

link a pivoting steel hook on a V-brake arm that the cable-guide "noodle" hooks into (see also "chain link").

locknut a nut that serves to hold the bearing adjustment in a headset, hub, or pedal, usually by jamming against another nut.

lockring a large, thin, circular locknut. On a bottom bracket, the outer ring that tightens the adjustable cup against the face of the bottom-bracket shell. On a freehub, the lockring holds the cogs on.

lock washer a notched or toothed washer that serves to hold surrounding nuts and washers in position.

master link a detachable link that holds the chain together. The master link can be opened by hand without a chain tool.

Mavic French bicycle-component company.

measuring caliper tool for measuring the outside dimensions of an object or the inside dimensions of a tube or hollow object by means of movable jaws.

mixte frame (see "step-through frame").

mounting bolt a bolt that mounts a part to a frame, fork, or component (see also "pivot bolt").

needle bearing steel cylindrical cartridge with rod shaped rollers arranged coaxially around the inside walls.

nipple a thin nut designed to receive the end of a spoke and seat it in a hole in a rim.

noodle curved cable-guide pipe on a V-brake arm that stops the cable housing and directs the cable to the cable anchor bolt on the opposite arm.

outer wire (see "cable housing").

outer wire stop (see "cable stop").

Park Tool bicycle tool manufacturer.

pedal platform the foot pushes on to propel the bicycle.

pedal overlap the overlapping of the toe with the front wheel while pedaling.

Pedro's bicycle tool and lubricant company.

pin spanner a V-shaped wrench with two tip-end pins to fit into holes in a lockring; often used for tightening the adjustable cup of the bottom bracket or other lockrings.

pivot a pin about which a part rotates through a bearing or bushing. Found on brakes and derailleurs.

pivot bolt a bolt on which a brake or derailleur part pivots.

preload (bearings) to adjust the bearings to rotate freely without end play in the axle. This allows them to turn most freely once loaded.

Presta valve thin, metal tire valve that uses a locking nut to prevent air from escaping out of the inner tube or tire.

quick-release 1) the tightening lever and shaft used to attach a wheel to the fork or rear dropouts without using axle nuts. 2) a quick-opening lever and shaft pinching the seatpost inside the seat tube, in lieu of a wrench-operated bolt. 3) a quick cable release on a brake. 4) a fixing mechanism that can be quickly opened and closed, as on a brake cable or wheel axle. 5) any anchor bolt that can be quickly opened and closed by a lever.

quill the vertical tube of a stem for a threaded headset system that inserts into the fork steering tube. It has an expander wedge and bolt inside to secure the stem to the steering tube.

quill pedal a pedal with a cage supporting the foot on only the top side, and whose cage plate is a single continuous piece that curves up to a point at the outboard end of the pedal to protect the side of the foot from being scraped on the road (Fig. 13.2). This type of pedal is meant to be used with a toeclip. The cage offset toward the top and curved upward at the outer end also serves to increase pedaling clearance when the rider leans the bike over when riding around a corner, as well as eliminating the excess weight of cage plates extending downward where they would never be used because of the toeclip on the top. A quill pedal will generally also have a tab on its trailing cage plate so that the rider can flip the pedal upright with the toe of the shoe in order to slide the foot into the toeclip.

race a circular track on which bearings roll freely.

rear triangle the rear part of the bicycle frame, including the seatstays, the chainstays, and the seat tube.

rebound damping the diminishing of speed of return of a spring by hydraulic or mechanical means.

rim the outer hoop of a wheel to which the tire is attached.

Ritchey an American bicycle-component and bicycle company.

rotor the brake disc attached to a wheel hub for a disc brake system.

Rotor a bicycle-component company.

saddle (or seat) a platform made of leather and/or plastic upon which the rider sits.

saddle rails the two metal rods supporting the saddle; the seatpost is clamped to these rods.

Schrader valve a high-pressure air valve with a spring-loaded air-release pin inside. Schrader valves are found on some bicycle inner tubes and air-sprung suspension forks as well as on adjustable rear shocks and automobile tires and tubes.

sealed bearing a bearing enclosed in an attempt to keep contaminants out (see also "cartridge bearings").

seat (see "saddle").

seat angle the acute angle formed by the centerline of the seat tube and the horizontal.

seatpost the tube (inserted into the frame) that supports and secures the saddle.

seatstay a frame tube on a bicycle connecting the seat tube or the rear shock to the rear dropout (and hence to the rear-hub axle).

seat tube the frame tube to which the seatpost (and, usually, the cranks) are attached.

sew-up tire (see "tubular tire").

shim a thin element inserted between two parts to ensure that they are the proper distance apart. On bicycles, a shim is usually a thin washer and can be used to space a disc-brake caliper away from the frame or fork or to space a bottom-bracket cup away from the frame's bottom-bracket shell.

Shimano Japanese bicycle-component company and maker of Dura-Ace and Ultegra component lines as well as SPD (pedals) and STI (shifting system).

sidepull cantilever brake (see "V-brake").

skewer 1) a long rod. 2) a hub quick-release. 3) a shaft passing through a stack of elastomer bumpers in a suspension fork.

Slime tire sealant consisting of chopped fibers in a liquid medium that can be injected inside a tire or inner tube to flow to and fill small air leaks.

snapring (see "circlip").

socket a cylindrical tool with a square hole in one end to mount onto a socket-wrench handle and with hexagonal walls inside the opposing end to grip a bolt head or nut to turn it.

socket wrench a cylindrical wrench handle with a ratcheting square head extending at right angles to the handle onto which sockets or other wrench bits for turning bolts or nuts are installed. Also called "socket-wrench handle" or simply "wrench handle."

spacer on a bicycle, generally a thick washer, cylindrical in shape, intended to space two parts farther apart. Spacers can be found between the headset and the stem and between the stem and the top cap on a threadless steering tube, or between the upper bearing cup and the top nut on a threaded steering tube. Spacers may also be used to space a bottom-bracket cup away from the frame's bottom-bracket shell.

spanner a wrench, in primarily British parlance.

spider a star-shaped piece of metal that connects the right crankarm to the chainrings.

spline one of a set of longitudinal grooves and ridges designed to interlock two mechanical parts together.

spokes metal rods that connect the hub to the rim of a wheel.

spring an elastic contrivance that, when compressed, returns to its original shape by virtue of its elasticity. In bicycle-suspension applications, the spring used is normally either an elastic polymer cylinder, a coil of steel or titanium wire, or compressed air.

spring preload the initial loading of a spring so that part of its compression range is taken up prior to impact.

sprocket a circular, multiple-toothed piece of metal that engages a chain (see also "cog" and "chainring").

SRAM American bicycle-component company. Owner of Sachs, Avid, RockShox, and Truvativ bicycle-component companies.

standover clearance (or standover height) the distance between the top tube of the bike and the rider's crotch when standing over the bicycle.

star nut (or star-fangled nut) a pronged nut that is forced down into the steering tube and anchors the headset top-cap bolt to adjust a threadless headset.

steering axis the imaginary line about which the fork rotates.

steering tube the vertical tube on a fork that is attached to the fork crown and that fits inside the head tube and swivels within it by means of the headset bearings. A steering tube can be threaded or threadless, meaning that the top headset cup can either screw onto the steering tube or slide onto it, and the stem can either 1) insert inside the steering tube and clamp with an expander wedge (threaded) or 2) clamp around the steering tube (threadless). Also called "steerer" or "fork steerer."

stem connection element between the fork steering tube and the handlebar. An archaic word for stem is "gooseneck."

stem length the distance between the center of the steering tube and the center of the handlebar measured along the top of the stem.

step-through frame (or women's frame or girl's bike or mixte frame) a bicycle frame with a steeply up-angled top tube connecting the bottom of the seat tube to the top of the head tube. The frame design is intended to provide ease of stepping over the frame and ample standover clearance.

STI (Shimano Total Integration) an integrated brake/shift lever manufactured by Shimano.

straddle cable short segment of cable connecting two brake arms together.

straddle-cable holder (see "yoke").

threaded headset a headset whose top bearing cup and top nut above it screw onto a threaded steering tube.

threadless headset (see "Aheadset").

three-cross a pattern used by wheel builders that calls for each spoke to cross three others in its path from the hub to the rim.

thumb shifter a thumb-operated shift lever attached on top of the handlebars.

tire bead the edge of the tire that seats down inside the rim. The bead's diameter is held fixed to established standards by means of a strong, stretch- and tear-resistant material—usually either steel or Kevlar. These strands alone are also referred to as the "bead."

tire lever a tool to pry a tire off the rim.

tire sealant (see "Slime").

toe overlap (or toeclip overlap) (see "pedal overlap").

top cap the round top part of a headset that has a bolt passing through it that screws into the star nut to apply downward pressure on the stem to properly load and adjust the headset bearings on a threadless steering tube.

top tube the frame tube that connects the seat tube to the head tube.

torque the rotational analogue of force. Torque is a vector quantity whose magnitude is the length of the radius from the center of rotation out to the point at which the force is applied, multiplied by the magnitude of the force directed perpendicular to the radius. On bicycles, we are primarily interested in 1) the tightening torque applied to a fastener (this value can be measured with a torque wrench—see Appendix E) and 2) the torque applied by the rider on the pedals to propel the rear wheel and hence the bicycle.

torque wrench a socket-wrench handle with a graduated scale and an indicator to show how much torque is being applied as a bolt is being tightened.

Torx key a tool with a star-shaped end that fits in the star-shaped hole in the head of a Torx bolt.

triple a term used to describe the three-chainring combination attached to the right crankarm.

Truvativ a bicycle -component manufacturer. Subsidiary of SRAM.

tub, tubular (see "tubular tire").

tubular rim a rim for a tubular tire. A tubular rim is generally double-walled and concave on top. It is devoid of hook sides that constrain the beads of a clincher tire.

tubular tire a tire without a bead. The tube is surrounded by the tire casing, which is sewed together on the bottom. A layer of cotton tape is usually glued over the stitching, and rim cement is applied to the base tape and the rim to bond the tire to the rim (also called "tubular," "sew-up," and in British parlance, "tub").

twist shifter a cable-pulling derailleur control handle surrounding the handlebar adjacent to the hand grip;

it is twisted forward or back to cause the derailleur to shift (see also "Grip Shift").

V-brake (sidepull cantilever brake) a cable-operated cantilever rim brake consisting of two vertical brake arms that can pivot on frame- or fork-mounted bolts when pulled together by a horizontal cable. A brake pad is affixed to each arm, and there is a cable link and cable-guide pipe on one arm and a cable anchor on the opposite arm.

vise a device, usually mounted on a workbench, with opposed jaws operated by a screw to hold objects.

Vise Grip brand name for adjustable clamping pliers.

Vise Whip an adjustable tool made by Pedro's for holding a cog when removing a casette; replaces a chain whip. Designed by the author of this book.

welding the process of melting two metal surfaces in order to join them.

wheel base the horizontal distance between the two wheel axles.

wheel dish (or wheel dishing) (see "dish" or "dishing").

wheel-dishing tool (see "dishing tool").

wheel-retention tabs integral or separate fixtures at the fork ends designed to prevent the front wheel from falling out if the hub quick-release lever of axle and nuts are loose.

wheelset a pair of wheels for the front and rear of the bicycle.

wild goose chase (see "chase, wild goose").

women's frame (see "step-through frame").

wrench a tool having jaws, a shaped insert, or a socket to grip the head of a bolt or a nut to turn it. In British parlance, a "spanner."

yoke the part on a cantilever or V-brake attaching the brake cable to the straddle cable.

Zinn author of this book; not to be confused with Zen.

APPENDIX E
TORQUE TABLE

One of the single biggest sources of mechanical problems (and breakage) is the overtightening or under tightening of fasteners, particularly on lightweight equipment. It is great to have a "feel" for what is tight enough, but many people either do not have this sense or overestimate their sensitivity to it; "feel" should only supplement torque measurement. With some parts, particularly today's superlight stems and handlebars, it is important to tighten them to their exact torque specification to prevent them from breaking while riding, which would result in an immediate and terrifying loss of control. Even experienced mechanics, with their sense of feel well developed from years of practice, sometimes overtighten the small bolts on lightweight stems.

That said, I do recommend that you try to develop that feel for bolt tightness. For small bolts, choke up on the wrench or hex key so you can tell more easily how hard you are twisting it. (Torque = Force × Radius; when you choke up on the wrench you reduce the radius at which you apply force, so that you have to apply more force to get the same torque on the bolt. That in turn makes you aware of the effort it takes.) When you think the bolt is tight enough, check the tightness with a torque wrench to calibrate your sense of feel.

There is also a danger in undertightening fasteners. The handlebar in an undertightened stem clamp can come loose and twist, or an undertightened brake cable can pull free when you yank hard on the brakes. Also, an undertightened bolt suffers more fatigue during use than one that is preloaded.

The standard method for calculating a torque specification is to load the fastener to 80 percent of its yield strength. This method works on rigid joints. High bolt preload ensures that the fastener is always in tension to prevent metal fatigue in the fastener. However, many

bike parts are not rigid, and high torques can overcompress or crush components. This is especially important when you are using parts of different brands, eras, or materials together, since a stem manufacturer's torque specification for a handlebar clamp may not have anticipated that a carbon handlebar would be used; what works for an aluminum bar can crush the carbon one. There is no springback in a rigid joint, but if parts flex under tightening (a handlebar is a good example), that flex may provide the preload that the bolt needs at a considerably lower torque setting than if it were bolted through solid steel parts.

Torque wrenches usually have a knob at the base of the handle to pull tension on an internal spring. Set the desired torque by twisting the knob and reading the torque setting on a vernier scale or in an indicator window. When the set torque is reached, the head of the wrench snaps over to the side or ratchets freely. An older style of torque wrench, called a beam wrench, has a needle arm parallel to the wrench shaft that moves across a scale. When using either type of torque wrench, hold it at the handle and pull smoothly.

You actually need two torque wrenches for working on bikes. Big ones cannot measure torques accurately for small bolts. Small ones have a limited capacity and cannot tighten a bottom bracket or crank bolt sufficiently.

Using a torque wrench is not a guarantee against a screwup; it simply reduces the chances of one. First, you must make sure that the torque setting you use is the one recommended for the bolt you are tightening. The torque table in this appendix includes a lot of bolts, but it obviously cannot include all bolts from all manufacturers, so if you can consult an owner's manual or find the correct torque on the manufacturer's website, do so. Also, manufacturers often adjust torque specifications following changes in design or materials, so always check the instruction manual for torque settings when possible, even if the bolt is listed in this chart.

Second, lubrication of the bolt, temperature, and a variety of other variables will affect torque readings as well. Bicycle specifications generally assume that the bolt threads have received lubrication or threadlock compound (which provides lubrication before it dries), but that the underside of the bolt head is dry. Lubricating under the bolt head allows the bolt to turn farther at the same torque setting than the same bolt without lubrication under the head, and it thus increases the tension on the bolt.

Third, the torque reading will depend on whether the bolt is turning or you are starting a stationary bolt into motion, since its coefficient of static friction will be higher than its coefficient of dynamic (sliding) friction. If you try to determine the torque of a bolt by checking the torque required to unscrew the bolt, you will have estimated a higher torque than the actual one, particularly if the bolt has been in place for some time and has corrosion or dirt around it. This may be the best you can do in some circumstances, but proceed with caution.

Fourth, the reading on the torque wrench assumes that the wrench head is centered over the bolt; the torque reading will be low if you have a radius multiplying the torque. For example, measuring tightening torque on a pedal axle (if not using a hex key in the hex hole in the axle end) requires a "crow's foot" 15mm open-end wrench attachment on a torque wrench. If extending straight out, the crow's foot creates an offset between the axle centerline and the tool head centerline, which multiplies the torque setting displayed on the wrench handle (i.e., it will make the wrench—the radius—effectively longer). The decimal by which you must multiply the torque reading on the wrench to determine the actual torque applied to the bolt will usually be imprinted on the crow's foot. You must use this torque multiplication factor if you have the crow's foot extending straight out from the torque wrench.

However, if you keep the crow's foot at 90 degrees from the torque wrench, provided the crow's foot is short relative to the length of the torque wrench, you can use the torque settings as is on the wrench (since the hypotenuse and the long side of the right triangle will be close to the same length).

Finally, torque wrenches are not 100 percent accurate, and their accuracy changes over time. Most torque wrenches can be calibrated; automotive parts stores and some hardware stores can do this for you. Ultimately, your feel and common sense are also necessary to ensure safety.

It will be worth your while to review 2-19 (in Chapter 2) to help you develop a feel for bolt tightness. Whether or not you have "the touch," a torque wrench is a wonderful thing, as long as you know how tight the bolt is supposed to be.

Listed below are tightening torque recommendations from many component manufacturers. Where there is only a maximum torque listed, you can assume the minimum torque should be about 80 to 90 percent of that number.

Most torques are for steel bolts; where possible, aluminum and titanium bolts are described as such in the table. Note that it is particularly important to use a copper-filled lubricant like Finish Line Ti-Prep on titanium bolts to prevent them from binding and galling; the same goes for installing any bolt into threads in a titanium component or bike frame.

CONVERSION BETWEEN UNITS

Table E.1 is in inch-pounds (in-lbs), foot-pounds (ft-lbs), and Newton-meters (N-m) (the latter being the one I find easiest to use, since the numbers tend to be nice, round one- or two-digit numbers).

Unit Conversion Factors for Table E.1

- Divide in-lbs settings by 12 to convert to foot-pounds (ft-lbs).
- Multiply in-lbs settings by 0.113 to convert to Newton-meters (N-m).
- Multiply kgf-cm settings by 0.098 to convert to Newton-meters (N-m).

BOLT SIZES

- **M5 bolts** are 5mm in diameter and take a 3mm or 4mm hex key (except on derailleurs, which often take a 5mm hex key or an 8mm box wrench).

- **M6 bolts** are 6mm in diameter and generally take a 5mm hex key.
- **M7 bolts** are 7mm in diameter and generally take a 6mm hex key.
- **M8 bolts** are 8mm in diameter and generally take a 6mm hex key.
- **M10 bolts** are 10mm in diameter and on bikes will likely take a 5mm or 6mm hex key (rear derailleur mounting bolt).

The designation M in front of the bolt size number means millimeters and refers to the bolt shaft size, not to the hex key that turns it; an M5 bolt is 5mm in diameter, an M6 is 6mm, and so on, but there may be no relation to the wrench size. For example, an M5 bolt usually takes a 4mm hex key (or in the case of a hex-head style, an 8mm box-end or socket wrench), but M5 bolts on bicycles often accept nonstandard wrench sizes. M5 bolts attach bottle cages to the frame, and while some accept a 4mm hex key, many have a rounded "cap" head and take a 3mm hex key or sometimes a 5mm hex key. The M5 bolts that clamp a front derailleur around the seat tube or that anchor the cable on a front or rear derailleur also take a nonstandard hex key size, namely a 5mm. And M5 disc-brake rotor bolts often take a Torx T25 key. Conversely, the big single pinch bolts found on old stems usually take only a 6mm hex key, but they may be M6, M7, or even M8 bolts.

Generally, tightness can be classified in four levels:

1. Snug (10–30 in-lbs, or 1–3 N-m): small setscrews, bearing preload bolts (as on threadless headset top caps) and screws going into plastic parts need to be snug.

2. Firmly tightened (30–80 in-lbs, or 3–9 N-m): this refers to small M5 bolts, like shoe cleat bolts, brake- and derailleur-cable anchor bolts, derailleur band clamp bolts, small stem faceplate, or stem steerer clamp bolts. Some M5 and M6 seatpost clamp bolts need to be firmly tightened.

3. Tight (80–240 in-lbs, or 9–27 N-m): wheel axle nuts, old-style single-bolt stem bolts (M6, M7, M8), and some seatpost binder bolts and seatpost saddle clamp bolts need to be tight.

4. Really tight (280–600 in-lbs, or 31–68 N-m): crankarm bolts, pedal axles, cassette lockring bolts, and bottom-bracket cups are large parts that need to be really tight. The load on them is so high that they will creak or loosen if they are not tight enough.

TABLE E.1 ——ROAD BIKE FASTENER TORQUE TABLE*

* See "Unit Conversion Factors for Table E.1" on page 436.

GENERAL TORQUE SPEC FOR STEEL BOLT THREADED INTO AN ALUMINUM PART	IN-LBS	N-M	FT-LBS
M5 bolt	60	7	5
M6 bolt	120	14	10
M6 bolt clamping a carbon part	100	11	8
M7 bolt	180	20	15
M8 bolt	220	25	18

BOTTOM BRACKETS AND CRANKS	IN-LBS		N-M		FT-LBS	
	MIN	MAX	MIN	MAX	MIN	MAX
Bontrager square-taper (Sport) crankarm fixing bolts, M8	320	372	36	42	27	31
Bontrager ISIS (Select, Race) crankarm fixing bolts, M15		480		55		40
Bontrager GXP (Race Lite, Race X Lite) crankarm fixing bolt		480		55		40
Bontrager chainring fixing bolt, steel	70	95	8	11	6	8
Bontrager chainring fixing bolt, aluminum	50	70	6	8	4	6
Campagnolo/Fulcrum Power Torque and Ultra-Torque crank fixing bolt	372	531	42	60	31	44
Campagnolo/Fulcrum Power Torque and Ultra-Torque external bearing cups		310		35		26
Campagnolo square-taper crankarm fixing bolt (M8 steel)	283	336	32	38	24	28
Campagnolo square-taper cartridge bottom-bracket cups		619		70	0	52
Campagnolo chainring fixing bolt		71		8		6
Easton external bearing cups	301	363	34	41	25	30
Easton left crankarm fixing pinch bolts (M5)		105		12		9
Easton chainring fixing bolt		40		4.5		3.3
FSA M8 steel crankarm fixing bolt	304	347	34	39	25	29
FSA M12 steel crankarm fixing bolt	434	521	49	59	36	43
FSA M14 steel crankarm fixing bolt	434	521	49	59	36	43
FSA M14 aluminum crankarm fixing bolt	391	434	44	49	33	36
FSA M15 steel crankarm fixing bolt	434	521	49	59	36	43
FSA M15 aluminum crankarm fixing bolt	434	521	49	59	36	51
FSA M18 bearing preload bolt, MegaExo	4	6	0.4	0.7	0.3	0.5
FSA M18 bearing preload bolt, BB90, BB86	6	13	0.7	1.5	0.5	1.1
FSA M5 pinch bolt, split aluminum crankarm, MegaExo	106	115	12	13	9	10
FSA M5 pinch bolt, split aluminum crankarm, BB90, BB86	97	133	11	15	8	11
FSA M17 crankarm fixing bolt, carbon crank, BB90, BB86	398	487	45	55	33	41
FSA M18 crankarm fixing bolt, carbon crank, MegaExo	398	487	45	55	33	41
FSA BB30 crankarm fixing bolt	345	434	39	49	29	36
FSA steel Allen chainring fixing bolt	80	106	9	12	7	9
FSA aluminum Allen chainring fixing bolt		87		10	0	7
FSA aluminum cartridge bottom-bracket cups	347	434	39	49	29	36
FSA MegaExo bottom-bracket cups	345	434	39	49	29	36

Continues >>

TABLE E.1 — ROAD BIKE FASTENER TORQUE TABLE, CONTINUED

BOTTOM BRACKETS AND CRANKS, CONT.	IN-LBS		N-M		FT-LBS	
	MIN	MAX	MIN	MAX	MIN	MAX
Kogel bottom-bracket cups	305	435	35	50	25	36
Race Face X-Type crankarm fixing bolt	363	602	41	68	30	50
Shimano square-taper crankarm fixing bolt (M8 steel)	305	391	34	44	25	33
Shimano crankarm fixing bolt (Octalink/Hollowtech)	305	435	35	50	25	36
Shimano left crankarm bearing preload cap (Hollowtech 2)	4	6	0.5	0.7	0.3	0.5
Shimano Hollowtech 2 left crankarm fixing pinch bolts (M5)	88	132	12	14	7	11
Shimano chainring fixing bolt, steel	70	95	8	11	6	8
Shimano square/Octalink cartridge bottom bracket cups	435	608	50	70	36	51
Shimano integrated-spindle (Hollowtech 2) bearing cups	305	435	35	50	25	36
Shimano loose-ball-bearing bottom bracket fixed cup	609	695	69	79	51	58
Shimano loose-ball-bearing bottom bracket lockring	609	695	69	79	51	58
SRAM/Truvativ steel chainring bolts	106	124	12	14	9	10
SRAM/Truvativ aluminum chainring bolts	71	80	8	9	6	7
SRAM/Truvativ GXP left crank bolt	416	478	47	54	35	40
SRAM/Truvativ GXP self extractor cup (16mm hex key)	106	133	12	15	9	11
SRAM/Truvativ GXP external bearing cups	301	363	34	41	25	30
SRAM/Truvativ Howitzer ISIS external bearing cups	301	363	34	41	25	30
Trek tandem eccentric	75	100	8	11	6	8
Truvativ ISIS cartridge bearing cups	301	363	34	41	25	30
Truvativ M8 crank bolts, square taper	336	372	38	42	28	31
Truvativ M12 crank bolts, ISIS	381	425	43	48	32	35
Truvativ M15 crank bolts, ISIS	381	425	43	48	32	35
Truvativ self-extractor cup, ISIS or square taper (10mm hex key)	106	133	12	15	9	11
Zinn Zinn-tegrated left crank bolt	400	450	45	51	33	38
Zinn Zinn-tegrated self extractor cup (10mm hex key)	106	133	12	15	9	11
Zinn Zinn-tegrated external bearing cups	301	363	34	41	25	30
BRAKES	**IN-LBS**		**N-M**		**FT-LBS**	
	MIN	MAX	MIN	MAX	MIN	MAX
SIDEPULL CALIPERS						
Caliper fixing bolt onto Trek, LeMond, or Klein carbon seatstays	55	60	6	7	5	5
Campagnolo caliper fixing bolt (to frame or fork)		88		10		7
Campagnolo cable-fixing bolt		44		5		4
Campagnolo brake-shoe fixing bolt		71		8		6
FSA caliper fixing bolt (to frame or fork)	70	86	8	10	6	7
FSA cable-fixing bolt	53	69	6	8	4	6
FSA brake-shoe fixing bolt	44	60	5	7	4	5
Shimano pad retainer bolt	1	2.5	0.1	0.3	0.1	0.2
Shimano cable-fixing bolt	53	69	6	8	4	6

Continues >>

TABLE E.1 — ROAD BIKE FASTENER TORQUE TABLE, CONTINUED

BRAKES, CONT.	IN-LBS		N-M		FT-LBS	
	MIN	MAX	MIN	MAX	MIN	MAX
Shimano brake-shoe fixing bolt	44	60	5	7	4	5
SRAM brake lever mounting bolt	53	70	6	8	4	6
SRAM caliper fixing bolt (to frame or fork)	70	86	8	10	6	7
SRAM cable-fixing bolt	53	69	6	8	4	6
SRAM brake-shoe fixing bolt	44	60	5	7	4	5
SRAM pad retainer bolt	4.5	9	0.5	1	0.4	0.7
CANTILEVERS AND V-BRAKES						
Avid brake shoe fixing bolt	26	44	3	5	2.2	3.7
Avid cantilever brake mounting bolt	44	61	5	7	3.7	5.1
Avid straddle wire carrier cable pinch bolts	26	44	3	5	2.2	3.7
Avid split-clamp flat handlebar lever dual mounting bolts	28	36	3	4	2	3
brake arm mounting bolt, M6	40	60	5	7	3	5
brake cable-fixing bolt, M5	50	70	6	8	4	6
cantilever brake pad fixing bolt	70	78	8	9	6	7
flat handlebar brake lever clamp bolt, M6	50	70	6	8	4	6
flat handlebar brake lever clamp — slotted screw	22	26	2.5	2.9	1.8	2.2
Shimano V-brake leverage adjuster bolt	9	13	1.0	1.5	0.8	1.1
straddle cable yoke fixing nut	35	43	4	5	3	4
Trek, Fisher, Klein spec for brake arm mounting bolt, M6	70	85	8	10	6	7
V-brake pad fixing nut	50	70	6	8	4	6
DISC BRAKES						
Shimano banjo bolt, 3mm hex	44	60	5	7	3.7	5.1
Shimano banjo bolt, 4mm hex	70	87	8	10	6	7
Shimano bleed nipple or bleed screw	32	52	4	6	3	4
Shimano cable-fixing bolts	52	70	6	8	4.4	5.9
Shimano caliper mounting bolts	52	70	6	8	4.4	5.9
Shimano disc-brake adapter bolts	52	70	6	8	4.4	5.9
Shimano sleeve nut	44	60	5	7	3.7	5.1
SRAM caliper mounting bolts	44	62	5	7	3.7	5.1
SRAM rotor mounting bolts		55		6.2		4.5
TRP cable anchor bolt	44	61	5	7	3.7	5.1
TRP caliper mounting bolts	53	69	6	8	4.4	5.9
TRP disc-brake adapter bolts	53	69	6	8	4.4	5.9
TRP hydraulic hose compression nut	35	53	4	6	2.9	4.4
TRP rotor bolts	35	53	4	6	2.9	4.4
TRP straddle cable yoke pinch bolt	22	26	2.5	3	1.8	2.2
TRP Parabox hydraulic master cylinder clamp	53	69	6	8	4.4	5.9
TRP Parabox master cylinder setup pin	7	8	0.8	1	0.6	0.7

Continues >>

TABLE E.1 — ROAD BIKE FASTENER TORQUE TABLE, CONTINUED

DERAILLEURS AND SHIFTERS	IN-LBS		N-M		FT-LBS	
	MIN	MAX	MIN	MAX	MIN	MAX
barrel adjuster mounting screw to frame down-tube shifter boss	13	18	1.5	2.0	1.1	1.5
Campagnolo braze-on type front-derailleur mounting bolt, M5		62		7		5
Campagnolo front-derailleur band clamp bolt, M5		44		5		4
Campagnolo front-derailleur cable-fixing bolt, M5		44		5		4
Campagnolo rear-derailleur cable-fixing bolt, M5		53		6		4
Campagnolo rear-derailleur mounting bolt, M10		133		15		11
Shimano front-derailleur cable-fixing bolt, M5	44	60	5	7	4	5
Shimano braze-on type front-derailleur mounting bolt, M5	44	60	5	7	4	5
Shimano rear-derailleur cable-fixing bolt, M5	35	52	4	6	3	4
Shimano rear-derailleur mounting bolt, M10	70	86	8	10	6	7
Shimano rear-derailleur pulley center bolts, M5	27	34	3	4	2	3
SRAM front-derailleur cable-fixing bolt, M5	35	45	4	5	3	4
SRAM braze-on type front-derailleur mounting bolt, M5	35	44	4	5	3	4
SRAM bolt-on front-derailleur adapter band for braze-on type, M5	27	35	3	4	2	3
SRAM front-derailleur band clamp bolt, M5	44	62	5	7	4	5
SRAM rear-derailleur cable-fixing bolt, M5	35	45	4	5	3	4
SRAM rear-derailleur mounting bolt, M10	70	85	8	10	6	7
SRAM rear-derailleur pulley center bolts, M5		22		3		2
Trek spec for front-derailleur clamp bolt, M5	25	35	3	4	2	3
DUAL CONTROL LEVER						
Campagnolo Ergopower fixing bolt (to handlebar)		88		10		7
Shimano STI fixing bolt (to handlebar)	35	43	4	5	3	4
Shimano Dura-Ace 7900 STI lever fixing bolt (to handlebar)	53	71	6	8	4	6
Shimano Dura-Ace 7900 lever nameplate screw		2		0.2		0.1
SRAM DoubleTap fixing bolt (to handlebar)	53	70	6	8	4	6

HEADSETS	IN-LBS		N-M		FT-LBS	
	MIN	MAX	MIN	MAX	MIN	MAX
Aheadset bearing preload, M6 top cap bolt		22		2		2
FSA bearing preload, top cap		43		5		4
FSA expander plug for carbon steerer	53	80	6	9	4	6.5
Trek bearing preload, top cap		35		44		3
Trek expander plug for carbon steerer	78	85	9	10	6.5	7

HUBS, CASSETTES, AND QUICK-RELEASE SKEWERS	IN-LBS		N-M		FT-LBS	
	MIN	MAX	MIN	MAX	MIN	MAX
bolt-on steel skewer		65		7		5
bolt-on titanium skewer		85		10		7
Campagnolo cassette cog lock ring, steel		442		50		37
Campagnolo cassette cog lock ring, aluminum, for 11-speed cogs		354		40		29

Continues >>

TABLE E.1 — ROAD BIKE FASTENER TORQUE TABLE, CONTINUED

HUBS, CASSETTES, AND QUICK-RELEASE SKEWERS, CONTINUED	IN-LBS		N-M		FT-LBS	
	MIN	MAX	MIN	MAX	MIN	MAX
locknut on quick-release axle	87	217	10	25	7	18
Mavic cassette cog lock ring		354		40		30
nutted front hub		180		20		15
nutted rear hub		300		34		25
Shimano freehub cassette body fixing bolt	305	434	35	50	25	36
Shimano cassette cog lock ring	261	434	30	50	22	36
Shimano hub quick-release lever closing	79	104	8.8	11.8	7	9
Shimano fixed-gear locknut	250	300	28	34	21	25
Shimano single-speed freewheel	250	300	28	34	21	25
Trek spec for front axle nuts (bolt-on hubs)	180	240	20	27	15	20
Trek spec for rear axle nuts (bolt-on hubs)	240	300	27	34	20	25

MISCELLANEOUS	IN-LBS		N-M		FT-LBS	
	MIN	MAX	MIN	MAX	MIN	MAX
fender to frame bolts, M5	50	60	6	7	4	5
Trek spec for rack or fender strut bolts to frame or fork	20	25	2	3	2	2
Trek spec for rear-derailleur hanger bolt	50	70	6	8	4	6
Trek spec for water bottle cage bolts, M5	20	25	2	3	2	2
water bottle cage bolts, M5	25	35	3	4	2	3
Campagnolo EPS V2 battery mounting bolt		25		2		2
Campagnolo EPS V2 bottle cage nut to battery bolt		15		1.2		1

PEDALS AND SHOES	IN-LBS		N-M		FT-LBS	
	MIN	MAX	MIN	MAX	MIN	MAX
Campagnolo pedal axle to crankarm		354		40		29
Crank Bros. pedal axle to crankarm	301	363	34	41	25	30
Crank Bros. shoe cleat fixing bolt, M5	35	44	4	5	3	4
pedal axle into FSA carbon crankarm	257	301	29	34	21	25
pedal axle into Truvativ ISIS or square taper crankarm	186	301	21	34	15	25
pedal axle into Truvativ GXP crankarm	416	478	47	54	35	40
Shimano pedal axle to crankarm	307		35		26	
Shimano shoe cleat fixing bolt, M5	44	51	5	6	4	4
Shimano shoe spike, M5		34		4		3
Speedplay Frog spindle nut	35	40	4	5	3	3
Time pedal axle to crankarm		310		35		26
toeclips to pedals, M5	25	45	3	5	2	4
Trek spec for pedal axle to crankarm	350	380	40	43	29	32

Continues >>

TABLE E.1 — ROAD BIKE FASTENER TORQUE TABLE, CONTINUED

SEATPOSTS AND SEAT BINDERS	IN-LBS		N-M		FT-LBS	
	MIN	MAX	MIN	MAX	MIN	MAX
Bontrager seatpost with bolt across seatpost head	120	130	14	15	10	11
Bontrager Select seatpost, M6 bolt		120		14		10
Bontrager Race, Race Lite, Race X Lite, Race XXX Lite, M6 bolt		150		17		13
Campagnolo seatpost single saddle rail clamp bolt	159	195	18	22	13	16
Campagnolo seatpost binder pinch bolt, carbon seatpost		88		10		7
Deda saddle rail clamp bolt		195		22		16
Easton EC90, EC70, EA70 saddle rail clamp bolts		100		11		8
Easton EC90 Zero, EC70 Zero saddle rail clamp bolts		55		6		5
FSA M5 seatpost rail clamp bolts (steel)		78		9		6
FSA M6 seatpost rail clamp bolts (steel)		106		12		9
FSA M7 seatpost rail clamp bolts (steel)		146		17		12
ITM K-Sword M6 (for GWS system)	88	97	10	11	7	8
ITM K-Sword Special Bolts (saddle clamp bolt)	88	97	10	11	7	8
ITM Forged Lite All series (alu, alu-carbon, carbon), M7	62	71	7	8	5	6
Oval Concepts M6 saddle rail clamp bolts		133		15		11
Ritchey dual saddle rail clamp bolt: WCS, New Pro, M6		165		19		14
seat collar bolt, M5	40	60	5	7	3	5
seat collar bolt, M6	60	80	7	9	5	7
seatpost saddle rail clamp bolt, M8	175	345	20	39	15	29
seat-tube clamp binder bolt, M6	105	140	12	16	9	12
Selcof saddle rail clamp bolt, M6		71		8		6
Selcof saddle rail clamp bolt, M8		177		20		15
Thomson saddle rail clamp bolt, M6		60		7		5
Trek Madone seat mast cap clamp bolt	44	62	5	7	4	5
Trek Madone saddle rail clamp bolt on seat mast cap	124	142	14	16	10	12
Trek spec for single bolt using 6mm hex key	150	250	17	28	13	21
Trek spec for single bolt using 5mm hex key	80	125	10	14	7	10
Trek spec for double bolt using 4mm hex key	45	60	5	7	4	5
Trek spec for binder bolt for aluminum seatpost	85	125	10	14	7	10
Trek spec for binder bolt for carbon-fiber seatpost	65	80	7	9	5	7
Truvativ M6 two-bolt	53	62	6	7	4	5
Truvativ M8 single bolt	195	212	22	24	16	18
two-piece steel seatpost saddle rail clamp bolt	175	345	20	39	15	29
two-piece seat binder bolt, M6	35	60	4	7	3	5

Continues >>

TABLE E.1 — ROAD BIKE FASTENER TORQUE TABLE, CONTINUED

STEMS	IN-LBS		N-M		FT-LBS	
	MIN	MAX	MIN	MAX	MIN	MAX
3T M5 bolts (front clamp, steerer clamp)		44		5		4
3T M6 bolts (single steerer clamp)		130		15		11
3T M6 bolts (two-bolt front clamp plate)		130		15		11
3T M8 bolts (single handlebar clamp)		220		25		18
NOTE: 3T specs also apply to Cinelli stems through 2006						
bar end for flat handlebar M6 clamp bolt	120	140	14	16	10	12
Bontrager M8 steerer tube clamp bolts		200		23		17
Bontrager M7 stem bolts (6mm hex key)		150		17		13
Bontrager M6 stem bolts (5mm hex key)		120		14		10
Bontrager M5 stem bolts (4mm hex key)	46	60	5	7	4	5
Cinelli M5 steel bolts (4mm hex key)		62		7		5
Cinelli M6 steel bolts (5mm hex key)		80		9		7
Deda M5 steel bolts (bar clamp, steerer clamp)		71		8		6
Deda M5 titanium bolts (bar clamp, steerer clamp)		70		8		6
Deda M6 bolts (bar clamp, steerer clamp)		85		10		7
Deda M6 old-model hidden steerer clamp bolt		130		15		11
Deda M8 bolts (quill expander)		85		10		7
Dimension two-bolt steerer tube clamp, M6	80	90	9	10	7	8
Dimension one-bolt handlebar clamp, M8 bolt	205	240	23	27	17	20
Easton M5 bar & steerer clamp bolts	44	70	5	8	4	6
FSA M5 titanium bolts (use Ti prep!)		68		8		6
FSA M5 steel steerer clamp bolts (4mm hex key)		53		6		4
FSA M6 steel bolts		104		12		9
FSA M8 steel bolts		156		18		13
ITM M8 bolts (single-bolt clamp or expander)	150	160	17	18	13	13
ITM M7 bolts	106	120	12	14	9	10
ITM M6 bolts (bar clamp, steerer clamp)	88	105	10	12	7	9
ITM M5 bolts (bar clamp, steerer clamp) 2 front bolts	62	70	7	8	5	6
ITM M5 bolts (bar clamp) 4 front bolts	35	44	4	5	3	4
ITM aluminum M6 bolts in magnesium stem	44	53	5	6	4	4
LOOK stems, all bolts		44		5		4
Modolo M5 handlebar clamp bolt (4mm hex key)		62		7		5
Modolo M5 steerer clamp bolt (4mm hex key)		71		8		6
Oval Concepts titanium M5 faceplate bolts for alloy bars		84		10		7
Oval Concepts titanium M5 faceplate bolts for carbon bars		49		6		4
Oval Concepts M6 faceplate bolts for carbon bars		53		6		4
Oval Concepts titanium M6 clamp bolts for alloy steerers		84		10		7

Continues >>

TABLE E.1 — ROAD BIKE FASTENER TORQUE TABLE, CONTINUED

STEMS, CONT.	IN-LBS		N-M		FT-LBS	
	MIN	MAX	MIN	MAX	MIN	MAX
Oval Concepts titanium M6 clamp bolts for carbon steerers		53		6		4
Oval Concepts M6 clamp bolts for alloy steerers		93		11		8
Oval Concepts M6 clamp bolts for carbon steerers		58		7		5
PRO M5 handlebar clamp bolt (4mm hex key)		35		4		3
PRO M5 steerer clamp bolt (4mm hex key)		44		5		4
Profile M6 steel bolts (5mm hex key)		80		9		7
RaceFace M5 steel bolts (4mm hex key)	61	79	7	9	5	7
Ritchey WCS M5 faceplate bolts for alloy bars	26	44	3	5	2	4
Ritchey WCS M5 faceplate bolts for carbon bars		35		4		3
Ritchey M5 steerer clamp bolts (4mm hex key)		44		5		4
Ritchey WCS M6 clamp bolts for alloy steerers	52	86	6	10	4	7
Ritchey WCS M6 clamp bolts for carbon steerers		78		9		7
Salsa SUL two-bolt face plate bar clamp, M6	120	130	14	15	10	11
Salsa one-bolt handlebar clamp, M6 bolt		140		16		12
Salsa one-bolt steerer tube clamp, M6 bolt	100	110	11	12	8	9
single stem handlebar clamping bolt, M8	145	220	16	25	12	18
Syntace M5 steel bolts (4mm hex key)		53		6		4
Thomson Elite, X2, X4 steerer clamp bolts, M5		48		5		4
Thomson X4 handlebar clamp bolts, M5		35		4		3
Trek spec for handlebar clamp bolts on forged stems	150	180	17	20	13	15
Trek spec for handlebar clamp bolts on welded stems	100	120	11	14	8	10
Trek spec for handlebar clamp bolts with carbon handlebar		100		11		8
Trek spec for stem angle adjustment bolt	150	170	17	20	13	14
Trek spec for stem expander bolt	175	260	20	29	15	22
Trek spec for stem steerer clamp bolts	100	120	11	14	8	10
Trek spec for tandem stoker stem extension adjustment bolt	120	140	14	16	10	12
Trek spec for tandem stoker stem seatpost clamp bolt	100	120	11	14	8	10
Truvativ M5 bolts-handlebar	40	50	5	6	3	4
Truvativ M6 bolts-handlebar	50	60	6	7	4	5
Truvativ M6 bolts-steerer	70	80	8	9	6	7
Truvativ M7 bolts-steerer	110	120	12	14	9	10
wedge expander bolt for quill stems, M8	140	175	16	20	12	15
HANDLEBARS						
Bontrager aluminum handlebar, M7 stem face plate bolt		150		17		13
Bontrager aluminum handlebar, M6 stem face plate bolt		120		14		10
Bontrager aluminum or carbon handlebar, M5 stem face plate bolt		60		7		5
Bontrager carbon handlebar, M6 stem face plate bolt		100		11		8

Continues >>

TABLE E.1 — ROAD BIKE FASTENER TORQUE TABLE, CONTINUED

	IN-LBS		N-M		FT-LBS	
STEMS, CONT.	MIN	MAX	MIN	MAX	MIN	MAX
AERO HANDLEBARS						
3T Bio Arms handlebar clamp/armrest bolts, M8		177		20		15
3T extension clamp bolts, all models, M6		133		15		11
3T New Ahero armrest offset arm mounting bolt, M5		80		9		7
3T New Ahero armrest bolts, M6		106		12		9
3T Sub-8 and Mini Sub-8 armrest bolts, with riser, M5		71		8		6
3T Sub-8 and Mini Sub-8 armrest bolts, without riser, M5		44		5		4
Oval Concepts A900 extension clamp bolts, M5		51		6		4
Oval Concepts A900 base bar clamp bolts, M5		51		6		4
Oval Concepts A700 extension clamp bolts, M6		84		10		7
Oval Concepts SLAM extension clamp bolts, M5		71		8		6
Oval Concepts SLAM handlebar clamp bolts, M5		71		8		6
Oval Concepts armrest bolts, carbon bars, M5		62		7		5
Oval Concepts armrest bolts, aluminum bars, M5		88		10		7
VisionTech armrest bolts, M5		70		8		6
VisionTech extension clamp bolts, M6		88		10		7
Trek spec for armrest bolts, M5		45		5		4
Trek spec for extension clamp bolts, M6		60		7		5

BIBLIOGRAPHY

Barnett, John. *Barnett's Manual: Analysis and Procedures for Bicycle Mechanics*. Brattleboro, VT: Vitesse Press, 1989, VeloPress, 1996.

Brandt, Jobst. *The Bicycle Wheel*. Menlo Park, CA: Avocet, 1988.

Compton, Tom. www.analyticcycling.com, 1998.

Dushan, Allan. *Surviving the Trail*. Tumbleweed Films, 1993.

Editors of *Bicycling* and *Mountain Bike* magazines. *Bicycling Magazine's Complete Guide to Bicycle Maintenance and Repair*. Emmaus, PA: Rodale Press, 1994.

Muir, John and Gregg, Tosh. *How to Keep Your Volkswagen Alive: A Manual of Step-by-Step Procedures for the Complete Idiot*. Santa Fe, NM: John Muir Publications, 1969, 1994.

Murphy, Brett. *Science Behind the Magic—Drivetrain Compatibility*. Art's Cyclery Blog: www.artscyclery.com, 2015.

Pirsig, Robert. *Zen and the Art of Motorcycle Maintenance*. New York, NY: William Morrow & Co., 1974.

Schraner, Gerd. *The Art of Wheelbuilding*. Denver, CO: Buonpane, 1999.

Taylor, Garrett. *Bicycle Wheelbuilding 101, a Video Lesson in the Art of Wheelbuilding*. Westwood, MA: Rexadog, 1994.

Van der Plas, Robert. *The Bicycle Repair Book*. Mill Valley, CA: Bicycle Books, 1993.

Zinn, Lennard. *Zinn & the Art of Mountain Bike Maintenance, 5th Edition*. Boulder, CO: VeloPress, 2010.

Zinn, Lennard. *Zinn & the Art of Triathlon Bikes*. Boulder, CO: VeloPress, 2007.

Zinn, Lennard. *Mountain Bike Performance Handbook*. Osceola, WI: MBI, 1998.

INDEX

ILLUSTRATION INDEX

ABOUT THE AUTHOR

LENNARD ZINN is a bike racer, frame builder, and technical writer. He grew up cycling, skiing, whitewater rafting, and kayaking as well as tinkering with mechanical devices in Los Alamos, New Mexico. After receiving his physics degree from Colorado College, he became a member of the U.S. Olympic Development (road) Cycling Team. He went on to work in Tom Ritchey's frame-building shop and has been producing custom road, triathlon, and mountain frames, as well as custom cranks and stems, at Zinn Cycles since 1982.

Zinn began writing for *VeloNews* in 1989 and is now the senior technical writer for *VeloNews* magazine and a columnist for velonews.com. Other books by Zinn are *Zinn & the Art of Mountain Bike Maintenance* (VeloPress, 5th ed. 2010), *Zinn & the Art of Triathlon Bikes* (VeloPress, 2007), *Zinn's Cycling Primer* (VeloPress, 2004), *Mountain Bike Performance Handbook* (MBI, 1998), and *Mountain Bike Owner's Manual* (VeloPress, 1998).

ABOUT THE ILLUSTRATORS

A former mechanic and bike racer, **TODD TELANDER** devotes most of his time these days to artistic endeavors. In addition to drawing bike parts, he paints and draws wildlife and landscapes for publishers, museums, design companies, and individuals. You can see more examples of his work on his website, www.toddtelander.com.

MIKE REISEL is a graphic designer who spends most of his time art directing magazines, riding his bike, and ignoring the pleas to lubricate his drivetrain.

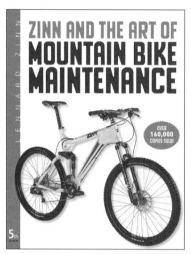